OZONE AND CHLORINE DIOXIDE TECHNOLOGY FOR DISINFECTION OF DRINKING WATER

Ozone and Chlorine Dioxide Technology for Disinfection of Drinking Water

Edited by J. Katz

NOYES DATA CORPORATION
Park Ridge, New Jersey, U.S.A.
1980

Library of Congress Catalog Card Number: 80-13134
ISBN: 0-8155-0809-3
Printed in the United States

Published in the United States of America by
Noyes Data Corporation
Noyes Building, Park Ridge, New Jersey 07656

Library of Congress Cataloging in Publication Data

Main entry under title:

Ozone and chlorine dioxide technology for disinfection of drinking water.

(Pollution technology review ; 67) (Chemical technology review ; 164)
Bibliography: p.
Includes index.
1. Water--Purification--Ozonization.
2. Water--Purification--Oxidation. 3. Water--Purification--Chlorination. 4. Drinking water.
I. Katz, Janet, 1934- I. Series.
II. Series: Chemical technology review ; 164.
TD461.095 628.1'662 80-13134
ISBN 0-8155-0809-3

Foreword

Disinfection is the most important stage in the treatment of drinking water supplies, because it removes or inactivates pathogenic organisms present in the water; it has been practiced since early times. Archeological findings reveal that the ancient Egyptians purified their water by first boiling and then dipping heated copper rods into it, but it was not until the twentieth century that large-scale disinfection practices with chlorine were developed. In 1908 the first chlorination treatment facility was installed in the city of Chicago's water supply. Chlorination has worked well, but it was discovered in the early 1970s that the use of chlorine does pose potential health risks due to the formation of carcinogenic halo-organic compounds, particularly trihalomethanes, as by-products in the chlorination process.

Consideration is now being given to the replacement of chlorine, totally or in part, by alternate technologies. Extensive federal regulations governing the allowable levels of chlorinated by-products have caused water treatment personnel to seek other disinfecting agents. Two of the best known and most practical of these agents are ozone and chlorine dioxide. Both are strong oxidizing agents, and both have experienced wide use outside of the U.S., particularly in Europe and Canada. Both have the advantage that, unlike chlorine which must be transported to a plant in liquid form under high pressure, they can be generated on-site, however neither leaves the same disinfectant residual level that chlorine does.

This book is composed of two reports which review pertinent international technology on ozonation and chlorine dioxide usage as well as microbiological and environmental aspects of their use. Significant advances noted in the engineering of water supply treatment equipment and systems were identified and discussed in detail. Attention was paid to on-site data collection for design of systems and to procedures for water quality analysis. Special emphasis was placed on documentation of the current knowledge of oxidation products of ozone and chlorine dioxide application.

Cost analyses are also included and, while data from different sources may vary, the information used in costing and the comparisons evolved will be of value to the reader.

This book is based on the following reports:

Ozone, Chlorine Dioxide, and Chloramines as Alternatives to Chlorine for Disinfection of Drinking Water—State-of-the-Art, compiled by J.M. Symons and written by J.K. Carswell, R.M. Clark, P. Dorsey, E.E. Geldreich, W.P. Heffernan, J.C. Hoff, O.T. Love, Jr., L.J. McCabe and A.A. Stevens, all of the U.S. Environmental Protection Agency, Office of Research and Development, Cincinnati, Ohio, for presentation at the Second Conference on Water Chlorination: Environmental Impact and Health Effects, Gatlinburg, Tennessee, October 31–November 4, 1977.

An Assessment of Ozone and Chlorine Dioxide Technologies for Treatment of Municipal Water Supplies, by G.W. Miller, R.G. Rice, C.M. Robson, R.L. Scullin, W. Kuhn and H. Wolf of Public Technology, Incorporated, Washington, D.C., prepared for the U.S. Environmental Protection Agency, Office of Research and Development, Municipal Environmental Research Laboratory, Cincinnati, Ohio, August 1978.

The Table of Contents provides easy access to the information contained in this book and is organized in such a way as to serve as a subject index.

In order to keep the price of this large and extensive book to a reasonable level, it has been reproduced by photo-offset directly from the original reports and the cost savings passed on to the reader. Due to this method of publishing, certain portions of the reports may be less legible than desired.

Contents and Subject Index

PART 1: DISINFECTANT ALTERNATIVES

Part 1

Disinfectant Alternatives

The material in Part 1 is based on: *Ozone, Chlorine Dioxide, and Chloramines as Alternatives to Chlorine for Disinfection of Drinking Water—State-of-the-Art,* compiled by J.M. Symons and written by J.K. Carswell, R.M. Clark, P. Dorsey, E.E. Geldreich, W.P. Heffernan, J.C. Hoff, O.T. Love, Jr., L.J. McCabe and A.A. Stevens, all of the U.S. Environmental Protection Agency Office of Research and Development, Cincinnati, Ohio, for presentation at the Second Conference on Water Chlorination: Environmental Impact and Health Effects, Gatlinburg, Tennessee, October 31-November 4, 1977.

OZONE, CHLORINE DIOXIDE AND CHLORAMINES AS ALTERNATIVES

TO CHLORINE FOR DISINFECTION OF DRINKING WATER

State-of-the-Art

Compiled by James M. Symons, Chief, Physical and Chemical Contaminants Removal Branch (P&CCRB), Water Supply Research Division (WSRD), Municipal Environmental Research Laboratory (MERL)

Written by:

J. Keith Carswell, Research Engineer, P&CCRB, WSRD, MERL
Robert M. Clark, Systems Analyst, P&CCRB, WSRD, MERL
Paul Dorsey, Research Assistant, P&CCRB, WSRD, MERL
Edwin E. Geldreich, Chief, Microbiological Treatment Branch (MTB), WSRD, MERL
W. Paul Heffernan, Research Pharmacologist, Laboratory Studies Division, Health Effects Research Laboratory (HERL)
John C. Hoff, Research Virologist, MTB, WSRD, MERL
O. Thomas Love, Jr., Research Engineer, P&CCRB, WSRD, MERL
Leland J. McCabe, Research Epidemiologist, Field Studies Division, HERL
Alan A. Stevens, Research Chemist, P&CCRB, WSRD, MERL

All in the Office of Research and Development, U.S. Environmental Protection Agency, Cincinnati, Ohio

INTRODUCTION

Although free chlorine is a good disinfectant, the problems of the formation of halogenated organics, particularly chloroform and related trihalomethanes, during its use have caused the entire subject of drinking water disinfection to be reviewed. Three other disinfectants are used by various water utilities throughout the world; ozone, chlorine dioxide, and chloramines. If any disinfectant or combination of disinfectants is to replace free chlorine as the most commonly used disinfectant, several criteria must be met, as follows:

It must be easily generated and be in widespread use;

It must be a good biocide;

It must provide an easily measured residual;

It must produce fewer undesirable by-products than does free chlorine; and

It must be cost effective.

The water supply research program, in support of a possible Agency regulation of trihalomethane concentrations, is currently evaluating the disinfectants ozone, chlorine dioxide, and chloramines as free chlorine alternatives with respect to these characteristics. This document will be a progress report on this evaluation. The report will be organized by sections, covering the individual criterion noted above. Each section will contain a summary review of the literature and a summary of research findings to date (Fall 1977) based on in-house and extramural projects. Because research is continuing in most areas, many of the studies discussed are unfinished and will be presented as progress reports. Detailed papers will be published in the technical literature on these studies as they are completed. In total, however, this report will summarize the Agency's knowledge on the subject of disinfectant alternatives at this writing.

Although this paper focuses on the possibility of changing disinfectants as a solution to the disinfection by-product problem, by-product formation is as much a result of the organic content of the water being disinfected as it is the result of chlorine being used. Therefore, another, and possibly preferable, option, which is beyond the scope of this paper, would be to reduce concentrations of disinfectant by-products by chlorinating water that had been treated, possibly with granular adsorbants, to remove the natural organic materials with which chlorine reacts. This would have the additional benefit of reducing the concentration of synthetic organics in finished waters as well as limiting the biological nutrients being carried into distribution systems.

GENERATION AND USAGE OF ALTERNATE DISINFECTANTS

Based on a review of the literature and a research grant sponsored by the Water Supply Research Division (WSRD) the following information is presented to demonstrate that the three alternate disinfectants considered in this paper meet the first criterion of suitability as a replacement for chlorination, that generation technology exists and that their use is sufficiently widespread that they should be considered "common practice."

OZONE

Generation

Lawrence and Cappelli in their recent review article[1] list three methods used for ozone generation:

a. silent electric discharge in air or oxygen

b. electrolysis of water

c. ultraviolet radiation of air or oxygen

Of these three, only (a) is used commercially because of its higher generation efficiency (maximum of 5-6 weight percent ozone).

In the usual generation process, low pressure, clean dry air (or oxygen) is passed between large area electrodes separated by an air gap and a dielectric (usually glass) across which an alternating potential of 15,000 volts (or greater) is maintained. In this type of ozone generator, the production efficiency depends on the rate of gas flow, the applied voltage and the cooling water (or air) temperature.

Mignot[2] has reviewed in detail the numerous methods used for contacting ozone with the liquid to be treated. These methods include:

a. Liquid dispersion in an atmosphere of ozonized air.

b. Fine orifice injection of ozonized air at the bottom of a co- or countercurrent contact chamber, with or without contact media and/or performated septa.

c. An injector operating on the total flow of water to be treated to aspirate low pressure ozonized air into a contact chamber and create mixing turbulence.

d. Similar to (c), with the injector operating on a portion of the total water flow

e. Over-ozonation of a portion of the water to be treated with mixing with the remaining non-ozonated portion

f. Porous media injection of ozonated air into a single or multiple (series) contact chamber with co- or counter-current flow

g. Injection of ozonized air by porous diffusers in a pressurized contact chamber with recovery of the residual ozonized air for re-injection at the chamber entrance (the Torricelli system).

h. Turbine diffusion of ozonized air without recirculation of the water to be treated.

i. Same as (h), with recirculation of the water to be treated

Selection of an appropriate method of contacting is a very important aspect in the design of an ozone treatment system.

Usage

Lawrence and Cappelli[1] have listed a number of common uses of ozone in drinking water treatment. They include:

a. Disinfection

b. Precipitation of iron and manganese

c. Destruction of sulfites

d. Destruction of many surfactants

e. Removal of turbidity

f. Oxidation of taste, odor and color producing compounds

g. Oxidation of other organic compounds

Hill[3], in reviewing the development of water and wastewater ozonation, lists several facts on the early use of ozone for drinking water treatment:

a. First permanent ozone installation -- 1893 at Oudshoon, Holland (Rhine River)

b. Other early ozone installations -- 1901 at Wiesbaden, Germany; 1902 at Paderborn, Germany; 1906 at Nice, France

c. By 1916 there were 49 water treatment plants (26 in France) using ozone

d. By 1940, there were 119 plants, with 90 in France and 14 in Italy.

The International Ozone Institute has summarized what was then (1976) known of the use of ozone for drinking water treatment in the United States.[4] This information includes:

a. Two water treatment plants are using ozone in operation. Whiting, Indiana began using ozone in 1941 for taste and odor control. Strasburg, Pennsylvania has been using ozone since 1973 for disinfection.

b. Monroe and Bay City, Michigan have ozonation facilities under construction.

c. New York City, Boston, Washington D.C., Louisville, and the State of Vermont have conducted research and/or pilot plant studies on ozonation in drinking water treatment.

d. Haverhill, Massachusetts and Laconia, New Hampshire have conducted pilot plant studies.

e. Vineland, New Jersey has received State approval to conduct a pilot-scale demonstration project

f. Boca Raton, Florida is considering an ozonation study.

In 1976, the WSRD funded a Research Grant titled: "Status of Ozonation and Chlorine Dioxide Technologies for Treatment of Municipal Water Supplies."[5] This project is a fact finding, state-of-the art survey of municipal water treatment practices involving the use of ozone and chlorine dioxide in the United States and selected foreign countries.

Thus far this study has developed some preliminary information on the use of ozone in other countries. Nineteen water treatment plants (18 in Quebec, 1 in the Northwest Territory) have been identified in Canada.

In Europe, Asia and Africa, a total of 1034 water treatment plants using ozone have been identified through a search of the literature and through contact with the manufacturers of ozone generation equipment. The vast majority of these plants are in France (577), with West Germany (247) and Switzerland (135) each having substantial numbers. Nineteen other countries have so far been identified as having at least one ozone facility.

CHLORINE DIOXIDE

Generation in Aqueous Solution

Chlorine dioxide, like ozone, is usually generated and used on site, although products referred to as "stabilized" chlorine dioxide are available on the Market. Chlorine dioxide is usually generated in aqueous solution for water treatment, as gaseous chlorine dioxide generation is not well suited to this application.[6] White[7] notes that chlorine dioxide would not be available for use in water or wastewater treatment, if it were not for the availability of solid sodium chlorite. This chemical was first made availble in commercial quantities by Matheson (now Olin-Matheson) in 1940.

Two methods are commonly used to generate chlorine dioxide for water treatment:

a. From chlorine and sodium chlorite

$$Cl_2 + H_2O \dashrightarrow HOCl + HCl \qquad \text{Eq.(1)}$$

$$HOCl + HCl + 2NaClO_2 \dashrightarrow 2ClO_2 + NaCl + H_2O \qquad \text{Eq. (2)}$$

This method is commonly used where a gas chlorinator is available.

b. From sodium hypochlorite and sodium chlorite

$NaOCl + HCl \rightarrow NaCl + HOCl$ Eq.(3)

$HCl + HOCl + 2NaClO_2 \rightarrow 2ClO_2 + 2NaCl + H_2O$ Eq.(4)

In either method, the addition of a mineral acid can be used to control the reaction stoichiometry, and thus eliminate an excess of either free chlorine (prevent trihalomethane formation) or chlorite (toxicity problems)[6].

In the case where a gas chlorinator is used, the chlorinator effluent must be at pH 3.5 or lower, and contain at least 500 mg/ℓ chlorine, if complete reaction of the sodium chlorite is to be effected.[8] To achieve this condition, the usual practice is to set the chlorine-technical sodium chlorite ratio at 1:1 to provide excess chlorine. If excess chlorine is not desired, the chlorine-technical sodium chlorite ratio is set at the theoretical 1:3 ratio and acid is added to maintain the pH at 3.5 or lower.[8]

Although not as common in practice, chlorine dioxide can also be generated from sodium chlorite and acid. The following equation illustrates this, using sulfuric acid as the acid.[9]

$10\ NaClO_2 + 5H_2SO_4 \rightarrow 8\ ClO_2 + 5\ Na_2SO_4 + 2HCl + 4\ H_2O$ Eq.(5)

This reaction has the advantage of producing "chlorine-free" chlorine dioxide easily.

Usage

White[7] reports the first known use of chlorine dioxide as a water treatment process in the United States occurred at Niagara Falls, New York in 1944. Chlorine dioxide was used to control phenolic tastes and odors from industrial wastes as well as tastes and odors from algae and decayed vegetation. The 1963 Inventory of Municipal Water Supplies[10] showed that only 8 of 11,590 water treatment plants in the Inventory used chlorine dioxide.

Information is still being gathered on the WSRD extra-mural project[5] and a final report is not due until early in 1978. Some information on the use of chlorine dioxide is available, however. It should be considered preliminary in nature at this writing (Fall 1977).

Chlorine Dioxide Use in the United States

103 facilities have been identified as using or have used chlorine dioxide

- 27 - Georgia
- 24 - Ohio
- 18 - Pennsylvania
- 8 - Michigan

Many of the treatment facilities in Georgia and Pennsylvania are thought to be small ground water systems with well-head chlorine dioxide disinfection prior to distribution. Most of the United States facilities are using chlorine dioxide for taste and odor control (phenols), for iron and manganese oxidation or for final disinfection.

Chlorine Dioxide Use in Canada

Ten facilities have been identified as using chlorine dioxide. All are in the Province of Ontario and all use chlorine dioxide for taste and odor control. Three of these plants are said to have "severe" taste and odor problems.

Chlorine Dioxide Use in Europe

In Europe, chlorine dioxide is used in several thousand water treatment facilities, most of them of very small size. Countries where extensive use is known include: West Germany, Switzerland, France, and to a lesser extent, Belgium. Italy and The Netherlands also are known to have a small number of chlorine dioxide facilities. At most locations, chlorine dioxide is used for final disinfection prior to distribution. The principal reason for its extensive use is that it does not impart a chlorinous taste to the water. When completed, the final report from the WSRD research grant[5] will provide more extensive information on the use of both ozone and chlorine dioxide in drinking water treatment.

CHLORAMINES, COMBINED CHLORINE, OR CHLORAMINATION

Generation

Combining ammonia with chlorine to form chloramines has been variously called the chloramine process, chloramination and combined residual chlorination. The usual goal in this water treatment process is to form monochloramine according to the reaction:

$HOCl + NH_3 \rightarrow NH_2Cl + H_2O$ (monochloramine) Eq.(6)

Reaction conditions to optimize production of monochloramine include:

a. A pH range of 7 to 8.

b. A chlorine-to-ammonia ratio of 5:1 (by weight), or less. The natural free ammonia content of the water, if any, should be included in determining this ratio. The preferred ratio, according to White[7], is 3:1 (by weight), although widely varying ratios are found in practice.

At higher chlorine-to-ammonia ratios, or at lower pH values, other cahloramine species will form according to the reactions:

$NH_2Cl + HOCl \rightarrow NHCl_2 + H_2O$ (dichloramine) Eq. (7)

$NH_2Cl + HOCl \rightarrow NCl_3 + H_2O$ (trichloramine, or nitrogen trichloride) Eq.(8)

The production of these latter species is related to the familiar chlorination "breakpoint" phenomenon. Their presence may contribute to taste and odor problems in the finished water.

Ammonia may be added to the water before or after the addition of chlorine. In the former case, pre-ammoniation can prevent the formation of tastes and odors resulting from the reaction of chlorine with phenols or other substances. In the latter, post-ammoniation permits a free chlorine residual to exist within the treatment plant, followed by the establishment of a lasting chloramine residual in the distribution system. The ammonia-chlorine process finds its widest use in this application, according to White[7].

Usage

One of the first small scale and subsequently full scale practical applications of combined chlorine as a disinfectant in water treatment was in Ottawa, Ontario (Canada). Joseph Race, chemist and bacteriologist for the City of Ottawa experimented with the material as a disinfectant during the winter of 1915-1916 because "the price of bleach advanced to extraordinary heights."[11] About the same time the City of Denver, Colorado (USA) was successfully applying a combination of chlorine and ammonia to control nuisance organisms in their distribution system.[12] By 1918, chloramines were being used in the Catskill system of the water supply for New York City because in previous use chlorine had disasterous effects on trout.[13] White[7] further reports that chloramines were most popular between 1929 to 1939 at which time their popularity decreased because of the discovery of breakpoint chlorination along with the inability to purchase ammonia during World War II.

The 1963 Inventory of Municipal Water Supplies[10] shows only 308 of the 11,590 water treatment plants in the Inventory using an ammonia-chlorine process. The current unpublished inventory lists about 40,000 water supplies, however no differentiation on the type of disinfectant used is recorded.

SUMMARY

These three disinfectants are judged to be sufficiently common in practice to be seriously considered as alternates to chlorination. Other disinfectants such as silver, iodine, pH, heat, ionizing radiation, permanganate, ultra-violet radiation, bromine chloride, ferrate, and so forth are recognized to exist but will not be discussed in this paper as their use for treating drinking water is minimal at the present time (Fall 1977).

MICROBIOLOGICAL ASPECTS OF DISINFECTION

The primary reason for the use of disinfectants in potable water treatment is to kill or inactivate pathogenic microorganisms that may be present. The material presented in this section describes the relative merits of chlorine, ozone, chlorine dioxide and chloramines as disinfectants from the microbiological standpoint, the second basic criterion of an acceptable disinfectant.

COMPARATIVE DISINFECTION EFFICIENCY

In considering the relative merits of the various chemicals used for disinfection of potable water that are reviewed in this paper, the major factor used to compare the different agents is their disinfection efficiency. This may be expressed in terms of relative disinfection concentrations needed to attain the same disinfection rate or relative disinfection rates produced by the same concentration of disinfecting agent. Because of the lack of standardization of methods by which disinfection research has been conducted, the variation in resistance shown by different isolates of the same microorganism species and variation in resistance between species, only generalized statements about the comparative efficiency of disinfectants can be made. In this context the disinfecting agents under consideration may be ranked as follows in order of decreasing efficiency[7].

$$O_3 \ (>) \ ClO_2 \ (>) \ HOCl \ (>) \ OCl^- \ (>) \ NHCl_2 \ (>) \ NH_2Cl \qquad \text{Eq.(9)}$$

ozone (>) chlorine dioxide (>) hypochlorous acid (>) hypochlorite ion (>) dichloramine (>) monochloramine

This order is derived from the results of numerous disinfection investigations and provides a general overview of the comparative efficiency of various disinfectants for inactivation of microorganisms.

A further complicating factor, particularly with attempts to rank the free chlorine and combined chlorine disinfectants, results from changes in chemical species that occurs with changing pH, temperature, and concentration. For instance, at pH 9.5 - 10 the usual pH for investigation of OCl^- disinfection, HOCl is present in low concentrations and because of its high relative germicidal efficiency compared to OCl^-, some of the inactivation measured may be caused by the HOCl rather than OCl^-. Similarly, in the case of chloramines, an equilibrium state exists as follows:

$$NH_3 + HOCl \rightleftharpoons NH_2Cl + H_2O$$

$$NH_2Cl + HOCl \rightleftharpoons NHCl_2 + H_2O \qquad \text{Eq. (10)}$$

Thus, a low concentration of HOCl may be present and may contribute to chloramine disinfection activity observed.[7] This equilibrium is further influenced by temperature, pH, the presence of other chemical species, and disinfectant concentrations present. Because of the effect of increasing temperature on the rate at which chemical reactions occur, the overall effect of increasing temperature is one of increasing disinfection rates, although the effect may not be the same for different disinfectant species. For instance, the relative effective ratio of OCl^- to HOCl for destruction of <u>Endamoeba histolytica</u> cysts was found to be 1 to 150 at 3°C and 1 to 300 at 23°C. That is, HOCl was 300 times as effective as OCl^- at 23°C , but only 150 times as effective as OCl^- at 3°C.[7]

Although such effects usually are slight, Scarpino <u>et al.</u>[14] in discussing some unusual disinfection results, speculated that a borate buffer system used in one of their studies may have had a major effect on the $HOCl \rightleftharpoons OCl^-$ equilibrium, may have suppressed ionization, or may have resulted in the formation of previously undescribed viricidal forms. Using this buffer system, their results indicated that OCl^- was seven times as effective as HOCl against poliovirus type 1.

In the same buffer system, HOCl was 50 times as effective as OCl^- in inactivating *Escherichia coli*. Because no further investigations of this phenomenon have been published, the reason for these findings is unknown.

Rather than the linear ranking of Equation (9) shown above, division of these disinfectants into two groups based on their disinfection efficiency is probably a better approach. One group, consisting of those agents that are considered very powerful disinfectants, includes ozone, chlorine dioxide, and HOCl. The other group, consisting of weaker disinfectants includes OCl^-, monochloramine and dichloramine.

CHEMICAL BASIS FOR DISINFECTANT ACTIVITY

Chemical disinfectant activity is dependent on contact between the disinfecting agent and the microorganism surface followed by reaction with the surface and/or penetration of and reaction with vital internal constituents resulting in death or inactivation of the microorganism. In the case of the disinfectants under consideration the reactions that take place are oxidative in nature. Thus, oxidation potential is an important factor with regard to disinfecting capabilities. Not all chemicals having a high oxidation potential are good disinfectants, however. For example, Morris[15] pointed out that gaseous oxygen, O_2 has a higher oxidation potential than many effective disinfectants. The statement of Peleg[16] "...substances that do not possess a high oxidation potential will not be germicidally active. Conversely, chemical species that have high oxidation potential may [or may not] possess germicidal potential", reiterates this point. Therefore, chemicals with a high oxidation potential cannot be assumed *a priori* to be good disinfectants.

Another factor involved in disinfectant activity relates to the chemical dissociation of the disinfecting agent and the presence of a charge on the germicidally active portion of the molecule. An ion such as OCl^- is a relatively poor disinfectant because the presence of the negative electrical charge prevents its approach to and diffusion through cell membranes of microorganisms. All three of the very effective germicides, ozone, chlorine dioxide, and HOCl have high oxidation potentials and also are undissociated. The germicidally reactive ozone chemical species is not known, but it may be the OH (hydroxy radical) or the HO_2 (hydroperoxyl radical). Other species such as O_3^- (ozonide radical), O_2^- (superoxide radical), and O^- (oxide radical) have been demonstrated, but their role in germicidal activity is not known[16]. Most experimental results indicate that the disinfection efficiency of ozone varies with pH and that it is somewhat more efficient at a lower pH (6.0) than at a higher pH (10.0). This is probably because of the greater stability of ozone in water at lower pH's.[17]

COMPARATIVE RESISTANCE OF MICROORGANISMS

Based on the results of many previous studies, the inherent resistance of different kinds of microorganisms also may be ranked.[7] As in the case of comparative ranking of disinfectants, the figures are not precise, but do provide a general impression of relative resistance. The rankings in order of decreasing resistance are as follows:

protozoan cysts > enteroviruses > enteric bacteria Eq. (11)

In a currently funded WSRD research grant at the University of Illinois involving an intensive examination of the chlorine resistance of 6 different enteroviruses, their resistance to HOCl (pH 6.0) was found to vary by a factor of five in the time required for a four log reduction.

The time required for a four log reduction of these same viruses using OCl^- (pH 10.0) ranged from 10 to 150 times that required for equivalent reduction by HOCl.[18] These results indicate that changing the pH had different effects on the chlorine resistance of different viruses.

Further indication of this phenomenon is shown in chlorine dioxide disinfection studies. In contrast to chlorine, chlorine dioxide does not react with water to hydrolyze, although it does disproportionate to ClO^-_2 (chlorite), and ClO^-_3(chlorate). The chlorine dioxide molecule remains in the same form over a pH range of at least 4 to 8.4.[19] The rate at which E. coli is inactivated by chlorine dioxide increases, however, as the pH increases within this range.[19] A similar effect with polio virus was shown by Cronier et al.[20] They showed that chlorine dioxide inactivated poliovirus 1 over three times as fast at pH 9.0 as at pH 7.0. Thus, both E. coli and poliovirus 1 are much more sensitive to chlorine dioxide at a slightly alkaline pH. Moffa et al.[21] in studying the disinfection of storm water, found a diphasic inactivation curve when using chlorine dioxide. Although chlorite and chlorate concentrations were not measured, they attributed the second slower phase of inactivation to the disappearance of chlorine dioxide and the appearance of chlorite. They concluded therefore that chlorite had "--weak bactericidal and viricidal properties--."

Differences in disinfection resistance between different isolates of the same virus type have also been shown. Kelly et al.[22] showed that the time required for a three log reduction of several different poliovirus 1 isolates varied from 3 to 6 minutes during exposure to 0.2 to 0.3 mg/ℓ of free chlorine. Recent studies by Bates et al.[23] indicate that progeny of poliovirus 1 repeatedly cycled through chlorine disinfection are somewhat more resistant to chlorine than the original parent stock virus. The resistant progeny required about 2.5 times the length of exposure required by the parent stock for inactivation to the same level.

The existence of extremely resistant polio 1 viruses has also been reported recently[24]. At free chlorine levels of about 0.5 mg/ℓ and pH 7.1, these isolates showed a decline of about one log in 2 minutes with little, if any, further decrease in titer after 30 minutes exposure. Thus, these viruses apparently show a complete resistance to chlorine for upwards of 30 minutes, rather than merely a difference in time required for complete destruction. Isolates of these resistant viruses have been obtained and further experiments are in progress in the laboratories of Dr. Engelbrecht of the University of Illinois and Dr. Sharp of the University of North Carolina, both funded by WSRD research grants.

Aggregation or clumping of viruses has long been considered to be a factor involved in increased viral resistance to disinfectants. Until recently, however, no evidence of this has been available. Through a series of WSRD research grants at the University of North Carolina, evidence of this phenomenon and information on the factors involved in virus aggregation and deaggregation have been obtained.[25-29] Although the effects are complex and appear to be different with different viruses, some general information has been developed from these studies as follows:

1. Although differences in halogen resistance of up to 300 fold between aggregated and dispersed viruses have been shown, the differences thus far shown are merely differences in rates of inactivation. No evidence of complete resistance to inactivation because of aggregation has been found.

2. pH and metallic cations are important determinants with regard to aggregation. Aggregation does not appear to be related to virus isoelectric point characteristics.
3. Induced aggregates of some viruses are quite stable although those of other viruses can be dispersed easily. Natural virus aggregates appear to be more stable than induced aggregates.

Of the most resistant forms of microorganisms of concern, protozoan cysts, only Endamoeba histolytica has been extensively studied with regard to disinfection. Most of this research has been conducted using chlorine in various forms. Newton and Jones[30] showed that ozone was a more effective cysticidal agent than chlorine. Although chlorine dioxide would be assumed to be an effective cysticidal agent, research findings on this point could not be found. The resistance of E. histolytica cysts to chlorine is such that White[7] states, "The best protection against this disease (amebiasis) is removal of the organisms from water supplies by filtraiton. They are large enough to be trapped by sand filters, and require high doses of free residual chlorine (10 ppm) and long contact time (one hour or more) to effect disinfection." According to Palin[34], the range of free residual chlorine recommended by the National Research Council for insuring cyst destruction within a 30 minute contact time ranged from 2 mg/ℓ to 70 mg/ℓ depending on the pH and temperature of the water being treated. Despite this and the lack of universal filtration of water supplies originating from source waters that may contain E. histolytica cysts, waterborne outbreaks of amebiasis are rare in the United States.

Outbreaks of giardiasis, a disease caused by Giardia lamblia, a protozoan that also forms cysts, have, however, been occurring with increasing frequency in recent years. In contrast to E. histolytica, a method for determining viability of G. lamblia cysts has not yet been developed. Although they are assumed to be highly resistant to disinfection, little is known about the resistance of these cysts to various disinfectants. A WSRD funded research grant at the University of Oregon Medical School in Portland on development of methods for determining viability is in progress and current results indicate that a method for determining cyst viability in vitro will soon be available for use in disinfection studies on this microorganism.[32]

PROTECTIVE EFFECTS OF PARTICULATE MATTER

As indicated previously, for disinfection to be effective, contact must occur between the disinfectant and the microorganisms to be inactivated or killed. Because source waters may contain a variety of inorganic and organic particulates, some of which may originate as fecal wastes, and because of the tendency of microorganisms, particularly viruses, to adsorb to various kinds of particles, removal of particulate matter, usually measured by light scattering, has been considered very important during water tratment. Although the major reason for removal of particulate matter (turbidity) has been the concern about its interference with disinfection, little direct evidence of such interference was available until recently.

Previously laboratory studies had established that enteric bacteria and enteroviruses ingested by aquatic nematodes found in some water supplies were protected against very high doses of chlorine[33]. In another instance, persistence of coliforms in a water supply system containing a substantial level of residual chlorine was attributed to ingestion and survival of the coliforms in small crustaceans present in the water[34]. More recent studies indicate that virus adsorbed onto surfaces of particles such as clay remain exposed to the disinfectant and disinfection rates are affected only slightly, if at all.[35,36] Further, viruses precipitated with aluminum phosphate, or flocculated with calcium carbonate or alum or either unprotected or are protected to only a minor degree.[36-38] Viruses associated with cell debris, however, are well protected from inactivation, showing survival for up to 50 minutes at HOCl concentrations of 1.5 - 2.0 mg/ℓ[37]. Protection of coliforms associated with primary sewage effluent solids for at least 60 minutes at HOCl concentrations of 0.5 mg/ℓ also has been shown[38].

All of these studies have been conducted using free chlorine as the disinfectant. Similar studies sponsored by WSRD research grants are in progress at the Unversity of Cincinnati using chlorine dioxide and at the University of Maine using ozone and studies using chloramines are planned. Based on the comparative disinfection efficiency of various disinfectants, one might speculate that the same order of efficiency would be maintained in inactivating microorganisms associated with particulate matter. The efficiency of ozone, in particular, is difficult to predict because of its extremely reactive nature and consequent rapid disappearance after addition to water.

INFLUENCE OF WATER QUALITY

The organic content of water influences the disinfectant demand of water. Waters with lower organic content require the application of lower doses of disinfectant to achieve good disinfection. Figures 1 and 2 show that the lower organic content of the effluent from granular activated carbon beds resulted in a reduced requirement for disinfectant, either ozone or chlorine dioxide, when compared to a typical filter effluent. This has important implications when considering the problem of by-products discussed later in this paper.

STABILITY DURING WATER DISTRIBUTION

In addition to their disinfection efficiency as defined above, another important consideration is the stability of these agents over a period of time and their ability to remain as residual disinfectants

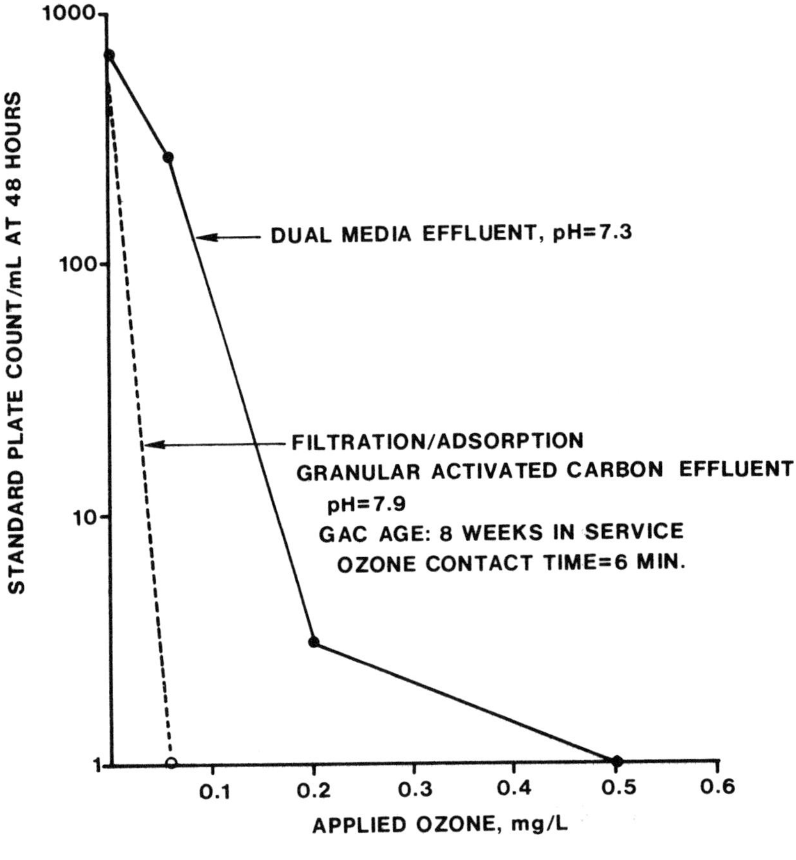

Fig. 1. POST DISINFECTION WITH OZONE

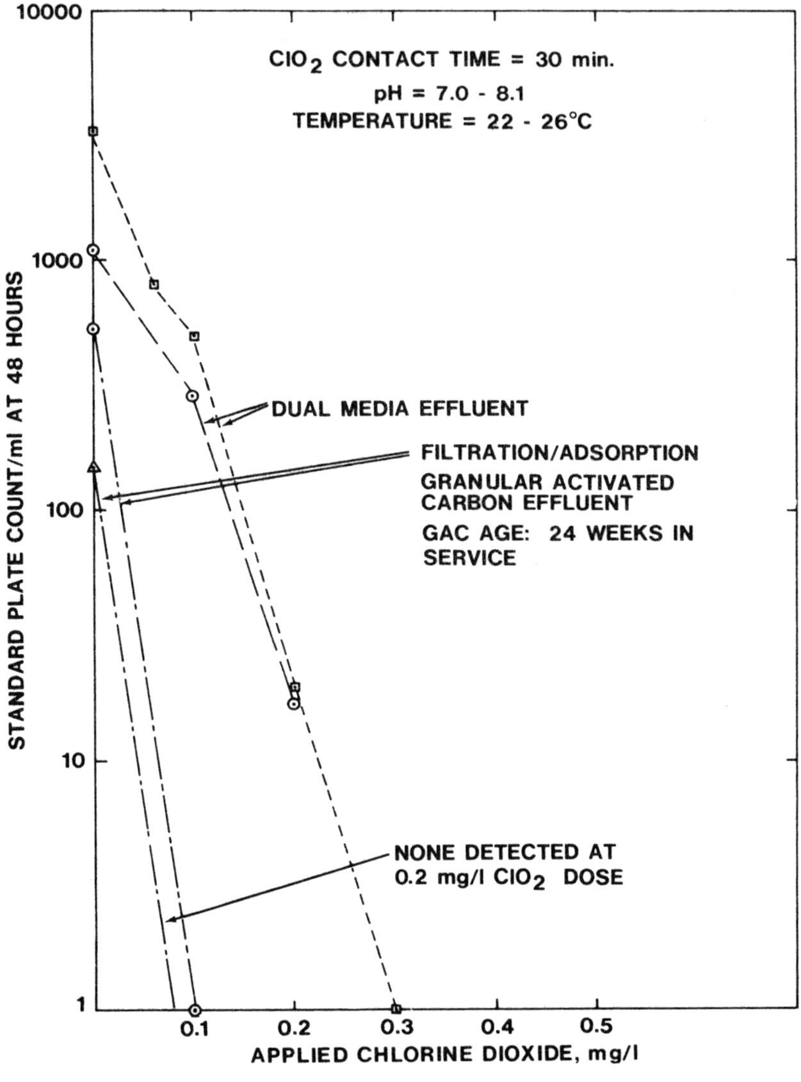

Fig. 2. POST DISINFECTION WITH CHLORINE DIOXIDE

to prevent growth of microorganisms in storage and distribution systems after initial disinfection during water treatment. Although the major concern regarding the regrowth of microorganisms is related more to harmful effects on the distribution system and aesthetic effects such as taste, odor and appearance rather than to potential health hazards, regrowth is not a trivial problem. Further, the maintenance of a biocidal residual to the consumers tap not only keeps the system clean and protects against some cross-connection contamination, but also its sudden disappearance is a rapid indication of distribution system problems. On the other hand, the application of chlorine can depress the pH and aggravate corrosion problems in some instances.

The level of cross connection protection attained by maintenance of a disinfectant residual has been the subject of much conjecture. A WSRD funded research grant in progress at the Johns Hopkins University is providing some information on this.[39] The results of batch studies in which E. coli, polio 1 and f2 bacteriophage mixed with sewage were added in proportions of up to 5 percent by volume to water containing various levels and types of chlorine residual indicated that both free and combined chlorine residual at equivalent concentrations eventually gave comparable degrees of inactivation of E. coli. A free residual at pH 6, however, provided inactivation in 2 minutes whereas an equivalent concentration of combined residual required two hours to inactivate E. coli. Neither free residual nor combined residual at pH 8 consistently inactivated polio virus or f2 bacteriophage to the lower sensitivity limit of the test.

The distribution system problems associated with the use of combined clorine residual or no residual have been documented in several instances.[40,41a,41b] In these cases the use of combined chlorine is characterized by an initial satisfactory phase in which chloramine residuals are easily maintained throughout the system and bacterial counts are very low. Over a period of years, however, problems may develop, including increased bacterial counts, dropoff of chloramine residuals, increased taste and odor complaints, and reduced main carrying capacity. Conversion of the system to free chlorine residual results initially in an increase of taste and odor problems resulting from sloughing off of accumulated organic material and problems in maintaining a free chlorine residual. Following this phase, a free chlorine residual is established throughout the system, bacterial counts are lower and taste and odor complaints decline.[40] Of the disinfectants under consideration, a residual level can be maintained with all except ozone, which, because of its high reactivity, disappears very rapidly.

SUMMARY

In considering the overall biocidal or germicidal advantages and disadvantages of these disinfecting agents as alternatives to free chlorine, which is currently advocated, from a microbiological point of view chlorine dioxide would be best. It is highly efficient, is most effective in the pH range in which water treatment is practiced, and can be maintained as a residual disinfectant in distribution systems. Chloramines have disadvantages with respect to their much lower disinfection efficiency, possible failure to completely inactivate pathogens present, their ineffectiveness in providing a barrier to cross connection contamination, and their failure to prevent regrowth in distribution systems in some cases. Ozone, although extremely efficient as a disinfectant, cannot be maintained as a residual in distribution systems. In the future, the effectiveness of these three alternate disinfectants and the influence of particulates and mixing intensity on disinfection will continue to be studied in both in-house and extramural bench-, and pilot- and field scale projects sponsored by WSRD.

MEASUREMENT OF DISINFECTANT RESIDUALS

In order to properly control the use of a disinfectant, the residual concentration of the disinfectant must be easily measurable, and where distribution system protection is needed, this residual must be maintained and monitored, the third criterion of an acceptable alternate to chlorine.

FREE CHLORINE

For measurement of free chlorine residuals ($HOCl+OCl^{-}+Cl_2$) the 14th edition of Standard Methods[42] lists and describes in detail eight methods of residual analysis: Two iodometric methods, Phenylarsine oxide -- Amperometric Titration, Stabilized Neutral Orthotolidine (SNORT), DPD Ferrous Titrimetric, DPD Colorimetric, Leuco Crystal Violet, and Syringaldazine (FACTS). The relative usefulness of each method plus known possible interferences in the analyses are described in the reference. Each method is based on the oxidizing capacity of aqueous chlorine and measures in some quantitative way the oxidation products of the reaction of the chlorine residual with the method reagent materials.

OZONE

Because all of the alternate disinfectants under consideration are also oxidizing agents (ozone, chlorine dioxide, chloramines) some version of most of the above methods is applicable to the analysis of each of the alternates as well. Indeed, Standard Methods[42] lists an iodide-iodine titrimetric method as the quantitative approach to ozone residual measurement where concentrations are over one mg/ℓ. An amperometrically determined end point extends the lower range of this analysis.[43]

CHLORINE DIOXIDE

Iodimetry, a modified orthotolidine, and amperometric titration are recommended for chlorine dioxide measurements[42]. Recently, Miltner[44] reviewed the work of Palin[45,46], Adams, et al.[47] and Hong and Rapson[48] and ultimately applied the DPD titrimetric procedure to the measurement of aqueous chlorine dioxide during his work describing the prevention of formation of trihalomethanes by use of this alternate disinfectant.[49]

CHLORAMINES

Amperometric titration is considered a standard of comparison[42] for free or chloramine residuals, although some of the other chlorine methods such as the DPD methods are also applicable to the determination of chloramines.

COMBINATION OF DISINFECTANTS

Thus, the measurement of chlorine and alternate disinfectant residuals *per se* is not a particularly difficult analytical task where considerations are given to the interferences and limitations described in each of the methods referenced. The measurement of individual disinfectants is only slightly more complicated when disinfectants are used in combination with one another. Commonly encountered examples of this are chloramines in the presence of chlorine (chlorine used with ammonia or where ammonia or organic amines are already present when chlorine is applied), chlorine in the presence of chlorine dioxide (chlorine dioxide is often produced by the action of excess chlorine on chlorite), and chlorine in the presence of ozone. Even these three examples have relatively straightforward analytical solutions:

Free Chlorine with Chloramines[42]

Free chlorine (HOCl, Cl_2 or OCl^-) is measured by either of the DPD methods or by amperometric titration followed by sequential measurement of iodine liberated by the reaction of the respective mono- and dichloramine with added iodide under specified pH conditions.

Chlorine with Chlorine Dioxide

After a thorough investigation of published methods, Miltner[44] selected the DPD approach as most reliable and convenient for measuring chlorine dioxide in his work. When DPD is used (either titrimetric or colorimetric) two separate measurements are necessary to determine chlorine and chlorine dioxide concentrations, and two approaches to this are possible.[44] Malonic acid can be added to an aliquot of the sample to react with residual chlorine prior to a chlorine dioxide measurement[45,46] and then two DPD measurements, one before and one after malonic acid addition, will yield total chlorine dioxide-plus-chlorine and chlorine dioxide respectively. Chlorine is obtained by difference. The other approach is to purge the chlorine dioxide from solution by gas stripping leaving the chlorine in solution[48]. In this approach the chlorine is measured directly and the chlorine dioxide determined by difference. In either case the analysis for these two species in simple combination is relatively uncomplicated.

Ozone with Chlorine

According to Dailey and Morrow[43], phenyl arsine oxide--amperometric titration of iodine liberated upon the reaction of both chlorine and ozone with added iodide followed by a similar titration of a second aliquot previously treated with ammonia will yield the concentrations of total ozone-plus-chlorine and chlorine alone respecitvely. The ozone concentration is obtained by difference. This approach depends upon the reaction of ozone and chlorine with ammonia to form nitrate (unreactive) and monochloramine (reactive with iodide) respectively.

Complex Combinations

From the above, the conclusion might erroneously be reached that the problems of disinfectant residual measurement are fairly well solved.

This is not necessarily the case, however. Where multiple disinfectants are used, where more than two or three active species need to be quantified, or possibly where certain organic materials are present (see next section), the number of analytical steps and techniques as well as the calculations necessary become much greater and the procedure more complex. For example, Miltner[44] attempted to measure free chlorine, chlorine dioxide, and all expected inorganic reaction products resulting from the action of these species with the organic materials in the source water. The goal was to generate a material balance of inorganic chlorine and oxychlorine species to aid in assessment of toxicological hazard (to be discussed later) from products (for example, chlorite and chlorate) possibly resulting from the widespread application of chlorine dioxide as a disinfectant alternative to chlorine.

Miltner was able to use a combination of five procedures, two required in triplicate, to construct a material balance of free chlorine, chlorine dioxide, chlorite, chlorate and chloride in a few simple cases in rather dilute solutions. Follow up work conducted over a wide concentration range[50] indicated that many problems with this total analysis still exist, especially with the measurements of chlorate and the relatively small contributions of chloride above background by the applied disinfectant.

Additionally, because of the differences in disinfecting power of each of the free chlorine species, HOCl and OCl^- (see previous section) distinguishing between them under any given set of treatment conditions rather than simply measuring the "free residual" might be desirable. This represents a different problem from that posed by the need to measure oxychlorine species of different oxidation states. Johnson[51] has developed a membrane electrode that claims promise, but it has not yet been generally accepted as a practical tool in the drinking water industry.

SUMMARY

If such differentiation of oxidizing species as in the two examples cited above are finally deemed to be unnecessary, disinfectant residual measurements are likely to remain relatively simple when proper consideration is given to potential interferences. Future work by WSRD will focus on the possible interferences of organic compounds when measuring chlorine dioxide residuals.

BY-PRODUCTS AND END PRODUCTS OF DISINFECTION

The occurrence of trihalomethanes in finished drinking water has been demonstrated to be widespread and a direct result of chlorination practices at the water treatment plant.[52-54] Because of the health significance of these trihalomethanes (discussed later in this section) an alternate disinfectant is being sought. The fourth criterion any alternate must meet is the creation of less undesirable by-products and end-products than chlorine.

CHEMISTRY

Free Chlorine

Chloroform results from the generalized reaction:

Free Chlorine + "Precursor" --> Chloroform[52-56] Eq.(12)

This occurs to some extent in any water treatment plant where chlorination for disinfection is practiced and natural organic (humic) material is present to act as precursor. The reaction is not instantaneous and occurs over a period of a few days until either chlorine or precursor is exhausted. In the presence of natural bromide, the reaction products include some mixed-halogen trihalomethane species (bromodichloromethane, dibromochloromethane) and bromoform.[52] This occurs in most chlorinated drinking water[54], even where bromide concentrations in the source water are small. Iodine-containing species have also been observed[57], presumably because of the presence of natural iodide. Because the chemical reactions for formation of these bromine- (and iodine-) containing trihalomethanes are probably mechanistically similar to that for formation of chloroform,[58] the trihalomethanes, including chloroform, can be discussed as a group.

Because trihalomethane formation is not an instantaneous reaction, as long as a free chlorine residual is present, trihalomethane concentrations will increase in the water as it flows through a water treatment plant (unless removed during treatment) usually reaching some value higher than that which would be observed if an analysis for trihalomethane species was performed immediately after sampling at the first point of chlorination. (A more complete discussion of these concepts has been presented elsewhere.[59].) Further, the consumer is likely to receive water with trihalomethane concentrations higher than those leaving the plant because the reaction proceeds in the distribution system. Additionally, not only are the concentrations of trihalomethanes time dependent, but the rate of reaction is dependent on pH, precursor concentration, nature of precursor(s), temperature, and to some degree, free chlorine concentration early in the chlorination process.[55,60,61] Finally, the ratio of chloroform to other trihalomethanes is highly dependent on the bromide content of the source water.[62,63] Typical concentrations of trihalomethanes formed range from < 1 μg/ℓ to a few hundred μg/ℓ.[54]

In addition to the trihalomethanes, numerous other compounds have been found in chlorinated waters and their presence attributed to the chlorine applied under those circumstances. For example, Rook[64] has characterized more than twenty by-products of chlorination of Rotterdam storage reservoir water. Jolly[65] and Glaze[66] have reported the presence of a large number of carbon-substituted chlorinated compounds in chlorinated waste waters. Nitrogenous compounds react with chlorine during both drinking water and wastewater disinfection to yield the nitrogen-substituted chloramines. Currently, Morris of Harvard University has a WSRD Research Grant to study the influence of pH and time on the reactions of chlorine with nitrogenous compounds likely to be found in natural waters. In addition to the chlorine demand reactions, the production of chloroform in these reactions is also being determined. Phenols have long caused a problem with odors in chlorinated waters by forming the highly odorous chlorophenols.[67] Generally speaking, however, the concentrations of these compounds, if formed upon chlorination of drinking water, are much lower in finished drinking water than are the concentrations of the trihalomethanes at locations where a free chlorine residual is maintained. Individual chlorinated compounds other than the trihalomethanes are not generally found at concentrations in finished drinking water higher than found in the corresponding source water. Additional confirmation on this is being sought through a thorough study of the occurrence of halogenated hydrocarbons in the Passaic River watershed and the rate of change of these (increase or decrease) concentrations through the water treatment process at Little Falls, New Jersey. The work is being carried out by J.V. Hunter at Rutgers, The State University under a WSRD research grant.

The recent findings concerning the carcinogenicity of chloroform[68] has stimulated EPA efforts to aid in reducing human exposure from the drinking water source to these chlorination by-products[69]. Any efforts to reduce trihalomethane concentrations should be expected to result in a presumably beneficial decrease in formation of the other chlorinated by-products. For example, Sontheimer et al.[70] reported an increase in total organic chlorine concentration, as concentrated by adsorption on ground granular activated carbon and measured by a pyrolysis technique, of 186 μg/ℓ after breakpoint chlorination of Ruhr River water, although the chloroform concentration only rose 6 μg/ℓ and the total trihalomethane concentration only increased 14 μg/ℓ. Insights as to the reduction in organo-chlorine load obtained as a consequence of control of trihalomethane formation may be gained as a result of understanding the mechanism of chlorination reactions of humic substances leading to trihalomethane production. This is currently under detailed review through a WSRD research grant to R.F. Christman, et al., at the University of North Carolina.

Ozone

In-house research by EPA's Water Supply Research Division[71] has shown that disinfection level (<5 mg/ℓ) ozone doses, under the conditions tested, will not:

a. form trihalomethanes (i.e., oxidize chloride to chlorine with subsequent reaction with trihalomethane precursors)

b. reduce the concentrations of trihalomethanes already formed prior to ozonation

c. reduce the concentration of trihalomethane precursors as measured by the trihalomethane formation potential[59]

At the present time plans have not been made to attempt to identify other by-products from ozonation in-house, although Chian of the University of Illinois has found that upon ozonation, 2-propanol was converted to acetone that in turn was oxidized to acetic and oxalic acids. Trace amounts of formaldehyde and formic acid were also detected in the ozonated acetone solution. Further ozonation of acetic acid resulted in the formation of glyoxylic acid that was oxidized readily to oxalic acid. Extramural research is now being sponsored by the WSRD on ozone oxidation by-products and end-products. A research grant to Glaze at North Texas State University is investigating the use of ozone with and without ultraviolet radiation in drinking water treatment. The removal of selected compounds (trihalomethane precursors, trihalomethanes, polycyclic aromatic hydrocarbons (PAH's), polychlorinated biphenyls (PCB's) and others) and the oxidation/photolysis products formed are being studied over a 3-year period. At this time (Fall 1977) preliminary results are not available.

Chlorine Dioxide

Some results of work by Miltner[49] in the WSRD laboratory to monitor trihalomethane production with time after chlorine dioxide, chlorine, and chlorine dioxide with chlorine were separately applied to pilot plant settled water are shown in Figure 3. The upper curve represents the action of chlorine alone; the curve coincident with the abcissa represents the application of chlorine dioxide alone; and the curve between these two represents the action of chlorine to form trihalomethanes in the presence of chlorine dioxide. These curves demonstrate two important points: (a) chlorine dioxide does not cause the formation of trihalomethanes and (b) chlorine dioxide plus excess chlorine (as is often the case in water treatment) results in the formation of lower concentrations of trihalomethanes than does the same amount of chlorine alone. Therefore, with consideration only to minimizing trihalomethane formation, chlorine dioxide is a possible alternative to chlorine as a disinfectant in water treatment. Although this is the case, consideration must still be given to other by-products that may be formed by chlorine dioxide treatment.

Considerable evidence exists that chlorine dioxide reacts with organic material during water treatment and therefore is likely to produce organic by-products:

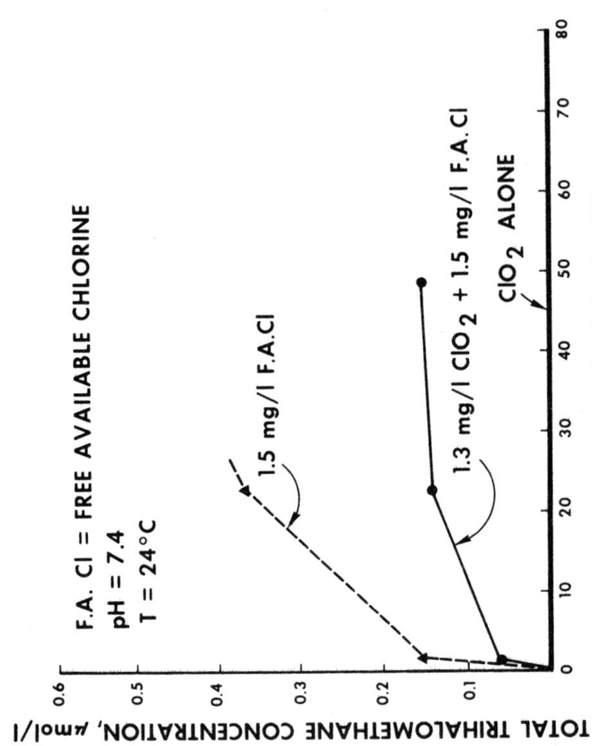

FIGURE 3. TRIHALOMETHANE FORMATION BY ClO_2 AND EXCESS FREE AVAILABLE CHLORINE, ERC PILOT PLANT SETTLED WATER (FROM MILTNER, R. J., 1976)[49]

1. Because chlorine dioxide is a good disinfectant (see earlier section), some reaction is taking place between the cell components of the organism and the chlorine dioxide.

2. Even though chlorine dioxide does not react with ammonia most waters exhibit a chlorine dioxide demand similar to that of chlorine.

3. At applied chlorine dioxide concentrations higher than those encountered in drinking water treatment, identifiable by-products have been isolated.[72]

4. Chlorine dioxide destroys phenolic compounds when the oxidant is used for taste and odor control in water supplies.[7]

5. Most importantly, as was shown earlier, the presence of chlorine dioxide reduces the formation of trihalomethanes by chlorine. This and other evidence obtained by Miltner[49] indicated that chlorine dioxide reacts with natural humic acid. This is not surprising because chlorine doxide is effective for reducing color in drinking water supplies.[73]

Before an informed decision can be made as to the relative desirability of free chlorine and chlorine dioxide with regard to the formation of organic by-products a detailed investigation and identification of the products formed during disinfection with both chlorine and chlorine dioxide was considered necessary and was initiated in the WSRD laboratory. The possible organic by-products arising from the use of chlorine dioxide as a disinfectant in drinking water was first considered on the basis of the existing literature, followed by an on-going laboratory study to determine whether extrapolations from the literature describing work where concentrations of oxidant and organic material were generally high were valid.

A brief review of the pertinent literature was recently presented by Stevens[74] although a much more extensive and complete review is available elsewhere.[72] Briefly, the literature describes chlorinated and non-chlorinated derivatives including acids, epoxides, quinones, aldehydes, disulfides, and sulfonic acids, that are products of reactions carried out under conditions vastly different from those experienced at water treatment plants.

Stevens[74] reviewed only a few reaction types that were suspected to be important in water treatment practice. Considered as reacting species were saturated aliphatic hydrocarbons, olefins, amines, and aromatic compounds. Based on the literature, chlorine dioxide would not be expected to react with saturated aliphatic hydrocarbons in aqueous media. Chlorine dioxide was reported to react with olefins to produce a variety of chlorinated and non-chlorinated products: aldehydes, epoxides, chlorohydrins, dichloro-derivatives, and chloro- and unsaturated-ketones. Amines react with chlorine dioxide in a manner very different from that of chlorine, chlorine dioxide does not produce chloramines. In general, the respective aldehydes are formed from amines in the order of reactivity: tertiary > secondary > primary[72].

The best known reactions of aqueous chlorine with aromatic compounds in the water treatment field are those that occur with phenols.[67] Chlorine reacts rapidly with phenol to form mono-, di-, and tri-chloro derivatives. These compounds are highly odorous and are slowly decomposed by excess chlorine. Other phenolics and substituted aromatics can also be chlorinated.[75] The formation of chlorophenols by chlorine treatment is one of the chief reasons that chlorine dioxide has been used in drinking water disinfection applications.[7]

According to the literature, chlorine dioxide usually does not form odorous compounds with phenol, but through a complex mechanism, forms the quinones and chloroquinones, and when in excess, oxalic and maleic acids.[72] Chlorine dioxide treatment of phenols can cause chlorine substitution or ring cleavage or both depending on the phenol reacted and the conditions of the reaction. The chlorinated products will generally be of different structure and are either less odorous or are formed in much lower yield than those formed by chlorination. For example, vanillin reacts at pH 4 with chlorine dioxide to give the non-chlorinated β-formymuconic acid monomethyl ester. Veratryl alcohol, however, produces 4,5-dichloroveratrole[72,76] when reacted with chlorine dioxide, which is one of the products expected from chlorination of veratryl alcohol.[77]

Some of the more detailed investigations of the reactions of chlorine dioxide with phenols (ring cleavage or ring retention with or without chlorine substitution) have been accomplished by Glabisz and Paluch at the Polytechnic University, Szczecin, Poland.[78] Glabisz states that, at least at concentrations of one mg/ℓ and above, the character of the reaction products of phenols falls into two groups. The first is the group in which the ring structure is retained and the end products are quinones. This group is made up of <u>para</u>-dihydric and monohydric phenols that are not <u>para</u> substituted. The second group is characterized by those phenols which undergo ring cleavage to give carboxylic acids as end products. Examples include <u>para</u> alkyl phenols and ortho or <u>meta</u> dihydric phenols.

In general, monohydric phenols reacting with chlorine dioxide undergo chlorination along with oxidation, and those of the first group form the chloroquinones, as well as chlorophenols. Glabisz[78] considers this to be somewhat similar to the reaction of chlorine with these phenols. Although chlorine dioxide tends to favor oxidation over chlorination, the relative amount of chlorinated versus oxidized products depends on the relative amounts of both chlorine dioxide used and phenols present. Excess chlorine dioxide favors oxidation.

Because of the potential for undesirable by-product formation resulting from chlorine dioxide disinfection of drinking water, an in-house investigation was begun at the WSRD Laboratory to determine if by-products of the type predicted by the literature, where reactions described were carried out at generally higher concentrations, would prevail in the drinking water disinfecting conditions. This work is being carried out in two phases:

1. The search of purgeable compound gas chromatographic data (for example from trihalomethane analyses) for differences between chlorine dioxide treated- and untreated waters.

2. Development and use of a more elaborate analytical scheme to detect products of more diverse nature, specifically those expected from reactions of phenolic compounds.

The semi-quantitative results of the first phase have already been briefly described[74] where C_2 through C_8 aldehydes were noted to increase in concentration after treatment of a natural water with chlorine dioxide. In that work, no other dramatic differences were observed between treated and untreated samples with regard to compounds amenable to the purge-and-trap[76] type of gas chromatographic analysis used.[54,57,79]

Phenol was selected as the model compound for the beginning of the second phase primarily because of the polyphenolic nature of humic materials[80] (see Figure 4) (trihalomethane precursors) that make up a large fraction of the organic material present in natural waters where trihalomethane formation is a problem. This work is not yet complete, but chlorophenols have been produced at high (5/4) molar ratios of phenol to chlorine dioxide. Lower ratios (5/14, 1/14) do not produce chlorophenols, but hydroquinone was isolated from all reaction mixtures. Other organic by-products such as oxalic and maleic acids, and 2,6- and 2,5-dichloro-p-benzoquinone were not immediately identifiable, although total organic carbon concentration data indicate that the phenol is not completely converted to carbon dioxide. To date (Fall 1977) no gas chromatographable compounds in this category have been identified in chlorine dioxide-treated natural waters, and humic- and fulvic acid solutions that were not present in the untreated sample (detection limits estimated to be in the range of 5 - 10 μg/ℓ as phenol).

Some circumstantial evidence now seems to indicate formation of poly quinone-like material from phenol as well as from at least one source of aqueous humic material, however. The DPD colorimetric method[42] was used to determine the chlorine dioxide demand of a highly colored natural water. An interference equivalent to approximately 0.3 mg/ℓ of chlorine dioxide was observed below the point on the demand curve where residual concentration began to rise as dose increased. A similar effect was observed after chlorine dioxide treatment of phenol solutions (see above). The compound p-benzoquinone, a model for possible organic byproducts of chlorine dioxide reactions and the oxidized form of hydroquinone (see above) was also found to produce a color with DPD without chlorine dioxide treatment, so an alternative test was sought for the determination of chlorine dioxide that might not be subject to quinone interference. The FACTS

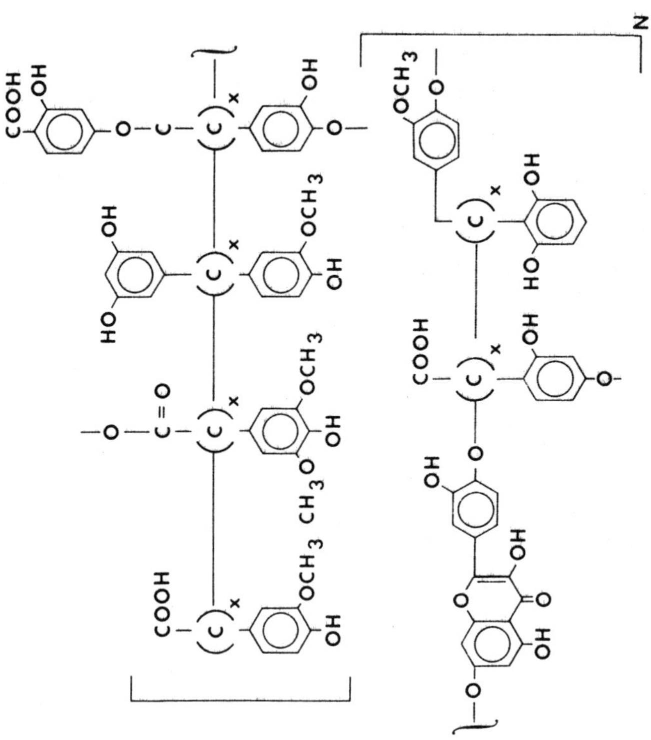

FIGURE 4. A PROPOSED HUMIC STRUCTURE (FROM CHRISTMAN & GHASSEMI,)[80]

test, using syringaldazine with spectrophotometric measurement at 530 nm[42] was found to be a suitable replacement for DPD. This reagent did not give a color with p-benzoquinone. In a chlorine dioxide demand determination on a second colored water, however, a water similar to but from a different source than that used previously, the syringaldazine method gave a higher blank (reagents + water, no chlorine dioxide) reading than the DPD method. Although the curves compared well when background was subtracted, neither curve showed the 0.3 mg/ℓ "false positive" seen with the previous sample.

A chlorine dioxide demand test on a second sample of the first highly colored raw water gave results similar to the first test reported above and again showed "false positive" concentrations below the "breakpoint" of the demand curve without the DPD test. Chlorite was found to be major inorganic product of chlorine dioxide treatment of this water (see Table I).

TABLE I

Chlorine Dioxide Treatment of a Colored Natural Water

ClO_2 Dose mg/ℓ	ClO_2 Residual mg/ℓ	ClO_2^- Conc. mg/ℓ	Contact Time Hours	23 Hour, Chlorine Demand mg/ℓ
13.13	1.19	6.33	23	11.94

The presence of chlorite did not cause this apparent chlorine dioxide residual using the DPD method. The same effect has not been observed on other raw water sources or treated humic- or fulvic acid solutions. This general problem requires considerably more work and will be the subject of a future more detailed report.

Inorganic by-products may also be of concern. As noted above evidence obtained at the WSRD laboratory indicates that a large portion of the applied chlorine dioxide remains as chlorite rather than chloride although accurately determining what percentage of this results from simple disproportionation in contrast to being a reduction product from organic reactions (see previous section on measurement of disinfectant residuals) has not been possible.

Chloramines

During the 1975 National Organics Reconnaissance Survey (NORS)[54] 10 of the 80 utilities sampled disinfected with chloramines. The sum of the concentrations of the trihalomethanes found, expressed as ΣTHM, in the finished water of these supplies ranged from 1 to 81 μg/ℓ with an average of 19 μg/ℓ. For comparison, those utilities that were using breakpoint chlorination had THM concentrations ranging from 1 to 472 μg/ℓ in the finished water with an average concnetration of 72 μg/ℓ. A similar study conducted in Ontario, Canada[81] on 48 water utilities included 10 that were disinfecting with combined chlorine, but the report does not contain sufficient details on treatment practices to assess the impact of combined chlorine on trihalomethanes. Stevens et al.[55] in a laboratory experiment, applied both free and combined chlorine for varying contact times to samples of untreated Ohio River water. Figure 5 shows the ΣTHM formation in the chlorinated sample reaching 160 μg/ℓ after 72 hours, yet the companion sample having combined chlorine formed only 16 μg/ℓ ΣTHM during the experiment.

The reason some of the water utilities in the NORS study[54] using chloramines had relatively high ΣTHM's is probably because of free chlorine being used prior to the addition of ammonia. For example, in the Kansas City Water Works the ΣTHM concentration at the time of the NORS was 34 μg/ℓ [54] This is not surprising because free chlorine is dosed about 5 hours before the ammonia is added. Another example of this effect is shown by Tuepker[82] for the St. Louis County Water Company. There, like Kansas City, the water is softened and chlorinated with approximately an eight hour contact time before ammonia is added. In that situation the chloroform concentration averaged 49 μg/ℓ over a 5 month sampling period.

According to Lambert et al.[83] trihalomethane production begins only when monochloramine, and presumably dichloramine, is destroyed by free chlorine. Further, once water is ammoniated and the disinfectant exists as combined chlorine the trihalomethane reaction essentially stops. This is shown by the plateau on the combined chlorine curve in Figure 5. Finally, the trihalomethane levels reaching the consumer should be similar to the trihalomethane concentrations leaving the treatment plant when combined chlorine is used as the disinfectant. Tuepker[82] has shown this to be the case where he reported no measurable differences between the trihalomethane concentrations leaving the treatment plant and those measured in the distribution system. As noted previously, free chlorine, on the other hand, continues to react with precursors in the distribution system and consumers in outlying reaches can receive water with trihalomethanes considerably higher than the treatment plant effluent concentrations. The use of combined chlorine during water treatment is currently being investigated in a WSRD Research Grant to Professor Leland Harms, South Dakota School of Mines and Technology at the Huron, South Dakota Water Treatment Plant[84] and at the Louisville Water Treatment plant as part of the EPA project with the Ohio River Valley Sanitation Commission[85].

HEALTH EFFECTS

Chlorine

Toxicology

Concern over the occurrence of chloroform and other halogenated

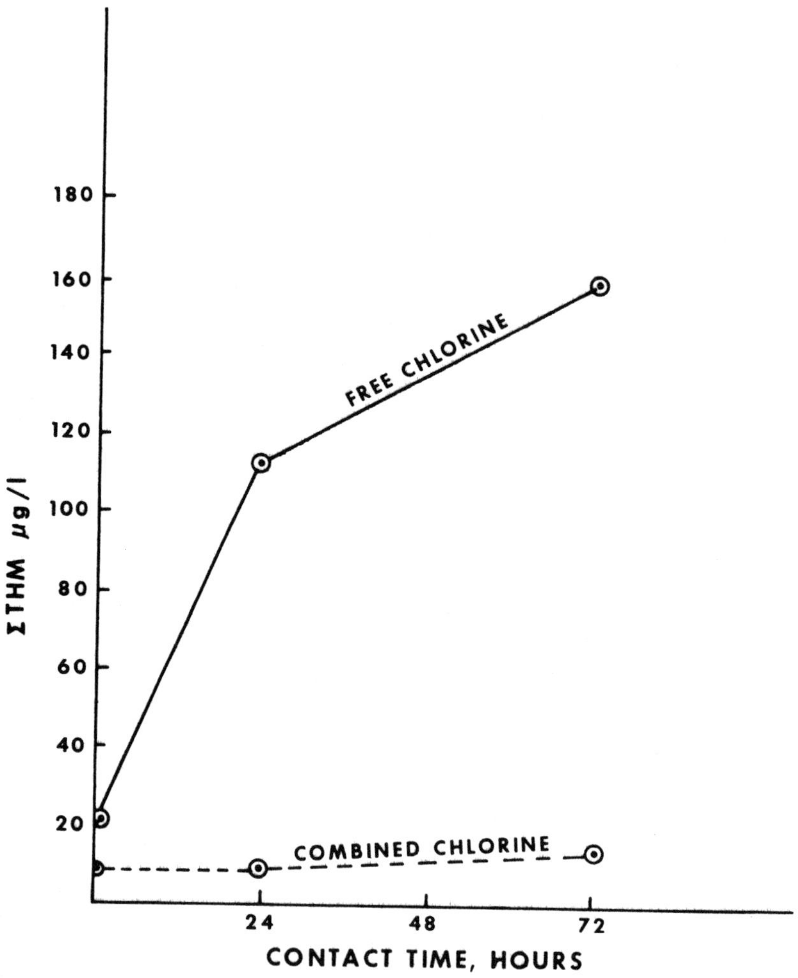

FIGURE 5: COMPARISON OF FREE AND COMBINED CHLORINE ON THE FORMATION OF TRIHALOMETHANES IN OHIO RIVER (AFTER STEVENS, ET AL[55])

hydrocarbons in drinking water, primarily as a result of chlorine disinfection, has precipitated investigations to evaluate the consequence of exposure to these compounds on human health.

Chloroform has been used medicinally for anestheia, as a cough suppressant, a carminative, a flavoring agent, and as a local irritant in liniments. Chronic ingestion of doses of 23-37 mg/K have been reported to produce some reversible hepatotoxic effects while ingestion of 0.3 to 0.96 mg/K for 1 to 5 years resulted in no observed hepatotoxicity[86].

Chloroform is rapidly absorbed from the gastrointestinal tract and other trihalomethanes may be expected to be absorbed and metabolized in a similar manner. Human exposure to chloroform may result from its presence in ambient air, from industrial exposure and in drinking water. A Threshold Limit Value (TLV) of 10 ppm in air had been established to protect workers[87] and long term clinical studies using a cough suppressant containing chloroform noted some reversible hepatotoxicity when about 2 grams of chloroform had been consumed per day for 10 years.[88] A short term, few animal study with mice gave an indication that chloroform might be carcinogenic in that species.[89]

In mid-1977 the National Institute of Occupational Safety and Health issued a "Revised Recommended Chloroform Standard" (unnumbered and undated) reducing the TLV to 2 ppm and its use in drugs, cosmetics and food packaging has been banned by the Food and Drug Administration[90] because the potential risk associated with its use are greater than its benefits.

Laboratory studies have produced LD_{50} toxicity data following acute oral exposure to chloroform at levels of 120 to 1350 mg/K in rats, male mice and male and female dogs.[91,92] No teratogenicity was determined in an oral exposure study with rats and rabbits.[93] Evidence of the animal carcinogenicity of chloroform has been confirmed by several studies that have been extensively reviewed in the Report of the National Academy of Sciences (NAS) titled "Drinking Water and Health."[94] They have found chloroform capable of producing malignant and metastatic neoplasms in at least one feeding study in mice and has produced tumors in both rats and mice in other studies. Utilizing the standard protocol for carcinogenic testing, a dose response relationship was found for epithelia tumors of the kidneys and renal pelvis in the rat and hepatocellular carcinomas in mice in studies by the National Cancer Institute[68], and the latency period for the carcinogenic effect decreased as the dose increased. Another study[95] showed hepatocellular carcinoma in femal rats and one strain of male mice. These studies established that chloroform is carcinogenic to the animals under those experimental conditions and therefore, presents a potential carcinogenic risk to humans.

The National Academy's Safe Drinking Water Committee[94] calculated the upper 95% confidence estimate of risk at 3×10^{-5} at an average two liters consumption of water with 20 μg/ℓ of chloroform for a lifetime. (one excess case of cancer for every 33,333 persons exposed for a lifetime). The NAS suggested that strict criteria be applied when limits for chloroform in drinking water are established. The use of maximum tolerable dose in the toxicological studies and the mathematical extrapolation to human population effects have an air of unrealness, but are based on a philosophy of safety using what data are available.

Before an alternative disinfectant for drinking water can be considered for use, the potential toxicity of not only the disinfectant itself, but also the potential toxicity of the by-products, both inorganic and organic, of the reactions between the disinfectant and water constituents, must be evaluated. For example, sodium hypochlorite was reported to be a weak base substitution mutagen in Salmonella typhimurium.[96]

Epidemiology

When the first nationwide data were available from NORS[54] on chloroform concentrations, they were compared with cancer mortality (all sites) in the cities served by the supplies sampled and a significant correlation noted. These data have continued to be analyzed in more sophisticated ways. An association with bladder cancer and the trihalomethanes other than chloroform is the most solid finding.[97]

Data on water quality for carrying out epidemiological studies have been lacking and other classifications of exposures have had to be used in order to proceed with an evaluation of the problem. The studies of the Environmental Defense Fund[98] comparing Parishes (counties) were the first to create a public interest. A recent paper[99] used this technique of comparing mortality of counties served mostly by surface or ground water in Ohio. A more elaborate study of this type[100] was conducted by the University of North Carolina for all the counties in the Ohio River Valley. These studies found that some aspects of cancer death rates were higher in counties where most of the people used surface water as compared to counties where most of the people used ground water. The assumption in these studies was that surface waters are more heavily contaminated than ground waters or that more chloroform is produced when surface waters are chlorinated.

Another general classification can be made of drinking water -- chlorinated or not chlorinated. Three studies have used this dichotomy as a basis for comparison. The Johns Hopkins study[101] covered a small population for which many variables were controlled and only obtained suggestive results. The University of California at Los Angeles study[102] demonstrated the extreme mobility of people in that area and was not definitive. The Columbia University study[103] had a population of adequate size without excess mobility and uncovered some interesting results. Males living in the chlorinated drinking water areas of Erie, Rensselaer, and Schenectady Counties and females living in the chlorinated drinking water areas of Erie and Schenectady Counties are at a greater risk of gastrointestinal and urinary tract cancer mortality than are individuals living in non-chlorinated drinking water areas.

The epidemiological studies are suggestive that the toxicological studies are reasonable, but may have in this case underestimated the cancer risk. Studies with more definitive water quality data and a more rigorous design are now being conducted.

Ozone

Ozone is a powerful oxidant and as such, can be expected to oxidize many compounds present in surface waters. Whether these oxidation products are more harmful or less harmful than the reaction products of chlorine is not known at this time (Fall, 1977). Numerous studies, summarized below, have been performed to evaluate the potential toxicity of ozonated water and wastewater, the majority of such studies concerning themselves with the formation of carcinogenic, mutagenic, and teratogenic substances from the ozonolysis of organic contaminants.

Comparative toxicity studies of ozonated and chlorinated waters with a defined chemical composition of organic materials such as pesticides or detergents indicated chlorine treated water exhibited a higher toxicity than ozone treated water, but the greater toxicity to the cell cultur system was speculated to be because of the residual chlorine rather than oxidative products.[104]

Cytochrome P-450 reduction, as an indication of liver necrosis, was evidenced in one strain of mouse subjected to intraperitoneal injections of the products of ozonation, but was not apparent after injections with the products of chlorination, however, the results were reversed with a different strain of mouse.[104] In studies conducted to determine if ozone would decrease the cytotoxicity of a synthetic military hospital wastewater containing methanol, ethanol, isopropanol, acetone and acetic acid, ozone produced products were more toxic to test cells than non-ozonated waste.[105] Reversion of histidine dependent mutant strains of *Salmonella typhimurium* by the products of ozonation of organic compounds bearing various functional groups showed that ethanol and a mixture of ethanol, acetaldehyde, and hydrogen peroxide were mutagenic and the ozonation of benzidine resulted in a form that no longer required metabolic activation for mutagenicity. Positive results were not obtained in similar bioassays for samples of unconcentrated secondary treated municipal effluent that were a) untreated, b) chlorinated, c) chlorinated/dechlorinated, d) chlorobrominated, or e) ozonated.[106]

Another study currently (Fall 1977) being conducted by Dr. Gruener in progress at the Hebrew University in Jerusalem showed these preliminary results on reused water. A reverse osmosis concentrate of reused water was ozonated and then tested in the *Salmonella* mutagenicity test. The same water sample was checked before ozonation and found to be negative. Water ozonated for 15 minutes was mutagenic in the two strains used: TA100 (base pair substitution) and TA98 (frameshift). No activation by microsomal fraction was needed to cause mutations in TA100. In the strain TA98, although an effect existed in the absence of the liver fraction (about three times the control value), a four-fold increase in the number of colonies occurred compared to the control without the liver fraction in the presence of phenobarbital induced liver fractions; the number of colonies was 9-10 times higher in the presence of methylcholanthrene induced- and Aroclor 1254 induced liver fractions. Clearly, the toxicity problems are so complex that it cannot be stated at this time if ozone is better or worse than chlorine from the point of view of by-product formation.

Chlorine Dioxide

Although chlorine dioxide possesses many desirable characteristics as a disinfectant it is, none-the-less, a powerful oxidant capable of oxidizing and chlorinating many compounds with which it comes in contact. The properties of the reaction by-products as well as chlorite and chlorate ion (end products), - the reduction products of chlorine dioxide oxidation, are being thoroughly evaluated by the Health Effects Research Laboratory (HERL) both in-house and through extramural research .

Toxicity

Considerable disagreement exists among literature sources concerning the toxicity of chlorine dioxide[107] and the majority of reports are concerned with exposure in air, principally from the paper pulp industry, describing symptoms of local contact irritation and two cases of illness including one fatality from exposure to 19 ppm inside a bleach tank.[108]

The effects, on rats, of chronic administration of chlorine dioxide in drinking water[109] showed no discernable toxic effects at 10 mg/ℓ or less over a two year period although rats consuming 100 mg/ℓ showed a higher mortality rate at the end of the two year period. Enger[110] reported that laboratory workers, consuming one liter of water containing 0.7 mg chlorine dioxide, experienced a feeling of weakness and gastrointestinal troubles. Musil[111] supposed that the symptoms were caused by: a) the chlorine dioxide itself, b) the final products from the reaction between chlorine dioxide and substances in the water, or c) unreacted chlorite from the generation of the chlorine dioxide.

The toxicity of chlorite ion arising from the reaction of chlorine dioxide and organic matter or the disproportionation of chlorine dioxide in water, similarly, is not well understood. One of the reported advantages of chlorine dioxide disinfection is the ability to maintain a residual in the distribution system.[112] Conversely, Myhrstad and Samdal[113] found that the chlorine dioxide dosed was completely converted to chlorite within 35 minutes in water from a system using microstraining and ozonation for treatment. The potential for chlorite ion to be present in water exists and its toxicity must be evaluated.

Studies have determined the LD_{50} of chlorite ion, for the rat, to be 140 mg/K[111] and for sodium chlorite, 182 mg/K[114]; 208 mg/ℓ and 50.6mg/ℓ for bluegill and rainbow trout, respectively[115] and for mallard ducks, 100 mg/K.[116] Rats exposed to 8 mg/ℓ chlorite or less in drinking water had no discernible toxic effects over a two year period.[109]

Hopf[117] proposed that the toxicity of chlorite might be approximately the same as chlorate and reported symptoms of a "weak furry feeling" on the tongues of subjects who drank sodium chlorite concentrations of more than 0.3 mg/ℓ.

Hemolytic Anemia

The propensity of certain chemicals and drugs to induce hemolytic anemia has been a subject of much interest for almost a century. The majority of these substances are aromatic compounds containing amino-, nitro-, or hydroxy groups, although a number of inorganic compounds such as hydroxylamines, nitrites, chloramines and chlorates are also active. Such compounds usually cause red cell injury without evidence of specific toxicity to other cells or tissues. The hemolytic process is characterized by two distinctive features in the affected red cells; a) the appearance of brownish or greenish derivatives of hemoglobin, including methemoglobin and, b) the formation of water-insoluble stainable granules called Heinz Bodies within the red cell.[118]

Methemoglobin is an oxidation product of the normal blood pigment, hemoglobin, in which the ferrous iron is oxidized to the ferric state and is unable to combine reversibly with oxygen and carbon dioxide, causing a shift in the oxygen dissociation curve thereby hindering the transfer of oxygen from the blood to the tissues. Abnormal amounts of methemoglobin will result in a functional anemia with cyanosis.[119] Musil[111] claimed that chlorite ion, as an oxidizing agent, carries the oxidation of hemoglobin to methemoglobin *in vivo*. He recommended a concentration of 0.0 mg chlorite per liter because of the hazard of methemoglobinemia in nursing babies. He further recommended that prior to the use of chlorine dioxide, other methods of treatment, such as activated carbon, should be considered. Based on these observations, the Norwegian Health Authority has recommended the absence of chlorite in drinking water.

The repeated administration of methemoglobin producing chemicals may cause hemolytic anemia, the exact mechanism of which is unknown as the chemicals themselves may not be directly hemolytic. Data dervied from experiments conducted at HERL indicated that *in vitro*, chlorite and chlorine dioxide will oxidize hemoglobin to methemoglobin and, in the intact human red cell, oxidizes hemoglobin with corresponding increases in the incidence of spicule cells of the acanthrocyte and stromatocyte type, morphological abnormalities that are associated with hemolytic disorders, after certain "threshold" levels are exceeded.

Continuous monitoring of the methemoglobin formed in cats from a single oral dose of sodium chlorite resulted in a dose response that indicated that chlorite was absorbed into the blood within 10 minutes, reached a methemoglobinemic maximum in 1-1/2 to 2 hours and was reduced at a rate of 16 percent per hour.

Although methemoglobin formation from a single dose of a compound does not necessarily indicate toxicity, an increasing degree of *in vivo* lysis of red cells was noted at doses above 20 mg/K and cats given 500 mg/ℓ chlorite in water exhibited a sharp decrease in hemoglobin content and packed cell volume after 2 weeks exposure, although methemoglobin was not detected. Similarly, rats given chlorite in water for 30 days resulted in significant ($p<0.05$) increases in red cell count at doses of 0.9 to 1.6 mg/K/day; decreases in hemoglobin and packed cell volume at doses of 11-16 mg/K/day; and decreased hemoglobin, packed cell volumes and red cell count at doses of 28-50 mg/K/day. Although methemoglobin was not detected in any of these animals, probably because the "threshold" level was not exceeded, a single dose of 50 mg/K resulted in excess of 50 percent of the animal's hemoglobin being oxidized to methemoglobin within 30 minutes.

Exposure of erythrocytes to mild oxidative stress converts hemoglobin to methemoglobin, a reaction that is reversible because metabolic pathways for the reduction of methemoglobin is present in normal erythrocytes. Further oxidative stress, probably involving the oxidation of -SH groups of the globin, causes irreversible damage with the formation of mixed disulfides between glutathione and the hemoglobin. Additional damage to the erythrocyte membrane from further exposure to oxidative stress has been postulated to lead to the destruction of the erythorocyte[120].
In the HERL laboratory, the erythrocyte survival rate of cats drinking 100 mg/ℓ chlorite in water (1.6 mg/K/day) was found to be reduced by

23.6 percent and the survival rate in cats drinking 500 mg/ℓ (5 mg/K/day) was reduced 27.2 percent, as compared to control drinking chlorinated tap water, ($p<0.05$).

Conversely, in other experiments conducted by HERL[121] in which an assessment of the effects of chlorine dioxide in drinking water on the hematopoietic system of African Green Monkey was made by exposing male and female monkeys to 20 and 200 mg/ℓ chlorine dioxide (avg. 1.72 and 8.96 mg/K/day, respectively) for 13 weeks, no statistical deviation from normal hemoglobin, hematocrit, red cell count, reticulocyte, methemoglobin or osmotic fragility values occurred. A series of other hematological parameters, for example, Heinz Bodies, differential count, red cell and leukocyte morphology were within normal limits and serum bilirubin and serum enzymes monitored throughout the study showed no abnormalities or statistically significant trends, indicating a stable homeostatic response.

Animal response data is difficult, at best, to extrapolate to the human. Species differences exist in the response to methemoglobinizing agents. The cat, for example, is generally recognized as the species most susceptible to methemoglobinemia, and is followed by man, dog, rat , rabbit and monkey in decreasing order of susceptibility to aromatic amine methemoglobin formation.[122] Segments of the human population also possess variations in methemoglobin susceptibility and sensitivity to oxidative stress to the hematopoietic system. Hemolytic anemia is frequently a serious problem among patients undergoing long-term hemodialysis. Newborn infants are all susceptible to drug induced hemolytic anemias[123] caused by an abnormality that persists for only a week or two. Genetically altered hemoglobins may confer abnormal sensitivity to nitrites and other oxidizing agents that cause methemoglobinemia. The abnormal pigment, hemoglobin H, is unusually sensitive to oxidation. The life span of these affected erythrocytes is about 40 days instead of the usual 120 days.

The enzyme deficienty that results in familial idiopathic methemoglobinemia has been known for a long time. Individuals with this condition are deficient in the ability to reduce methemoglobin. Studies have shown that erythrocytes with decreased glucose-6-phosphate dehyrogenase activity, an enzyme deficiency present in 13 percent of U.S. Black males, are more vulnerable to oxidant induced hemolysis[124] which may begin with precipitation of hemoglobin within the cell.[125] To continue the research in this area, studies are being initiated to determine, first, if animals deficient in glucose-6-phosphate dehydrogenase activity are more susceptible to oxidative stress caused by exposure to chlorite and chlorine dioxide in water.

In addition to these controlled laboratory studies, HERL is presently investigating the feasibility of conducting epidemiological studies on populations exposed to chlorine dioxide in their drinking water. Although numerous treatment plants use chlorine dioxide for periodic taste and odor control, few use it consistently or in concentrations exceeding 1 - 2 mg/ℓ. Appropriate sites are being sought at this time. Also, a community water supply in Massachusetts utilized chlorine dioxide exclusively during World War II. A retrospective mortality study in which the rates and causes of death and the incidence of neonatal jaundice for this community could be compared with similar areas utilizing conventional chlorine disinfection is being considered.

<u>Chloramines</u>

Only two reports of combined chlorine toxicity were found in the literature. Chloramines have been found responsible for two epidemics of acute hemolytic anemia characterized by Heinz Bodies in dialyzed uremic patients.[126] The effected patients had undergone dialysis in two hospitals that had recently adopted a reverse osmosis technique for purifying dialysis water. Patients in a third hospital, which utilized activated carbon filtered water, were not affected. Also monochloramine has been shown to be a weak mutagen when reversion of trpC to trp^+ in B. subtilis was used as an assay.[127]

Comparative Studies

A clinical trial is being formulated that will assess the human tolerance to each of the proposed alternate disinfectants, including chlorine, and some of the end products such as chlorite and chlorate. These trials will initially be conducted on normal healthy male subjects, followed by controlled exposure to levels of chemicals comparable to those used in water treatment practices to individuals possessing some of the deficiencies that may render them more susceptible to oxidative stress than the normal human population. The results of these studies will determine the absorption, distribution, and metabolism of these agents and provide a thorough assessment of human tolerance and effects, if any.

The comparative toxicity of by-products from the three proposed alternative disinfectants and chlorine is being evaluated by HERL in a project in which organic concentrates in aqueous solution obtained by a 100-fold reverse osmosis concentration technique, from undisinfected coagulated, settled and dual-media filtered Ohio River water and different samples of similar dual-media filtered water disinfected with chlorine, chlorine dioxide, ozone, or chloramine will be tested for cytotoxicity in both mammalian cells in culture and bacterial cells, bacterial mutagenesis, rat mortality, carcinogenic initiation/promotion, effects on DNA binding, and on Xenobiotic metabolic systems. In the first test, the concentrates from the four disinfectants were negative for bacterial mutagenesis and rat mortality and only the concentrate from the system disinfected with chloramines showed any significant bacterial cytotoxicity. At this time (Fall, 1977) the carcinogenic initiation/promotion studies are just starting.

SUMMARY

At the present time (Fall 1977) chlorine is known to produce trihalomethanes and possibly other chlorinated by-products. Ozone, chlorine dioxide, or chloramines on the other hand, either do not cause trihalomethane formation, or cause reduced quantities of trihalomethane to be produced. Evidence is growing, however, that as strong oxidants, both ozone and chlorine dioxide do produce non-trihalomethane by-products, and inorganic end products in the case of chlorine dioxide. Chloramines have not yet been studied in this regard. As yet, however, none of the organic by-products of ozone or chlorine dioxide have been demonstrated to be formed at concentration levels comparable to the trihalomethanes. Both in-house and extramural research is continuing in the area.

Chloroform has been demonstrated to be a carcinogen in some animal tests and epidemiological studies support the concept of some adverse effects from chlorination. Similar data are not currently (Fall, 1977) available for ozone, chlorine dioxide and chloramines. Preliminary studies both by EPA and others have not shown any definitive dramatic adverse effects from the three proposed alternates, but the completion of on-going research is needed to add confidence for the continued use of these alternatives. One possible result of these studies could be the recommendation of a maximum acceptable dose for any or all of these disinfectants.

COST OF DISINFECTION

If all other factors are equal, drinking water treatment unit processes should be selected on a minimum cost basis. In the case where the least expensive unit process has an unacceptable risk, however, a higher cost process may actually have a more favorable benefit/cost ratio than a minimum cost process and might be selected. The fifth criterion of a disinfectant alternate to chlorine should be cost-effectiveness. The costs of the alternate disinfectants are discussed below.

GENERAL CONSIDERATIONS

The costs in this section are intended for the development of planning estimates only and not for the preparation of bid documents or detailed cost estimates. Exact capital and operating costs are highly variable from location to location within the United States, even for plants of the same size and design. Variables -- such as local costs of land, materials and labor; state or regional differences in building codes; and existing facilities suitable for modification -- may accentuate the differences in treatment costs for similar plants to reduce chloroform and other trihalomethanes in drinking water.

These cost data are presented in such a way as to enable the planner to make adjustments to the reported costs when local information is available. For example, operation and maintenance costs can be reduced if the delivered cost of chemicals is less than the costs upon which the estimates were based. A cost index is used to provide a baseline for projecting costs and for estimating escalation because of inflation.

BASIS OF COST ESTIMATES

Much of the basic information utilized in this report was obtained from data developed by the Systems and Economic Evaluation Section (SEES) of EPA's Wastewater Research Division.[128,129] For example, computations for chlorine costs were performed using a computer program developed by SEES, but operational modifications were assumed in the analysis to reflect conditions unique to water supply.

Costs reported as Capital Costs include:

a. construction for site preparation,

b. plant construction,

c. legal, fiscal and administrative service,

d. interest during construction, and

e. start-up costs.

Costs reported as Operation and Maintenance Costs include:

a. chemical costs,

b. labor costs, and

c. operation and maintenance costs, such as utilities, annual replacement of expendable items, etc.

CHLORINE

In this analysis, a number of variables have been evaluated as to their effect on the cost of chlorination. Baseline or standardized values for a set of design parameters were assumed and the cost of chlorination was calculated. The parameters and their associated "standardized" levels are shown in Table II.

TABLE II

Design Parametes for the Cost of Chlorination

Design Parameter (Variable)	Level
Chlorine Dose	2 mg/l
Chlorine Contact Time	20 min.
Construction Cost Index	273.8
Cost of Chlorine	300 $/ton
Wholesale Price Index	194.6
Direct Hourly Wage Rate	$5.58/hr
Amortization Interest Rate	7%
Amortization Period	20 yr.
Capacity Factor	70 percent

Assuming design flow rates of 1, 5, 10, 100 and 150 mgd, at an operating to design ratio of 0.7 and the design parameter values in Table II, the unit costs for chlorination are shown in Table III.

TABLE III

Unit Chlorination Costs (¢/1000 gal)*

	Design Capacity				
Item	1 mgd	5 mgd	10 mgd	100 mgd	150 mgd
Capital Cost	2.19	0.88	0.62	0.26	0.24
Operating Cost	1.06	0.56	0.46	0.32	0.31
Total Cost	3.25	1.44	1.08	0.58	0.55

*Operating costs are calculated on 70% on design capacity.

OZONE

The standardized design parameters listed in Table IV were used as the basis for analysis. Because of increased biocidal activity the disinfectant dose of ozone was assumed to be one-half the chlorine dose used in Table II.

TABLE IV

Design Parameters for Ozonation Costs

Design Parameters (Variables)	Level
Ozone dosage	1 mg/ℓ
Ozone contact time	20 min.
Electrical Power Cost	$0.03/kw-hr
Construction cost index	273.8
Cost of Oxygen	$0.046/lb.
Wholesale Price Index	194.6
Direct Hourly Wage Rate	$5.58/hr
Amortization Interest Rate	7%
Amortization Period	20 yr
Capacity Factor	70 percent

Costs based on these parameters have been calculated for ozone generated from both air and oxygen. These costs are shown in Table V and VI.

TABLE V

Unit Costs for Disinfection with Ozone - Air (¢/1000 gal.)

	Design Capacity				
Item	1 mgd	5 mgd	10 mgd	100 mgd	150 mgd
Capital Cost	2.90	1.36	1.11	0.76	0.73
Operating Cost	2.75	1.05	0.77	0.40	0.38
Total Cost	5.56	2.41	1.88	1.16	1.11

Operating costs are based on 70% of capacity.

TABLE VI

Unit Costs for Disinfecting with Ozone - Oxygen (¢/1000 gal)

	Design Capacity				
Item	1 mgd	5 mgd	10 mgd	100 mgd	150 mgd
Capital Cost	4.46	1.50	1.08	0.61	0.58
Operating Cost	2.87	1.17	0.88	0.52	0.49
Total Cost	7.33	2.67	1.96	1.13	1.07

MGD x 43.81 = l/sec

CHLORINE DIOXIDE

In this cost analysis, the dose of chlorine dioxide was assumed to be half the dosage of chlorine to achieve equivalent disinfection results. Therefore, 1 mg/ℓ of chlorine dioxide would achieve disinfection results equivalent to those achieved by 2 mg/ℓ of chlorine, see Table II. If 0.5 mg/ℓ of chlorine is combined with 1.6 mg/ℓ of technical grade sodium chlorite, a 1 mg/ℓ dosage of chlorine dioxide will result. The equation below shows this relationship (assuming 80 percent pure $NaClO_2$ yields 1.3 mg/ℓ of reactive material):

	$2\ NaClO_2$	$+\ Cl_2$	$\rightarrow\ ClO_2$	$+\ 2\ NaCl$
Mol.Wt.	(181)	(71)	(135)	(117)
	1.3 mg/ℓ	0.5 mg/ℓ	0.9 mg/ℓ	0.8 mg/ℓ

Therefore, the chlorine feeding system and contact basins were estimated for a 0.5 mg/ℓ dosage of chlorine with the rest of the standardized values fixed at the levels shown in Table VII.

TABLE VII

Design Parameters for the Cost of Chlorine Dioxide

Design Parameters (Variable)	Level
Chlorine Dose	0.5 mg/ℓ
Contact Time	20 min.
Cost of Sodium Chlorite	$0.70/lb
Chlorine Cost	$300/ton
Construction Cost Index	273.8
Ratio of Sodium Chlorite Concentration to Chlorine Concentration	3.2
Wholesale Price Index	194.6
Direct Hourly Wage Rate	$5.56/hr
Amortization Interest Rate	7%
Amortization Period	20 yrs.
Capacity Factor	70 percent

TABLE VIII

Unit Cost of Disinfecting with Chlorine Dioxide (¢/1000 gal.)*

	Design Capacity				
Item	1 mgd	5 mgd	10 mgd	100 mgd	150 mgd
Capital Cost	1.90	0.76	0.51	0.22	0.20
Operating Cost	1.55	1.18	1.12	1.03	1.02
Total Cost	3.45	1.90	1.63	1.25	1.22

*Operating Costs are based on 70% of design capacity.

CHLORAMINES

The standardized levels for cost estimation are shown in Table IX. Assuming design flows of 1, 5, 10, 100 and 150 mgd and an operating to design ratio of 0.7 and the design parameter values shown in Table IX, the unit costs for chloramines are shown in Table X.

TABLE IX

Design Parameters for Chloramine Costs

Design Parameters (Variable)	Level
Chlorine Dose	0.5 mg/ℓ
Contact Time	20 min.
Construction Cost Index	273.8
Cost of Chlorine	$300/ton
Cost of Ammonia	$200/ton
Ratio of Chlorine to Ammonia Conc.	5.0
Wholesale Price Index	104.6
Direct Hourly Wage Rate	$5.58/hour
Amortization Rate	7%
Amortization Period	20 yrs.
Capacity Factor	70 percent

TABLE X

Unit Chloramine Costs (¢/1000 gal.)*

	Design Capacity				
Item	1 mgd	5 mgd	10 mgd	100 mgd	150 mgd
Capital Cost	1.70	0.62	0.42	0.17	0.15
Operating Cost	0.63	0.25	0.19	0.10	0.10
Total Cost	2.33	0.87	0.61	0.27	0.25

*Operating costs are calculated on 70% of design capacity.

SUMMARY

Unit costs for five disinfection processes have been developed and are summarized below in Table XI. Costs for chlorination, chlorine dioxide and ozonation are developed on the assumption of equivalent disinfecting capability. The costs for chloramines are calculated on the assumption that the values shown in Table IX yield 3 mg/ℓ of monochloromine, which is not equivalent in disinfecting power to the other disinfectants. Therefore to obtain equivalent biocidal activity chloramines must be used with some other disinfectant, such as chlorination or in high quality waters where high biocidal activity is not needed.

TABLE XI

Summary of Unit Costs for Disinfectants (¢/1000 gal.)

	Design Capacity				
Item	1 mgd	5 mgd	10 mgd	100 mgd	150 mgd
Chlorination 2 mg/ℓ	3.25	1.44	1.08	0.58	0.55
Cost of Ozone - Air 1 mg/ℓ	5.65	2.41	1.88	1.16	1.11
Cost of Ozone - Oxygen 1 mg/ℓ	7.33	2.67	1.96	1.13	1.07
Chlorine Dioxide 1 mg/ℓ	3.45	1.90	1.63	1.25	1.22
Cost of Chloramine 3 mg/ℓ*	2.33	0.87	0.61	0.27	0.25

*Not equivalent to the other disinfectants in biocidal activity.

To change from one disinfectant to another represents a minimal investment. This is illustrated by Table XII. The data in Table XII is derived from the assumption that a typical family of 4 consumers would use 117 thousand gallons of water in the home per year (80 gal/capita/day - a generous figure). Using data from Table XI an annual cost for disinfection per family is presented in Table XII, based on system size.

TABLE XII

Cost to a Family of Four in Dollars/Year for the Disinfection Unit Process

	Design Capacity				
Item	1 mgd	5 mgd	10 mgd	100 mgd	150 mgd
Chlorination	$3.80/yr	$1.68/yr	$1.26/yr	$0.68/yr	$0.28/yr
Ozonation - Air	$6.61/yr	$2.82/yr	$2.20/yr	$1.36/yr	$1.29/yr
Ozonation - Oxygen	$8.58/yr	$3.12/yr	$2.29/yr	$1.32/yr	$1.25/yr
Chlorine Dioxide	$4.04/yr	$2.22/yr	$1.91/yr	$1.46/yr	$1.43/yr
Chloramines*	$2.73/yr	$1.02/yr	$0.71/yr	$0.32/yr	$0.29/yr

*Not equivalent in biocidal activity to the other disinfectants

Therefore the annual cost per family for any disinfectant is small, and the cost increment for changing disinfectants is minimal.

DISCUSSION

Choosing the optimum or ideal disinfectant is a difficult task as each of the candidates discussed in this paper has advantages and disadvantages: See Table XIII. Other disinfectants are currently (Fall 1977) in minor use and were not discussed in this paper. In this country, historically, the disinfectant of choice was chlorine. Chorine was inexpensive and very effective in controlling waterborne disease and although it sometimes imparted taste and odor to water, it did not produce any obvious effect on people's health. Subsequent studies showed humans had a high tolerance to chlorine itself.[130] Later chloramines were substituted in some cases to prevent taste and odor, although some biocidal activity was sacrified. Outside of the United States the disinfectants ozone and chlorine dioxide were widely used. Although slightly more expensive, they were effective, did not impart a taste and odor to treated water, and did not cause any obvious effects on the health of consumers. In practice, the disinfectant dosage was often less than with chlorine, which reduced the additional cost.

Currently (Fall 1977) the situation is changing. Improved analytic techniques for organic chemicals have shown that chlorine produces at least one carcinogenic by-product, and epidemiological studies indicate some adverse effect on health may be caused by drinking chlorinated water. Therefore, the U.S. Environmental Protection Agency is considering establishment of limits for the concentrations of disinfection by-products, particularly trihalomethanes. Should such regulations be promugulated, they will cause changes in drinking water treatment practices in this country.

The concentrations of trihalomethanes can be controlled either by changing disinfectants such that less trihalomethanes are produced or by treating the water, for example with granular absorbants, such that chlorine could still be used, but trihalomethane concentrations would be reduced. Under these latter circumstances, the benefits of chlorination could be maintained without the corresponding undesirable organic by-products being produced and no alternate disinfectant would be required. This approach would have the additional benefit of removing synthetic organics found in raw water as well as controlling concentrations of biological nutrients entering the distribution system. Another treatment approach would be removal of the trihalomethanes themselves by treatment.

The details of the final regulation, if promulgated, will influence which of the above choices are available as alternate treatment procedures. If fairly high concentrations of trihalomethanes are permitted, free chlorine can continue to be used in many circumstances, possibly followed by ammonia to form chloramines. If the permissible concentrations of trihalomethanes are low, either now or in the future, the use of chlorine will be limited only to waters with low concentrations of precursors, either naturally or after treatment, and alternate disinfectants or treatment for removal of trihalomethanes will have more widespread use. Further, if in an attempt to limit the formation of all disinfection by-products, the regulation were to contain a total organic carbon concentration or disinfectant demand limit, chlorine could have a continuing role in disinfection in some instances.

Considering only the disinfectants themselves, their major characteristics are presented in Table XIII.

TABLE XIII

SUMMARY OF DISINFECTANT CHARACTERISTICS

	Chlorine	Ozone		Chlorine Dioxide	Chloramines
Use & Generation	Widespread in U.S. Off-site generation	Widespread outside U.S. On-site generation		Widespread outside U.S. On-site generation	Limited use On-site generation
Biocidal Activity	Affected by pH HOCl - third in activity OCl^- - weak	Little pH influence Most active		Little pH influence Second in activity	Affected by pH $NHCl_2$ - Fourth in activity NH_2Cl - Fifth in activity
Measur. & Persist. of Residual	Eight methods[42] HOCl & OCl^- not easily distinguished Persists	Titration method[42] Does not persist		DPD & FACTS Iodimetry, Mod. OT Amperiometric titration[42] ClO_2 & ClO^-_2 can be distinguished Certain by-products may interfere with DPD Persists - slowly changes to ClO^-_2 & ClO^-_3	Amperometric titration DPD Persists
By-products and End-products					
Chemistry	Trihalomethanes other chlorinated compounds, some of which are odorous	Some, but not yet well characterized No trihalomethanes		Some, but not yet well characterized No trihalomethanes ClO_2 & ClO^-_3 are end products	Unknown No trihalomethanes if no chlorine present
Health	$CHCl_3$ a carcinogen epidemiology supports some adverse effect	None positive yet, Still under study		None positive yet, Both by-products and end-products under study	None positive yet, Studies beginning
Total Cost Cents/1000 gal.					
	2 mg/ℓ	1 mg/ℓ		1 mg/ℓ	3 mg/ℓ*
		Air	Oxygen		
1 mgd	3.25	5.65	7.33	3.45	2.33
10 mgd	1.08	1.88	1.96	1.63	0.61
100 mgd	0.58	1.16	1.13	1.25	0.27
150 mgd	0.55	1.11	1.07	1.22	0.25

*Not equivalent in biocidal activity to the other disinfectants.

Because many on-going studies are not complete, the choice of a disinfectant alternative to chlorine must be made on the basis of incomplete information, if circumstances dictate that an alternate disinfectant is the proper approach. Although each of the alternatives have advantages and disadvantages, at the present time (Fall 1977), four alternatives are worthy of consideration: 1) chlorine dioxide; 2) ozone plus chlorine dioxide or chloramines to maintain a residual; 3) chlorine followed shortly by ammonia to stop the trihalomethane formation reaction; or 4) chlorine dosed at a concentration less than the chlorine demand followed by chlorine dioxide or chloramines to maintain a residual. Each of these options would reduce trihalomethane concentrations, would provide adequate initial disinfection and would provide a biocide* in the distribution system. Although theoretically attractive, chloramines are such weak biocides, that any option based on their use should be evaluated prior to recommendation, something that has not yet been accomplished for all situations.

Whether required by a regulation or not, another prudent step would be to limit the dose of any disinfectant, thereby reducing the potential for the formation of any undesirable by-products and end-products regardless of the disinfectant used. In a given situation, if the disinfectant demand is much greater than 1 mg/ℓ, treatment of the water prior to disinfection to reduce the disinfectant demand, see Figures 1 and 2, would be generally beneficial. As an extension of this concept, as noted above, chlorine could continue to be used as the disinfectant if the water being chlorinated were properly treated with a granular adsorbant to remove organic matter.

*Currently (Fall 1977) some State regulatory agencies specify a "chlorine" residual in the distribution system, something that must be taken into consideration where applicable. Another possible constraint to use of an alternate disinfectant is Section 141.21(h) of the National Interim Primary Drinking Water Regulations that discusses the substition of "chlorine" residual measurements for some required microbiological contaminant samples.

Although controlling disinfectant demand is a prudent concept, the problem of what to use as a raw water disinfectant, where one is needed or mandated, remains. The control of growths, such as fresh water clams, in raw water transmission lines, the prevention of slime growths in flocculating and settling basins or the destruction of ammonia without producing undesirable by-products and end-products is a practical problem that has yet to be completely evaluated. Periodic shock chlorination or the use of small doses of chlorine, ozone or chlorine dioxide may be effective for controlling microorganism development.

SUMMARY

Providing a wholesome drinking water to the consumer is the primary goal of all water purveyors. Effective killing or inactivating of pathogenic organisms through disinfection is the most important process in drinking water treatment and distribution. Because chlorination has been so effective in controlling waterborne disease over the past 70 years in this country, any departure from current practice must be made cautiously to avoid any deterioration in the microbiological quality of the Nation's drinking water. All other criteria by which to judge any new treatment proposal is secondary to the maintenance of a water free of all pathogens.

Several options exist for disinfecting water without by-product contamination. Treatment of water to remove by-product precursors or treatment to remove formed by-products are possibilities. This paper has focused on a third possibility, replacement of chlorine as the disinfectant.

Although all of the on-going research cited in this paper is not complete, enough is known at this time (Fall 1977) to suggest that "chlorine-free" or "low-chlorine-concentration" chlorine dioxide either as the primary disinfectant or in combination with ozone as the primary disinfectant, preferably used in waters with a disinfectant demand of about 1 mg/ℓ or less, would be two possible acceptable alternates to chlorine. Although not the only possible suggestions, these two meet all of the five criteria established at the beginning of this paper as necessary for any alternative to chlorine as a disinfectant.

ACKNOWLEDGEMENTS

The authors and the compiler of this paper wish to thank Mr. Gordon G. Robeck for his helpful suggestions during review and Mrs. Maura M. Lilly for typing the several drafts and the final paper.

REFERENCES

1. Lawrence, J. and Cappelli, F.P., "Ozone in Drinking Water Treatment: A Review.", The Science of the Total Environment, 7, 2, 99-108, (Mar 1977).

2. Mignot, J., "Practices for Contacting Ozone with Liquids to be Treated.", In: Proceedings of the Second International Symposium on Ozone Technology, Montreal, Canada,(May 11-14, 1975), 15-46.

3. Hill, A.G., "The Development of Ozonation: Application to Water and Wastewater Treatment.", Department of Chemical Engineering, Louisiana Tech. University, Ruston, Louisiana, (Dec 1976), mimeo, 66 pp.

4. OZONews--, 3, 1, Jan 1976, International Ozone Institute, Syracuse, N.Y.

5. Miller, G.W., "Status of Ozonation and Chlorine Dioxide Tech nologies for Treatment of Municipal Water Supplies.", EPA Research Grant No. R804385-01, Washington, D.C., (In Progress).

6. Ward, W.J., "Chlorine Dioxide-- A New Selective Oxidant/Disinfectant for Wastewaters.", In: Proceedings of the Forum on Ozone Disinfection, Chicago, Illinois, (June 2-4, 1976), 382-393.

7. White, G.C., "Handbook of Chlorination.", Van Nostrand Reinhold, New York, N.Y., (1972), 744 pp.

8. Anon., "Treatment of Water Supplies with Chlorine Dioxide.", Olin Corporation, Stamford, Conn., (undated), mimeo, 10 pp.

9. Gall, R.J., "Chlorine Dioxide-An Overview of its Preparation, Properties and Uses.", Hooker Chemicals and Plastics Corporation, Niagara Falls, N.Y., (1976), 36 pp.

10. Anon., "Statistical Summary of Municipal Water Facilities in the U.S.--Jan. 1, 1963.", PHS Publ. No. 1039, U.S. Dept. of Health, Education and Welfare, Public Health Service, Division of Water Supply and Pollution Control, Washington, D.C. 20201. (1965).

11. Race, J., "Chlorination and Chloromines.", JAWWA, 5,3,79, (Mar 1918).

12. Anon., "Chloramine at Denver Solves Aftergrowth Problem.", Eng. News-Record, 210, (Aug. 2, 1917).

13. Hale, F.E., "The Chloramine Process as Applied to the Catskill (ESOPUS) Water.", JAWWA, 6, 6, 804, (June 1919).

14. Scarpino, P.V., Berg, G., Chang, L.L., Kahling, D., and Lucas, M. "A Comparative Study of the Inactivation of Viruses in Water by Chlorine.", Water Research, 6, No. 8, pp. 959-965, (Aug. 1972).

15. Morris, J.C. "Disinfectant Chemistry and Biocidal Activities.", In: Proc. National Specialty Congress on Disinfection., Amherst, Mass., (1970), pp 609-633.

16. Peleg, M., "The Chemistry of Ozone in the Treatment of Water.", Water Research, 10, No. 5, pp. 361-365, (May, 1976).

17. Farooq, S., Chian, E.S.K., and Engelbrecht, R.S., "Basic Concepts In Disinfection with Ozone.", J. Water Pollution Control Federation, 49, No. 8, pp. 1818-1831, (Aug. 1977).

18. Engelbrecht, R., "Virus Sensitivity to Chlorine Disinfection of Water Supplies.", EPA Res. Grant No. R803346, Univ. of Illinois, (In Progress).

19. Benarde, M.A., Israel, B.M., Olivieri, V.O., and Granstrom, M.L., "Efficiency of Chlorine Dioxide as a Bactericide.", Applied Microbiology, 13, No. 5, pp. 776-780, (Sep. 1965).

20. Cronier, S., Scarpino, P.V., Zink, M.L., and Hoff, J.C., "Destruction by Chlorine Dioxide of Viruses and Bacteria in Water.", Abstr. Ann. Meeting American Society for Microbiology, (1977), p. 238.

21. Moffa, P.E., and Smith, J.E., "Bench-scale high-rate disinfection of Combined Sewer Overflows with Chlorine and Chlorine Dioxide.", EPA Report No. S-80240, Adv. Waste Treat. Res. Lab, NERC Cincinnati, Ohio, (Nov. 1974), 181 pp.

22. Kelly, S. and Sanderson, W.A., "The Effect of Chlorine in Water on Enteric Viruses.", American Journal of Public Health, 48, No. 10, pp. 1323-1334, (Oct. 1958).

23. Bates, R.C., Shaffer, P.T.B., and Sutherland, S.M., "Development of Polio-virus Having Increased Resistance to Chlorine Inactivation.", Applied and Environmental Microbiology, In Press, 34:6 (Dec. 1977).

24. Shaffer, P.T.B., Metcalf, T.G. and Sproul, O.J., "Chlorine Resistant Viruses from Drinking Water.", Presented at the National Environmental Engineering Conference of the American Society of Civil Engineers, Vanderbilt University, Nashville, TN, July 11, 1977.

25. Sharp, D.G., Floyd, R., and Johnson, J.D., "Initial Fast Reaction of Bromine on Reovirus in Turbulent Flowing Water.", Applied and Environmental Microbiology, 31, No. 2, pp. 173-181, (Feb. 1976).

26. Floyd, R., Johnson, J.D., and Sharp, D.G., "Inactivation by Bromine of Single Poliovirus Particles in Water.", Applied and Environmental Microbiology, 31, No. 2, (Feb. 1976).

27. Sharp, D.G., "Virus Particle Aggregation and Halogen Disinfection of Water Supplies.", EPA Report No. 600/2-76-287, Municipal Environ. Res. Lab, Cincinnati, OH, (Dec. 1976) pp. 59.

28. Floyd, R., and Sharp, D.G., "Aggregation of Poliovirus and Reovirus by Dilution in Water.", Applied and Environmental Microbiology, 33, No. 1, pp. 159-167, (Jan. 1977).

29. Young, D.C., and Sharp, D.G., "Poliovirus Aggregates and Their Survival in Water.", Applied and Environmental Microbiology, 33, No. 1, pp. 168-177, (Jan. 1977).

30. Newton, W.L. and Jones, M.G., "The Effect of Ozone in Water on Cysts of Endamoeba histolytica.", American Journal of Tropical Medicine, 29, No. 5, pp. 669-681, May 1949).

31. Palin, A.T., "Water Disinfection - Chemical Aspects and Analytical Control.", In: Disinfection Water and Wastewater. J.D. Johnson, Ed., Ann Arbor Sci., Ann Arbor, Mich., 1970, 425 p.

32. Meyer, E.A. "Determination of Giardia Cyst Viability.", EPA Res. Grant No. R804898, Univ. of Oregon Health Sciences Center, (In Progress).

33. Chang, S.L., Berg, G., Clarke, N.A., and Kabler, P.W., "Survival and Protection Against Chlorination, of Human Enteric Pathogens in Free-living Nematodes Isolated from Water Supplies.", American Journal of Tropical Medicine, 9, No. 1, pp. 136-142, (Jan. 1960).

34. Tracy, H.W., Camarena, V.M., and Wing, F., "Coliform persistence in Highly Chlorinated Waters.", Journal of the American Water Works Association, 58, No. 9, pp. 1151-1159, (Sep. 1966).

35. Boardman, G.D., "Protection of Waterborne Viruses by Virtue of Their Affiliation with Particulate Matter.", Ph.D. Thesis., Univ. of Maine, 196 p., (Dec. 1976).

36. Stagg, C.H., Wallis, C., and Ward, C.H., "Inactivation of Clay-Associated Bacteriophage MS-2 by Chlorine.", Applied and Environmental Microbiology, 33, No. 2, pp. 385-391, (Feb. 1977).

37. Symons, J.M., and Hoff, J.C., "Rationale for Turbidity Maximum Contaminant Level.", In: Proc. AWWA Technol. Conf., Atlanta, Georgia, (Dec. 1975), pp. 2A-4a, 1-15.

38. Hoff, J.C., "Relationship of Turbidity to Disinfection of Potable Water.", USEPA Office of Water Supply, Symposium on Eval. of Microbiol. Stand. for Water, (1977), In press.

39. Kruse, C.W. and Olivieri, V.P., "Biological Evaluation of the Benefits of Maintaining a Chlorine Residual in Public Water Systems," EPA Res. Grant No. R804307, Johns Hopkins Univ. (In Progress).

40. Brodeur, T.P., Singley, J.E., and Thurrott, J.C., "Effects of a Change to Free Chlorine Residual at Daytona Beach, Florida.", In: AWWA Technol. Conf. Proceedings, San Diego, Calif., (Dec. 1976), pp. 3A-5, 1-12.

41a. Vendryes, J.H., "Experiences with the Use of Free Residual Chlorination in the Water Supply of the City of Kingston, Jamaica.", In: Proc. AIDIS Cong. of Washington, D.C., (1962).

41b. Buelow, R.W. and Walton, G., "Bacteriological Quality vs. Residual Chlorine," JAWWA, 63, No. 1, 28-35 (Jan. 1971).

42. Anon., "Standard Methods for the Examination of Water and Waste-Water." 14th ed., Amer. Pub. Health Assn., New York, N.Y., (1975).

43. Dailey, L., and Morrow, J., "On Stream Analysis of Ozone Residual,". In: Proceedings, First International Symposium on Ozone for Water and Wastewater Treatment, Vol. I, Rice, R.G., and Browning, M.E., Eds., International Ozone Institute, (1975).

44. Miltner, R.J., "Measurement of Chlorine Dioxide and Related Products" In: Proceedings AWWA Water Quality Technology Conf., Dec. 6-7, 1976, San Diego, Calif. Paper No. 2A-5; AWWA, Denver, Colorado, 1977.

45. Palin, A.T., "Methods for the Determination, in Water, of Free and Combined Available Chlorine, Chlorine Dioxide and Chlorite, Bromine, Iodine and Ozone, Using Diethyl-p-phenylene Diamine (DPD)." Jour. Inst. Water Engr., 21, 537, (1967).

46. Palin, A. T., "Analytical Control of Water Disinfection with Special Reference to Differential DPD Methods for Chlorine, Chlorine Dioxide, Bromine, Iodine and Ozone." Jour. Inst. Water Engr., 28, 139-154, (1974).

47. Adams, D., Carter, J., Jackson, D. and Ogelby, J., "Determination of Trace Quantities of Chlorine, Chlorine Dioxide, Chlorite and Chloramines in Water." Proc. Soc. Water Treatment and Exam., 15, 117, (1966).

48. Hong, C. and Rapson, W., "Analyses for Chlorine Dioxide, Chlorous Acid, Chlorite, Chlorate, and Chloride in Composite Mixtures." Can. Jour. Chem. 46, 2061, (1968).

49. Miltner, R.J., "The Effect of Chlorine Dioxide on Trihalomethanes in Drinking Water." M.S. Thesis, University of Cincinnati, August 1976.

50. Seeger, D.R., USEPA Cincinnati, personal communication, 1977.

51. Johnson, J.D., "Measurement and Persistence of Chlorine Residuals in Natural Water" In: Proceedings of the Conference on the Environmental Impact of Water Chlorination, CONF-751096, R.L. Jolley, Ed., Oak Ridge National Laboratory, Oak Ridge, Tennessee (1976).

52. Rook, J.J., "Formation of Haloforms During Chlorination of Natural Waters," Water Treatment and Examination, 23, Part 2, 234-243, (1974).

53. Bellar, T.A., Lichtenberg, J.J. and Kroner, R.C., "The Occurrence of Organohalides in Chlorinated Drinking Water," Jour. AWWA, 66, No. 12, 703, (Dec. 1974).

54. Symons, J.M., Bellar, T.A., Carswell, J.K., DeMarco, J., Kropp, K.L., Robeck, G.G., Seeger, D.R., Slocum, C.J., Smith, B.L., and Stevens, A.A., "National Organics Reconnaissance Survey for Halogenated Organics in Drinking Water," Water Supply Research Laboratory and Methods Development and Quality Assurance Laboratory, National Environmental Research Center, U.S. EPA, Cincinnati, Ohio, Jour. AWWA, 67, 634, (Nov. 1975).

55. Stevens, A.A., Slocum, C.J., Seeger, D.R., Robeck, G.G., Chlorination of Organics in Drinking Water. In: Proceedings, Conference on the Environmental Impact of Water Chlorination, CONF-751096, R.L. Jolley, Ed., Oak Ridge National Laboratory, Oak Ridge, Tennessee. See also: Jour. AWWA, 68. No. 11, 615, (Nov. 1976).

56. Love, O.T., Jr., Carswell, J.K., Miltner, R.J. and Symons, J.M., "Treatment for the Prevention or Removal of Trihalomethanes in Drinking Water," In: Symons, J.M., "Interim Treatment Guide for the Control of Chloroform and Other Trihalomethanes," Water Supply Research Division, Municipal Environmental Research Laboratory, U.S. EPA, Cincinnati, Ohio (June 1976) App. 3

57. Coleman, W.E., Lingg, R.D., Melton, R.G., Kopfler, F.C., "The Occurrence of Volatile Organics in Five Drinking Water Supplies Using Gas Chromatography/Mass Spectrometer (GC/MS)," In: "Identification and Analysis of Organic Pollutants in Water," 1st ed, Keith, L. H., Ed., Ann Arbor Science Publishers, Inc., Ann Arbor, Michigan, (1975) Chapter 21.

58. Gould, E.S., "Mechanism and Structure in Organic Chemistry," Holt, Reinhart & Winston, New York, (1964).

59. Stevens, A.A., Symons, J.M., "Measurement of Trihalomethane and Precursor Concentration Changes" Jour. AWWA, 69, No. 10, 546-554, (Oct. 1977).

60. Rook, J.J., "Haloforms in Drinking Water," Jour. AWWA, 68, No. 3, 168-172 (Mar. 1976).

61. Seeger, D.R., 1976, U.S. EPA, Cincinnati, personal communication

62. Bunn, W. W., Haas, B.B., Deane, E.R., and Keopfler, R.D., "Formation of Trihalomethanes by Chlorination of Surface Water," Environmental Letters, 10, No. 3, 205-214, (1975).

63. Moore, L., 1976, U.S. EPA, Cincinnati, personal communication.

64. Rook, J.J., "Chlorination Reactions of Fulvic Acids in Natural Waters," Environmental Science & Technology, 11, No. 5, 478, (1977).

65. Jolley, R.L., "Chlorine-Containing Organic Constituents, in Chlorinated Effluents," Jour. WPCF, 47, No. 3, 601-618, (1975).

66. Glaze, W.H., and Henderson, J.E., IV, Formation of Organochlorine Compounds from the Chlorination of a Municipal Secondary Effluent, Jour. WPCF, 47, No. 10, 2511-2515, (Oct. 1975).

67. Burttschell, R.H., Rosen, A.A., Middleton, F.M., and Ettinger, M.B., "Chlorine Derivatives of Phenol Causing Taste and Odor," Jour. AWWA, 51, 205-214, (Feb. 1959).

68. Anon., "Report on the Carcinogenesis Bioassay of Chloroform," Carcinogen Bioassay and Program Resources Branch, Carcinogenesis Program, Division of Cancer Cause and Prevention, National Cancer Institute.

69. Train, R.E., U.S. EPA News Release, March 29, 1976.

70. Sontheimer, H., Heilker, E., Jekel, M., Nolte, H. and Vollmer, F.H., "The 'Mulheim Process' - Experience with a New Process Scheme for Treating Polluted Surface Water, " Jour. AWWA, In Press.

71. "Interim Treatment Guide for the Control of Chloroform and Other Trihalomethanes," Water Supply Research Division, Municipal Environmental Research Laboratory, Office of Research and Development, U.S. Environmental Protection Agency, Cincinnati, Ohio 45268, June 1976.

72. Gordon, G., Kieffer, R.G., and Rosenblatt, D.H., "The Chemistry of Chlorine Dioxide: In: Progress in Inorganic Chemistry, Vol. 15, Lippard, S.J., Ed., Wiley - Interscience, New York, (1972), p. 201.

73. Black, A.P., and Christman, R.F. "Chemical Characteristics of Fulvic Acids," Jour. AWWA, 55, 897 (1963).

74. Stevens, A.A., Seeger, D.R., Slocum, C.J., "Products of Chlorine Dioxide Treatment of Organic Materials in Water." To be published in: "Proceedings of Workshop on Ozone/Chlorine Dioxide Oxidation Products of Organic Materials, Nov. 17-19, 1976, Cincinnati, Ohio.

75. Morris, J.C., "Formation of Halogenated Organics by Chlorination of Water Supplies," EPA-600/1-75-002, U.S. Environmental Protection Agency, Washington, D.C., (1975).

76. Dence, C.W., Gupta, M.K., and Sarkanen, R.V., "Studies on Oxidative Delignification Mechanisms, Part II. Reactions of Vanillyl Alcohol with Chlorine Dioxide and Sodium Chlorite.", Tappi, 45, 29, (1962).

77. Dence, C.W., and Sarkanen, K.V., "A Proposed Mechanism for the Acidic Chlorination of Softwood Lignin," Tappi, 43, 87, (1960).

78. Glabisz, U., "The Reactions of Chlorine Dioxide with Components of Phenolic Wastewaters - Summary. Monograph 44, Polytechnic University, Szczecin, Poland (1968).

79. Bellar, T. A., and Lichtenberg, J. J., "Determining Volatile Organics at Microgram-Per-Liter Levels by Gas Chromatography," Jour. AWWA, 66, No. 12, 739-744, (Dec. 1974).

80. Christman, R.F., and Ghassemi, M., "Chemical Nature of Organic Color in Water," Jour. AWWA, 58, 723, (1966).

81. Smillie, R.D., Nicholson, A.A., Meresz, O., Suholke, W.K., Rees, G.A.V., Roberts, K., and Fund, C., "Organics in Ontario Drinking Waters, Part 11, A Survey of Selected Water Treatment Plants," OTC 7703, AAN-7504, Ontario Ministry of the Environment, 1977.

82. Tuepker, J.L., "Sampling and Analysis of Chloro-Organics in the Distribution System," Proceedings 4th Annual AWWA Water Technology Conference, Dec. 1976, San Diego, California, AWWA, Denver, Colorado, 1977.

83. Lambert, M., Vilagines, R., Montiel, A., and Derreumaux, A., "A Comparative Study of Halomethanes Formation During Drinking Water Treatment by Chlorine or its Derivatives" Presented at the 96th Annual AWWA Conference, Anaheim, CA., May 8, 1977.

84. Harms, L. "Preventing Haloform Formation in Drinking Water," South Dakota School of Mines and Technology. EPA Research Grant No. R805149-01 (In progress).

85. Razor, W., "Organic Substances in the Ohio River and Associated Water Supplies," Ohio River Valley Water Sanitation Commission (ORSANCO) EPA Grant No. 804615-02 (In progress).

86. DeSalva, S., Volpe, A., Leigh, G., and Regan, T., "Long Term Safety Studies of a Chloroform-Containing Dentrifice and Mouth Rinse in Man," Food Cosmet. Toxicol., 13, 529-532 (1975).

87. Anon. "Criteria for a Recommended Standard -- Occupational Exposure to Chloroform," National Institutes of Occupational Safety and Health, HEW Publication No. (NIOSH) 75-114, U.S. Government Printing Office, Washington, D.C. (1974).

88. Wallace, C.J., "Hepititis and Nephrosis Due to Cough Syrup Containing Chloroform," Calif. Med., 73, 442 (1959).

89. Eschenbrenner, A., "Induction of Hepatomasin Mice by Repeated Oral Administration of Chloroform with Observations on Sex Differences," J. Nat. Cancer Inst., 5, 251-255 (1945).

90. Federal Register, 41, No. 126 (June, 1976). p. 26842

91. Kimura, E.T., Ebert, D.M. and Dodge, P.W. "Acute Toxicity and Limits of Solvent Residue for Sixteen Organic Solvents," Toxicol. Appl. Pharmacol., 19, 699-704 (1971).

92. Hill, R.N., Clemens, T.L., Liu, D.K., Vesell, E.S., and Johnson, W.D., "Genetic Control of Chloroform Toxicity in Mice," Science, 190, 159-161 (1975).

93. Thompson, D.J., Warner, S.D., and Robinson, B.V., "Teratology Studies on Orally Administered Chloroform in the Rat and Rabbit," Toxicol. Appl. Pharmacol. 29, 348-357 (1974).

94. Anon. Drinking Water and Health, A Report of the Safe Drinking Water Committee, Advisory Center on Toxicology Assembly of Life Sciences, National Research Council, National Academy of Sciences, Washington, D.C. (1977).

95. Roe, F.J.C., "Preliminary Report of Long-Term Tests of Chloroform in Rats, Mice and Dogs," Unpublished Report (1976).

96. Wlodkowski, T.V., Rosencranz, H.S. "Mutagenicity of Sodium Hypochlorite for Salmonella typhimurium" Mutation Res., 31, No. 1, 39-42 (1975).

97. Cantor, K.P. "The Epidemiological Approach to the Evaluation of the Health Effects of Water Chlorination" Presented at Conference on Water Chlorination: Environmental Impact and Health Effects, Oct. 31 - Nov. 4, 1977, Gatlinburg, Tennessee.

98. Page, T., Harris, R.H., and Epstein, S.S., "Drinking Water and Cancer Mortality in Louisiana", Science, 193, 55-57, (July 2, 1976).

99. Kuzma, R.J., Kuzma, D.M. and Buncher, C.R., "Ohio Drinking Water Source and Cancer Rates" Am. J. of Public Health, 67, No. 8, 725-729 (1977).

100. Salg, J., "Cancer Mortality Rates and Drinking Water in 346 Counties of the Ohio River Valley Basin," Report on purchase order No. 5-03-4528J U.S. EPA, Cincinnati, Ohio, mimeo. (1977).

101. Kruse, C., "Preliminary Report: Chlorination of Public Water Supplies and Cancer, Washington County, Maryland." Report on purchase order No. CA-6-99-3284 U.S. EPA, Cincinnati, Ohio, mimeo. (1977).

102. Spivey, G.H., Sloss, E., and Mah, R.A., "Cancer and Chlorination of Drinking Water (Los Angeles County)" Part II, Report on purchase order No. CA-6-99-3349J U.S. EPA, Cincinnati, Ohio, mimeo. (1977).

103. Alavanja, M., Goldstein, I., and Susser, M., "Case Control Study of Gastrointestinal Cancer Mortality in Seven Selected New York Counties in Relation to Drinking Water Chlorination." Report on purchase order No. CA-7-2805-J. U.S. EPA, Cincinnati, Ohio, mimeo. (1977). Also presented at Conference on Water Chlorination: Environmental Impact and Health Effects, Oct. 31 - Nov. 4, 1977, Gatlinburg, Tennessee.

104. Hartemann, J., "Biochemical Aspects of the Toxicity Involved by the Ozone Organic Oxidation Products in Water," presented at IOI Workshop, Cincinnati, Ohio, (Nov. 1976).

105. Kinman, R., et al., "Effects of Ozone on Hospital Waste Water Cytotoxicity." presented at IOI Workshop, Cincinnati, Ohio, (Nov. 1976).

106. Simmon, V.F., Eckford, S.L., "Methods for Evaluating the Mutagenic Activity of Ozonated Chemicals," presented at IOI Workshop, Cincinnati, Ohio, (Nov. 1976).

107. Haller, J.F. and Northgraves, W.W., "Chlorine Dioxide and Safety" TAPPI, 38, No. 4, 199-202, (1955).

108. Elkins, H.B., "The Chemistry of Industrial Toxicology," John Wiley and Sons, Inc., New York, N.Y. (1950) p. 87.

109. Haag, H.B. "The Effect on Rats of Chronic Administration of Sodium Chlorite and Chlorine Dioxide in the Drinking Water" Report to The Mathieson Alkali Works, Niagara Falls, N.Y. (Feb. 7, 1949).

110. Enger, M., "Treatment of Water With Chlorine Dioxide to Improve Taste," Gas und Wasser Fach, 14, 340-343 (Apr. 1, 1960).

111. Musil, J., Kontek, Z., Chalupa, J., and Schmit, P., "Toxicological Aspects of Chlorine Dioxide Application for the Treatment of Water Containing Phenol," Sf vys. Sk. Chem-Technol. Praze, 8, 327-345 (June 1963).

112. Augenstein, H.W., "Use of ClO_2 to Disinfect Water Supplies" JAWWA, 66, 716-717 (Dec. 1974).

113. Myhrstad, J.A. and Samdal, J.E., "Behavior and Determination of Chlorine Dioxide," JAWWA, 61, 205-208 (April 1969).

114. Sperling, F., "LD_{50} Determination of Sodium Chlorite" Report No. OM-B5 to Olin Mathieson Chemical Corp., Baltimore, Md. (Oct. 8, 1959).

115. Macek, K.J., "Acute Toxicity of Sodium Chlorite to Bluegill (Lepomis macrochirus) and Rainbow Trout (Almo gairdner:)" Report to Olin Water Services, Stamford, Conn., Dec. 1971.

116. Calandra, J.C. "Acute Oral Toxicity Study with Sodium Chlorite in Mallard Ducks." Report No. 1BT-J2118 to Olin Corporation, New Haven, Conn. (Jan. 9, 1973).

117. Hopf, W., "Experiments with Chlorine Dioxide Treatment of Drinking Water", Gas und Wasser Fach, 108, No. 30, 852-854, (1967).

118. Jandl, J.H., Engle, L.K. and Allen, D.W., "Oxidative Hemolysis and Precipitation of Hemoglobin, I. Heinz Body Anemias as an Acceleration of Red Cell Aging," J. Clin. Invest. 12, No. 39, 1818-1836, (Dec. 1960).

119. Davidson, I. and Henry, J.B., "Clinical Diagnosis by Laboratory Methods," 14th ed. Saunders, Philadelphia, PA (1962) p. 135.

120. Jaffe, E.R., "Introduction to Discussion of Red Cell Metabolism: Exp. Eye Res., 11, 306-309, (1971).

121. Tardiff, R.G., Bercz, J. "The Assessment of Chlorine Dioxide in Drinking Water on the Hematopoietic System of African Green Monkey." Project No. TAB-006 U.S. EPA, Health Effects Research Lab., Cincinnati, OH (Aug. 1977).

122. Spicer, S.S., "Species Differences in Susceptibility to Methemoglobin Formation," J. Pharm. Exp. Ther., 99, 185-194 (1950).

123. Goldstein, A., Aronow, L., Kalman, S., "Principles of Drug Action: The Basis of Pharmacology," 2nd Ed., John Wiley & Son, N.Y., N.Y., (1974). pp. 461.

124. Calabrese, E.J., Kojola, W.H. and Carnow, B.W., "Ozone: A Possible Cause of Hemolytic Anemia in Glucose-6-Phosphate Dehydrogenase Deficient Individuals," J. Tox. Environ. Health, 2, 709-712, (1977).

125. Kiese, M., "Methemoglobinemia: A Comprehensive Treatise," CRC Press Inc., Cleveland, Ohio (1974), p. 34.

126. Eaton, J.W., Kolpin, C.F. and Swofford, H.S., "Chlorinated Urban Water: A Cause of Dialysis Induced Hemolytic Anemia," Science, 181, No. 4098, 463-464 (Aug. 3, 1973).

127. Shih, K.L., Lederberg, J., "Chloramine Mutagenesis in Bacillus subtilis," Science, 192, No. 4244, 1141-1143 (June 11, 1976).

128. Patterson, W.L., and Banker, R.F., "Estimating Costs and Manpower Requirements for Conventional Wastewater Treatment Facilities for the Environmental Protection Agency," Black and Veatch, Consulting Engineers, Kansas City, Missouri (1971).

129. Eilers, Richard G. and Smith, Robert, "Executive Digital Computer Program for Preliminary Design of Wastewater Treatment Systems." NTIS-PB222765 (report NTIS-PB222764 (card deck), (November 1970).

130. Muegge, O.J., "Physiological Effects of Heavily Chlorinated Drinking Water," Jour. Am. Water Works Assn.. 48, No. 12, 1507-1509 (Dec. 1956).

Part 2

Technology Assessment

The material in Part 2 is based on: *An Assessment of Ozone and Chlorine Dioxide Technologies for Treatment of Municipal Water Supplies,* by G.W. Miller, R.G. Rice, C.M. Robson, R.L. Scullin, W. Kühn and H. Wolf of Public Technology, Incorporated, Washington D.C., prepared for the U.S. Environmental Protection Agency, Office of Research and Development, Municipal Environmental Research Laboratory, Cincinnati, Ohio, August 1978.

ACKNOWLEDGEMENTS

This report could not have been written without the assistance of many people and organizations throughout the United States, Europe, and Canada. The researching and writing of the report required the acquisition of information that had not previously been collected or well documented. It was only through the cooperation and assistance of many key individuals that this body of information was successfully put together. Unfortunately, not everyone who contributed can be mentioned here.

Principal among those persons are the forty-three or so water plant superintendents who, in most cases, gave much of their valuable time to our site visit team in order to assure our understanding of their water treatment processes. Their contribution is gratefully acknowledged.

The first persons whose contributions must be acknowledged are two excellent engineers, Ronald L. Scullin and Daniel H. Houck of the Public Technology, Inc. staff. Without their efforts, the "nuts and bolts" portions of the report would not have been accomplished.

Grateful acknowledgment is extended to Professor Dr. Heinrich Sontheimer of the Engler-Bunte Institute, University of Karlsruhe, Federal Republic of Germany. Professor Sontheimer spent many hours with the project team answering questions on European water treatment practices. He was also instrumental in securing the services of Dr. Wolfgang Kühn, one of his outstanding young research assistants, to work with PTI on this project.

Both of the large, vertically integrated water companies that operate in France made significant contributions in many different ways. They forwarded detailed cost information, filled out questionnaires, helped to arrange for site visits, accompanied the site visit teams to many plants in France, and also answered hundreds of detailed questions. Our thanks go to all of those people in Compagnie Générale des Eaux (CGE) and in the Société Lyonnaise des Eaux et de l'Eclairage (SLEE) who assisted us. Pierre Schulhof, Vice President of CGE and Jean-Jacques Prompsy, Director of Operations of SLEE receive our special thanks because they not only assisted the team personally, but provided the services of their key personnel as well.

Several manufacturers of ozone generating equipment spent a great deal of time discussing philosophy, history, and actual application of ozone. The contributions of Jacques Le Paulouë of Trailigaz Corporation, Jean Mignot of Degrémont Corporation, and Dr. Harvey Rosen of Union Carbide Corporation are gratefully acknowledged. Our thanks go also to André Gagnaux, formerly of Sauter Corporation in Switzerland, for his assistance.

In Canada, six individuals stand out as having given much time, effort, and information to the project team. Foremost among these persons is Ron

Larocque of Francis Hankin Ltd. in Montreal. Mr. Larocque accompanied the site visit team to several Canadian plants and provided substantial detailed information on ozone systems engineering and costs. The efforts of Olaf Skorzewski, Vice President of Degrémont-Canada, were also substantial and are acknowledged. Our thanks also go to Meyer Schwartz of Fenco Consultants in Toronto for the insights into ozone systems which he provided in the early days of the project. Lastly, the efforts of Dr. Donald MacGregor of the Department of National Health and Welfare in Canada and of Messrs. Ronald Lampron and Roland Mercier of the Government of Quebec cannot be overemphasized. These gentlemen were instrumental in getting questionnaires completed and returned for 18 plants in the province of Quebec that use ozone.

Several top researchers were instrumental in assessing various aspects of ozone and chlorine dioxide use. One of the most impressive persons that the site visit team had the pleasure of meeting was Dr. Willy Masschelein of the Tailfer (Brussels) waterworks. Dr. Masschelein, who wrote a review book on chlorine dioxide in 1969 (which is being updated, in English, for publication in 1978), is an acknowledged expert on ozone contacting and water treatment in general. His contributions were invaluable.

Another top scientist who assisted Dr. Wolf in determning the public health aspects of O_3/ClO_2 use is Dr. Cyril Gomella of SETUDE in Paris. Dr. Gomella's work in the field of bacterial and viral inactivation is well known. We thank him for his contribution.

In addition to Dr. Masschelein, two water plant personnel who were most cooperative and impressive must be mentioned. These are Dr. Dietrich Maier, Chief Chemist of the Bodensee Wasserversorgung in Sipplingen, Germany, and Maarten Schalekamp, Director of the Lengg Waterworks in Zürich, Switzerland. Dr. Maier is currently heading up an international committee on ozone monitors and instrumentation that will publish the first report on this subject in mid 1978.

The contributions of Jean Chedal of CGE, Paul Thaler Blue of Infilco-Degremont in Richmond, VA, and Maurice Paré of Degrémont deserve mention. Paré provided some of the best theoretical explanations of ozone energy demands and balances and control systems which we encountered. Paul Blue and Jean Chedal accompanied the site visit team to many plants and offered constant encouragement.

Grateful acknowledgment must also go to Paul Chapsal, Directeur of Trailigaz Corporation for his theoretical "lectures" on ozone production.

There are many stateside people who have contributed greatly. Foremost among these is J. Keith Carswell, our EPA project officer, who did an excellent job of keeping us on schedule, but at the same time letting the project take its own course. The contributions of Dr. Jay Lehr of the National Water Well Association and Elroy Spitzer of the American Water Works Association/Research Foundation are likewise gratefully acknowledged.

Finally, the cooperation of Dr. Morton Klein, President of the International Ozone Institute, and all IOI committees and members who contributed time and information is acknowledged.

SECTION 1

INTRODUCTION

The fact that ozone and chlorine dioxide are widely used for treating drinking water in Europe and Canada has been known for many years in the United States. Likewise, the fact that chlorine dioxide is used by many water treatment facilities in the U.S. for some purpose was known. Until recently, however a full knowledge of specific uses, applications, and experiences with these chemicals has not existed.

Interest in these so-called advanced treatment techniques and others has been stimulated by the results of numerous studies that have revealed the presence of potentially carcinogenic organic chemical compounds in the drinking water supplies of many U.S. cities. Contributing greatly to this increased interest by the U.S. water treatment community was the enactment of PL 93-523, the Safe Drinking Water Act of 1974. Stated simply, the Safe Drinking Water Act reflects updated knowledge and technology that has allowed us to learn more about the contents of our drinking water; this greater body of knowledge has necessitated the establishment of more stringent criteria for water treatment plants to meet in order to protect the public health. The Safe Drinking Water Act, among its many provisions, mandates the establishment of maximum contaminant levels for a number of microbiological, chemical, and physical substances and likewise mandates investigation of treatment technologies that will permit attainment of these contaminant levels.

Specific studies that have provided evidence of potential carcinogens in drinking water and thus provided impetus to investigation of technologies that can remove these organic materials are the National Organics Reconnaissance Survey of 1975, the 1976 National Cancer Institute (NCI) Report on the carcinogenicity of chloroform, and the National Organics Monitoring Survey (NOMS) of 1977. Each of these studies has contributed to the growing body of evidence and resulting concern regarding toxic substances in water.

Specifically, the National Organics Reconnaissance Survey (NORS) revealed the presence of halogenated organic compounds thought to be carcinogenic in the drinking water of all 80 U.S. cities surveyed. The NCI report confirmed the carcinogenicity of chloroform in rats and mice; chloroform was one of the five organics whose presence was determined in the NORS survey. The NOMS expanded on the NORS by including analyses for trihalomethanes in 113 cities throughout an entire year, and quantified a number of other synthetic organic chemicals found in the water. The NOMS demonstrated that trihalomethanes could form in finished water on the way to a customer's tap and that other organics, also potentially carcinogenic, such as certain bromine-containing trihalomethanes, could exceed the concentration of chloroform.

These events and others led to the current state-of-the-art assessment of the use of ozone and chlorine dioxide in municipal water supplies. Few details regarding the engineering, costs, public health aspects, effectiveness, or even purposes of use of ozone were known in the U.S. prior to the initiation of this study in June, 1976. Experiences with chlorine dioxide, though significantly greater, were not well documented.

Therefore, the objective of this study was to obtain as much information as possible regarding the actual use of ozone and chlorine dioxide in water treatment. The several objectives of the project were as follows:

- Documentation of data on known applications of O_3/ClO_2 for drinking water treatment;
- Determine how many plants worldwide are using these oxidants for some purpose;
- To obtain data on engineering designs, operating history and effectiveness, capital and operating costs, health and safety aspects, and materials of construction.

In addition, special emphasis was placed on the documentation of what is currently known about: 1) the use of ozone for removing dissolved organic materials; and 2) organic oxidation products which may be formed as a result of the use of ozone or chlorine dioxide. During the course of the study, the project team discovered a new treatment phenomenon, the use of ozone in front of granular activated carbon to produce "biological activated carbon" which is being used in modern European drinking water plants for removal of organic materials and ammonia. This is reported in detail in Section 13.

It should be emphasized that the purpose of this study was to gather information on the various aspects of these treatment technologies and report on them. It is not, nor was it meant to be, an in-depth analysis or technology assessment. The primary intent of this report is to characterize and convey objectively "real world", hands-on experiences and to provide a knowledge base from which a technology assessment or other studies can be initiated.

SECTION 2

CONCLUSIONS

GENERAL

Ozone and chlorine dioxide are well established, proven technologies in Europe. Use of ozone for treating drinking water is likewise well established in the Province of Quebec, Canada, the Soviet Union, and to a lesser degree, Japan. With proper engineering design and operator knowledge of the purposes and application methods, ozone can be used effectively for a number of purposes including iron and manganese oxidation, microflocculation, taste and odor control, organics removal, and viral inactivation and general disinfection. The project team has inspected plants using ozone that were beautifully engineered and operated. Similarly, plants were inspected in which ozone was being grossly misused due to the lack of sophistication of the designer and the low level of operator training and consequent appreciation of the technology.

Chlorine dioxide is, in general, not being used optimally in the various installations in the United States. Little consistency is evident in dosages or methods of application. There appear to be some questions regarding the proper usage of this oxidant.

This study has found that ozone is being used quite successfully for a varied number of purposes in western Europe. This fact tends to render inoperative the preconceived belief that ozone can be used only as a disinfecting agent and that it should be viewed as an alternative to chlorine for drinking water treatment. Europeans and Canadians do not view ozone simply as a disinfectant, but rather view ozonation as a treatment process capable of performing many useful functions, disinfection being one.

Ozone is being used currently by at least 1039 water treatment plants throughout the world. This is the number identified and located; the number is believed to be low.

Ozone Applications

Ozone is seldom used as a terminal step in Germany. Ozone causes chemical transformation of dissolved organic compounds in the water, making them more easily biodegradable and thus providing food for bacteria which can lead to regrowth in the water distribution system. Ozone, according to Professor Dr. Sontheimer, of the University of Karlsruhe, Germany, can be used as a terminal step only if the dissolved organic carbon concentration of water to be distributed is less than or equal to 0.2 mg/l. Also, ammonia cannot be present, otherwise regrowth of nitro-bacteria can result when ozonation is applied as the terminal step:

If ozone is used for iron and manganese oxidation, microflocculation, organics removal, color removal, or for taste and odor control, it is most effective if applied after a filtration step for removal of coarse particles in the water and before a second filtration step, either through sand or carbon, to remove the substances acted upon or produced by the ozonation.

Ozone Systems Engineering

European ozonation techniques evolved from country to country. As a result, national approaches to ozone system design are evident. The French, Germans and Swiss use ozone successfully, but each has a different approach to air preparation, generation, and contacting. The degree of sophistication of European control systems for ozonation is quite high. This level of control and obvious reliability of equipment has led to a rather casual approach to operation. That is to say that ozonation technology is established to the degree that operation presents no significant problems, other than proper initial training of plant personnel.

A wide variety of contactor designs is in use in Europe. Much of this probably is due to the national approaches to ozonation, but some is based upon energy expenditures and technical performance factors.

Ozone Costs

Economies of scale are realized in larger ozonation systems, those in the 500 pounds/day and greater generation range. Total capital costs for these systems, including housing, is $1000/lb of ozone generating capacity/day or less. Smaller units, less than 100 pounds per day, can cost up to $4,000/lb capacity/day. This includes the cost of air treatment, ozone generation, and contacting. Because most existing U.S. water plants fall in the smaller size range, the latter costs would be most applicable in this country. By contrast, the general trend in Europe is toward large, regional plants serving multiple communities, achieving significant economies of scale on ozonation and other treatment plant equipment and processes.

Operating costs depend on cost of electricity, energy demand of the system as designed, and average dosage. Total capital and operating costs of an ozone system, base on limited field cost data acquired, can range from 1.75¢/1000 gallons to about 4¢/1000 gallons.

Public Health Aspects of Ozone

Since ozone is seldom used as a terminal step in the water treatment process, no adverse health effects can be directly attributed to its use. Any discussion of the public health aspects of ozone, as currently used, cannot be divorced from chlorine or chlorine dioxide usage, except in rare cases. Chemical evidence obtained to date does not indicate any untoward health hazard to be associated with the use of ozone in water treatment.

Ozone can produce many new oxidized moieties and biodegradation products, but in general these are compounds of lesser, not greater toxicity, provided sufficient oxidation has been caused to occur. This general rule does not always hold because certain pesticides (parathion and malathion) form their respective oxons initially before degrading upon further oxidation. The oxons are more toxic than are the starting compounds.

Chlorine Dioxide

This survey has identified 84 plants in the United States and 10 plants in Canada that use ClO_2 for some purpose. Most of these plants use it for taste and odor control. Only one plant uses ClO_2 as a final disinfectant, Hamilton, Ohio. Chlorine dioxide use in Europe is widespread; the total number of plants using it is not yet determined, but more than 400 have been identified. A number of plants use ClO_2 for oxidizing organically complexed iron and manganese, but its predominant use in Europe is for supplying a residual for the distribution systems. It is applied in concentrations of less than 0.5 mg/l for this purpose.

Chlorine dioxide systems in Europe, for the most part, are automatically controlled, while those in the U.S. are manually controlled.

Chlorine dioxide is a stronger disinfecting agent than chlorine, does not react with organics* to form trihalomethanes, does not react with ammonia to produce chloramines,* and can provide a longer lasting residual in the distribution system than free residual chlorine. Chemical costs for chlorine dioxide average about 10 to 15 times those of chlorine at those U.S. plants reporting costs, depending on the feed ratios. In Europe, ClO_2 application is reported to cost about three times as much as chlorine.

Concern about the potential toxicity of the chlorite ion (ClO_2^-) was expressed by a number of European and U.S. plant managers whose plants are using ClO_2.

Chlorine dioxide usage for water treatment began in the United States in 1944. In reviewing G. C. White's classic text entitled "Handbook for Chlorination" which contains a chapter on chlorine dioxide, it appears that little more is known about ClO_2 now than was known a decade ago. The methods of generation, engineering, applications and methods of analysis have been well known for some time[2].

Biological Activated Carbon

A new treatment phenomenon known as biological activated carbon (BAC) was discovered during the course of the project. The Germans are using it effectively for removal of organics and ammonia at the new, full scale Dohne plant in Mülheim, near Düsseldorf. The French plant at Rouen-la-Chapelle has been operating the BAC process on full scale since January, 1976, also for ammonia and organics removal.

*Except in the presence of free chlorine.

BAC is a combination of physical/biological processes. It involves promotion of biological growth on activated carbon columns; the growth is facilitated through the transformation of non-biodegradable organic compounds into more biodegradable states by the use of pre-ozonation and provision of an oxygen enriched environment. Bacteria on the carbon then biologically regenerate the carbon, postponing the necessity for regeneration.

Biological activated carbon appears to be the most advanced treatment technique currently available for removal of organic compounds and ammonia in drinking water treatment. Use of BAC significantly increases the time period of activated carbon effectiveness, thus reducing the frequency of regeneration. BAC has eliminated the need for breakpoint chlorination for ammonia removal at the French Rouen and German Mülheim plants. In turn, this has reduced the amounts of organochlorine compounds formed initially, which formerly (at Mülheim) caused frequent regeneration of activated carbon columns, used for dechlorination. This process could make a substantial impact on lowering of activated carbon costs through elimination or reduction of expensive thermal regeneration requirements, as well as chemical costs for chlorine.

Organic Oxidation Products

Major conclusions of this phase of the study are as follows:

- Complete oxidation of dissolved organic materials to carbon dioxide and water in aqueous solution is rare, by means of any oxidant. In general, if an organic material is resistant to oxidation by ozone (the most powerful oxidant used in water treatment), it will also be resistant to oxidation by other (weaker) oxidants.

- Oxidation products formed by ozonation do not contain halogen atoms.

- Ozonation of certain pesticides has produced intermediates which are more toxic than the starting materials. Continued ozonation produces further oxidation products which are less toxic and more easily biodegraded.

- Oxidation products formed upon ozonation, and non-chlorinated oxidation products from chlorine dioxide are more biodegradable than are the starting organic materials.

- Oxidation of pesticides with ozone proceeds rapidly in clean water, but significantly slower when humic materials or soil particles are present. Thus, dissolved organics can be adsorbed onto humic or soil materials and be resistant to oxidation.

- In many cases, oxidation products from chlorine dioxide do not contain halogen atoms. This is not a general rule, however, because chlorinated compounds often are produced, even when chlorine dioxide is synthesized so as to contain no free chlorine.

- Both chlorinated and non-chlorinated organic oxidation products are formed with chlorine. For the most part, many of the non-halogenated oxidation products from chlorination also can be formed with chlorine dioxide and with ozone.

- When oxidants are used in drinking water treatment, it is important to know what oxidizable materials are present and to incorporate a sufficient quantity of oxidant to produce the desired end result.

SECTION 3

RECOMMENDATIONS

Biological activated carbon appears to have great potential for use in U.S. water treatment. Pilot projects on this treatment combination should be initiated to determine its effectiveness for organics and ammonia removal. Engineering and cost details of currently operating BAC plants in Europe should be investigated.

A thorough investigation of the potential savings in activated carbon regeneration costs resulting from extended life of the biological activated carbon also should be investigated.

An in-depth cost analysis of the use of ozone as a unit process capable of performing a number of useful functions is needed. Various combinations of air preparation, generation, contacting, and off-gas use or destruction systems should be studied.

Epidemiological studies of populations using drinking water treated with chlorine dioxide and ozone should be conducted. Research directed at determining whether or not the chlorite ion is toxic, and at what levels, would be a part of that study. Toxicity of residual ozone in water is of no import because in properly designed systems it is not present at the tap.

Technical assistance should be made available to those municipalities currently using chlorine dioxide in order to assure that it is being used properly and in safe concentrations. This could be provided by EPA, supported by expert assistance as needed.

SECTION 4

PROJECT DESCRIPTION

OBJECTIVES

The objectives of the project, listed earlier in Section 1, were to obtain as much information as possible regarding actual operating experiences with ozone and chlorine dioxide. Ozone has been used in hundreds of water treatment facilities in Europe, beginning in 1906. Canada opened its first plant in 1956 and now has 20 operating facilities. Chlorine dioxide has been used in a number of plants in the U.S. since the 1940s; the latest survey prior to this one was in 1964 and listed 37 operating facilities. Most of these were thought to use ClO_2 for taste and odor control. The use of chlorine dioxide in Europe likewise was known, but the scope of its use was unknown.

Thus the purpose of this project has been to obtain detailed information regarding the application and effectiveness of these two oxidants. Information has been obtained on: 1) engineering of system components and design; 2) operating history and effectiveness; 3) capital and operating costs; 4) public health aspects; and 5) special emphasis has been given to the effectiveness of ozone for removal of organic compounds in water and the products, toxic or otherwise, which may be formed as a result of its usage. Likewise, emphasis has been given to organic oxidation products of chlorine dioxide usage and, for comparative purposes, those of chlorine.

SCOPE OF WORK

The scope of work involved several distinct tasks. First, detailed questionnaires on various aspects of chlorine dioxide and ozone usage were developed for distribution to plants using one or both of the oxidants. Simultaneously, efforts were made through contact with ozone and chlorine dioxide manufacturers, professional associations such as the International Ozone Institute, the Deutscher Verein der Gas und Wasserfaches e.v.--(German Association of Gas and Waterworks) (DVGW), research institutes, and the literature to identify and locate plants using the oxidants. Once plants were identified and located, the detailed questionnaires were mailed and responses requested. Questionnaires were sent to all plants using ozone (20) and chlorine dioxide (10) in Canada, to plants thought to be using ClO_2 in the U.S.[103], and to hundreds of plants using one or both oxidants in various European countries.

Based on the questionnaire responses and personal contacts with knowledgeable persons in the field, representative plants were selected for visitation.

Twenty-three plants in four European countries were visited in May, 1977 by a site visit team consisting of a design engineer, a public health expert, an expert in ozone, a European water supply research expert, and the principal investigator. Seven plants in Canada which use ozone were visited in August, 1977. In September and October of 1977, 13 of 84 U.S. facilities using ClO_2 were visited.

Concurrent with the questionnaire and site visit activities, a comprehensive literature search was conducted. The primary goal of the literature search was to gather pertinent information on oxidation products formed through the usage of ozone or chlorine dioxide.

This report is the result of these several tasks. Details on each aspect of the study are included in later sections. Sample questionnaires are attached as Appendix A. A detailed bibliography is given in Appendix G.

SECTION 5

WATER TREATMENT PHILOSOPHIES

Prior to discussion of the results of the study, it is necessary to decribe the water treatment philosophies of various countries. In the United States, we often speak of a "conventional" water treatment process. This usually means coagulation, sedimentation, filtration, and chlorination. This is not the case in Europe. Drinking water treatment practices vary from country to country, and often from city to city.

The most striking difference between U.S. and European practices is the concentration on bacteriological quality of water in the U.S. as an indicator of safety. The Europeans, while concerned with bacteriological and virological safety of water, are much more concerned with chemical contamination. Europeans are brought up with the understanding that when there are any unnatural tastes in water (especially chlorine tastes) the water is contaminated. Americans are brought up with the understanding that when chlorine cannot be tasted, the water may be contaminated. Therein lies one of the major reasons for the two basic approaches to treating water supplies. Americans use relatively heavy dosages of chlorine in the water to assure bacteriological (but not necessarily chemical) safety; Europeans reduce the chlorine demand of water (insuring chemical safety) so that the residual chlorine finally used to maintain bacteriological safety in the distribution systems will be so small as to be tasteless in the water.

The following describes in some detail the treatment philosophies of France, Germany, Switzerland, and Canada.

FRANCE

Since 1906, the city of Nice has been disinfecting its drinking water supplies solely with ozone, and is credited with being the longest ozone-using municipal water supply system in the world. Over the years, three Nice water treatment plants have been constructed (all using ozone for disinfection), but in 1970 these three plants were combined into a single plant which incorporates the latest water treatment technologies (including continued use of ozonation) appropriate to its mountain stream water supply. Today in France, 593 water supply systems are using ozone for a variety of purposes.

In France, only 40% of the water plants are owned and operated by private companies. The remaining 10% are both owned and operated by large, vertically integrated water companies.

Many of the plants which are owned and/or operated by private companies are the largest ones. For instance, the Paris suburbs are mostly served by three large plants, Choisy-le-Roi (on the Seine), Neuilly-sur-Marne, and Méry-sur-Oise. Each of these plants was pilot tested, designed, constructed, equipped, and is now operated by Compagnie Générale des Eaux (CGE) for the Paris suburbs.

CGE is one of two large companies located in Paris that is in the business of water supply (Soc. Lyonnaise des Eaux et de l'Eclairage -- SLEE -- is the other), which means that they are involved in the operation of all facets of the system, similar to the manner in which large oil companies operate. This "turnkey" approach has the advantage that it tends to insure that the best treatment process and all necessary and ancillary equipment will be installed, since the designer, constructor, equipment supplier, and operator all are part of the same business entity.

Great emphasis is placed on designing a process to produce a finished water of high quality. One plant process was pilot tested for three years before design and installation were begun. It is normal for CGE and SLEE to pilot test several processes for 12 to 18 months before settling on a final design.

There is no such thing as a "conventional" process employed in France, since the raw waters contain pollutional loads that vary greatly. In several plants located along the Seine and Marne rivers, as many as nine different treatment steps are employed.

A typical process is the one used at the newest plant serving the Paris suburbs, Neuilly-sur-Marne (Figure 1).

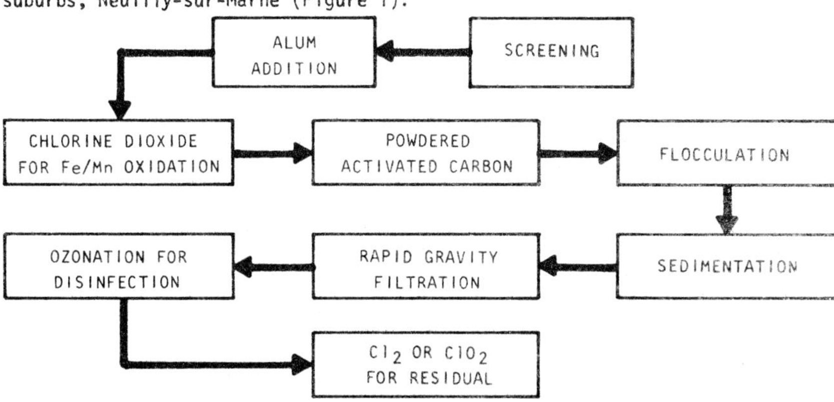

Figure 1. Process flow diagram of Neuilly-sur-Marne water treatment plant.

Even though the selling price of water to individual consumers is significantly higher than in the U.S., there is little concern shown by consumers regarding the high prices. One reason could be that the French use much less water than do Americans. French plants normally are designed based on an assumed per capita consumption of 250 l/day (66.1 gpd). Average water consumption for four plants visited in the Paris suburbs was 64 gallons/-person/day (gpcd).

Considerable emphasis is placed on achieving viral and bacterial inactivation. This is one of the reasons for the use of ozone. Two now famous papers by Coin, et al.[4] provide the basis for the French theory of contacting. This theory, which is practiced widely, is that to achieve virtually complete viral inactivation, a 0.4 mg/l residual of ozone must be maintained for at least 4 minutes following satisfaction of the initial ozone demand.

A final point of significance is the aversion of the French consumer to the taste of chlorine. One of the unwritten laws of French water treatment is that the finished product must have no odor and no undesirable taste.

FEDERAL REPUBLIC OF GERMANY

The predominant philosophy in West Germany is that the best source of drinking water supply is unpolluted groundwater. If well conditions prove unobjectionable chemically and bacteriologically, the groundwater is used without any treatment and without chlorination. According to Kühn[3], there are still hundreds of European waterworks which provide water without any chlorination and yet do not have any problems with bacteria in their distribution systems. Among such waterworks are large cities such as Munich and Karlsruhe. A common practice in many of these waterworks is to saturate the water with pure oxygen before it enters the distribution system in order to maintain a high dissolved oxygen level[26].

When water authorities are forced to use surface waters as a source, which occurs in the densely populated regions along the Rhine and Ruhr rivers in northern Germany, every effort is made to treat the water to a quality identical to that of a pure, unpolluted groundwater.

Because of the historically good experiences using groundwater, efforts are made to use treatment procedures that include passage of water through the ground. This practice of drawing water from wells located 50 to 250 meters from the bank of the river is known as river sand bank filtration. Groundwater is extracted in vertical and horizontal wells from large 10 to 30 meter thick water-bearing diluvial gravel and sand sediments. In the Düsseldorf area, the ratio of bank filtrate to groundwater in the water to be treated is about 3 to 2, varying with the water level of the river and changes in the water table.

The effect of natural treatment by ground passage between the Rhine River and the wells results in a 65 to 75% reduction in the amount of dissolved organic carbon (DOC)[5]. During ground passage, numerous physical, chemical and biological processes occur. Much turbidity and associated organic and inorganic substances, such as pesticides, heavy metals and bacteria, are retained in the gravel and sandstone sediment. The resulting filtrates combined with groundwater are clear and have low concentrations of bacteria[6].

At one time, river sand bank filtration was the only treatment applied to surface waters. The goal of treatment is to produce a water which has a DOC of less than 2 mg/l and which will require no more than 0.3 to 0.5 mg/l of chlorine for disinfection purposes[3]. As the level of pollution increased over the years, due to increasing population density and industrial pollution, physical/chemical treatment became necessary in order to achieve this goal. Thus, 40 years ago, ozone and activated carbon processes were used for the first time in Germany.

Concurrently, breakpoint chlorination was installed at those plants having high ammonia contents in their raw waters. Breakpoint chlorination involves chemical oxidation of ammonia nitrogen to nitrogen gas with chlorine, and requires the use of 12 grams of chlorine per gram of nitrogen. This means that very large amounts of chlorine and chlorinated products were present in the processed water. Since German drinking water regulations limit residual chlorine concentrations to a maximum of 0.3 to 0.5 mg/l, granular activated carbon was installed at those plants using breakpoint chlorination for the purpose of dechlorinating.

Today, both ozone <u>and</u> activated carbon are now used in many plants in Germany. When the combination was first used in full scale facilities in Düsseldorf during the early 1960s, the application was mainly for iron and manganese oxidation (ozone) and taste and odor removal (granular activated carbon). Over the years, the Germans followed the efficiency of activated carbon in removing dissolved organics, and have found that the ozone/activated carbon combination was removing more dissolved organics than could be expected simply on the basis of ozone oxidation alone or by activated carbon adsorption alone. In addition, ammonia was found to be removed by passage through the preozonized granular activated carbon.

This prompted detailed investigations at the Bremen water works over a three year period beginning in 1969[7] from which it was discovered that this synergistic organics removal was caused by the bacterial activity present in the carbon columns. Bacterial activity was optimized by preozonation and was found to be effective in removing ammonia from the waters as well. Most significantly, granular activated carbon columns in waters which were devoid of chlorinated organics and which were subjected to preozonation, were found to be active for at least two or three years, <u>without requiring regeneration</u>. This concept, now termed Biological Activated Carbon (BAC) is discussed in detail in Section 13.

Ozone is seldom used as the terminal treatment step in Germany. This is because oxidation of organic compounds with ozone leads to transformations of such organic matter, making many non-biodegradable compounds biodegradable. This can cause an increase in bacterial growth in the distribution system. Therefore, an additional filter plus final disinfection by chlorine or chlorine dioxide should follow ozonation. However, ozone can be used safely as a terminal step if the Dissolved Organic Carbon (DOC) of the finished water is less than or equal to 0.2 mg/l.

Also, according to Sontheimer[5], ozonation should be preceded by a filtration step or a flocculation and sedimentation step in order to remove as much particulate matter as possible. In other words, ozone should not be used to accomplish what a simpler process can better and more cheaply accomplish. This approach is endorsed at the majority of European plants using ozone.

Finally, chlorine dioxide or chlorine is used in very small dosages to provide residuals in the distribution systems and to prevent bacterial regrowths. The concentrations added are usually limited to 0.3 mg/l and never exceed 0.6 mg/l.

There are two reasons for not using high concentrations of final disinfectants. First, if the water has been treated properly, many of the dissolved organic compounds and other reactants will have been removed and the disinfectant dosages needed to provide residuals are very low. Secondly, the Germans believe that the water treatment processes also should be designed to remove bacteria prior to final disinfection. This can be done either by sand filtration or ozonation. The belief is that since chlorine is harmful to bacteria, it may likewise be harmful to humans.

SWITZERLAND

In Paris along the Seine, Oise and Marne Rivers and in Germany along the Rhine and Ruhr Rivers, sophisticated treatment steps are required to cope with the high concentrations of pollutants in the surface water supplies. Ozone in concentrations of 3 to 6 mg/l and activated carbon sometimes are necessary to treat the waters to qualities suitable for consumption. In Switzerland, however, the raw waters, which are taken from lakes fed by mountain streams, or from the streams themselves, are of very high quality. The TOC is usually 2.0 mg/l or less. Many countries would like to have finished waters of such high quality as are many Swiss raw waters.

Swiss waters are basically glacial in origin, from very high altitudes, and generally are not contaminated with organic compounds. For example, Schaffhausen requires an ozone dosage of 0.4 mg/l, which produces an ozone residual of 0.3 mg/l, and this residual lasts for 24 hours. Long term stability of this ozone residual is due in part to high water quality, but also to low water temperature.

Swiss people will not accept excessive chlorination in their water because of tastes. They are used to very cold water, free of undesirable tastes.

The Swiss began using ozone for drinking water treatment at their very small and remote plants. Automatic controls became a necessity immediately, because the communities are sometimes snow bound for months and manual operation and maintenance is not possible. Modern Otto plate generators manufactured by Sauter Corporation of Switzerland are used in many of these small plants.

The next stage of ozone use in Switzerland was to treat lake waters. The first large Swiss lake water application was at St. Gallen in 1957-58, whose water supply is the Lake of Constance. The process involved filtration, chlorination and distribution.

Nearby, however, gas was being produced from coal, and coal processing drainages were being discharged to the Lake. Over a short period of time, phenolic compounds built up in the St. Gallen water supply and these imparted bad tastes and odors when treated by the normal chlorination methods. This was acutely felt in St. Gallen, where there is a fairly large food processing industry, and at one time, several days worth of production had to be destroyed because of contamination with chlorophenols.

Ozone did not produce these tastes when tested on St. Gallen's water supply, so St. Gallen was the first large Swiss lake water treatment plant to install ozone. During the next 5 to 6 years, all Swiss Lake of Constance plants changed over to ozone.

In Germany, ozonation of Lake of Constance water came later, but their initial efforts were always conducted using ozone as the terminal step for sterilization.

Using ozone as the final treatment step, the Swiss found bacterial regrowth, plus slime growth, in the distribution systems. Swiss systems are 100% metered and many repairs had to be made to these water meters because of slime growths. The Lake Director thought this was caused by the ozone, and he became very disenchanted with its use. However, when small quantities of chlorine dioxide were added to the ozonized waters, bacterial regrowth and slime growths ceased. Today, chlorine dioxide is used as a post-treatment at Swiss plants on the Lake of Constance, the Lake of Zürich, and others.

During 1963-65, German scientists conducted a 2 year pilot study to compare the then standard Lake of Constance treatment process with ozonation and microstraining (ozonation and sand filtration after microstraining). After only 6 months of this study (at the City of Konstanz treatment plant) it became clear that ozone must be applied between pre-filtration and post-filtration.

Plant personnel at the large German waterworks at the Bodensee Wasserversorgung (Sipplinger Berg), agree today that it would be a gross misapplication to use ozone as a terminal step for lake of Constance water.

The modern, fully automated plant at Kreuzlingen (Lake of Constance) using ozone and activated carbon is the result of this historical move toward the use of ozone. Pressure filtration occurs prior to ozonation followed by a bed of activated carbon; chlorine dioxide is then applied as the terminal treatment step.

Today, there are 150 ozone installations in operation in Switzerland[8], in a country approximately twice the size of the state of New Jersey.

Ozone plus activated carbon facilities have been installed in some Swiss waterworks for safety reasons, to guard against accidental oil spills. In plants on the Bodensee (Lake of Constance), for instance, 40% of the capital costs of the ozone and activated carbon systems were paid by an international oil company that constructed an oil pipeline under the lake. The 40% figure was arrived at because this was calculated to be the excess ozone generation capacity needed to provide safe water should the oil pipeline ever rupture or leak.

The salient characteristics of Swiss water treatment are the high quality of the raw water and the "safeguard" treatment processes employed.

CANADA

There is no national water treatment philosophy in Canada. In fact, drinking water standards are not national mandates, but rather are recommended standards. Each province decides whether to adopt the Canadian Drinking Water Standards and Objectives prepared in 1968 by the Joint Committee on Drinking Water Standards.

The Joint Committee, made up of representatives from each province, developed these standards in 1968, recognizing the desirability of having uniform drinking water standards. The current standards will be updated in 1978 and are expected to reflect some of the provisions contained in the U.S. law, the Safe Drinking Water Act of 1974 (PL 93-523). However, only one province in Canada (Quebec) is expected to adopt or follow closely the Primary Drinking Water Standards which went into effect in the U.S. on June 24, 1977.

Because Canadian provinces may differ in philosophy and treatment standards, it is difficult to pinpoint any particular set of characteristics. The two most populous provinces, Quebec and Ontario, are the only two known to use ozone or chlorine dioxide to any major extent. There are currently 20 plants in Canada, 19 in the province of Quebec, the other in the Northwest Territories, that use ozone. There are ten plants in Canada which use chlorine dioxide, all in the province of Ontario.

The history of ozone use in Quebec can be traced back to 1949 when an engineer having difficulties in eliminating taste and odors from drinking water decided to learn about ozone experiences in Europe. He consequently made a trip to European water treatment plants using ozone, was convinced by what he observed, and decided to try the process on his next project. So in 1956 the first filtration plant using ozone was built in Ste. Therèse in Quebec; the plant had a capacity of 3 mgd[9].

Whereas ozone was first used for taste and odor purposes and to improve appearance, the work of Coin, Hannoun, and Gomella[4] provided impetus to its adoption through their work which demonstrated its effectiveness in poliovirus inactivation.

French Canadians soon recognized the multiple advantages of ozonation: taste and odor removal, improved appearance and taste, viral inactivation, and, unknown in the 1950s and 1960s, use as an effective oxidant for organic chemicals. Nadeau and Pigeon[9] point out that ozone should not be compared with chlorine, since chlorine only sterilizes, whereas ozone acts as a sterilizing agent and removes tastes and odors.

Chlorine dioxide use in Ontario, as far as can be determined, is a relatively recent occurrence, and is used to remove taste and odors resulting from phenolic substances.

In Ontario, all surface water supplies are required to have continuous turbidity and chlorine residual monitoring and recording equipment. Chlorination equipment is required at all ground water sources. Approximately 85% of the total population of Ontario is provided with water from municipal waterworks. In 1975, Ontario had 442 municipal waterworks.

There are approximately 1800 municipal and private waterworks systems in Quebec. About 92% of Quebec's population obtains its water from surface supplies, 7% from groundwater supplies, and 0.5% to 1% from a combination of both. About 20% of the population drink water which is chlorinated only. According to the "National Inventory of Municipal Waterworks and Wastewater Systems" from which these data were taken, Quebec's population is presently provided with a wide distribution of good quality drinking water.

Canada has become concerned recently about the concentration of trihalomethanes in drinking water. A national survey was carried out in seventy municipalities in an attempt to relate levels of trihalomethanes to type of raw water supply, water treatment and total organic content. Of four trihalomethanes investigated, chloroform was found in the highest concentrations, ranging from 0 to 121 μg/liter. As in the recent U.S. EPA study (NOMS), trihalomethane levels in water in the distribution systems were found to be significantly higher than in water sampled in the treatment plants. A report entitled "National Survey for Halomethanes in Drinking Water" has been published recently by the Health Protection Branch of the Ministry of National Health and Welfare.

SECTION 6

OZONE

GENERAL OVERVIEW OF OZONATION

General History[10]

In 1785 Van Marum, a Dutch philosopher, noticed that the air in the neighborhood of his electrostatic machine acquired a characteristic odor when subjected to the passage of a series of electric sparks. In 1801, Cruickshank observed the same odor in the gas formed at the anode during the electrolysis of water. In 1840, Schonbein reported the odor as being due to a new substance to which he gave the name ozone, derived from the Greek word "ozein", meaning to smell. Schonbein also was the first to suggest that ozone may occur naturally in the atmosphere.

The present construction of electric discharge ozone generators evolved from the apparatus originally designed by Werner von Siemens in 1857 in Germany. Brodie (England) and Berthelot (France) also designed early ozone generators. Siemens' first ozonizer essentially consisted of two coaxial glass tubes, the outer coated externally and the inner coated internally with tin foil, air feed gas being passed through the annular space. Brodie substituted water as the electrode material in place of tin foil, and Bertholet used sulfuric acid.

The Siemens type of ozone generator has been developed commercially into a form suitable for commercial production of ozone, and today most of the ozone generating systems installed in drinking water treatment plants are of this type. Glass tubes are coated internally with a metal dielectric and the individual tubes are cooled by means of water.

Modifications to the original Siemens generator have been made, primarily dealing with the method of cooling, and today many different types of generators are available. All operate on the same general principal, corona discharge, which requires high voltages and/or high frequencies, thus creating considerable heat. In turn, this requires that the generators be cooled in order to maximize ozone production yields and minimize power consumed.

Characteristics of Ozone

Ozone itself is an unstable gas which boils at minus 112°C at atmospheric pressure, is partially soluble in water (about 20 times the solubility of oxygen) and has a characteristic penetrating odor, readily detectable at concentrations as low as 0.01 to 0.05 part per million. It is a powerful oxidant, having an oxidation potential of 2.07 volts, and therefore should be considered a dangerous material, capable of oxidizing many types of organic materials, including human body tissue. At the relatively low concentrations of ozone produced by commercial generation equipment (1 to 3% in air; 2 to 6% in oxygen) no explosive hazard exists, but mixtures of ozone concentrated to 15 to 20% or higher in air can be explosive. In aqueous solution, ozone is relatively unstable, having a half-life of about 20 to 30 minutes in distilled water at 20°C. At lower temperatures the half-life of ozone in water is extended considerably. If there are oxidant-demanding materials present in solution, the half-life of ozone in such solutions will be shorter.

On the other hand, ozone in air (especially under dry conditions) is much more stable than in water. The half-life of ozone in the ambient atmosphere has been measured by the U.S. Environmental Protection Agency and determined to be on the order of 12 hours.

History of Ozone Use in Water Treatment

The earliest experiments on the use of ozone as a germicide were conducted by de Meritens in 1886 in France, who showed that even dilute ozonized air will effect sterilization of polluted water. A few years later (1891), the bactericidal properties of ozone were reported by Frölich from pilot tests conducted at Martinikenfeld in a plant erected by the German firm of Siemens & Halske. In 1893, the first drinking water treatment plant to employ ozone was erected at Oudshoorn, Holland. Rhine River water was treated with ozone, after settling and filtration. Siemens & Halske next built treatment plants at Wiesbaden (1901) and Paderborn (1902) in Germany which employed ozone.

A group of French doctors studied the Oudshoorn plant and its ozonized water and, after pilot testing at St. Maur (in Paris) and at Lille, a 19,000 cu m/day (5 mgd) plant was constructed at Nice, France (the Bon Voyage plant) in 1906, which employed ozone for disinfection. Because ozone has been used continuously at Nice since the Bon Voyage plant began operating in 1906, Nice is referred to as "the birthplace of ozonation for drinking water treatment".

Full scale water treatment plants then were constructed in several European countries. As of 1916 there were 49 treatment plants having a total capacity of 319,200 cu m/day (84 mgd)[11] in operation, and 26 of these were in France. By 1940 this number had increased to 119 water treatment plants, located as follows[12]:

90	France
14	Italy
5	Belgium
4	England
3	Rumania
2	USA
1	Russia
119	

Today there are more than 1000 drinking water treatment plants throughout the world employing ozone for one or more purposes (see Table 1).

In Canada, the first ozonation plant was built in Ste-Therèse, in the Province of Quebec in 1956. Today Canada has 20 ozonation plants (19 in Quebec), with 3 more being built. The largest ozone system in the world for drinking water plant currently is under construction in Montreal.

In the USA there are currently four operating ozone drinking water treatment plants, with one other under construction. Since 1941, Whiting, Indiana has been ozonizing about 7 mgd of water for taste and odor control, followed by chloramination for residual. In 1973, Strasburg, Pennsylvania constructed a 110,000 gal/day plant using ozone for disinfection, with no residual disinfectant being employed.

During the National Organics Reconnaissance Survey, conducted by EPA in early 1975, water from the Strasburg plant contained lower concentrations of trihalomethanes than any other city examined. Waters of Whiting, Indiana contained the second lowest amounts of trihalomethanes.

Monroe, Michigan recently has started up a new, 18 mgd plant employing ozone for taste and odor control. Bay City, Michigan also has started up a new 30 mgd treatment plant using ozone for taste and odor. Saratoga, Wyoming will be starting up a 3 mgd plant by mid-1978, also using ozone for taste and odor control.

Current Extent of Ozone Usage

In conducting the current state-of-the-art analysis, considerable data were assembled as to the location of drinking water treatment plants using ozone from the published literature, from manufacturer brochures and from personal conversations with people who have been involved with the design and installation of ozonation systems for a number of years. From all of these sources the approximate number of currently operational drinking water treatment plants using ozone for one or more purposes was determined to be

1039 (+200,-50) throughout the world. Location of these plants by country is given in Table 1.

TABLE 1. OPERATIONAL PLANTS USING OZONE -- 1977

Country	Number of Plants
France	593
Switzerland	150
Germany	136
Austria	42
Canada	23*
England	18
The Netherland	12
Belgium	9
Poland	6
Spain	6
USA	5
Italy	5
Japan	4
Denmark	4
Russia	4
Norway	3
Sweden	3
Algeria	2
Syria	2
Bulgaria	2
Mexico	2
Finland	1
Hungary	1
Corsica	1
Ireland	1
Czechoslovakia	1
Singapore	1
Portugal	1
Morocco	1
Total	1039

* Includes expansions. Actual number of operating plants in Canada = 20, with 3 more under construction.

Two comments are applicable to the number of plants determined. First, the number is high because many manufacturers list each expansion of a particular plant as a separate installation (some ozonation plants have undergone as many as three expansions). Some plants have ceased operating or have been consolidated into other plants. This would also make the number high. On the other hand, it was not possible to contact all ozone generator manufacturers, and the smaller manufacturers are known to have many existing ozonation installations for drinking water. On balance, it is believed that the actual number of plants currently using ozone to be 100 to 200 higher than the 1039 number.

Although France has the greatest number of plants using ozone, the plant with the largest ozone gneration capacity for treating drinking water is Moscow, Russia. This 1,200,000 cu m/day (317 mgd) plant, built in 1969, generates 200 kg/hour of ozone (10,582 lbs/day.

Montreal, Canada currently is constructing a 2,300,000 cu m/day (608 mgd) plant in two stages. When completed in 1980 the Charles-J. des Baillets Montreal plant will be capable of generating 300 kg/hour of ozone, and at that time will be the largest drinking water treatment plant in the world using ozone.

Applications of Ozone in Drinking Water Treatment

Because ozone is a powerful oxidant and because many contaminants in raw water supplies are oxidizable, ozone can be used for many specific applications. The major applications are listed in Table 2. Although the early uses for ozone were predominantly for disinfection (bacterial kill and viral inactivation), today oxidative applications account for a significantly increasing number of applications.

TABLE 2. APPLICATIONS OF OZONE FOR WATER TREATMENT

Bacterial Disinfection
Viral Inactivation
Oxidation of Soluble Iron and/or Manganese
Decomplexing Organically-Bound Manganese (Oxidation)
Color Removal (Oxidation)
Taste Removal (Oxidation)
Odor Removal (Oxidation)
Algae Removal (Oxidation)
Removal of Organics (Oxidation)
such as Pesticides
Detergents
Phenols
Removal of Cyanides (Oxidation)
Suspended Solids Removal (Oxidation)
Preparation of Granular Activated Carbon for Biological Removal of Ammonia and Dissolved Organics

In recent years, multiple uses for ozonation in the same water treatment process have been developed. For example, if ozone is applied for, say, color removal, near the end of the treatment process, a significant amount of disinfection also will be obtained. The conjunctive use of contactor off-gases from the primary purpose ozone contacting chambers can be effective in such treatment processes. These off-gases (which contain 5 to 10% of the ozone charged to the primary purpose contact chambers) can be recycled to the initial steps of the treatment process to oxidize iron and manganese, to aid in flocculation of suspended solids, or simply to destroy excess ozone while performing useful work. The reuse of contactor off-gases will not eliminate the need for ozone destruction facilities (see Section 8 -- Engineering Aspects of Ozonation).

The most recent application of ozonation is before granular activated carbon beds or columns. This treatment, which can be accomplished by using primarily contactor off-gases containing ozone and air or oxygen, supplies considerable dissolved oxygen to the water, oxidizes dissolved organics to generate more biodegradable organic compounds, and provides an oxygen and dissolved carbon environment conducive to aerobic bacterial proliferation in the activated carbon. In turn, these aerobic bacteria biologically remove dissolved organic carbon compounds as well as ammonia. The phenomenon is termed "Biological Activated Carbon" and is discussed in detail in Section 13 of this report.

In the future, it appears likely that, as the technology of treating drinking water supplies with ozone becomes more widely understood, the application of ozone will increase. One of the most significant points for the water treatment engineer to learn about ozonation is the multiple application aspect. The use of ozone for more than one purpose can effectively reduce the cost of ozone installations charged to disinfection.

For example, ozone can be installed for a single purpose, say disinfection, at a cost of X dollars. If this is the only purpose for ozone, then contactor off-gases must be disposed of safely. This can be done catalytically or thermally, but at some additional (Y dollars) cost. Therefore, total ozonation system cost is: $(X + Y).

Instead, the contactor off-gases can be recycled to the front of the water treatment process and utilized for a second water treatment purpose, such as the oxidation of iron and manganese, or as a flocculation aid, or for color removal, organics oxidation, etc. The additional cost for this second ozonation step (Z dollars) can be of the same order of magnitude as the cost for ozone destruction (Z = Y). Using some of the off-gas ozone to perform useful work can be more energy cost-effective than destroying excess ozone. Costs for utilizing off-gas ozone also would be smaller than if preozonation were the only ozonation purpose.

Such a dual application of ozone could even require a small amount of additional ozone generation capacity, which might not be economical if only the initial ozonation process step were to have been considered without later disinfection. As part of a dual ozonation process, the relative costs of ozonation for the disinfection step will be lowered. In addition, the initial use of ozone can replace chemical usage and lower chemical costs for the total plant process.

Generation of Ozone

Nature generates ozone continuously by means of sunlight acting upon oxygen in the atmosphere, or intermittently by lightning passing through the air. Man simulates this natural process in generating ozone by passing high voltage electrical discharges, high or low electrical frequencies, or high energy radiation through air or oxygen. Ozone is also generated unintentionally by man as a by-product during operation of electrical power generation equipment, electrostatic precipitators, welding equipment, electrostatic copying machines, ultraviolet lights, and a variety of other electrical devices.

Commercial quantities of ozone are generated on-site and as needed in equipment which includes gas preparation, electrical supply, ozone generators. Generation of ozone is energy intensive, with some 90% of the power supplied to the generator being utilized to produce light, sound, and primarily heat, rather than the desired ozone. Thus minimizing electrical power requirements is a prime target of the ozone generator manufacturer.

A more detailed discussion of the factors affecting ozone generation by the corona discharge technique is given in Section 8. In general, however, type, thinness and surface area of the dielectric medium used, discharge gap between electrodes, quality of the dielectric medium (no pinholes or misalignment), pressure, temperature, rate of flow of gas through the ozone generator, composition (air vs. oxygen), and moisture content of the feed gas are among the most important factors. Rosen[13] gives an excellent discussion of the theory of generation of ozone.

Ambient air or recycled oxygen-rich gas contains moisture which, if allowed to remain in the feed gas during generation of ozone will: 1) reduce the yield of ozone per kwhr of electrical energy applied: and 2) form nitric acid, which can result in severe corrosion of some generator components and downstream ozone handling equipment. For those reasons, air to be fed to the generator should be adequately dried. This is the function of the gas handling equipment mentioned earlier. In modern ozone generating systems, air is dried to a dew point of at least minus 40°C, and preferably to minus 60°C or below.

When dry air is used to generate ozone, a mixture of approximately 1% ozone in air is produced for the lowest power expenditure consistent with generation ozone at a reasonably rapid rate. Increasing the rate of flow of feed gas through the ozone generator at constant power will <u>increase</u> the amount of ozone generated per unit time and per unit of electrical energy applied, and will <u>reduce</u> the concentration of ozone in the output gas mixture. Reducing the rate of feed gas flow will <u>increase</u> the concentration of ozone per unit of electrical power applied, will <u>decrease</u> the amount of ozone produced per kwhr of power applied and will <u>decrease</u> the amount of ozone produced per unit time. Proper choice of parameters for the particular ozonizing job at hand will guide the user to produce the optimum amount of ozone for minimal expenditure of power.

Generation of ozone by electrical discharge produces heat, and heat causes decomposition of ozone in the product gas. Thus it is important that the heat generated in producing ozone be removed as quickly and efficiently as possible. This is normally accomplished by water, air, or heat transfer fluid cooling of the dielectric.

In recent years, ozone generator efficiency has been increasing. It has been noted that the Bon Voyage plant of Nice in 1906 originally required 73 kwhr/kg of ozone generated[14]. By contrast, the City of Montreal's

Charles-J. des Baillets drinking water treatment plant, which is under construction and will house one of the largest ozone generation capabilities in the world (6750 kg/day) when completed in 1980, is projected to produce ozone for 23.6 kwhrs/kg of ozone generated[15].

Use of oxygen as the feed gas rather than air will allow production of essentially double the amount of ozone per kwhr of electrical energy (2% ozone in oxygen). Thus, capital requirements for ozone generation will be lowered by using oxygen. It should be recognized however, that the gases now discharging from the ozone generator will be 98% oxygen and 2% ozone. Therefore, to minimize loss of oxygen, contactor off-gases, still very rich in oxygen, should be recycled to the ozone generator (after removing contaminants and water). Gas handling equipment will normally be required, therefore, when oxygen is used as the feed gas.

At the present time, only one of the 1039 municipal water treatment plants is known to be using oxygen to generate ozone (Duisburg, Germany). The Brussels, Belgium plant (Tailfer) is installing oxygen facilities so as to be able to use oxygen during summer periods of high ozone dosage requirements. The use of oxygen in drinking water treatment plants for generating ozone currently is very small, but worthy of future consideration.

Contacting of Ozone

Because ozone is only slightly soluble in water at the partial pressure at which it is generated and applied, contacting ozone with water involves bringing bubbles of ozone-containing air or oxygen into intimate contact with the water. Mass transfer of ozone from the gaseous bubbles occurs across the gas/liquid interface into the water.

Factors which affect the mass transfer of ozone in liquids, and which themselves are affected by design and operation of the contactor system, include:

- the miscibility with water and ozone demand of substance(s) to be ozonized
- concentration of ozone in the gas (with regard to mass transfer)
- method and time of contact
- bubble size
- pressure and temperature
- presence of interfering substances

Generally speaking, there are two major categories of reaction which ozone undergoes in solution: 1) those which are so rapid that they are limited by the rate of mass transfer of ozone into solution; and 2) those which are slower than the mass transfer rate, thus are limited by the reaction kinetics of the material to be ozonized. Reaction rates of such constituents as bacteria, nitrites, hydrogen sulfide, phenols, unsaturated organic compounds, etc., are very rapid, and are limited only by the rate at which ozone is "mass transferred" into solution.

Other materials react with ozone very slowly, such as acetic acid, urea, saturated aliphatic alcohols, ammonia and oxalic acid. Even in the presence of large excesses of ozone, the rates of these reactions are very slow, and thus are "reaction rate limited".

Therefore, in designing an ozone contacting system, it is important to minimize the amount of ozone required for the specific purpose for which the ozone is to be used. If bacterial disinfection only is desired, a contactor which causes rapid mass transfer of ozone should be used. For oxidation of biorefractory organic materials, the rate of ozone mass transfer is less important than maintaining a specific concentration of ozone for a longer contact period.

These two situations are illustrated graphically in Figure 2. The upper curve shows that a high ozone dose for a short period of time uses the minimum amount of ozone for maximum mass transfer. On the other hand, the lower curve shows that low ozone dosages for long periods of time provides the minimum amount of ozone for reaction rate controlled reactions.

Basically, there are only six different types of gas/liquid contacting systems[16]. These are:

- Spray towers (liquid sprayed into gas)
- Packed beds
- Bubble plate or sieve plate towers (an intermediate situation between 1 and 2)
- Mechanical mixers
- Injectors
- Diffusers

The most common ozone contacting systems are based on some method for dispersing gas bubbles within a liquid. Generally there are two ways of accomplishing this:

- Gas is introduced initially into the fluid as bubbles of the desired size for optimum ozone dispersion into the liquid, or
- A massive bubble or gas stream is disintegrated into the fluid.

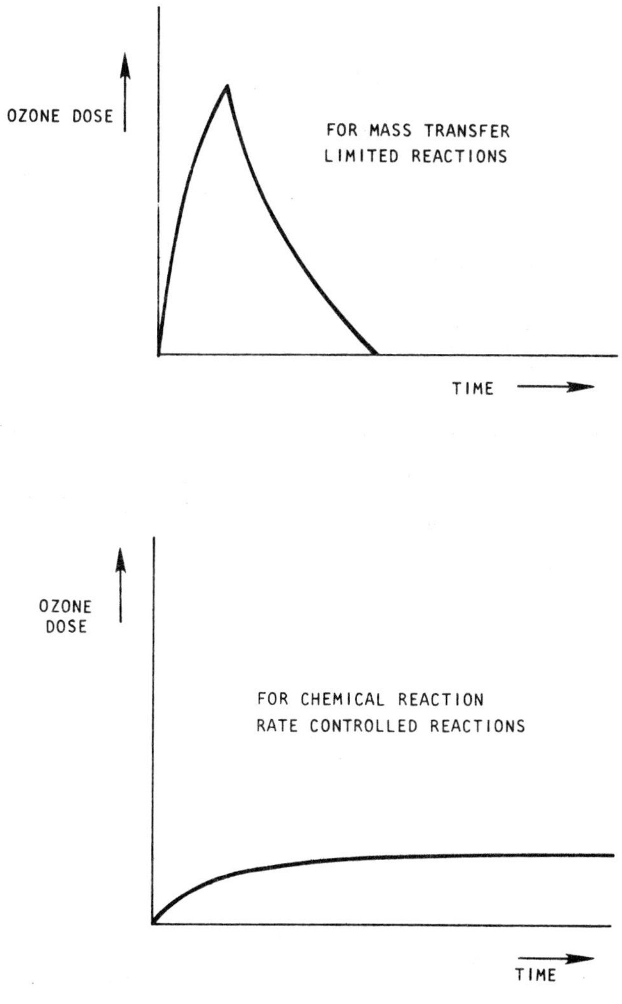

Figure 2. Ozone dosage vs. time for two types of reaction: mass transfer rate controlled and chemical reaction rate controlled.

These two techniques are exemplified by the diffuser (or sparger) and the injector (or eductor) contactors, respectively. Both of these types of contactor were developed originally for treating potable water supplies, and have been utilized throughout the world for this purpose for many years. Many variations of each type have evolved.

With the diffuser, ozone is added at the bottom of the contact chamber, through a porous medium (ceramic, Teflon, stainless steel, etc.), and the gas bubbles rise through the water which is passed cocurrently or countercurrently through the chamber. Many installations utilize multiple diffuser chambers, alternating liquid flow first cocurrently then countercurrently with the gas stream. Diffusers can be operated with little energy being added, are especially useful when large volumes of water are being passed through the plant by gravity flow, and also are flexible in terms of changing flow rates. Additional power costs to effect contacting thus are minimized. However, contacting chambers to house the diffusers are rather large.

Injectors require added energy, the simplest involving pumping of water to be ozonized rapidly past a small orifice, through which the ozone is forced into the liquid under pressure or drawn into the liquid by the vacuum created by rapid flow of water past the orifice. In those installations in which water already is being pumped about the plant, injectors are especially suitable, because of their smaller size and high mass transfer rates. However, injectors are relatively inflexible in terms of changing flow rates.

Detailed discussions of the theory of contacting have been published by Nebel,[17]; Sherwood & Pigford[18] and Treybal[19]. For purposes of this report, however, it is sufficient to point out that all contactors have their advantages and their limitations (Table 3), and it is necessary to consider the specifics of the ozonation job at hand, with respect to material(s) to be ozonized, volume of liquid to be treated, etc. A pilot study should be conducted with the contactor(s) selected, and the following parameters should be studied over practical and economic limits[17,20,21,22]:

- Liquid head or total pressure at initial point of gas/liquid contact,
- Distribution of ozone throughout the contact period,
- Total contact time necessary,
- Concentration of ozone applied,
- Volume ratio of liquid to gas,
- Ozone residual.

Once these parameters have been determined satisfactorily, design of the fullscale ozone contacting system can proceed with a high degree of confidence.

Specific ozone contacting systems will be discussed in more detail in Section 8.

TABLE 3. TYPES OF OZONE CONTACTORS

CONTACTOR TYPE	ADVANTAGES	DISADVANTAGES
SPRAY TOWERS	- HIGH RATE OF MASS TRANSFER - UNIFORM O_3 IN GAS PHASE	- REQUIRES HIGH ENERGY TO SPRAY LIQUID - SOLIDS CAN PLUG SPRAY NOZZLES - SHORT CONTACT TIME
PACKED COLUMNS	- WIDE GAS/LIQUID OPERATING RANGE - SMALL SIZE AND SIMPLICITY	- EASILY CHANNELLED & PLUGGED
PLATE COLUMNS	- SAME AS PACKED COLUMNS, BUT NO CHANNELING AND BROADER GAS/LIQUID OPERATING RANGE	- EASILY CLOGGED, BUT EASIER TO CLEAN - BEST SUITED FOR VERY LARGE INSTALLATIONS
SPARGERS	- REQUIRES ONLY GRAVITY FEED; NO ADDED ENERGY - WIDE GAS/LIQUID OPERATING RANGE ALLOWS INTERMITTENT OPERATION	- INTERMITTENT FLOWS MAY CAUSE PLUGGING OF POROUS MEDIA - LONGER CONTACT TIMES REQUIRE LARGER HOUSINGS - TENDENCY TO VERTICAL CHANNELING OF GAS BUBBLES, REDUCING CONTACT EFFICIENCY
AGITATORS, SURFACE AERATORS, INJECTORS, TURBINES, STATIC MIXERS	- HIGH DEGREE OF FLEXIBILITY - SMALL SIZES - INTIMATE CONTACT & GOOD DISSOLUTION	- REQUIRE ENERGY - NARROW GAS/LIQUID OPERATING RANGES - CANNOT ACCOMODATE SIGNIFICANT FLOW CHANGES (INJECTORS AND STATIC MIXERS), THEREFORE REQUIRE MULTIPLE CONTACTOR STAGES

Ozone Analysis and Control

The most generally used technique for determing ozone (actually total oxidants) in water is by the starch-potassium iodide technique. Recently, however, many other colorimetric techniques have proved to be successful as well. Instrumental techniques include ultraviolet adsorption and amperometric titration[23].

Process control of some types of modern ozonation systems can be achieved readily by monitoring residual ozone in the water just after the contacting chamber exit. This technique is employed in many modern European drinking water treatment plants, and is planned for the Charles-J. des Baillets plant being constructed in Montreal[24]. The analyzer is coupled with the ozone generators so that if the level of dissolved ozone drops below a pre-determined level, the generators are signaled to increase production.

By thus instrumenting, programming, and monitoring dissolved ozone after contacting, the normally varying loadings of ozonizable materials in water treatment plants can be handled on an automated basis using minimum quantities of ozone.

In an exactly similar manner, the amount of ozone present in ambient plant atmospheres can be monitored by another analyzer, or series of analyzers. In the event of an ozone leak from the generators or their associated downstream piping, the ozone sensors will react by signaling the ozone generators to be turned off and sound an alarm. Once electricity ceases to pass through the ozone generator, ozone generation ceases immediately.

Operational Experiences With Ozone

A large portion of this report is based on findings and observations made during site visits to twenty municipal water treatment plants in Europe and seven plants in Canada which use ozone. The 20 plants in four European countries were visited in May 1977 by a team of scientists and engineers. In August 1977, this team inspected seven plants in the province of Quebec, Canada.

Plants were selected on the basis of variability and uniqueness of application, size, and ozone treatment train variability. The site visit team inspected plants in Europe ranging in size from Annet-sur-Marne (25,000 cu m/day), and the large plant in the Federal Republic of Germany, Sipplinger Berg (650,000 cu m/day).

Plants containing ozone equipment from each of the major European manufacturers were inspected. These included Trailigaz and Degrémont of France, Gebruder Hermann and Demag of Germany, and Kerag and Sauter Corporation of Switzerland. Plants were inspected in Canada supplied with Canadian made equipment (Degrémont Infilco Ltd.). Also plants using U.S. manufactured equipment were observed in Europe (Welsbach) and Canada (Welsbach and PCI ozone.

The only operating municipal plant (Duisburg, Germany) using oxygen as the starting material for ozone production was visited. Five plants in the Düsseldorf area which practice river sand filtration and which place heavy emphasis on ozone plus activated carbon were inspected. The newest concepts in German drinking water practice were viewed at the Dohne plant in Mülheim near Düsseldorf. In southern Germany, a large plant which uses ozone primarily for microflocculation, followed by activated carbon, was seen (Langenau).

Rouen-la-Chapelle, located about 70 miles northwest of Paris, uses two stage ozonation, practices river sand bank filtration, and is the first French plant to use biological activated carbon (BAC). This process began operating at Rouen in early 1976. Two plants in southern France using ozone as the terminal step in the process, Clarifont in Toulouse and Super Rimiez in Nice, were visited. Several plants having highly sophisticated control systems were visited. The most notable ones in this category were Neuilly-sur-Marne, (600,000 cu m/day) in the Paris suburbs and Kreuzlingen on the Bodensee (Lake of Constance) in southern Germany.

Ozone systems that had been in operation for a number of years were also of interest. Two such plants, Choisy-le-Roi in Paris and the Holtausen plant in Dusseldorf have had their current systems in operation for more than 10 years.

A complete list of European plants listed in the order visited is shown in Table 4. Contrasts among plants are covered in detail in Section 7.

Six of 19 operating plants in the province of Quebec, using ozone, were visited. A seventh, the new Charles-J. des Baillets plant in Montreal, scheduled to go on-line in 1980, also was visited. The largest ozone plant in Canada, Quebec City (218,000 cu m/day), was inspected. The other five visited were Sherbrooke, Pierrefonds, Laval-Chomedey, St. Denis, and Ile Perrot. Pierrefonds and Sherbrooke are the two newest ozone facilities in Canada and thus represent the most recent efforts in North American practices of the technology. St. Denis is a very small plant (27,000 cu m/day, 7.2 mgd) that has PCI ozone equipment. Ile Perrot is an older small system that has a Welsbach ozonator. The Chomedey plant, located in the City of Laval, is rated at 39 million Imperial gallons per day (177,990 cu m/day). One side of the plant, the older side, uses powdered activated carbon for taste and odor control; the newer side uses ozone for taste and odor removal.

The selection of plants to visit in Canada was based on the same criteria as in the case of Europe. The team wanted to see systems made by U.S. manufacturers (Welsbach and PCI), small systems, Otto plate systems, and newer systems using French technologies. Additionally, if any plant used ozone as a terminal step, the conditions were to be observed and evaluated. Plants chosen exhibited one or more of these criteria. Table 5 gives pertinent information for each of the Canadian ozone plants located in the province of Quebec.

SUMMARY OF DATA FROM EUROPEAN OZONE QUESTIONNAIRES

General

Eleven hundred ninety-two (1192) questionnaires were mailed to municipal water plants in western Europe in mid-1977. These questionnaires (see copy in Appendix D) asked for detailed information on various aspects of ozone usage. Questionnaires were mailed to plants in France, the Federal Republic of Germany (FRG), Switzerland, Austria, The Netherlands and Great Britian. One questionnaire was sent to Brussels (Tailfer) for review; it was completed and returned.

The number of questionnaires mailed to those water plants thought to be using ozone is actually much smaller than 1192. In Germany (FRG), the

TABLE 4. EUROPEAN PLANTS INSPECTED BY SITE VISIT TEAM

Name	Location	Design Capacity (cu m/day)	Ozone Generator Manufacturer	Ozone Generator Type
Choisy-le-Roi	Paris suburbs	800,000	Trailigaz	Tube
Morsang-sur-Seine	Paris suburbs	150,000	Degrémont	Tube
Rouen-la-Chapelle	NW of Paris	30,000	Trailigaz	Tube
Aubergenville	NW of Paris	100,000	Welsbach	Tube
Neuilly-sur-Marne	Paris suburbs	600,000	Trailigaz	Tube
Annet-sur-Marne	Paris suburbs	25,000	Trailigaz	Tube
Clairfont	Toulouse	110,000	Trailigaz	Tube
Super-Rimiez	Nice	90,000	Trailigaz	Tube
Tailfer	Brussels	260,000	Trailigaz	Tube
Holthausen	Düsseldorf	192,000	Hermann	Tube
Flehe	Düsseldorf	88,000	Hermann	Tube
Am Staad	Düsseldorf	144,000	Hermann	Tube
Dohne	Mülheim	48,000	Trailigaz	Tube
Wuppertal	Wuppertal	168,000	Hermann	Tube
Wittlaer III	Duisburg	48,000	Demag	Tube
Lengg	Zürich	250,000	Kerag	Tube
Kreuzlingen	Kreuzlingen (Switzerland)	34,560	Sauter	Otto Plate
Konstanz	Konstanz (Germany)	50,000	CEO (Trailigaz)	Otto Plate
Sipplinger Berg	Sipplingen (Germany)	650,000	Hermann	Tube
Langenau	near Ulm (Germany)	198,700	Degrémont	Tube

German Association of Gas and Waterworks mailed out 835 questionnaires to waterworks, most of which do not use ozone; thirty-one of the 136 plants in Germany (FRG) thought to be using ozone have responded to date.

Thus, the number of water plants (using ozone) that received the questionnaire totalled approximately 400.

Questionnaire mailings and responses by country are as follows:

Country	Mailed	Received	Total Municipal Plants Using Ozone
France	300	63	593
Germany (FRG)	835	31	136
Great Britain	15	6	18
The Netherlands	10	7	12
Austria	11	5	42
Switzerland	20	9	150
Belgium	1	1	9
Total	1192	122	960

The total number of ozone water plants by country is shown in Table 1. Total number of plants listed for western Europe is 999. With responses numbering 122, this means that the response rate for all western Europe plants using ozone to date is 12.2%.

The following pages summarize questionnaire results by country. Tables giving details on each plant responding to the survey are included in Appendix D.

Great Britain

Fifteen questionnaires were mailed to British plants using ozone and six responses were received. There are 18 water plants in Great Britain known to be using ozone. The plants responding averaged 105,000 cu m in size, and ranged from 5,000 cu m/day to 450,000 cu m/day (Watchgate Plant). The primary application of ozone in Great Britain is for color removal. Each of the six plants responding indicated ozone use for this purpose. Other uses indicated include: bacterial disinfection (2 plants); iron and manganese oxidation (2 plants); taste and odor removal (1 plant); and viral inactivation (1 plant).

Ozone usage is relatively new in Great Britain. The oldest of the six ozone facilities which responded is the Loch Turret treatment works which installed ozone in 1967, but is not now operating.

Treatment processes vary. Ozone is used after a filtration step in each of the six plants. In four of the plants microstraining is the filtration method used. Chlorine is used as a final disinfectant in each of the six plants. In 3 of 6 plants chlorination follows ozonation with no filtration step in between.

Ozone generation equipment in 5 of the 6 plants reporting was supplied by Trailigaz of France. Five of the six plants have tube type, water cooled ozone generators. Ozone contacting in three plants is by porous tubes (diffusers). Other methods of contacting are: injector (2 plants); and submerged turbine (1 plant).

Off-gas treatment is practiced in 5 of the 6 plants. The methods vary from destruction by passing through wet activated carbon (2 plants), recycling (1 plant), thermal destruction (1 plant), and catalytic destruction (1 plant).

Power consumption for ozone generation and application (including air preparation, generation, contacting, and off-gas treatment) averaged 29.7 kwh/kg of ozone produced. Consumption ranged from 23 kwhr/kg to 37 kwhr/kg.

Ozone dosage averaged 2.37 mg/l for the four plants that responded to that portion of the questionnaire. It should be noted that chlorine dioxide and ozone are not used jointly in all British plants. Chlorine dioxide is often used as a distinct unit process, added stepwise to insure a better residual.

The Netherlands

Questionnaires were mailed to the 10 major plants in the Netherlands using ozone. Seven were returned. Results of the survey show that there are 12 water plants in the Netherlands which use ozone.

The prevalent water treatment philosophy is that physical and biological treatment processes are best. Chemicals and disinfecting agents are used only in cases where they are unavoidable. Chlorine dioxide currently is not used. Chlorine is used as a terminal disinfecting agent although many plants have plans to switch to ClO_2 in the future.

The plants responding to the questionnaire are relatively large, averaging 49,500 cu m/day. The range in size was 1000 cu m/day to 120,000 cu m/day. Five of seven facilities have used ozone for five years or less. The Noordendijk Plant in Dordrecht has used ozone for nine years. Average age of the ozone facilities is 3.7 years.

The primary application of ozone appears to be for color removal (6 plants), taste removal (6 plants), and odor removal (4 plants). Other applications are for bacterial disinfection (2 plants), viral inactivation (2 plants), organics removal (3 plants) and as a filtration aid (1 plant). Ozone dosages range from 0.23 mg/l to 5 mg/l. Average dosage for seven plants was 2.59 mg/l.

Each of the seven plants reporting has tube type, water cooled ozone generators. Kerag, a Swiss manufacturer, furnished the equipment for 4 of 7 plants.

Contacting in 5 of 7 plants is accomplished by means of submerged turbines; one plant uses an injector and one uses porous tubes for dispersion.

Power consumption ranges from 17 to 20 kwh/kg of ozone produced at the Houdsweg plant to 50 kwh/kg of ozone produced at the Engelse Werks plant. Power consumption averages 29.85 kwh/kg at the six plants reporting data.

Contactor off-gas treatment is by thermal destruction at 4 plants and thermal destruction plus a catalyst at another. One plant uses filtering through wet activated carbon for ozone destruction. The seventh plant does not treat the off-gas.

Chlorine is used as a final disinfectant in 3 of the 7 plants; a fourth applies 0.2 mg/l sodium hypochlorite. Ozone is used as the terminal step in the Engelse Werk plant. In the other two plants, sand filtration is the terminal step and immediately follows the ozonation process.

River sand bank filtration is practiced at two of the seven plants.

Austria

There are 42 municipal water plants in Austria currently using ozone. Questionnaires were mailed to the 11 largest water companies in Austria which serve 50% of the Austrian people. Five questionnaires were returned from water companies in Salzburg, St. Pölten and Linz.

Ozone has been used in the five facilities an average of 6.6 years, with the oldest plant having used ozone for 12 years. Plant size ranges from 4,000 cu m/day to 58,000 cu m/day, average size being 21,000 cu m/day.

Bacterial disinfection is the main purpose of ozonation. Other applications indicated are for viral inactivation (2 plants), color removal (1 plant), and organics removal (1 plant).

Ozone dosages are relatively low, ranging from 0.06 mg/l to 1.2 mg/l, averaging 0.48 mg/l for the five plants reporting data. The low dosage can be attributed to the high raw water quality. The water sources for each of the plants are deep wells.

Ozone generators in three plants are tube type, water cooled units. In the other two plants, plate type water cooled units are used. Contacting is by injection in three plants. The other two plants use a "schwimmbegaser", which has been described to the project team as a surface aeration device which induces turbulence and thus mass transfer of gas into liquid.

Contactor offgases are handled by dilution (2 plants), partial reinjection (1 plant), filtration through wet activated carbon (1 plant) and are not treated in the fifth plant.

Power consumption for ozone production, contacting, and off-gas treatment averages 34.4 kwh/kg of ozone produced. Consumption ranges from 16 kwh/kg at the two St. Pölten plants to 55 kwh/kg at the Salzburg City Waterworks.

The only oxidant other than ozone used in the five Austrian plants is chlorine; in the form of sodium hypochlorite at one plant. In four of the five plants, ozonation is the only treatment of the water. The fifth plant applies ozone and then adds 0.8 mg/l of sodium hypochlorite.

Switzerland

There are approximately 150 water plants in Switzerland which use ozone. Many of these are very small plants. Questionnaires were not mailed to all plants, but rather to 20 representative waterworks which serve more than one million people. Nine of 20 questionnaires have been completed and returned.

The raw water for the nine plants comes from wells or glacial supplied lakes. Thus the raw water quality is very high and the ozone dosage is very low, ranging from 0.3 to 1.5 mg/l. For the nine plants reporting, the average dosage is about 0.7 mg/l.

Ozone is used for several purposes, among them bacterial disinfection (7 plants), viral inactivation (7 plants), odor removal (7 plants), taste removal (6 plants) and organics removal (4 plants).

Five of the nine plants have been using ozone for more than 10 years; average time of usage is about 12 years.

The plants reporting data range in size from about 2,500 cu m/day (Waterworks Alstatten) to the Lengg Plant in Zürich which has a capacity of 250,000 cu m/day. Average plant size for the nine plants is about 65,600 cu m/day.

Five of nine plants have tube type, water cooled ozone generators. The other four have plate type, water cooled generators. Contacting is accomplished by a number of methods: submerged turbines (4 plants), injector (3 plants), porous tubes (2 plants).

Power consumption ranges from 17 to 81 kwh/kg. Average power consumption for ozone generation and contacting for the five plants reporting data is about 33.5 kwh/kg.

Contactor off-gases are not treated at all in five plants; two dilute the off-gases before venting to the atmosphere; two thermally destroy the ozone.

Six of nine plants use chlorine dioxide as a terminal step. It is estimated that 80% of the Swiss plants which use ozone also use ClO_2. Chlorine dioxide is applied in very small amounts. The Lengg plant in Zürich adds only 0.06 mg/l. Chlorine is used as a disinfectant at two plants. Ozone is the only treatment applied at the small (2500 cu m/day) Altstatten plant.

Four plants reported the use of granular activated carbon (GAC) directly following ozonation.

Germany (Federal Republic of Germany)

There are approximately 136 municipal waterworks in Germany which use ozone. Thirty-one of these waterworks responded to a questionnaire mailed out by the DVGW (German Association of Gas and Waterworks).

Ozone usage in Germany, based on site visits conducted in May 1977, is more varied than in any other country. Purposes of ozonation, dosage, methods of contacting, and the number of manufacturers supplying equipment do not fall into a consistent pattern as is the case in other countries. Power consumption also varies greatly.

The average period that ozone has been installed in the German plants responding is 7.6 years. The plant that has been using ozone for the longest period is the Neustadt-Sisch plant, which installed ozone facilities in 1959.

The average plant size is about 54,100 cu m/day. This number can be misleading, however. Many small plants of less than 1,000 cu m/day use ozone as their only treatment. Ozone is also used at the large German water treatment plant, the Bodensee Wasserversorgung (Lake of Constance) which has a capacity of 650,000 cu m/day.

In Section 5, it was noted that the Germans try to use groundwater as a source whenever possible and that their water treatment philosophy is to try to restore waters to the quality of a pure, unpolluted groundwater. The questionnaires received substantiate this stated philosophy. Nineteen of the 31 plants responding use groundwater as then raw water source, 6 practice river sand bank filtration, and 4 take their water from the Lake of Constance. Only three plants (Düren, Dohne and Langenau) uses river water directly without ground passage first (Dohne uses ground passage after treatment).

Ozone is used for many purposes in Germany. Twenty-four of the 31 plants indicate its use for organics removal, taste (13 plants), viral inactivation (8 plants), iron oxidation (7 plants), manganese oxidation (6 plants), odor removal (7 plants), turbidity reduction (6 plants), and color removal (5 plants). Ozone dosages range from 0.15 mg/l at the Diez/Lahn plant to 5.7 mg/l at the Osterode plant.

Ten different manufacturers of ozone generators supplied equipment to the 31 plants. Twenty-six facilities have tube type, water cooled generators while 5 have plate type, water cooled generators.

Contacting of ozone with the water stream is accomplished mainly by means of injection (21 of 31 plants). Other methods are submerged turbine (2 plants), packed column (3 plants), and porous tubes (1 plant). Wuppertal uses spray towers. Three plants did not specify contacting method in their responses.

Power consumption for ozone generation, air preparation, contacting, and off-gas treatment appears to be higher at some German plants than in other countries surveyed. Interestingly, the lowest power consumption cited (15 kwh/kg) is at Duisburg, the only known municipal water plant that produces ozone from oxygen.

No offgas treatment is practiced at 14 of the 31 plants; six plants indicate some form of treatment, but do not specify the method. Destruction of ozone on wet activated carbon is practiced in 4 plants, 3 use catalytic destruction, 1 plant recycles ozone and 1 plant uses a combination of activated carbon and a catalyst. Thermal destruction is not practiced at any of the 31 plants.

For final disinfection, 9 plants use ClO_2, 8 use Cl_2 and 2 use sodium hypochlorite. Nine plants apply ozone as their only treatment step. In 3 other plants, ozone is the only oxidant used.

Ten of 31 plants use granular activated carbon as an adsorbent. In every case, it is used after the ozonation step.

France

Ozone is used in approximately 600 French water plants. Questionnaires were mailed to 300 of these plants. A total of 63 questionnaires has been received.

Ozone usage for water treatment began in France in 1906. Thus, one would expect many of the plants responding to the questionnaire to be older plants. While there were 10 plants that have used ozone for more than 10 years (the oldest having used ozone for 52 years), the average period of use is about 8 years. There are many new ozone plants in France, which substantiates the findings earlier in this Section that ozone use has doubled in the past decade.

The average size of plants responding to the questionnaire was 29,100 cu m/day. Sizes ranged from 350 cu m/day to 240,000 cu m/day.

Unlike Germany, most of the plants use surface water as a raw water source. Forty-three of 63 plants indicated use of surface water while only 8 use groundwater as a source.

Ozone is used for many purposes, but the primary purposes are bacterial disinfection (59 plants), viral inactivation (35 plants), taste removal (31 plants), and organics removal (24 plants). Other indicated uses include color removal (18 plants), turbidity reduction (10 plants), iron (7 plants) and manganese (5 plants). Ozone dosages range from 0.15 mg/l to 10 mg/l.

Only 3 manufacturers supplied ozone generation equipment to the 63 plants reporting. Most of them were equipped by the large French companies, Degrémont (32 plants) and Trailigaz (24 plants). Most of the plants use tube type, water cooled generators. Only nine indicated usage of plate type, water cooled units.

Contacting is accomplished primarily by porous plate diffusers, this method being used in 44 plants. Injectors are used in 10 plants, packed columns in 3 plants and spray towers in 2 plants. The remaining plants did not report data.

Forty-two of 63 plants vent off-gases from the contactor directly to the atmosphere. Nine plants thermally destroy the off-gases, 3 recycle and 2 dilute before discharging.

Power consumption overall is fairly consistent for the 33 data points reported. Average power consumption is 31.3 kwh/kg for the total ozonation subsystem.

Chlorine is used as a final disinfecting agent in 26 plants. Chlorine dioxide is used in 13 plants. In twenty-two of 63 plants, ozone is the only oxidant used and often it is the terminal step.

The classic French water treatment process is remarkably similar to the U.S. "conventional" process. The process used in many plants involves prechlorination, coagulation, sedimentation, filtration, ozonation, and use of chlorine or chlorine dioxide as a terminal treatment step. The only marked difference is the ozonation step. This allows small quantities of chlorine to be used as the residual disinfectant. Normally less than 1 mg/l of chlorine is added, or less than 0.6 mg/l of ClO_2, to provide a protective residual for the distribution system.

SUMMARY OF DATA FROM CANADIAN OZONE QUESTIONNAIRES

Eighteen questionnaires were distributed in Canada, and all were completed and returned. These comprise all of the plants in Canada using Ozone with the exception of a plant at Frobisher Bay in Northwest Territories, and the Chomedey Plant at Laval. Data for the latter, obtained at the time of the plant visits, are given in Table 5 which summarizes all of the plant data from Canada.

The philosophy of ozone usage in Canada evolved from a need to deal with largely seasonal taste and odor problems, plus continuous disinfection needs with surface water supplies[9]. Ozone has been viewed in Canada as an alternate for two processes: chlorine disinfection and activated carbon treatment. Nadeau and Pigeon[9] argued that, when the costs of the latter two processes combined are compared to ozone costs, they are comparable, particularly when the side benefits of decolorization and superior appearance, plus improved taste and odor of ozonated waters are taken into consideration. Chlorine still is used in the majority of plants to provide a disinfectant residual for preventing bacterial regrowth in the distribution system.

Energy consumption for ozone treatment reported by the Canadian plants falls mainly in the 20 to 30 kwh/kg range. One old plant, Ile Perrot, reported 45 kwh/kg; whereas one of the newest Canadian plants, Pierrefonds, reported 18 kwh/kg. Off gas destruction is not normally practiced in Canada. Only one plant, Rivière du Loup, reported using catalytic off gas destruction.

TABLE 5. QUEBEC PLANTS USING OZONE

Name	Flow	U.S.	Ozone Generator			Unit cap.	Instld. Cap.	Yr. O_3
	cu m/day	mdg	Manufacturer	Type	No. Units	kg O_3/hr.	kg O_3/hr	start-up
1. Terrebonne	15,120	4	CEO	Otto	2	0.6	1.2	1963
2. Rivière-du-Loup	15,120	4	PCI	Tube	1	1.5	1.5	1977
3. Lac Etchemin	2,383	.63	CEO	Otto	1	0.3	0.3	1966
4. Roberval	7,714.5	2.04	Welsbach	Tube	1	1.8	1.8	1977
5. *Quebec City	218,000	57.6	CEO	Otto	12	1.25	15.0	1969
6. St. Eustache	1,060	3.0	CEO	Otto	3	-	-	1957
7. Oka (Deux Montagnes)	2,192	0.3	CEO	Otto	1	-	0.3	1958
8. *Sherbrooke	98,862	26.0	Degrémont Infilco	Tube	4	2.36	9.44	1977
9. Laval (Pont-Viau)	95,850	25.3	CEO	Otto	12	0.6	7.2	1957/62/68
10. Laval (Ste Rose)	11,410	3.0	CEO	Otto	3	0.66	2.0	1961/68
11.*Ile Perrot	6,800	1.8	Welsbach	Tube	1	0.3	0.3	1963
12.*Pierrefonds	95,500	25.2	Trailigaz	Tube	2	3.0	6.0	1976
13. L'Assomption	9,988	2.64	CEO	Otto	1	-	-	1966
14. Repentigny	22,730	6.0	Trailigaz	Tube	1	1.9	1.9	1973
15. Drummondville	54,000	14.3	CEO	Otto	3	0.57	1.71	1966
16.*St. Denis sur Richelieu	27,300	7.2	PCI	Tube	2	0.58	1.16	1972
17. L'Epiphanie	4,500	1.2	CEO	Otto	1	-	-	1962
18. Buckingham	18,900	5.0	Degrémont Infilco	Tube	1	1.2	1.2	1976
19.*Laval (Chomedey)	176,900	46.8	Welsbach	Tube	2	-	-	1973

*Plants visited by survey team.

Porous diffusers and injectors are the most widely used form of ozone contacting. Only one plant, Ile Perrot, uses submerged turbines. Contact times, where given, ranged from 5 to 20 minutes with most reporting in the 2 to 10 minute range. The Roberval plant doses with ozone twice, utilizing ozone for both coagulation and disinfection.

A summary table of the Canadian ozone questionnaire responses is given in Appendix D.

SECTION 7

OZONE SITE INVESTIGATIONS

EUROPEAN PLANT VISITATIONS

Table 6 lists some of the pertinent features of the 20 European plants visited by the site survey team. As is evident, ozone is used for numerous purposes and is applied by many methods. In this section we will discuss the most striking points of similarity and differences in the purposes for which ozone is being utilized and in the methods of application in the various drinking water treatment plants visited. Detailed discussions of engineering and cost aspects will be addressed in later sections.

France

Eight plants were visited in various locations in France (see Figure 3.) Six plants visited use river water, but the raw water quality varies considerably. Two plants use well water which is partially affected by river water. Two plants visited are located on the Marne River (Annet -- upstream of Paris and Neuilly -- Paris suburbs), four on the Seine (Morsang -- upstream of Paris, Choisy-le-Roi -- Paris suburbs, Aubergenville and Rouen-la-Chapelle -- both downstream of Paris), one on the Garonne in southwestern France (Toulouse, Clairfont plant) and one on a mountain stream at Nice, southeastern France. The Rouen and Aubergenville plants use well water.

All eight French plants visited use ozone for disinfection. Most of these plants also use ozone for oxidation of organics, two additionally use ozone for color removal and taste and odor control. One employs two stage ozonation for all these purposes, plus preparation of Biological Activated Carbon for ammonia and organics removal.

Raw water qualities of the Marne and Seine rivers upstream of Paris are higher than in the Paris suburbs. The large Aschères sewage treatment plant for Paris is located on the Seine, and Seine river water quality west of Paris is lower still. In fact, French public health officials prohibit raw Seine river water from being taken directly into drinking water treatment plants located downstream of the Aschères plant. Instead, wells are dug into the river banks and water is drawn from these wells.

This technique of drawing groundwater instead of highly polluted river water is used at the Aubergenville and la Chapelle plants along the Seine, and at all other drinking water treatment plants on the Seine west of Paris. Ground filtration of highly polluted river waters allows considerable

TABLE 6. Pertinent Features of European Water Treatment Plants Visited

Plant, Location, Water Source & Year Ozone Installed	Design Capacity (cu m/day)	Ozone Generation Capacity (kg/day)	Treatment Process	Purpose(s) of Ozonation	Type of Contacting
Annet-sur-Marne, France, Marne River, east of Paris (1973)	25,000	162	-ClO_2 presterilization, $FeCl_3$, flocculation, neutralization with soda, activated carbon, filtration, decantation, ozonation, post-chlorination	disinfection & organics	4 chambers; diffusers in 2,3 & 4; turbine in #1
Choisy-le-Roi, Paris, France Seine River (1968)	800,000	2760	-ClO_2, $AlCl_3$, powdered activated carbon, soda, coagulation, flocculation, decantation, sand filtration, dechlorination with bisulfite, ozonation, ClO_2	disinfection & organics, color & tastes & odor	4 chambers; diffusers in 1, 2 & 3
Morsang-sur-Seine, France, Seine River, east of Paris (Stage 1-1970)	150,000 being expanded to 1 MM	317	(Line #1) -75,000/day, breakpoint chlorination (for NH_3), $Al_2(SO_4)_3$ + silica + powd act. carbon, flocculation, rapid decantation, sand filtration, dechlorination with SO_2 (if excess Cl_2 remains) or add Cl_2 (if insufficient residual), pH correction with soda, ozonation, Cl_2 (Line #2) -75,000/day, breakpoint chlorination, $Al_2(SO_4)_3$ + silica (no powd. act. carbon) flocculation, rapid decantation, sand filtration. Now split stream into halves: -a) ozonation, GAC, Cl_2 -b) GAC, ozonation, Cl_2	disinfection & organics removal	2 chambers, both countercurrent; diffusers
Rouen-la-Chapelle, France, Seine River, west of Paris, Uses deep well water	50,000	137	-preozonation, sand filtration, GAC filtration, post-ozonation, chlorination	pre: Fe & Mn, phenols & detergents, oxygenate water to develop aerobic nitrifying bacteria post: disinfection tastes, odors & color	pre: turbine post: 2 chamber diffusers; counter-then cocurrent
Aubergenville, France, Seine River, west of Paris (1961), Uses deep well	100,000	276 (Welsbach 550 tube generator)	-$Al_2(SO_4)_3$ + silica + powd. act. C, coagulation, flocculation, phosphate addition, aeration, nitrification (to remove 0.5-3 mg/l NH_3), sand filtration, ozonation, chlorination	disinfection & organics	diffusers, 2 countercurrent chambers, 60% O_3 in #1, 40% O_3 in #2

(continued)

TABLE 6. (continued)

Plant Name	Ozone Dosage	Energy Demand, Kwh/kg	Contactor off-gas Treatment	Type of Generators	Conjunctive Use of Cl_2 or ClO_2 In Pretreatment	Conjunctive Use of Cl_2 or ClO_2 As Post-Treatment
Annet-sur-Marne	2 mg/l (0.4 mg/l after 7 min) in 3 & 4	NA	15 min in chamber #1, then atmosphere	tube type, water cooled Trailigaz	1 mg/l ClO_2 CIFEC ClO_2 system since 7/76	0.5 mg/l ClO_2 in summer; 0.4 mg/l Cl_2 in winter (lower cost)
Choisy-le-Roi	3–5 mg/l (0.4 mg/l after 10 min)	24		tube type, water cooled Trailigaz	1.5 mg/l ClO_2 to destroy organic Mn complexes	0.4–0.5 mg/l ClO_2 at least 6 mos of the year. Chlorine during the balance (up to 0.5 mg/l) (using CIFEC ClO_2 generator since June 1976)
Morsang-sur-Seine	2–4 mg/l (0.4 mg/l after 10 min)	37.5	thermal or discharge	tube type, water cooled, Degremont	3–6 mg/l Cl_2 for 0.5 mg/l NH_3	0.2–05 mg/l Cl_2 to attain 0.02–0.03 mg/l in distribution system
Rouen-la-Chapelle	pre: 0.7 mg/l, 3 min contact time post: 1.4 mg/l (0.4 mg/l after 12 min contact time)	NA	post contactor off-gases to pre contactor with added O_3, then discharge to atmosphere	tube type, water cooled Trailigaz	—	0.4–0.5 mg/l Cl_2
Aubergenville	0.5 mg/l in both chambers for 6 min (total = 12 min). Dosage varies in first chamber	NA	thermal destruction	tube type, water cooled, Welsbach	—	0.2 mg/l Cl_2 added @ pumping

TABLE 6. (continued)

Plant, Location, Water Source & Year Ozone Installed	Design Capacity (cu m/day)	Ozone Generation Capacity (kg/day)	Treatment Process	Purpose(s) of Ozonation	Type of Contacting
Düsseldorf, Germany, Holthausen plant (1964). Bank filtered, Rhine River water	192,000	1840	river bank filtration, ozonation, holding tank, anthracite filtration, GAC, ClO_2	Fe & Mn & organics	partial injection into 1% of product water
Düsseldorf, Germany, Flehe plant (1964). Bank filtered, Rhine River water.	88,800	816	river bank filtration, ozonation, holding tank, anthracite filtration, GAC, ClO_2	Fe & Mn & organics	partial injection into 1% product water
Düsseldorf, Germany, Am Staad (early 1970's). Bank filtered, Rhine River water.	119,200	1056	river bank filtration, ozonation, holding tank, anthracite filtration, GAC, ClO_2	Fe & Mn & organics	partial injection into 1% of product water
Wuppertal, Germany (1967). Bank filtered, Rhine River water	168,000	360	river bank filtration, aeration (with preozonation), rapid sand filtration, GAC, ClO_2	pre: Fe + Mn major: organics	pre: with aeration device post: spray water into ozone/air mixture
Duisburg, Germany (1965), Bank filtered Rhine River water. Only plant generating ozone from "oxygen" (actually 60% O_2: 40% N_2)	65,000	672	river bank filtration, ozonation (10% of product water), ozonized water combined with raw water, sand filtration, GAC filtration (O_2 added before GAC to 10-12 mg/l), ClO_2, NaOH	Fe, Mn & organics	10% product water countercurrent in packed column. This is combined with raw water in second packed column, cocurrently
Mülheim, Germany, (Dohne plant) Ruhr River water (4/15/77)	48,000	192	preozonation, $Al_2(SO_4)_3$, lime, flocculation, sedimentation, filtration, ozonation, $Al_2(SO_4)_3$ + polyelectrolyte, flocculation, filtration, GAC, ground storage, chlorination	pre: flocculation & Mn major: organics	pre: turbine major: diffusers
Konstanz, Germany, Bodensee water (1965)	50,000	72	microstraining, ozonation, sand filtration, ClO_2	disinfection & organics	partial preinjection (similar to Düsseldorf)
Sipplingen, Germany (Sipplinger Berg plant) Bodensee (1970)	648,000	816	microstraining, ozonation, holding tank, rapid sand filter, Cl_2	disinfection & organics	packed columns, countercurrent

(continued)

TABLE 6. (continued)

Plant Name	Ozone Dosage	Energy Demand, Kwh/kg	Contactor off-gas Treatment	Type of Generators	Conjunctive Use of Cl_2 or ClO_2 In Pretreatment	Conjunctive Use of Cl_2 or ClO_2 As Post-Treatment
Holthausen	3-3.5 mg/l (5-7 min)	10-gas preparation 20-O_3 generators	GAC, but changing to catalyst	14-Herrmann horizontal tube, water cooled	--	0.1-0.3 mg/l ClO_2 made from excess Cl_2 + $NaClO_2$
Flehe	3-3.5 mg/l (5-7 min)	10-gas preparation 20-O_3 generators	GAC, but changing to catalyst	6-Herrmann horizontal tube water cooled	--	0.1-0.3 mg/l ClO_2 made from excess Cl_2 + $NaClO_2$
Am Staad	3-3.5 mg/l (5-7 min)	10-gas preparation 20-O_3 generators	GAC, but changing to catalyst	10-Herrmann horizontal tube, water cooled	--	0.1-0.3 mg/l ClO_2 made from excess Cl_2 + $NaClO_2$
Wuppertal	pre: unknown post: 1-2 mg/l total: 1-3 mg/l	35	pretreatment, then to atmosphere	8-Herrmann horizontal tube, water cooled	--	0.1-0.15 mg/l ClO_2 made from Cl_2 + $NaClO_2$
Duisburg	10-25 mg/l in packed column, 1.5-3 mg/l total. 8-10 min. total	13	recycled totally	4 DEMAG vertical tube, water cooled	--	0.2-0.25 mg/l ClO_2 made from Cl_2 + $NaClO_2$ to attain 0.1 mg/l at plant exit. Then NaOH to neutralize CO_2
Mülheim	pre: 1 mg/l major: 2 mg/l	-	to preozonation, then catalytic destruction	2 Trailigaz, horizontal tube, water cooled	--	0.2-0.3 mg/l Cl_2
Konstanz	1.1 mg/l (0.4 mg/l after 8 min. contact time)	-	GAC or thermally	3 VAR (Sauter) plate, water cooled	--	0.25-0.3 mg/l ClO_2 made from Cl_2 + $NaClO_2$
Sipplinger Berg	0.9-1 mg/l (2 min contact time)	29.5 total 2.5 gas preparation 17 ozone generators 10 contacting	GAC	6 Herrmann horizontal tube, water cooled	--	0.6 mg/l Cl_2

TABLE 6. (continued)

Plant, Location, Water Source & Year Ozone Installed	Design Capacity (cu m/day)	Ozone Generation Capacity (kg/day)	Treatment Process	Purpose(s) of Ozonation	Type of Contacting
Langenau, Germany, Danube River (50% max. during summer) + ground water	198,720	206	River Water: prechlorination at intake for slime control, breakpoint chlorination, $FeSO_4$, lime, preozonation, flocculation, filtration, ozonation, $FeSO_4$, filtration, GAC, mix with ground water and chlorinate	pre: flocculation & destruction of off-gases major: disinfection, make water blue, and organics	pre: turbines major: diffusers
Neuilly-sur-Marne, Paris suburbs, France, Marne River water (1974)	600,000	2640	prechlorination (ClO_2), $FeCl_3$, soda, flocculation, sedimentation, rapid sand filtration, ozonation, chlorination	disinfection & organics	4 chambers; diffusers in 2, 3 & 4, turbine in 1st
Toulouse, France (Clairfont Plant), on the Garonne River in southern France (1970)	110,000	216	ClO_2 + $Al_2(SO_4)_3$ + $AlCl_3$, flocculation, decantation, sand filtration, ozonation, pH adjustment + Na silicate 1-2 days/months at 1 mg/l	disinfection	2 diffuser chambers, counter - then cocurrent
Nice, France on Mediterranean, Super Rimiez plant, mountain stream water (1972)	90,000	216	microstraining, $Al_2(SO_4)_3$, coagulation, flash mixing, flocculation, sedimentation, rapid sand filtration, ozonation	disinfection & organics	4 chambers, diffusers
Brussels, Belgium, (Tailfer plant) 30 miles northeast of Brussels (1973). River Meuse water	195,000	456	H_2SO_4, Cl_2, ClO_2, coagulation, powd. act. C, settling & sedimentation, filtration, ozonation, chloramination	viruses, tastes & odors, color & organics	2 turbines
Zürich, Switzerland (Lengg plant) on Lake of Zürich, (1975)	250,000	780	pre-chlorination for molluscs, $Al_2(SO_4)_3$, lime, ozonation, GAC, slow sand filtration, ClO_2	disinfection, viruses, & organics	3 turbines
Kreuzlingen, Switzerland, on Lake of Constance (1977, but same process as was operated since 1957 at older plant). Also uses some groundwater which receives ClO_2 treatment only	32,400	96	prechlorination for molluscs (if necessary), $Al_2(SO_4)_3$, pressure filtration, ozonation, GAC (70 cm) and sand filtration, ClO_2	disinfection & organics	2 stacked diffusers

(continued)

TABLE 6. (continued)

Plant Name	Ozone Dosage	Energy Demand, Kwh/kg	Contactor off-gas Treatment	Type of Generators	Conjunctive Use of Cl_2 or ClO_2 In Pretreatment	Conjunctive Use of Cl_2 or ClO_2 As Post-Treatment
Langenau	pre: 0.2-0.3 mg/l major: 1-1.2 mg/l, 20 min. contact time	-	to preozonation then atmosphere	2 Degrémont horizontal tube water cooled	--	0.4 mg/l Cl_2
Neuilly-sur-Marne	2-4 mg/l (0.4 mg/l after 4 min)	21 at 600 Hz 19 at 50 Hz	to #1 contact chamber, then vent to atmosphere	4 Trailigaz horizontal tube, water cooled	ClO_2 for presterilization & oxidation (probably of complexed metals)	0.5 mg/l Cl_2 at reservoir prior to distribution system
Toulouse	1-8 mg/l (0.4 mg/l after 8 minutes)	43	thermally in diesel generator stacks	4 Trailigaz horizontal tube, water cooled	ClO_2 for taste & odors, color removal & predisinfection	pH adjustment w/ NaOH and Na silicate 1-2 days/mo. at 1 mg/l. No residual disinfectant added
Nice	0.4 mg/l after 6 min	-	dilute & discharge to atmosphere	4 Trailigaz horizontal tube, water cooled	--	--
Tailfer	1.5-2 mg/l, 8 min contact time	38 total 6 air pretreatment 20 O_3 generators 12 contacting	thermally	6 Trailigaz horizontal tube, water cooled	injected after pre-Cl_2 for organically complexed Fe & Mn; made under pressure (Cl_2 & $NaClO_2$)	0.25 mg/l post-ammoniation to produce $ClNH_2$
Lengg	1.25-3.0 mg/l (6-10 min)	12 total 8 O_3 generators	thermally	6 Kerag vertical tube, water cooled	1 mg/l Cl_2 to prevent molluscs	0.1 mg/l ClO_2, made from HCl + $NaClO_2$
Kreuzlingen	1.3 mg/l into #1; 0.1 mg/l in #2, if at outlet of #1. 5 min in # 2	-	dilute & discharge to atmos-	4 Sauter plate, water	Cl_2 available to prevent mollusc needed yet	0.2 mg/l ClO_2, made from HCl

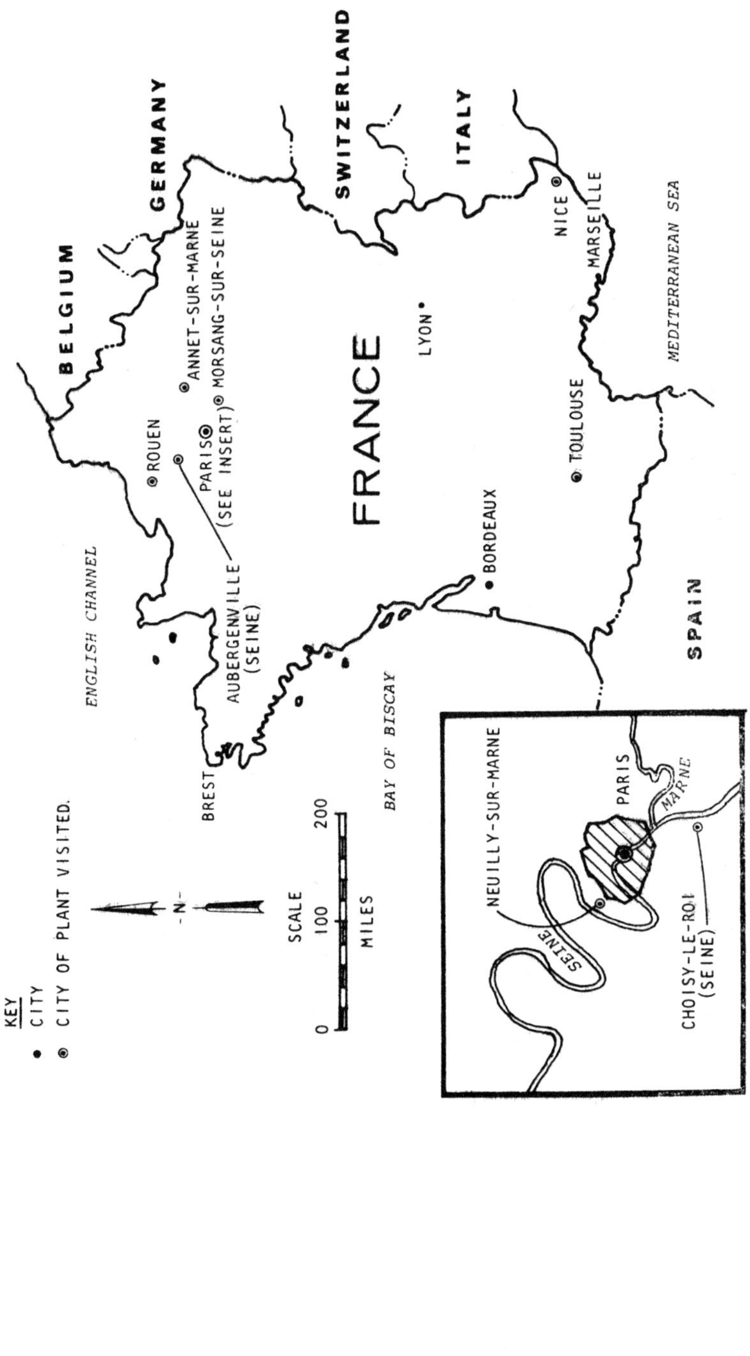

Figure 3. Location of French plants visited.

amounts of organic materials to be removed, both by filtration and by biological action. The concept of river sand bank filtration is practiced in Germany all along the Rhine River. Some 60 to 75% of the organic material in the raw river water, along with some ammonia, is removed by passage, through sand banks located along the river.

Ammonia content of river waters upsteam of Paris is low (0.005 mg/l at Annet-sur-Marne; 0.5 mg/l at Morsang-sur-Seine). In the Paris suburbs, the ammonia content is 0.43 mg/l at Choisy-le-Roi, but at Aubergenville is 0.5 to 3 mg/l (averaging 0.6 to 0.7) and at Rouen is 2 to 3 mg/l. At these last two plants, ammonia is removed biologically.

Water treatment processes at Annet, Choisy, Neuilly and Toulouse are similar, involving pre-chlorination with ClO_2, chemical addition ($FeCl_3$, $AlCl_3$, alum and/or silica), powdered activated carbon, neutralization with caustic soda, coagulation, sedimentation, decantation and rapid sand filtration -- all before ozonation.

Ozonation at these four plants is accomplished by means of porous diffusers in multiple contact columns. A minimum ozone residual of 0.4 mg/l is maintained and monitored continuously after 4 to 10 minutes of contacting time.

In the two chamber porous diffuser contactor at Toulouse (as well as at Morsang and Aubergenville), some 60% of the total ozone applied is fed to the first chamber (which has water containing the highest ozone demand), and the balance of the ozone feed is added to the second chamber. Using this two chamber technique, the first contacting chamber is referred to by the French as the "oxidation chamber", in which the initial ozone demand is satisfied. The second contact chamber is called the "residual chamber", wherein the requirement of maintaining 0.4 mg/l of residual ozone in solution for at least 4 minutes is attained in order to insure viral inactivation (bacterial disinfection is attained in lesser contact times).

Ozone contactor off-gases from the two chambers at Toulouse are sent to the hot exhaust stack of the diesel generators where ozone is destroyed thermally. At Morsang, the off-gases are sent to a cold stack where they can be destroyed catalytically or diluted with atmospheric air (10:1) and discharged to the atmosphere (the Morsang plant is in a rural area).

Choisy, Annet and Neuilly utilize four chamber contactor systems. Choisy has diffusers in all four chambers, and diffuses ozone into the chambers in staged amounts. Recovery of spent off-gases is not practiced at Choisy. Annet diffuses fresh ozone into the second and third chambers, and recovers spent off-gases in the fourth. The off-gases are injected using a turbine into the first chamber. Neuilly diffuses ozone into just three chambers, recovers spent off-gases in the fourth chamber, and reinjects the off-gases with a turbine in a compartment immediately preceeding the first diffusion chamber. Exhaust gases from the turbine chambers in both plants either are discharged to the atmosphere through a tall stack, or are destroyed by passage through wet granular activated carbon.

Ozone dosages at Annet-sur-Marne average 2 mg/l to attain the 0.4 mg/l residual after 7 minutes contact time. At Choisy, the ozone dosages required average 3 to 5 mg/l, and at Neuilly-sur-Marne the ozone dosages required average 2 to 4 mg/l. At Toulouse, 1.5 to 1.8 mg/l of ozone is required to attain the 0.4 mg/l residual after 8 minutes of contacting time.

Chlorination follows ozonation at the above mentioned plants (except for Toulouse), either through the use of chlorine or chlorine dioxide. However, only small amounts of these chlorine compounds are employed for residual -- usually 0.5 mg/l or less, depending upon the distribution system length, complexity and residence time. It is apparent that the objective of the water treatment processes at these plants is to produce water of high quality which minimizes the amount of chlorine or chlorine dioxide that must be employed. From the consumers point of view, the objective is to produce a bacterially and chemically safe potable water with no chlorinous taste.

At Aubergenville, the pretreatment steps involving alum, silica, powdered activated carbon and coagulation are followed by biological nitrogen removal. This is accomplished by addition of phosphate and aeration, then passage through a wet mass (sludge blanket) maintained in an upflow-filter. After sand filtration the water is ozonized in a two chamber contactor as described earlier, then it is chlorinated just before entering the distribution system.

Water supply for the City of Nice comes from a mountain stream and is of very high quality. The Super Rimiez plant is situated high above the city. Mountain stream water flows by gravity through microstrainers, then into the treatment plant, where alum addition, coagulation, flash mixing, coagulation, sedimentation, rapid sand filtration and ozonation processes are applied. No chlorination or other treatment is practiced after ozonation. Water is sent to underground reservoirs and flows into a three level distribution system into the various sections of the city. Water does not need to be pumped at Super Rimiez -- water flows into the plant, through the plant and throughout the distribution system entirely by gravity.

Although the Super Rimiez plant began operation in 1972, the previous three Nice plants (which were combined when Super Rimiez was constructed) all used the same treatment process involving terminal ozonation. The oldest Nice plant, Bon Voyage, first installed ozonation in 1906.

Toulouse's Clairfont plant (on the Garonne River) also does not use post-chlorination after ozonation. However, pH adjustment is employed after ozone treatment, and for 1 to 2 days/month, 1 mg/l of sodium silicate is added to protect the distribution system from corrosion.

The distribution systems at Toulouse and Nice are short and the average residence time of treated water in these systems is rather short (less than 12 hours). Because of the low temperature of the raw and treated waters (both plant raw waters originate in the mountains), the low carbon and ammonia contents, as well as the short residence time in the distribution systems, the French find no necessity for residual chlorine in the treated drinking waters of Toulouse or Nice. More than 70 years of experience attest to the safety of this practice at Nice.

The Morsang-sur-Seine plant (east of Paris) is planned eventually to have five individual treatment lines, each capable of treating 75,000 cu m/day. By May of 1977, two lines were operating and the third was being completed. The two operating lines employ breakpoint chlorination at the plant entrance, followed by alum. Line 1 then continues with poly-aluminum chloride then powdered activated carbon, flocculation, rapid decantation, sand filtration, chlorine residual adjustment, pH correction with soda, then ozonation (2 to 4 mg/l to maintain the 0.4 mg/l residual over 10 minutes of contact time), followed by chlorination for residual.

Line 2 receives breakpoint chlorination, alum, silica, flocculation, rapid decantation and sand filtration. This stream then is split into halves. Line 2A now receives ozonation, then granular activated carbon (GAC) then chlorination. Line 2B receives GAC, then ozonation and chlorination.

Details of the Morsang plant are included in a comprehensive paper by Richard & Fiessinger[25]. For purposes of this section, however, it suffices to say that Line #1 represents the process initially installed at Morsang. Line #2 represents two attempts to improve the water quality by incorporating granular activated carbon, before ozonation and after ozonation. To date, although the qualities of the waters from lines 2A and 2B are similar, there are some distinct advantages observed in the line in which ozonation precedes the GAC. These advantages are: 1) extended operating life of the carbon column; and 2) better performance efficiency of the carbon. Both of these advantages are attributed to beneficial bacterial growths in the carbon columns.

Finally at Morsang, a pilot process recently has been set up and is being studied in which breakpoint chlorination is eliminated, chemical treatment remains the same, and ozonation followed by granular activated carbon (biological activated carbon) is employed to remove ammonia and organics. Morsang is one of only two French plants known to be using or studying Biological Activated Carbon (see Section 13).

At Rouen-la-Chapelle, on the Seine, west of Paris, well water is preozonized to oxidize iron & manganese, oxidize organics (phenols and detergents particularly) and to prepare GAC for optimum biological growths. After preozonation, the water is sand filtered, filtered through Biological Activated Carbon, then post-ozonized (for disinfection, tastes, odors and colors), then chlorinated to obtain a residual.

The major portion of ozone at Rouen is used in post-ozonation. Ozone dosages in this stage average 1.4 mg/l to achieve the 0.4 mg/l ozone residual after satisfying the initial ozone demand and after 12 minutes of contacting in a two chamber diffuser apparatus. Off-gases from the post-ozonation contactor are recompressed and sent to the pre-ozonation turbine contactor, which has a three minute contact time. If necessary, make-up ozone is added at this point so as to provide an average pre-ozonation dosage of 0.7 mg/l. This dosage is sufficient to oxidize iron and manganese, to partially oxidize phenols and detergents, and to aerate the water so as to cause proliferation of aerobic bacteria and nitrifying bacteria in the sand filters and in the subsequent granular activated carbon filters.

The beneficial effect upon ozonation costs of a second ozonation step at the plant, utilizing some of the off-gases from the ozone contacting system, is significant. Installing ozone generation capacity for the single purpose of disinfection near the end of the treatment process, involves installing generation capacity for the amount of ozone required, plus some backup generation capability. If ozone is used only for a single process such as disinfection, then all contactor off-gases must be treated before discharge to the atmosphere -- at some cost increment.

If some of these contactor off-gases are utilized instead for a pretreatment purpose, say oxidation of iron and manganese, then the main additional cost involved will be that of compression (for conveyance), contacting and piping for the pretreatment step. Even if supplementary ozone must be generated for the pretreatment step, this small increased need for ozone generation capacity can be determined during pilot studies and purchased at the time of ozonation system installation. On the other hand, not all of the ozone in primary contactor off-gases will be utilized even in a reuse application such as just described. Therefore ozone destruction facilities still will be required. The amount of ozone to be destroyed after recycle will be less, however -- see Section 8.

Conversely, if the ozone has not been installed for disinfection, the installation of ozone as a pretreatment step to oxidize iron and manganese will require ozone generation capacity (with backup generation capacity) higher than in the case of using contactor off-gases.

U.S. water treatment engineers can conclude from the Rouen plant (and others like it which the site visitation team visited in other countries) that once having installed ozone for one specific purpose, its use for other purposes may be cost-effective and should be evaluated.

Germany

Nine plants were visited in Germany, seven of which use river water and two of which are located on the Lake of Constance (the Bodensee), (see Figure 4.) Five of the river plants are located on the Rhine in the Düsseldorf area, one is on the Ruhr river near Düsseldorf, and one is on the Danube river in southern Germany.

None of the plants on the Rhine River draw water directly from the river for processing in water treatment plants. Instead, river sand bank filtration is practiced. Wells are dug along the Rhine and water is drawn from these wells for the water treatment plants. The average time of passage of river water from the Rhine to these wells is 20 days. Because of the bacterial activity in the ground, some 60 to 70% of the total organic

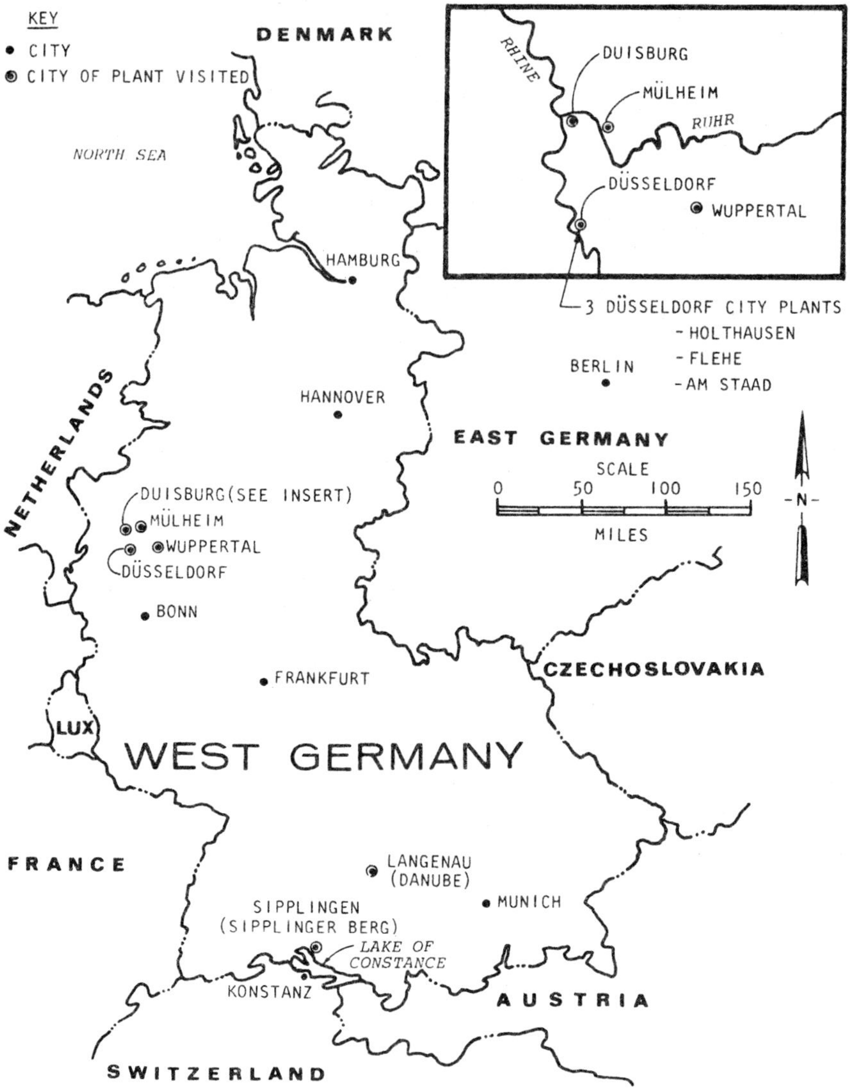

Figure 4 Location of German plants visited.

carbon contained in the Rhine river water is removed during ground passage. River sand bank filtration has been practiced along the Rhine for many years (at the Am Staad plant since the late 1800s) in the Düsseldorf area, and today is considered to be an essential first step in water treatment processing.

Waters of the Ruhr and Danube rivers in this part of Europe are not as polluted as those of the Rhine River, and river bank filtration is not considered necessary. Consequently, raw water is drawn directly from these rivers and treated in the two plants visited.

All plants on the Lake of Constance (German and Swiss) take water from the Lake for treatment without bank filtration. The Rhine River flows through the Lake of Constance, but the river is not highly polluted at this point.

All plants along the Rhine in the Düsseldorf area use ozone for oxidizing iron & manganese (which is present because of filtration through the sand banks) and for oxidation of organics. The three Düsseldorf city plants (Holthausen, Flehe and Am Staad) all use the same water treatment process. River bank filtered water is ozonized as it enters the plant. Lower valent iron (II) and manganese (II) are oxidized to the iron (III) and manganese (IV) states, at which states these metal ions quickly hydrolyze and precipitate. During ozonation, dissolved organic compounds also are oxidized, at least partially.

The ozonized waters then are sent to a holding tank for 20 to 30 minutes. The primary purpose of the holding tank is to allow the concentration of residual ozone to decay to nearly zero before the ozonized water comes into contact with the granular activated carbon. Otherwise, activated carbon will be oxidized by the ozone present, in a wasteful manner both of carbon and of ozone. During this holding time, residual ozone concentration decreases because of continued oxidation of dissolved organics. Any permanganate formed upon initial ozonization also is reduced to the Mn (IV) state by oxidizing dissolved organics. After this holding period, the water then is filtered through anthracite (to remove precipitated solids) then through granular activated carbon. Finally, small dosages of chlorine dioxide (up to 0.3 mg/l) are added for residual. Ozone contactor off-gases have been destroyed in the past by passage through wet granular activated carbon. However, all three plants have tested and now are installing catalytic ozone destruction units for this purpose.

The three Düsseldorf city plants are unique in that there is little reservoir capacity in the distribution system served by the plants for treated water (two small reservoirs store water for the highest elevations of the city). Therefore the three plants operate almost totally on demand. When demand is low, the plants operate at low levels. When demand is high, the plant operates at full capacity. Consequently, the three plants are sized to produce water at the peak demand rate.

The combination of ozonation followed by granular activated carbon was first installed in the Düsseldorf City plants in the late 1950s, and other German plants using this combination are generally patterned after what has come to be known as "the Düsseldorf process". Activated carbon was installed originally for taste and odor control, but now is used also for adsorption of chlorinated organic compounds, which are prevalent in the Rhine River, and which are only partially removed during river sand bank filtration, and for biological removal of ammonia. When activated carbon columns were first installed in these Düsseldorf plants, reactivation was required as infrequently as every two years. However, because of the buildup of chlorinated organic materials in the Rhine River water and their adsorption onto the activated carbon, today regeneration is required every 5 to 6 months[26].

It was observed that removal of organics by the combined ozonation/-granular activated carbon process at Düsseldorf was greater than expected if the expected removals by ozonation and by activated carbon were added separately. In addition it was found that ammonia levels were being reduced during passage through the granular activated carbon filters. These observations prompted studies which led to the development of the Biological Activated Carbon process recently installed at the Dohne plant in Mülheim (on the Ruhr River). Details of Biological Activated Carbon technology are presented in Section 13 of this report.

At the Wuppertal plant (on the Rhine near Düsseldorf), river bank filtered Rhine river water is aerated an preozonized, using off-gases from the ozone contactors employed later in the process. This oxidizes iron and manganese, partially oxidizes organics, and also serves to dispose of off-gas ozone. Any excess ozone in the preozonation off-gases is diluted with air during the aeration step and vented to the atmosphere. After pre-ozonation, the water is filtered, then ozonized, filtered through rapid sand, then through granular activated carbon and treated with chlorine dioxide. The sequence of primary ozonation, rapid sand and granular activated carbon filtration, all take place in a series of 16 units which house all three steps in a single structure.

At Wuppertal the ozone contacting system is one in which water to be ozonized is forced through narrow orifices into a gas atmosphere which contains ozone and air. The effect is much like that of spray drying, and the small droplets of liquid fall slowly to the bottom of the chamber, in constant contact with ozone-containing air.

The Duisburg plant employs a process similar to the Düsseldorf process, the major difference being in the method of ozone generation. Duisburg is the only known operational municipal plant that uses oxygen as the feed gas for generating ozone. As a result, the oxygen/ozone system operates in a closed loop, and only a single pump is used to move water through the total plant. Because of the closed loop design, all gases, including ozone contactor off-gases, are recycled through gas drying equipment to the ozone generator. Carbon dioxide (formed upon oxidation of organics) does not have a chance to escape from the system. Therefore, after carbon filtration, sodium hydroxide is added to adjust the pH of the treated water. A full description of the Duisburg plant is given in Appendix B.

Contacting in the closed loop system at Duisburg also is by means of a two stage partial injection system, but through two Raschig ring packed towers. In this case, 10% of the product water is treated with the total amount of ozone countercurrently through the first tower, called an "ozone washer". The output of this column then is passed upwards through the second Raschig ring packed tower, termed the "contactor".

Ozone contacting at the three Düsseldorf city plants (and at the City of Konstanz) is by partial injection. About 1% of the finished Rhine River water is subjected to direct ozonation. The total amount of ozone to be added is passed into a pipe through which 1% of the finished water is pumped. Within seconds after the ozone has mixed with the 1% water volume in the pipe, the ozonized water meets the remaining 99% as it is pumped down through a wider diameter pipe to the bottom of the contact chamber. From the bottom of the 10 meter deep contactor, water and ozone bubbles rise co-currently.

Ozone contacting is described in greater detail in Section 8.

The Dohne plant at Mülheim recently has installed a new water treatment process which involves ozonation and Biological Activated Carbon, and is similar to the process employed at the Rouen-la-Chapelle plant in France. Ruhr river water is pre-ozonized (partially with ozone-containing off-gases from the oxidation contactor used later in the process), then treated with alum, lime, then subjected to coagulation, sedimentation, then ozonation for oxidation of organics, more alum plus a polyelectrolyte, coagulation, filtration, granular activated carbon, ground storage, then chlorination for residual as the water enters the distribution system. Ozone-containing off-gases from the pre-ozonation step are destroyed by passage through a catalytic ozone destruction unit.

The older process at Mülheim began with breakpoint chlorination (for ammonia removal), followed by chemical addition, coagulation, sedimentation, filtration, granular activated carbon (for dechlorination) then ground storage and chlorination. As pollution in the Ruhr river has increased the ammonia level has increased, necessitating an increase in the amount of chlorine required during the initial treatment step (10 to 50 mg/l). In turn, this produced increasingly large quantities of chlorinated organics, which required increasingly more frequent reactivation of the GAC columns. Even with fresh activated carbon, unacceptably large quantities of chlorinated organics were noted in the product water.

Mülheim's new process, which became operational on April 15, 1977, eliminates the breakpoint chlorination step and effects ammonia removal in the Biological Activated Carbon. Thus, considerable quantities of chlorine are saved. In addition, rather than having to reactivate the GAC as often as every 2 to 3 months (by the old process), the anticipated life of the carbon columns before regeneration now is at least two years[27]. After 9 months of operation, the carbon columns at Dohne are performing as efficiently as when they were installed, and show no signs of requiring regeneration[28].

Long operating life of the activated carbon columns at Mülheim is to be expected on the basis of the 24 month operation of the biological activated carbon columns beds (without regeneration as yet) at Rouen-la-Chapelle in France.

Dusseldorf city plants use ozone dosages of 3 to 3.5 mg/l and contact times of 5 to 7 minutes. Wuppertal uses 1 to 3 mg/l total ozone dosage (pre- and post-ozonation), Duisburg adds 1.5 to 3 mg/l for 10 minutes, and Dohne (Mülheim) uses 3 mg/l (1 in pre-ozonation, 2 in post-ozonation) ozone dosages.

Contactors at Mülheim include a turbine for pre-ozonation and 2-chamber diffusers for post-ozonation. The high degree of agitation provided by the turbine has been found to be beneficial in causing improved flocculation[28].

In southern Germany, the water qualities are much higher than in the Düsseldorf area. Both the City of Konstanz and Sipplinger Berg plants are located on the Lake of Constance. Both treatment processes start with microstraining, followed by ozonation (for disinfection and oxidation of organics), sand filtration, then chlorine dioxide (at Konstanz) and chlorine (at Sipplinger Berg). The Sipplinger Berg plant incorporates a holding tank after ozonation. Off-gases containing ozone at Konstanz are destroyed thermally and/or by passage through wet granular activated carbon. At Sipplinger Berg, off-gases are passed through wet granular activated carbon for destruction of excess ozone.

The contactor at Konstanz is a partial injector, similar to the Düsseldorf contactors, whereas the Sipplinger Berg contactors are Raschig ring packed towers. Water flows downward through the packed columns while ozone-containing air flows upward. Ozone dosages at Konstanz average 1.1 mg/l and are 0.9 to 1 mg/l at Sipplinger Berg. Konstanz monitors an ozone residual of 0.4 mg/l after 8 minutes of contact, but Sipplinger Berg uses only a 2 minute contact time. However, the ozonized water is allowed to stand 30 minutes in a holding tank before sand filtration.

Sipplinger Berg is the largest municipal water treatment plant in Germany using ozone, with a design capacity of 648,000 cu m/day. Uniquely, this water, treated at the Lake of Constance, is sent 150 kilometers north to Stuttgart and other communities along the line. The highest amount of residual chlorine (0.6 mg/l) observed by the site visitation team at any of the plants visited is added here. The fact that residual chlorine still is present in the treated water when it reaches Stuttgart attests to the high quality of water produced by microstraining, ozonation and sand filtration. There is one chlorination station in the distribution system on the way to Stuttgart, but this is present only for safety reasons, and is rarely used. Sipplinger Berg plant personnel advise that without ozonation, the residual of chlorine could not be maintained in the distribution system without adding considerable chlorine at several points along the way to Stuttgart.

As pointed out in Section 5, all water treatment plants on the Lake of Constance have installed excess ozonation capacity and activated carbon capabilities to be used in the event of leakage of an oil pipe which runs through the Lake connecting Switzerland and Germany.

The plant at Langenau treats a combination of groundwater and Danube River water. The maximum amount of river water treated, however, occurs during periods of peak water demand during summer, and is a maximum of only one-half of the plant flow at that time. The amount of Danube River water treated during winter is minimal, and is necessary only when the groundwater volumes are insufficient to meet demands.

Raw Danube water is prechlorinated (1 mg/l) at the intake for slime control (the intake is 7 km from the plant), then treated by breakpoint chlorination (4 to 6 mg/l for ammonia), $FeSO_4$, lime, preozonation (using off-gases from the final ozonation), filtration, final ozonation, $FeSO_4$ addition, sand filtration, then granular activated carbon. At this point, treated river water is mixed with groundwater (which has not been treated to this point) and the combined stream chlorinated.

Pre-ozonation is employed to aid flocculation and to destroy contactor off-gases containing excess ozone. Post-ozonation is used for disinfection, organics oxidation and "to make the water blue" (the plant director stated that Langanau is the only place in Europe where the Danube is blue -- because of ozonation). The pre-ozonation contactors are high speed turbines and the post-ozonation contactors are two chamber porous diffusers. Ozone dosages are 0.2 to 0.3 mg/l in pre-ozonation (sometimes additional make-up ozone is required) and 1 to 1.2 mg/l for post-ozonation (10 minute contact time plus 10 minutes travel time to the filters). Off-gases from the pre-ozonation contactors are vented to the atmosphere (the Langenau plant is located in a rural area).

The Langenau plant has a number of unique features. First, the plant process is controlled <u>at every step</u> by turbidity monitors. Secondly, it is the only plant visited in which ferrous sulfate is added to the post-ozonized water. The purpose of this second iron addition is to provide an additional flocculation step for removing soluble organics which are oxidized to less soluble and/or more easily flocculated states during post-ozonation.

Granular activated carbon columns at Langenau are biologically active, but are only 1.5 meters deep, compared to 5 meters in the Düsseldorf area. Carbon reactivation times here range from 4 months to 2.5 years. This widely varying operational life of carbon is attributed to the relatively low amounts of chlorinated organics produced by the prechlorination and breakpoint chlorination steps. Ammonia levels in the Langenau raw water are much lower (0.2 to 0.5 mg/l) than in the Rhine or Ruhr Rivers near Düsseldorf. At Mülheim, breakpoint chlorination required 10 to 50 mg/l of chlorine as the old treatment process was practiced. At Langenau, only 4 to 6 mg/l is required presently. Nevertheless, a pilot Biological Activated Carbon program is being studied at Langenau with the objectives of replacing the breakpoint chlorination step, thus saving chlorination costs and prolonging the activated carbon life even further.

Finished water from Langenau is sent to Stuttgart, through a 350 km distribution system, and is treated with 0.4 mg/l of chlorine for residual at the plant reservoir.

Switzerland

Two plants were visited in Switzerland, both of which treat lake water (see Figure 5). The Lengg plant in Zürich is large (250,000 cu m/day), and the Kreuzlingen plant is small (30,000 cu m/day). Both of these plants incorporate ozonation plus activated carbon in their treatment processes.

Zürich's Lengg plant is situated on the Lake of Zürich, which is fed by mountain streams. Raw water from this lake is very high in quality, having a TOC of only 2 mg/l, for example. Water is prechlorinated (1 mg/l to prevent growth of molluscs), then treated with alum, lime, ozonation, granular activated carbon, slow sand filtration and chlorine dioxide.

Ozone is used at Lengg for disinfection, viral inactivation and oxidation of organics. Dosages are 1.25 to 3 mg/l (yearly average 1.5) and contact times are 6 to 10 minutes. Contactors at Lengg are turbines, and three of these are installed in each of the two treatment lines. Water to be ozonized passes through each contacting chamber, however only during periods of maximum flow are all three turbines utilized at the same time. Contactor off-gases are destroyed thermally.

Before distribution, Lengg water is treated with 0.1 to 0.15 mg/l of chlorine dioxide.

At Kreuzlingen, on the Lake of Constance (Bodensee) (which is fed by the Rhine River in its early stages), ozonation has been used since the late 1950s. The original treatment plant employing ozone was situated on the lake's edge, and in recent years began to sink into the lake. As a consequence, a new plant was constructed, and was placed in operation early in 1977.

The Kreuzlingen treatment process includes prechlorination for mollusc control alum, pressure filtration, ozonation (for disinfection and oxidation of organics), granular activated carbon, sand filtration, and chlorine dioxide for residual.

Ozone contacting at Kreuzlingen takes place in two chamber porous diffusers, but the two chambers are stacked one on top of the other. Water flows through the upper chamber then into the lower chamber. The geometry of the lower chamber is such that its off-gases collect at the top of the lower chamber beneath the inlet of the porous diffusers of the upper chamber. When the pressure of off-gases in the lower chamber exceeds the head of pressure in the upper chamber, these off-gases will pass into the upper chamber. Off-gases from the upper chamber are diluted with atmospheric air and discharged to the atmosphere.

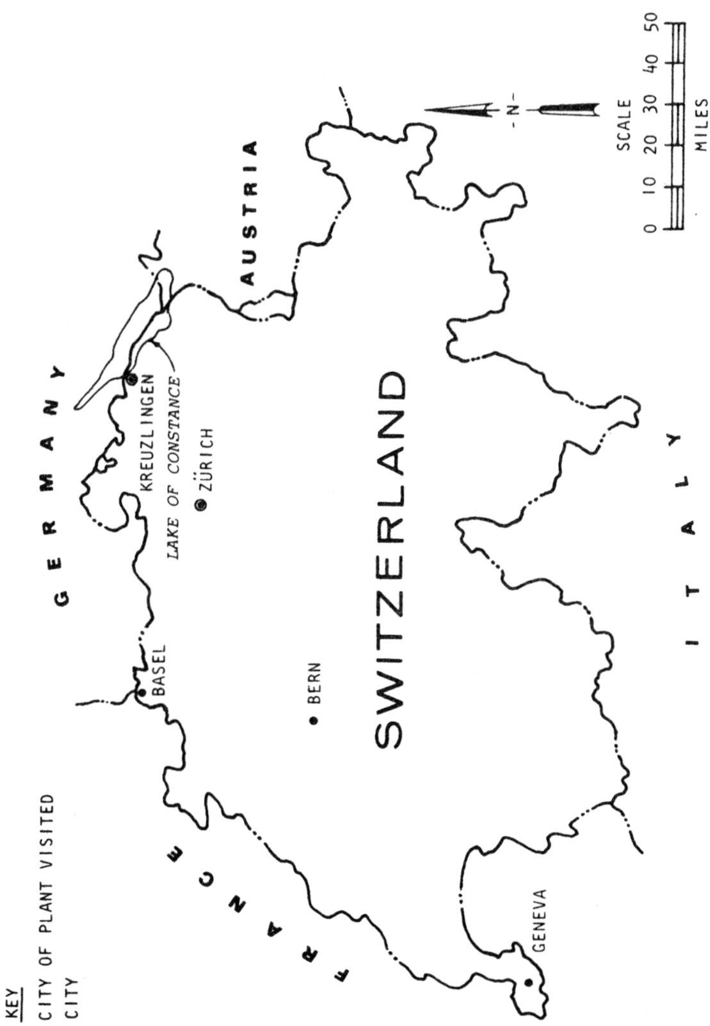

Figure 5. Location of Swiss plants visited.

Ozone dosages of about 1.3 mg/l are used in the first (lower) contacting chamber with contact times of 6 minutes. A residual ozone concentration of 0.4 mg/l is measured at the outlet of the lower chamber. The ozone concentration in the upper chamber is 0.1 mg/l, and the contact time in the upper chamber is 5 minutes (total contact time = 11 minutes).

The Kreuzlingen plant is designed to operate unattended. Finished water is stored in two reservoirs, and when demand lowers the level in one reservoir, the plant starts up. The plant superintendent lives in a house adjacent to the plant, and he prepares chemicals, checks operation of equipment and controls and provides maintenance services.

Belgium

Only one Belgian plant was visited, the Tailfer plant which serves Brussels. This plant treats Meuse river water with sulfuric acid, chlorine, chlorine dioxide (to decompose organic complexes of iron and manganese), coagulation, powdered activated carbon, settling, filtration, then ozonation (for viral inactivation, tastes & odors, color and oxidation of organics), followed by chloramination for residual.

Contacting at Tailfer is accomplished with two turbines per contactor. Dosages of ozone are 1.7 mg/l (yearly average) over 8 minutes contact time. Off-gases are destroyed thermally.

Tailfer currently is installing oxygen facilities so as to study the advantages of having higher ozone generation capacity during periods of peak ozone demand. When installed (late 1977 or early 1978), ozone will be generated from oxygen for a one year period to allow the ozone generation and associated processes to be studied on full scale in one of its treatment lines during all seasons. After this year of operation, Tailfer will decide upon whether to utilize this ozone generation technique during summer periods of highest ozone demand or to use oxygen continuously.

General Comments on European Plant Practices

With the exception of the Tailfer, Belgium plant, all plants visited had installed excess ozone generation capacity. This allows the generators to be operated routinely at less than their maximum production rate, thereby reducing deterioration of the generator dielectrics. Such practice allows maintenance to be performed on one generator while others are being operated. Maintenance of ozonation equipment is provided on a scheduled basis, rather than waiting until malfunction occurs unexpectedly.

Ozone demand varies seasonally, particularly as raw water temperatures change. In winter, ozone demands generally are lower than in summer. With those raw waters having relatively constant temperatures the year around, the ozone demands are also relatively constant the year around.

None of the plants visited had open reservoirs for treated water. The visitation team was repeatedly advised in all four countries visited and by representatives of other European countries not visited that there are no uncovered reservoirs for treated drinking water in Europe.

All plants visited are concerned about and prepared for accidental ozone leakages in the plants. All rooms where ozone can be present either are monitored for ozone in the air, or operators are instructed not to enter rooms when ozone can be smelled. Normally, plant personnel do not work in the rooms containing ozone generators or ozone contactors. These rooms all have exhaust fans which are programmed to start automatically when the ambient ozone monitors signal the presence of 0.1 mg/l (by volume), or greater, of ozone, or are turned on manually when ozone can be smelled by plant personnel.

Modern European plants have installed ozonation systems and entire treatment plant processes which are highly automated, and the degree of automation is not related to plant size. For example, there are many small water treatment plants in the Alps of Austria and Switzerland which are inaccessible throughout much of the year. The storage, handling and preparing of chemical solutions is impractical under these circumstances, and so the plants are designed with physical and/or mechanical treatment steps to operate unattended, with controls and warning alarms being installed in central facilities, such as the town hall. Filtration, ozonation and post-treatment are controlled automatically.

Operations at the Kreuzlingen plant (30,000 cu m/day), also are automated, even though the plant is not isolated in location. The plant superintendent visists the plant periodically to prepare solutions and check equipment operation, otherwise the plant is designed to operate unattended.

Some of the largest ozonation plants also are highly automated. At Neuilly-sur-Marne, in the suburbs of Paris, France, this 600,000 cu m/day treatment plant employs chemical and mechanical pretreatment on one side of the river Marne, and ozonation and post-chlorination on the other side. The entire plant process is operated and controlled by two plant operators who are stationed at the control panel around the clock. Analytical data obtained from the laboratory (automatic sampling) are fed to the control panel and adjustments in process parameters are made automatically.

Several water treatment plants visited contained one or more aquaria with trout to indicate the presence of toxic materials. The simplest trout "monitors" are aquaria through which product water is passed. If materials toxic to these sensitive trout are present in the finished water, the trout will die, or at least show visible signs of distress. These aquaria are located in areas of the plants which operating personnel frequent in conducting their duties. Plant personnel are trained to check the condition of the trout as often as they pass them.

A more sophisticated trout monitoring system has been developed by the Keuringsinstitut voor Waterleidingsbedrijven (KIWA -- the Netherlands Water Research Institute) and is operating at the Tailfer (Brussels, Belgium) plant. Another unit is to be installed at the Lengg plant in Zürich,

Switzerland shortly. This system uses the response of the trout to an electrical stimulus to monitor the raw water quality.

In the Netherlands trout monitoring system, fish are first conditioned to the raw water of the plant. After one week of conditioning, a trout is placed into a U-shaped, narrow raceway, through which the raw water flows at a constant rate. At the start of the raceway, provision is made for an electrical impulse across the channel. If the trout backs into this area, he triggers the electrical impulse and receives a small electrical shock. Upon receiving this shock, the trout swims to the other end of the raceway, and the time required to pass a specific point in the raceway is measured in relation to the time of the original electrical impulse. The raceway is narrow, so as to prevent the trout from turning around and swimming in the wrong direction upon receiving the electrical impulse.

To establish the swimming time of a new fish, each trout is impulsed several times and the times required for it to swim past the timing point are averaged to provide a calibration time. The new trout then is kept in the raceway for a one week period. During this time, he will occasionally back into the impulse area. When this occurs and the time to swim to the timing point is within the calibration range, nothing happens. But if the trout takes longer to swim to the timing point than his calibration time, this indicates that something has affected his ability to swim and an alarm rings.

There are four such trout raceways at the Tailfer plant. When one alarm rings, plant personnel are alerted that something may be amiss. If three of the four trout alarms ring, the plant is shut down and the raw water analyzed for toxic chemicals.

At the Sipplinger Berg plant in southern Germany, plant scientists have found that water fleas (daphnia) and a species of African electric eel can be used similarly for monitoring of toxic materials in the raw water. These chemists have also found that the rate of oxygen release from certain algae is greatly reduced in the presence of many water pollutants, and a monitoring procedure based upon this technique is being developed.

Overall, the most striking features of the European ozonation plants visited include the design of the water treatment processes to reduce oxidant-demanding materials to such low levels that either no residual disinfectant is required, or if one is required, levels of only 0.1 to 0.3 mg/l of chlorine, chlorine dioxide or chloramine generally are sufficient. Only in the case of Sipplinger Berg, which sends treated water 150 km north to Stuttgart, is a post-chlorination dosage as high as 0.6 mg/l necessary.

CANADIAN PLANT VISITATIONS

Table 7 lists some of the pertinent features of the six operational Canadian municipal water treatment plants visited by the site survey team. All of these treatment plants are located in the Province of Quebec (see Figure 6). The only other currently operating Canadian water treatment

TABLE 7. CANADIAN OZONE TREATMENT PLANTS VISITED

Plant, Location, Water Source & Year Ozone Installed	Design Capacity (cu m/day)	Ozone Generation Capacity (kg/day)	Treatment Process	Purpose(s) of Ozonation	Type of Contacting
Quebec City, St. Charles River, (1969)	218,000	360	Prechlorination, alum, lime, polyelectrolyte, coagulation, flocculation, sedimentation, filtration, ozonation, chlorination, pH adj.	tastes, viruses, viruses, organics (phenols), disinfection	4 injectors (4 min)
Laval (Chomedey), Rivière des Prairies (1959), river polluted with pulp & paper wastes	177,295 total 113,651 using ozonation 63,644 using Powd. Act. C.	(two 160 tube ozonators)	Prechlorination, alum, silica, lime addition, settling, filtration, ozonation, final chlorination.	disinfection, tastes, odors	3-chamber, porous diffusers Cocurrent in #1, Countercurrent in #2. #3 is detention tank.
Pierrefonds, Rivière des Prairies (1976)	95,500	250	Screening, intermittent prechlorination, caustic soda ammonium sulfate, sodium silicate, flocculation, settling, filtration, ozonation, pH correction, final chlorination and fluoridation. Chlorinate also in distr. system.	disinfection, odors	2-chamber, porous diffusers
St. Denis, Rivière Richelieu (1972)	27,300	28	Chlorination, alum, coagulation, settling, filtration, ozonation, chlorination, soda ash addn.	tastes, odors, disinfection	2-porous diffusers (10 min)
Sherbrooke, Lake Memphremagog (1977 plant replaced 15 yr old plant)	98,262	227	Occasional prechlorination, microstraining, ozonation, final chlorination	tastes, odors, disinfection	2-porous diffusers
Ile Perrot, Lake St. Louis (1963)	6,800	9	Chlorination, alum & silica, coagulation, settling, filtration, ozonation, lime, fluoridation	color, taste, odors, disinfection	1-submerged turbine (20 min)

(continued)

TABLE 7. (continued)

Plant Name	Ozone Dosage	Energy Demand, Kwh/kg	Contactor off-gas Treatment	Type of Generators	Conjunctive Use of Cl_2 or ClO_2 In Pretreatment	As Post-Treatment
Quebec	1.3 mg/l gives 0.4 mg/l residual O_3 after 4 min.	30	none	Trailigaz (plates) water cooled	1.2-2.4 mg/l to kill fly larvae (max. in summer)	1.2-4 mg/l provides 0.35 mg/l at plant exit and 0.10 mg/l 7 mi. from plt
Laval	Unknown -- 8 min total contact time		none	2 Welsbach horizontal tube, water cooled	120 lbs/day to maintain 0.3 mg/l residual on filters	250 lbs/day in 2 additions: 235 at entrance of reservoir, 15 at exit gives 0.15 mg/l residual at exit
Pierrefonds	2.0 mg/l to provide 0.4 mg/l after 4 min.	18 total 6-air prep. 12-O_3 generators	dilute and discharge to atm.	Trailigaz horizontal tubes, water cooled	0.8 mg/l summer for bacteria on filters	1 mg/l dose in reservoir gives 0.5 mg/l at exit, 0.1 in distr. system
St. Denis	2.2-2.3, 13 mins. contact time	19.0 (theor.) not measured	none	PCI-Ozone horizontal tube, water & air cooled	1-3 mg/l for bacteria	0.2-0.3 mg/l dosage gives 0.5 mg/l total oxidant (Cl_2+O_3) at exit
Sherbrooke	2.0 dosage, 8-13 mins contact, provides 0.4 mg/l	29.1 total 7.9 air prep. 21.2 O_3 generators	none	Degrémont horizontal tube, water cooled	prechlorinate to kill algae on microstrainers	1.4 mg/l in reservoir, 0.2 mg/l at exit (88 kg/day)
Ile Perrot	0.8-1.0, 20 min contact time	45 total 9-air prep. 21 O_3 generators 15 contactor	-	1-Welsbach horizontal tube, water cooled (0.8-1.0 mg/l dose)	1.3 mg/l to raw water at coagulation basin	none

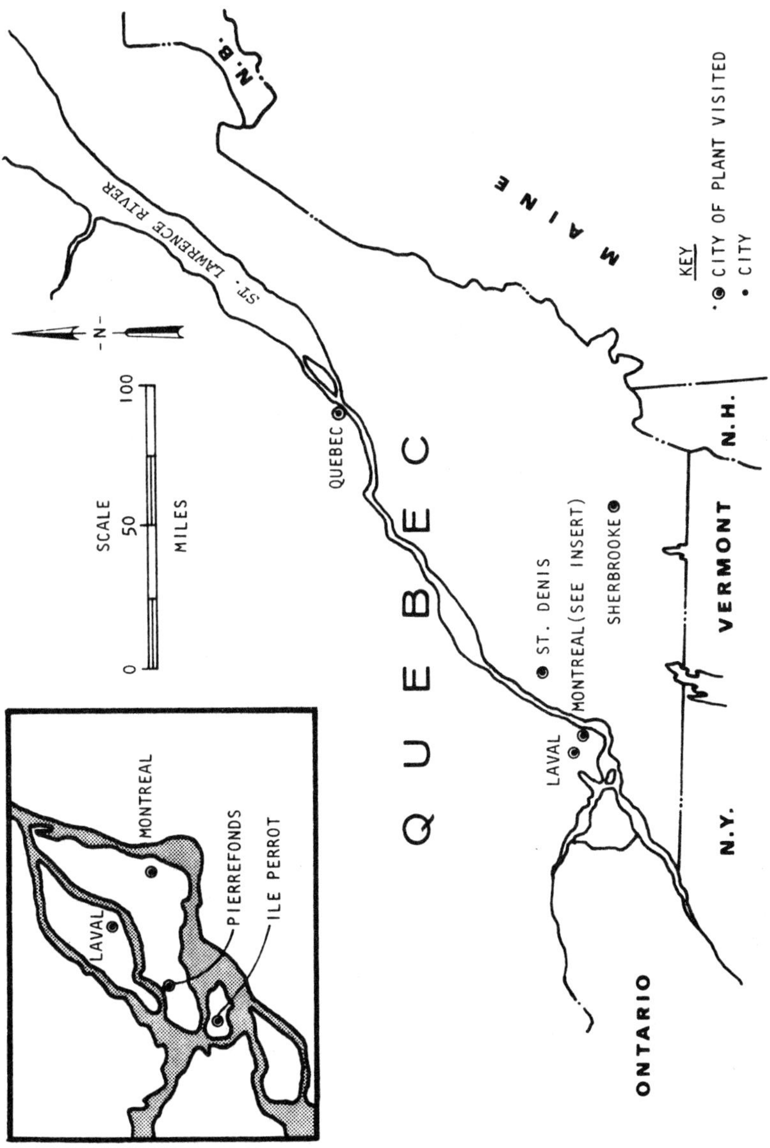

Figure 6 Location of Canadian plants visited

plant employing ozone is located in Frobisher Bay, Northwest Territories, on Baffin Island, near Greenland.

At the six Canadian plants visited, ozone is used primarily for disinfection and taste and odor control. However, color removal and organics removal are also important applications of ozone in Canada.

Four of the six plants use river water and two use lake waters. Although Canadian river waters are considered by local water treatment plant personnel to be "highly polluted", water taken directly from the source is treated, as opposed to river sand bank filtration practices before treatment along the lower Rhine river and drawing deep well water along the Seine downstream of Paris, France.

Age of the plants visited ranged from new to 14 years old. The oldest plant visited was Ile Perrot, on Lake St. Louis. The newest plants visited were Sherbrooke (mid-1977) and Pierrefonds (late 1976). Both of these newer plants are patterned after modern French water treatment plants, but with some modifications. The site of the new 500 mIgd (million Imperial gallons per day) (2,281,920 cu m/day) Charles-J. des Baillets plant currently being constructed to provide water for the city of Montreal also was visited. This is being constructed in two 250 mIgd (1,140,960 cu m/day) stages, the first of which will be operational in 1980. When fully completed in the mid-1980s, this plant will generate the world's largest amount of ozone per day for drinking water treatment (6750 kg/day). The des Baillets plant, as currently designed and under construction, represents the closest attempt to duplicate a modern French ozonation plant (Neuilly-sur Marne) in North America. When completed, the plant will be fully automated and computer controlled.

Water treatment capacities at the plants visited ranged from 6,800 cu m/day at Ile Perrot to 218,000 cu m/day at Quebec City (currently the largest Canadian water treatment plant using ozone). Three of the plants visited use ozonation systems manufactured in the United States.

An excellent discussion of the introduction of ozone technology into Canada, and in particular, into the Province of Quebec, is given by Nadeau and Pigeon[29]. The first ozonation plant for treating municipal drinking water in Canada was installed at Ste. Therèse in 1956. Today there are 19 operational plants in Quebec employing ozone (see Table 5) and one in Frobisher Bay. In addition, water treatment plants in St. John's Newfoundland (41,800 cu m/day -- 11 mIgd) and in Portage la Prairie (199,000 cu m/day -- (5 mIgd), Manitoba are under construction. Both should be operational in 1979. Several other Canadian municipal water treatment plants in Quebec are being designed which will employ ozonation.

Five of the six plants visited employ post-chlorination after ozone treatment, but in low dosages (0.5 to 2.5 mg/l) to provide residuals of 0.1 to 0.2 mg/l of chlorine in the local distribution systems. One plant visited (Ile Perrot) does not use chlorination after ozonation, but does adjust pH and adds fluoride.

In Quebec, 11 of the 19 plants use the older, Otto plate type ozone generators, which operate at slightly negative pressure and generate 1 kg of ozone/hour. The other plants use water cooled, tube type ozonizers. Nine of the plants have only a single ozone generator. Back-up ozone generation capacity is provided by installing excess generation capacity initially. This allows the generator to be operated routinely below its rated capacity. In turn, this results in longer operation times for the dielectrics without unexpected downtime of the ozone generators.

Quebec City

Twelve water-cooled, Otto plate type ozone generators are at this plant, and maintenance is performed every 18 months by one maintenance man over a two week period (per generator). Thus the plant programs one-third of a man-year for ozone generator maintenance. This involves changing gaskets, cleaning cooling water tubing, washing glass plates and restarting. This last step requires that dried air from the pretreatment system be passed through the cleaned and reassembled ozone generator for at least one hour to insure the presence of dried and cooled air in the generator. Electrical power then is turned on, but at the lowest setting. Gradually the voltage is increased (over a period of hours) to the normal operational level (about 50% of the maximum capacity).

The ozonation process of Quebec City is controlled by measuring a residual ozone concentration of 0.4 mg/l after 4 minutes contacting time. An ambient air ozone monitor recently has been installed in the ozone generator room, because of some leakage of ozone from the generators. This occurs when the Otto generators are operated above atmospheric pressure.

Contactor off-gases containing excess ozone are discharged to the atmosphere (the plant is situated in a rural location).

Until 1969, the St. Charles River (water source for Quebec City plant) received no treatment other than prechlorination (for fly larvae), primary screening and chlorination.

Laval (Chomedey)

Ozonation was introduced 20 years ago into this Chomedey plant, which at that time was the l'Abord-à-Plouffe plant. During expansion of this plant in 1961, the treatment process was converted back to conventional chlorination, but with powdered activated carbon added for taste and odor control. This 53,200 cu m/day (14 mIgd) activated carbon/chlorination plant still operates today at the Chomedey site using this process.

In 1973, however, an additional 95,000 cu m/day (25 mIgd) expansion was made at Chomedey, but as a second operational line which utilizes ozonation. Today at Chomedey, there are two operational plants on the same site, treating the same raw water for taste and odor control and disinfection by two different treatment processes (activated carbon/chlorination in one, ozonation/post-chlorination in the other). Therefore, Chomedey would appear to be a good installation at which to conduct a comparative study of these two processes, not only to compare their process efficiencies, but also for their effect upon production of halogenated organics as well as for cost-effectiveness. The study team was advised by Laval water treatment personnel that such a study is desired, but funds have not been available.

Two Welsbach ozone generators are at the plant, and although the generators themselves have provided satisfactory service, there have been problems with the air preparation units. When designed, no backup air compression capacity was installed. Therefore, when one of the air compressors malfunctions, insufficient air is dried and insufficient ozone is generated. Malfunction of the air compressors has occurred frequently at Laval, and the installation of larger air compression capacity to avoid this situation currently is being planned.

Ozone contacting at Laval is by means of injectors, which are similar to those used in the Düsseldorf and Konstanz, Germany plants. Contactor off-gases containing excess ozone are discharged to the atmosphere, although the consulting engineer is studying other disposal methods (thermal, catalytic or GAC destruction), since the plant is situated in an urban location.

Standby diesel power generators are on hand in case of electrical failure. These diesels can provide power for all of the Chomedey plant operations, except for ozone generation. Standby chlorination and powdered activated carbon will be used in the event of power failures. The presence of diesel power generators in Chomedey (and at other Canadian plants) parallels standby procedures in European ozonation plants.

Ozone generator tubes at Chomedey are individually fused, but the plant superintendent would prefer no fuses, because of economics. He advises that fuses (costing $30 each) fail much more frequently than tubes (costing $190 each). He feels that he would save more money by not having to replace fuses than he would lose in the cost of tube failures.

St. Denis

This 27,300 cu m/day plant currently uses two PCI Ozone tube type, water cooled generators. The first ozone generator was installed in 1972, and the second in early 1977. Both generators contain 6 dielectric tubes, and are rated at 32 lbs/day of ozone production. However, they are operated normally at production rates of 15 to 20 lbs/day. It is the opinion of the plant manager that alternate daily operation of the units is desirable from the standpoint of maintenance. Therefore, only one generator is used each day, for 18 hours of operation, and each generator is operated every other day. This causes greater than normal maintenance problems with respect to gaskets, fuses and the like, but no major maintenance problems.

Maintenance on the PCI Ozone generators is performed after every 2500 hours of use. Accurate records are kept during their operation, so that tube cleaning and other maintenance can be performed at the proper time intervals. Since 1972, the older generator has been cleaned only twice. No dielectric tubes have had to be replaced to date. As replacement tubes for these ozone generators cost $500 each, no spare tubes are stocked at St. Denis. Service from the supplier is very good, and in the event of tube failure, the plant superintendent feels that a replacement should be available within a day or two.

After ozonation, the plant laboratory measures a combined (ozone + chlorine) residual, which should be 0.5 mg/l. Usually, 0.3 mg/l of chlorine must be added to bring this combined residual up to 0.5 mg/l at the plant exit.

Monitors for detecting ozone in the ambient plant air are present in the rooms above the contacting chambers, but not in the ozone generation room. Contactor off-gases are discharged to the atmosphere through an exhaust pipe in the roof of the 3-story building. This exhaust pipe is pointed toward the ground, and underneath it, at ground level, is a box of flowers. Several of these flowers had been adversely affected by the waste ozone, but no other signs of damage were noticed in the discharge area. The plant is in a rural location.

Pierrefonds

This new plant went into operation in late 1976 and uses Trailigaz water cooled, tube-type ozone generators. The plant is patterned after Neuilly-sur-Marne, in the Paris, France suburbs (screening, prechlorination, chemical addition, decantation, filtration, ozonation, chlorination and fluoridation). Ozonation is controlled by maintaining 0.4 mg/l residual ozone after 4 minutes of contact time (2 mg/l dosage). Post-chlorination requires an average dosage of 1 mg/l to produce 0.5 mg/l at the plant exit. Residual chlorine in the distribution system averages 0.1 mg/l.

Analytical instruments monitor the ambient air for ozone in those plant areas where ozone can be present. Contactor off-gases are discharged to the atmosphere (the plant is located in a rural area).

Sherbrooke

This is the newest Canadian plant using Degrémont tube type, water cooled ozone generators. The 90-day commissioning phase of operation was completed in July, 1977. Air pretreatment at this plant uses a high pressure (100 psi) desiccant drying system. Ozone contacting is accomplished in two chambers, each of which contains porous diffusers. Water and ozone flow is counter-current in each chamber. Between the two chambers is a narrow passage which changes direction of water flow before the second chamber (similar to Degrémont contacting chambers seen in European treatment plants).

Two-thirds of the ozone generated is dosed into the first contacting chamber, and one-third to the second. Excess ozone in the contactor off-gases is vented to the atmosphere. There are dissolved ozone monitors at the outlets of both contact chambers. Residual ozone concentration of 0.4 mg/l is attained at the outlet of the first chamber, and this first ozone monitor controls the process. The second ozone monitor at the outlet of the second contact chamber signals an alarm if it measures less than 0.4 mg/l. Total contact time in both chambers is 8 to 13 minutes.

Since the plant break-in phase, the plant superintendent has been running the ozone generating systems at their maximum production capacity to study their operational performance capabilities. Starting January 1, 1978, the plant will be operated normally (ozone generation at less than rated capacity) and accurate records of operational parameters, costs, etc., will be maintained.

Contactor off-gases currently are discharged to the atmosphere, but the plant superintendent advised that he is going to study other disposal techniques. The plant is situated adjacent to the campus of the University of Sherbrooke, and the plant finished water reservoir is underneath the university track.

Ile Perrot

This small (6800 cu m/day) and older (1963) plant has been operating with a single Welsbach ozone generator since its installation. Each year, the generator is shut down and cleaned. Contacting is by means of porous diffusers and off-gases are discharged to the atmosphere.

City of Montreal

This plant is currently under construction, and is scheduled to begin operations in 1980. At that time the treatment capacity will be 250,000 mIgd (1,140,660 cu m/day), which is one-half of its ultimate capacity. The process will be an exact copy of the Neuilly-sur-Marne plant in the Paris suburbs, except for the cooling water system which will be modified because of differences in water temperatures. The process will be fully automated and computer controlled.

Montreal's Charles-J. des Baillets plant will treat St. Lawrence river water. At the time of site visit, 13 Trailigaz, Choisy 7500, water cooled, tube type ozone generators had been delivered, 12 for generation and one for backup. These generators operate at 600 Hz and can produce 27 kg/hr of ozone each.

Ozone contacting will be conducted in three-chamber contactors. Porous diffusers are present in chambers #2 and #3. Ozone will be introduced into these last two chambers through the diffusers, and off-gases will be injected (by turbines) into the first chamber.

An excellent description of the new Montreal plant has been published by Bouchard[15]. One year pilot studies leading to adoption of the planned process have been described in detail by Dellah[24].

Comparison of Canadian Plants With European Plants

All 20 Canadian plants generate ozone from air. No Canadian plants employ granular activated carbon after ozonation. All Canadian plants currently discharge excess ozone in contactor off-gases to the atmosphere, without using thermal, catalytic, or granular activated carbon destruction. None of the Canadian water treatment plants using ozone also use chlorine dioxide, either for pre- or post-treatment. Ozone contacting at all Canadian plants is by means of porous diffusers or injectors.

Canadian water treatment plants are designed by consulting engineering firms and usually the low bid equipment is purchased. This sometimes leads to cost cutting on the installation of controls and monitoring equipment. For example, at Laval--Chomedey, back-up air compression capability was not installed initially, which causes an insufficiency of ozone generation whenever an air compressor is down.

By contrast, modern European water treatment plants utilizing ozone, activated carbon, chlorine dioxide, and other water treatment processes provide back-up capabilities for these processes. When such European plants are designed, constructed, and operated by fully integrated water supply companies, the municipality merely contracts to purchase a given amount of water per unit time at a specified price. The capital and operating costs of these treatment plants thus are the concern of the water company, not of the municipality.

When an equipment supplier to the integrated water company is asked to bid on a designed treatment process, he is assured that the water company is sophisticated in the specification and use of that equipment (such as ozone generation). Controls and monitoring equipment usually are specified. On the other hand, when an engineering concern designs an ozonation system or chlorine dioxide system and does not have the operational experience with these processes to specify the installation of excess capacity, advanced analytical monitors, controls, etc., and is concerned only with installing the lowest price process design, costly operation and maintenance problems can develop early on during plant operations.

SECTION 8

ENGINEERING ASPECTS OF OZONATION EQUIPMENT AND PROCESSES

INTRODUCTION

Current U.S. waterworks design standards[21] and texts[22,23] provide little or no guidance for process applications of ozone in water treatment. Although this subject is well covered in ozone specialty literature[24,25,26] very little practical information has entered the waterworks trade publications. It is not surprising therefore, that the detailed guides and standards required to evaluate and design ozonation systems for any process application are not readily available. This lack of authoritative information discourages consideration of ozonation as a water treatment process and hampers the design of cost-effective, reliable, operative applications of ozone.

Ozone process selection and design criteria currently must be based on published information and on-site evaluation of operating facilities which are few in number in the United States. This section is intended to provide a basic background on the essential components of the ozonation process as applied for various water treatment purposes. It is by no means exhaustive and more detailed information may have to be obtained by visits to operational ozonation facilities and from knowledgeable equipment suppliers.

The ozonation system may be divided into four main parts: gas preparation, electrical power supply, ozone generation, and contacting (Figure 7). Associated with these basic components are additional subunits such as controls, instrumentation, cooling water and/or cooling air supply, as well as environmental provisions including noise control and ventilation. In view of the sparsity of U.S. experience with ozone, special concern must be given to materials of construction.

GAS PREPARATION

Feed-gas to the ozone generator must be nominally free of oil, dust and moisture. Oil will reduce the capacity of the desiccant driers by covering portions of the active surface of the adsorbent. Dust will collect in the ozonizer tubes, slowly reduce the efficiency of ozone generation, and require more frequent opening of the generators and cleaning of the tubes. Moisture in the feed gas will reduce the amount of ozone generated per unit power input, increase maintenance, and potentially decrease dielectric life.

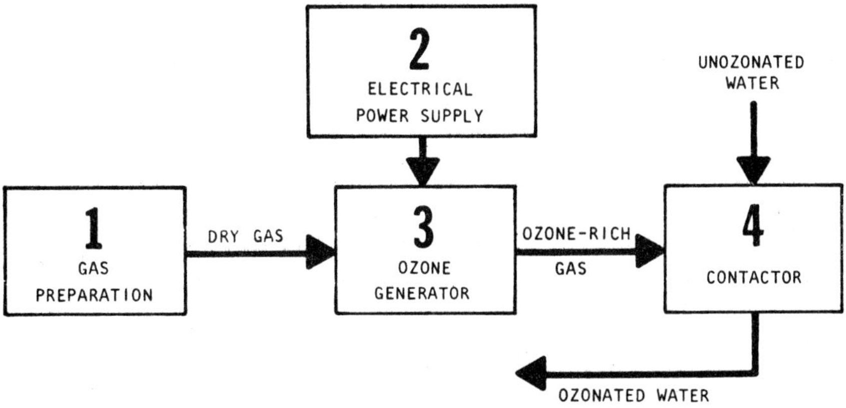

Figure 7. The four basic components of the ozonation process.

Factors to be considered in selection of a gas preparation system include the following:

1. Type of feed gas: air, high purity oxygen, oxygen-rich gas mixture.
2. Type of ozone generation system, including the gas pressure at which the generator is capable of operating.
3. Type of ozone contactor, including the pressure requirements of the gas diffusion system.
4. Temperature of plant water or other water to be used for cooling water purposes.
5. The capacity of the ozone generation system (a large system could justify a more complex system).

The objective of preparing gas is to provide a clean, dry and cool feed-gas for preparation of ozone. The cleaner and drier the gas, the less frequently the ozone generator will have to be maintained. The dryer and colder the gas, the more ozone will be produced per unit power input. Furthermore, the capacity of the generator will be constant when these factors are held constant.

Inlet Gas Filters

Air for the system is drawn from the outside through a duct or chimney. The air supply also may be drawn from within the building, apparently to maintain negative pressure within the ozone generation rooms. Dust and other particulate matter is filtered from the air through modular paper or fabric filters which remove material larger than five microns.

Inlet Gas Pressurization

Gas must be pressurized to draw gas through the inlet air filters and push it through the air preparation system to the ozone generation system. Depending upon the selected contactor design it may also be necessary to push the air through porous diffusers submerged under 4 to 6 meters of water. The following three types of gas pressurization were observed:

High Pressure (70 to 100 psi)--

This level of pressurization was required to enable use of a high pressure desiccant drier and the advantages of such a system are discussed in the Desiccant Gas Drier portion of this subsection. Compressors are required to provide this level of gas pressurization. This pressure level necessitates the provision of a gas pressure reducer prior to the ozone generator. This system was observed only at Sherbrooke, Quebec, and is illustrated by Figure 8.

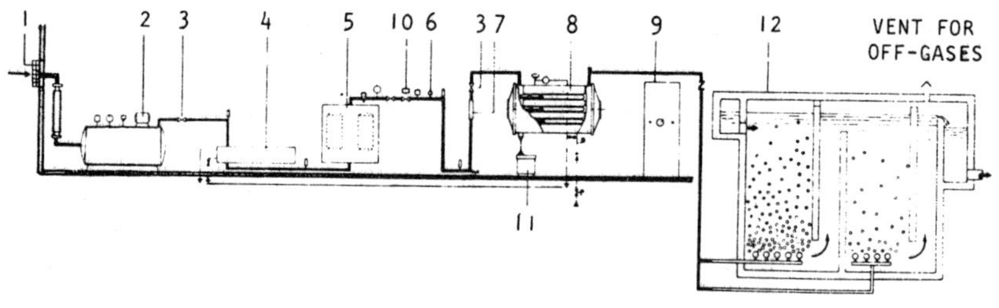

1. AIR FILTER
2. COMPRESSOR
3. AIR CONTROL VALVE
4. HEAT EXCHANGER/COOLER
5. DESICCANT DRIER
6. HYGROMETER
7. AIR FLOW METER
8. OZONE GENERATOR
9. SWITCHGEAR CONSOLE
10. PRESSURE REDUCING VALVE
11. TRANSFORMER
12. CONTACTOR

Figure 8. High pressure gas preparation system[36]

Intermediate Pressure (6 to 12 psi)--

This level of pressurization is required with contactor systems which discharge ozone-rich gas at the contactor bottom, such as through porous tubes, or with contactor systems which do not draw the ozonized gas from the ozone generator. Positive displacement or centrifugal blowers are commonly used for this application. This system is illustrated by Figure 9.

Low Pressure (0 to 8 inches H_2O)--

This level of pressurization is used with a contactor system that draws the ozone-containing gas from the ozone generator. Therefore, only sufficient pressure to draw air through the inlet filter and push it through the cooler and drier components of the air preparation system is required. This system is illustrated in Figure 10.

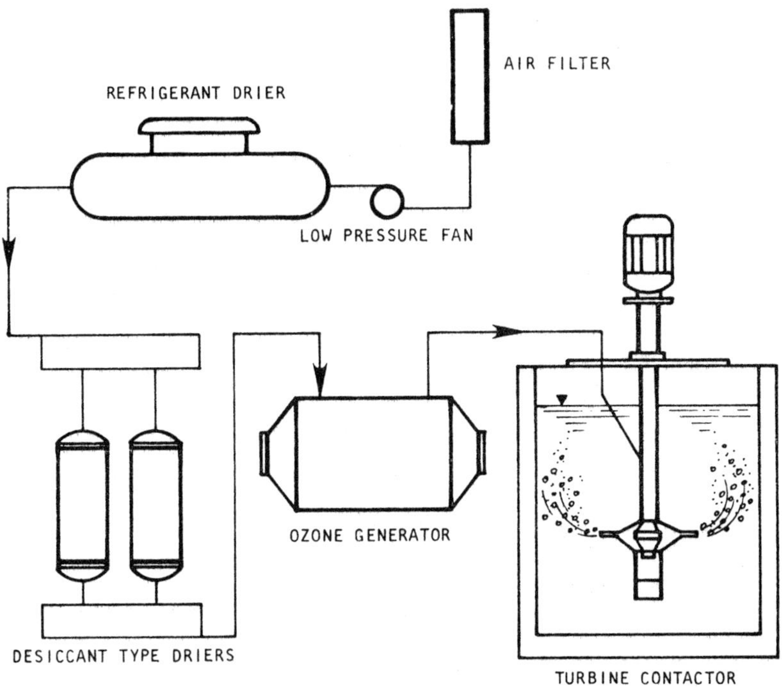

Figure 10. Low-pressure gas preparation system.

Some ozonation systems are essentially fixed capacity trains in which fixed capacity air pressurization units are used. These are primarily the Otto plate-type generators which are designed to operate at slightly negative

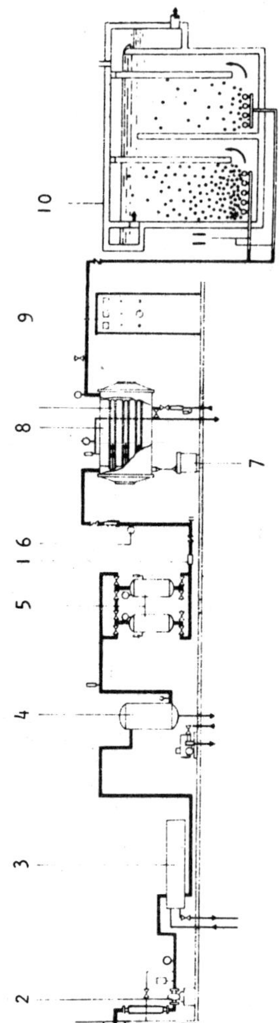

1. AIR FILTER
2. BLOWER
3. HEAT EXCHANGER/COOLER
4. REFRIGERANT DRIER
5. DESICCANT-TYPE DRIERS
6. HYGROMETER
7. STEP-UP TRANSFORMER
8. OZONE GENERATOR
9. SWITCHGEAR CONSOLE
10. CONTACTOR
11. OZONE SPARGERS

Figure 9. Intermediate-pressure gas preparation system[36]

or at atmospheric pressure. However, some form of gas flow control is provided with the majority of tube type ozone generators. This includes local, manual blower discharge control and remote manual control. Several installations, such as the Morsang and Aubergenville plants near Paris, are equipped with control systems which integrate air flow with ozone generation to maintain a constant ozone concentration in the gas.

Selection of the type of gas pressurization unit is important. Contamination of the gas by oil from the pressurization units must be considered unless "oil less" units are used. The gas temperature increase resulting from gas pressurization must be overcome in the subsequent cooling and drying steps of the air preparation system.

Considerable energy is consumed in the gas pressurization step and attention must be paid to control of noise generated by these units. The noise problem is addressed in the Occupational Safety portion of this section.

Water Cooled Heat Exchanger

Shell and tube heat exchangers are used as aftercoolers to compensate for the gas temperature increase resulting from gas pressurization by compressors in the high pressure gas preparation system and by blowers in the medium pressure system. The source of cooling water is normally pressurized, finished water. Cooling water generally is returned to the raw water supply but is sometimes returned immediately upstream of final disinfection. Design of the water cooled heat exchanger would be based on the gas temperature reduction required in the heat exchanger and on the temperature of the cooling water. The amount of temperature reduction possible in this unit is water temperature plus 10°C. Maximum gas temperatures emerging from these units appears to be approximately 25°C.

The simplicity of these units and lack of moving parts should render them relatively maintenance free. This inferred reliability should eliminate the need for redundant units in an ozone generation train or system. Few or no noise problems would be expected from this unit.

Refrigerant Gas Cooler

The refrigerant gas cooler is a major component of medium pressure and low pressure gas preparation systems. Gas temperature is reduced by this unit to approximately 5°C, which is just above the level of frost formation on piping and equipment. The primary function of the refrigerant cooler is gas drying. Lowering the gas temperature causes the condensation of 70% of the water vapor in the gas, thereby reducing the size and cost of the downstream desiccant drier. High pressure gas preparation systems do not normally use refrigerant gas coolers, as pressure swing desiccant driers are relatively insensitive to inlet gas temperatures.

Conventional refrigeration systems are used with this unit. The reliability and operational noise level of the compressors and other equipment incorporated into this unit must be evaluated in the design. The compressors of the refrigerant units at Pierrefonds, Quebec were being replaced while observed to be noisy. Stand-by refrigerant gas cooler units may be provided if the ozonation system is designed on the basis of an overall system rather than individual trains.

Oil Removal Device

An oil removal unit such as an oil "trap" or oil coalescer is recommended to protect the desiccant drier, particularly if an "oil free" gas pressurization unit is not used.

Desiccant Gas Drier

The desiccant gas drier is the key unit in all types of gas preparation systems. Weiner provides an excellent discussion of this method of gas drying[32]. Desiccant driers are necessary for reduction of the gas moisture content to a dew point (the temperature at which a given gas is saturated with water vapor and below which water vapor will start to condense and liquid appears) of minus 40°C to minus 60°C for low maintenance costs and high efficiencies for both the Otto plate and the tube type ozone generator units. Gas drying to a minus 40°C dew point is reportedly sufficient for the Lowther air-cooled plate unit.

Figure 11 is a chart illustrating the moisture contents associated with various dew points. When discussing gas dew points, it must be stated clearly whether the dew point is at line pressure or at atmospheric pressure.

A desiccant drier unit consists of two compartments in which the adsorbent in one compartment is being used for adsorption while the adsorbent in the other compartment is being reactivated and purged. The major difference among the various types of desiccant driers is the mode of adsorbent regeneration. These include heated desiccant regeneration and heatless desiccant regeneration.

Heat desiccant regeneration utilizes thermal swing desorption and operates at the low pressures utilized in the medium and low pressure gas preparation systems. The thermal swing desorption cycle is one in which temperature differential is the major factor causing desorption. Cooling generally is required before the adsorbent is ready for the subsequent adsorption step. This was the desiccant regeneration method generally observed during the inspection trip. Normally, this would be the unit of choice for ozone generation systems with a capacity in excess of 150 lbs of ozone per day. In this type of unit, adsorbent heating may be provided by means of an external heat source, or by an internal heat source. The latter would be considered for only the smallest of installations. A regeneration period of eight to twelve hours is customary with this type of unit. Figure 12 illustrates the thermal swing method of desiccant drier regeneration.

Heatless desiccant regeneration utilizes a pressure swing cycle in which desorption is carried out at a pressure lower than that of the adsorption step, and in which the pressure differential is the major factor bringing about desorption. These units normally are used in high pressure air preparation systems. Regeneration cycles of one to two minutes may be expected with this type of unit. (A drier cycle is defined as the time required for the unit to pass through one drying period and one regeneration period). Pressure swing driers are significantly smaller than the thermal swing driers, making the high pressure systems appropriate for modular, skid mounted ozone generator systems. Also, the pressure swing driers do not require refrigerant coolers or regeneration heaters, thereby resulting in a simpler system which should be easier to operate and maintain. Pressure swing desiccant driers were observed at two Quebec plants, St. Denis and Sherbrooke. This type of drier normally would be considered for installations with ozone generation capacities up to 150 pounds per day. Figure 13 illustrates the pressure swing method of desiccant drier regeneration.

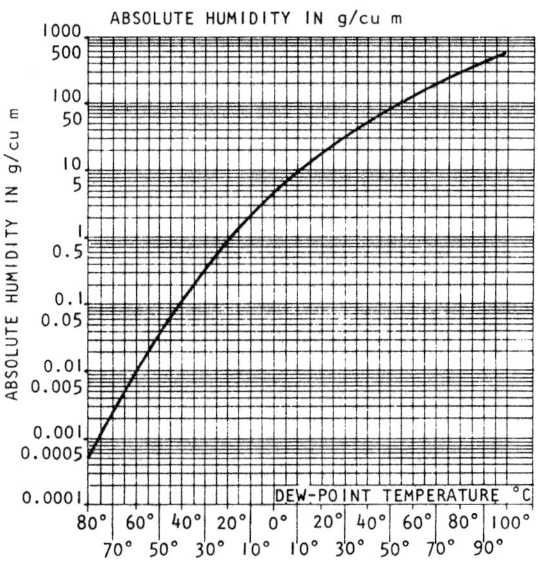

Figure 11. Absolute humidity of atmospheric air at saturation point as a function of dew point[36].

Silica gel and activated alumina are the two most commonly used desiccants (moisture adsorbers) although molecular sieves (crystalline zeolites) are promoted for similar applications. Molecular sieve material also is used as an initial layer in some desiccant dryers which contain activated alumina or silica gel. Advantages are claimed for each adsorbent. For example, silica

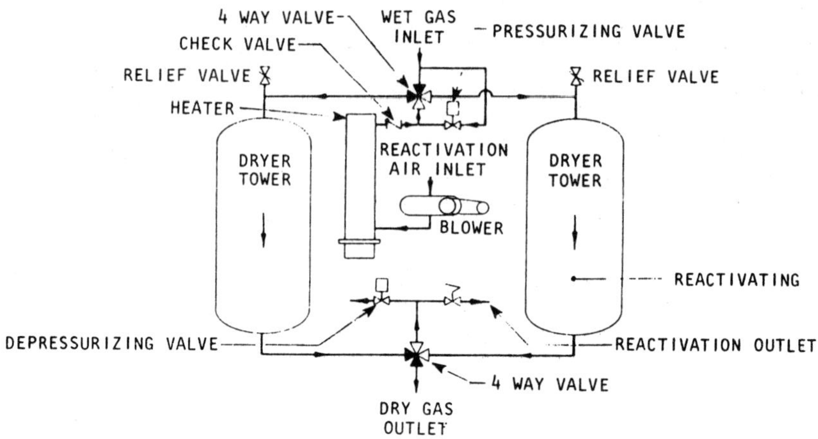

Figure 12. Thermal-swing desiccant drier[33,39]

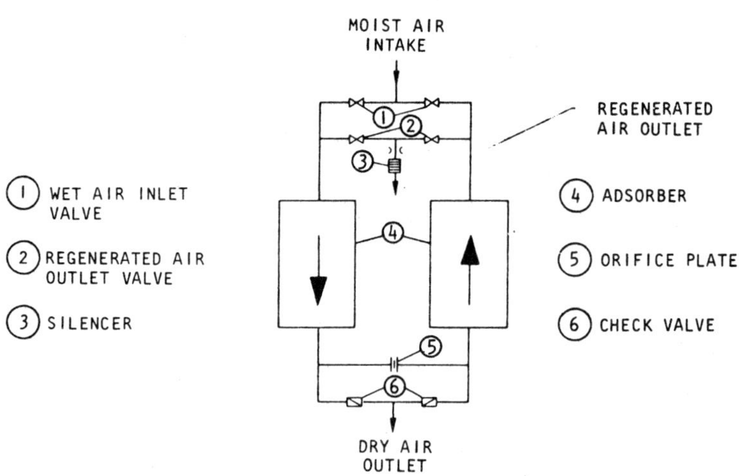

Figure 13. Pressure-swing desiccant drier[39].

gel is claimed to be more fragile than activated alumina and is subject to decomposition in the presence of water. However, the Federal Republic of Germany silica gel installations reported no operational problems based on long term (more than 10 years) operational experience.

An interesting feature of the Kreuzlingen plant in Switzerland was the provision of view ports at two levels of the silica gel desiccant drier units. The plant operator therefore could visually determine the condition of the desiccant by observing the color of the silica gel (which was impregnated with a color changing indicator).

Pressure Reducing Valve

The 70 to 100 psi operating pressure associated with the high pressure air preparation system requires the provision of a pressure reducing valve prior to discharge of the dried gas to the ozone generator. This is because the normal operating pressures of the Otto plate type ozone generator are slightly negative or close to atmospheric pressure. The tube type generator operates up to approximately 9 psi while the Lowther, air cooled plate ozone generator operates up to 12 psi. While operational experience demonstrates the reliability of these units, a pressure switch normally is provided at the ozone generator as a back-up to the pressure reducing valve.

Desiccant Drier After-Filter

A filter normally is required to capture adsorbent dust escaping from the desiccant drier. It may be argued that purging of any "fines" from the drier during the start-up of the system would eliminate the need for the after-filter. However, consideration of the numerous regeneration cycles involved in operation of the desiccant drier would appear to confirm the need for this unit. A filter unit capable of removing particles as small as one micron is recommended. Modular paper or fabric filter media normally would be used.

Ozone-Rich Gas Compressor

Application of a low pressure gas preparation system in conjunction with a positive pressure diffusion system (see Ozone Contacting portion of this section) will require pressurization of the ozone-rich gas discharged from the ozone generator. The Kreuzlingen, Switzerland ozonation system is operated in this manner using a low pressure gas preparation system, Otto plate ozone generator, and positive pressure diffuser contactor. The concept apparently is based on successful application elsewhere in Switzerland. Selection of a mode of pressurization must take into consideration that ozone decomposes at elevated temperatures. A liquid-ring compressor apparently serves the required purpose.

Ozone contactor off-gases similarly must be pressurized if they are to be recycled to a positive pressure diffusion system. Systems of this type were observed at Rouen-la-Chapelle, Annet-sur-Marne and Neuilly-sur-Marne in France and is to be installed at the Montreal, Quebec plant. The pressurization blowers used for this application are said to be made of aluminum.

ELECTRICAL POWER SUPPLY

A critical element in the overall ozonation system is the electrical power supply. Descriptions of this portion of the system are sparsely covered in the literature and it is generally isolated and well hidden in the plants themselves.

The power input to the ozone generator is expressed by the following empirical equation:

$$W = 4fC_g e_o(e_m - C_a/C)e_o$$

Where:

- W = power
- f = frequency
- C_g = capacity of dielectric (insulator)
- C_a = capacity of discharge gap
- $C = (C_a C_g)/(C_a + C_g)$
- e_o = discharge potential across gap
- e_m = peak value of voltage across electrodes

For a given ozonator, C_g, C_a and e_o are constant and independent of voltage and frequency. Therefore, power consumption is directly proportional to frequency and peak voltage (above a minimum threshold voltage). Therefore, two ways of controlling the ozone production of a given unit are to vary the voltage or the frequency.

Diagrammatically, three frequently used electrical power supply control systems are shown as follows[40]:

A. Fixed line low frequency (60 Hz), variable voltage[40]

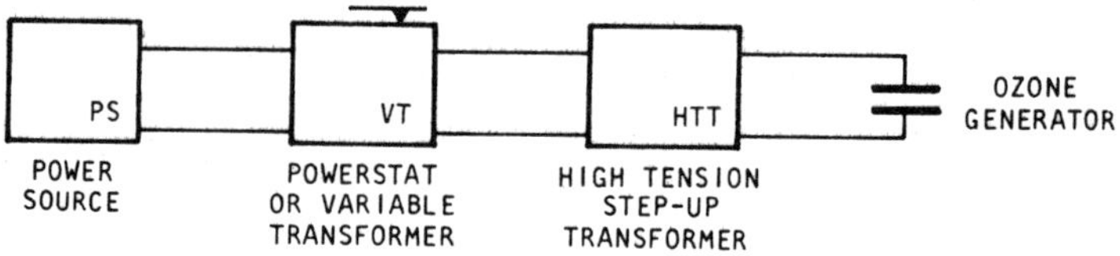

B. Fixed medium frequency (600 Hz), variable voltage[40]

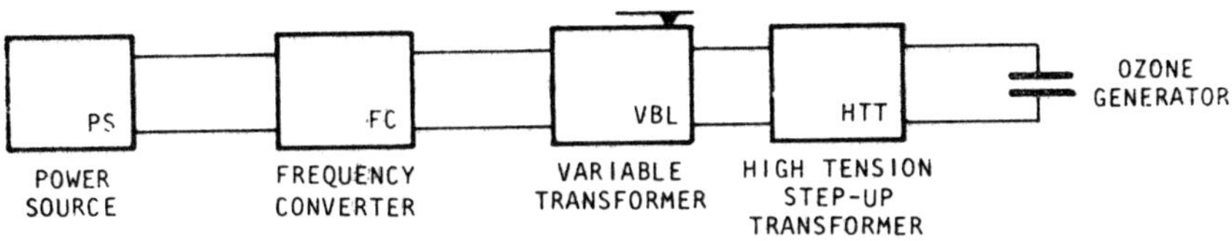

C. Fixed voltage variable frequency[40]

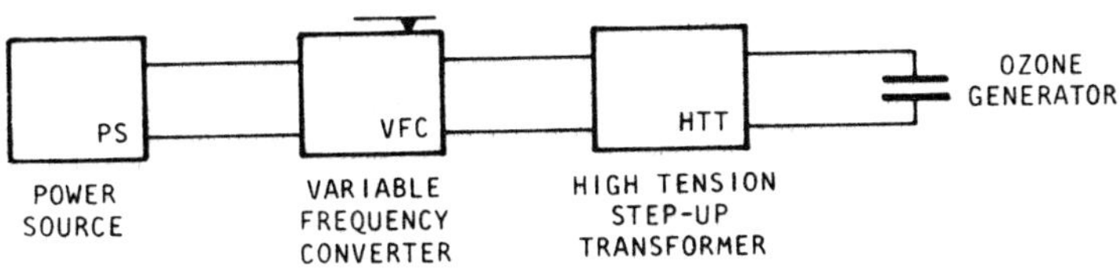

Typical controlled voltage and controlled frequency electrical power supply systems schematics are shown in Figures 14 and 15[40]:

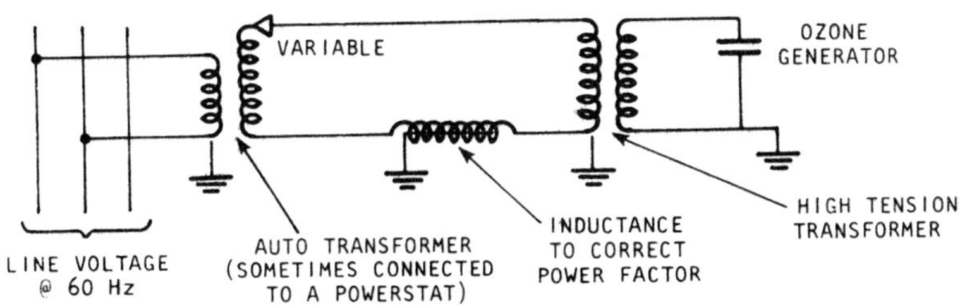

Figure 14. Typical controlled voltage, 60 Hz, electrical power supply system schematic[40].

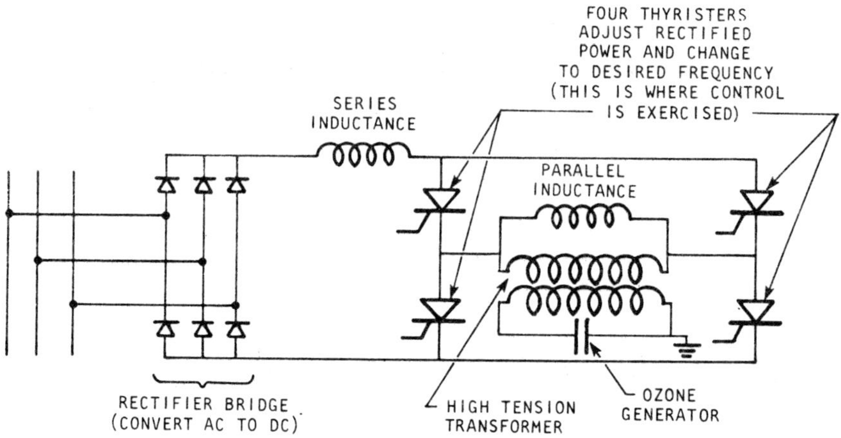

Figure 15. Typical variable frequency electrical power supply [40].

A constant frequency of 60 Hz (or 50 Hz in Europe) typically would be used for ozone generation. Production, therefore, is controlled by varying the voltage to the ozonator. Voltage may be varied by applying the line current to one of a series of tappings on the primary side of the transformer, thereby changing the transformer ratio and the resultant secondary voltage. Otherwise, a variable autotransformer may be used to feed the primary side of the transformer with a constantly variable current.

Ozone production requirements in excess of 10 kg of ozone per hour may justify the use of variable frequency ozone generation control. The use of a frequency converter to increase the frequency from 60 Hz to levels up to 2,000 Hz at the primary side of the transformer is becoming more common because of the availability of solid state electronic control systems. Ozone generation at 600 Hz produces nearly double the amount of ozone from the same generator operated at 50 Hz, but at a higher power consumption per unit weight of ozone produced.

The following table shows typical operational ranges of the various units.

	VOLTAGE (Volts)	FREQUENCY (Hertz)
OTTO PLATE	7,500 to 20,000	50 to 500
SIEMENS TUBE	15,000 to 20,000	50 to 600
LOWTHER PLATE	9,000	60 to 2,000

No guidance can be provided regarding the reliability of the various electrical power units. Very little information is available on the quality required to provide a high degree of efficiency and reliability. A number of problems with dry-type transformers were noted at startup of the ozonation systems at the Chomedey Plant of Laval, Quebec, Canada. Ten separate transformer failures occurred before the dry-type transformers were converted to oil cooling through immersion in an oil bath. Problems with transformers at the Belmont Plant in Philadelphia, Pennsylvania (not operating) are described in the literature[40].

The ozone generator supplier should be responsible for providing the electrical power supply subsystem in view of the dependence of the ozone generation on the power supply. Transformers should be units especially designed for ozone service, and the ozone generator supplier should provide a record of successful tranformer performance at other ozonation installations. In view of the need to substitute other oils for the non-flammable polychlorinated biphenyls (PCBs) previously used, the system designer must consider the possiblility of transformer fires should flammable oils be employed. In any case, drainage of the transformer room must not cause oil contamination of the product water. Dry type, or air-cooled, transformers also have operational problems, and these units are less common in ozonation service.

OZONE GENERATION THEORY

The silent electrical (corona) discharge currently is considered to be the only practical method of generating ozone in plant scale quantities; hence, it is the only mode of ozone generation considered for this report.

Any individual ozone generating element essentially consists of a pair of electrodes separated by a gas space and a layer of insulation (dielectric) such as glass. Oxygen-containing gas is passed through the empty space and high voltage alternating current is applied. The ozone generator acts as a leaky condenser which allows energy to pass through. A corona discharge occurs across the gas space. A portion of the oxygen in the feed gas is converted to ozone in the following manner:

$$O_2 + e^- \longrightarrow 2[O] + \text{heat}$$

and

$$2[O] + 2O_2 \longrightarrow 2O_3$$

Overall, the reaction is $3O_2$ + electrical energy $\rightleftharpoons$ $2O_3$ + heat. It is important to realize that the reaction is an equilibrium reaction, meaning that the reverse reaction is occurring at the same time as the forward reaction. This is the major reason that yields of ozone are so low (i.e., 1% to 3% concentrations of ozone are generated in air; 2% to 6% concentrations are generated in oxygen). In fact, at temperatures above 40°C the reverse reaction is dominating and very low yields of ozone are obtained.

Since some 90% of the electrical energy fed to the ozone generator is lost as radiation, sound and mainly heat, it is essential that efficient cooling of the electrodes be provided, so as to minimize the ozone decomposition and thereby maximize the yields of ozone generated. Therefore there is a cost tradeoff involving the energy penalty incurred in cooling the generator to conserve the ozone generated.

The basic ozone generator with the required electrical power supply is shown below[40].

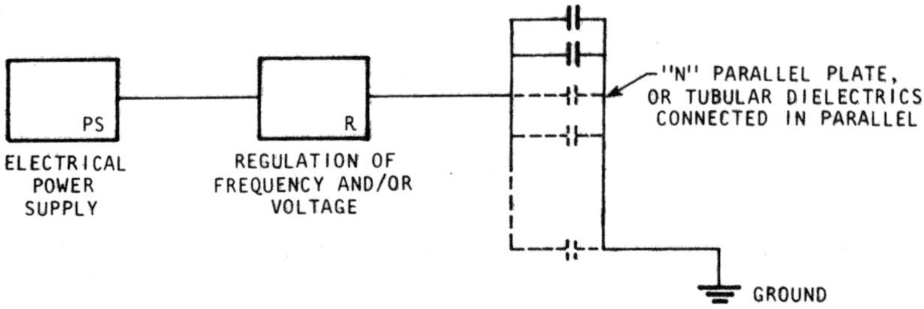

Prior to the application of power, the ozone generator is a group of capacitors connected in parallel. A single capacitive element (tube or pair of parallel plates) is shown below.[40]

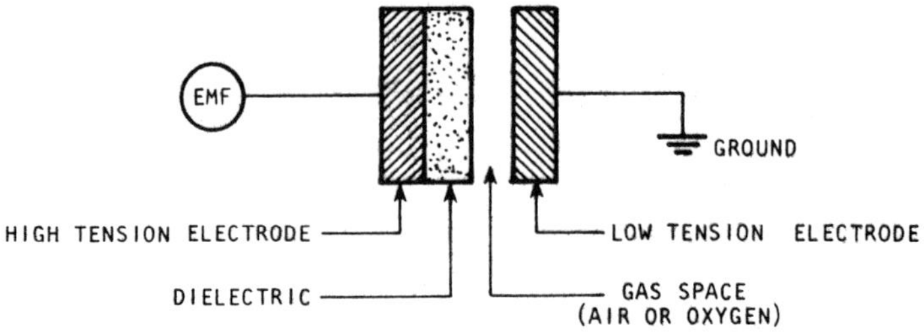

The electrical properties of this single capacitive element are shown in the following sketch which represents the system before the ionization potential of the air or oxygen in the gas space is reached.[40]

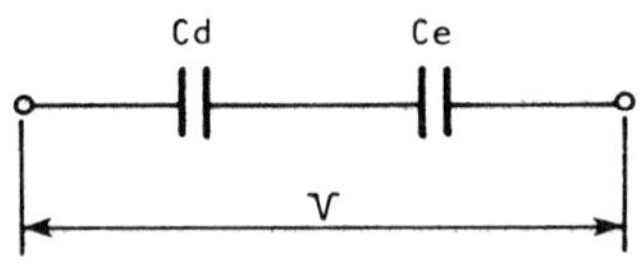

Cd: CAPACITANCE OF GLASS DIELECTRIC,
Ce: CAPACITANCE OF THE FEED-GAS (AIR OR O_3),
V: POTENTIAL DIFFERENCE ACROSS DIELECTRIC AND DISCHARGE GAP,
Re: RESISTIVE ELEMENT

As the ionization level of the gas space is reached, the gas space conducts electricity and becomes a resistive element as illustrated below.[40]

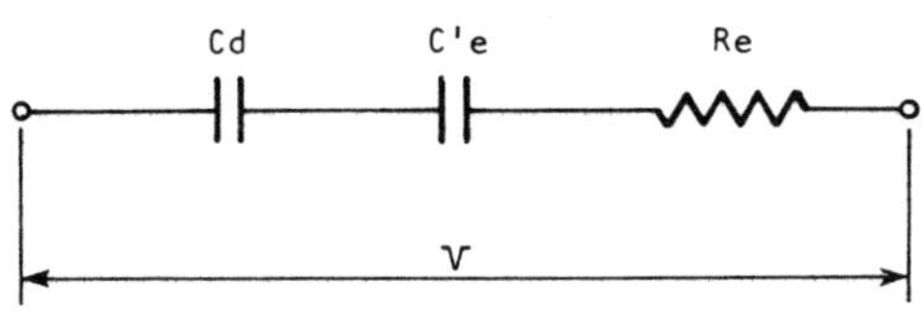

The electrical characteristics of the ozone generator change. As alternating current power is applied to the circuit, an alternating cyclical variation in the circuit occurs when the ionization level of the air or oxygen in the gas space is traversed.[40]

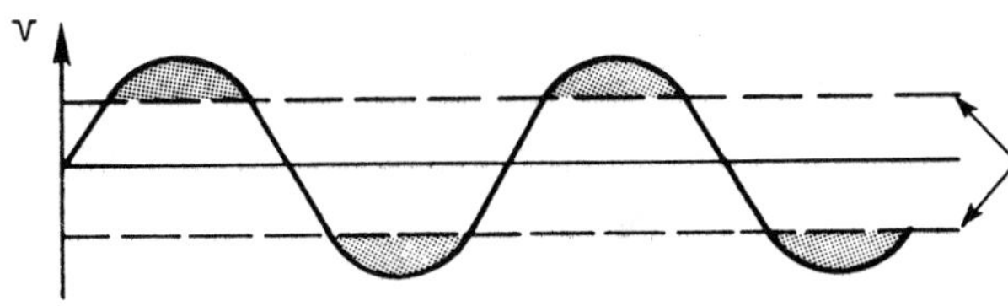

The gas (air or oxygen) remains partially ionized and the voltage drops below the ionization potential. The complicated waveform for the potential difference across the gas space (Ue) is shown below.

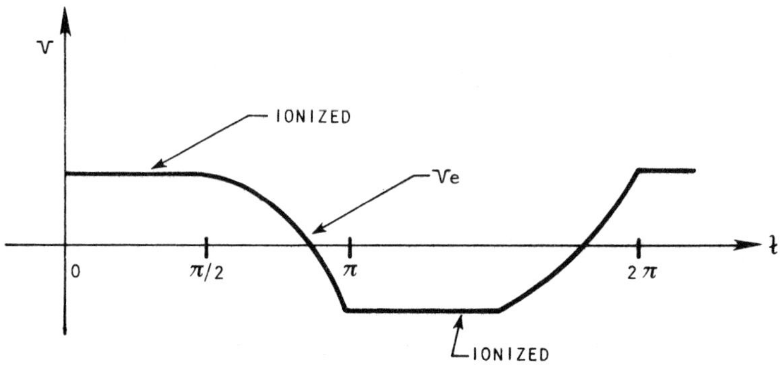

For a conventional ozone generator

$$\frac{Q}{A} \;\alpha\; \frac{FEV^2}{D}$$

where,

Q = ozone production

A = electrode surface area

F = frequency

E = dielectric constant

V = applied voltage

D = distance between electrodes

For a given type of ozone generator (and therefore a given electrode configuration) the ionization voltage is a function of gas pressure and the discharge gap:

$$V_i \;\alpha\; pg$$

where,

V_i = ionization voltage

p = pressure

g = discharge gap

Ozone is generated during the period that the oxygen-containing gas is present. Free electrons resulting from ionization are accelerated by the applied voltage and frequently collide with oxygen molecules. Collisions are more frequent when the feed gas is high purity oxygen because the density of the oxygen molecules is greater.

If the excited electron possesses sufficient energy at collision, the oxygen molecule is split into oxygen units. Each oxygen unit is available either to combine with an oxygen molecule (thus forming O_3) or to recombine with another unit to reform O_2.

The gas (air or oxygen) is being ionized continuously and deionized within the gas space by the alternating current. The free electrons, travelling from one electrode to the other as the polarity reverses, collide with oxygen molecules. Clearly, increasing the frequency of operation will increase the number of potential collisions. Ionization voltage normally is in the neighborhood of 10,000 volts; however, the voltage required to cause ionization would be greater for wider discharge gaps and less for narrow discharge gaps. Figure 16 illustrates the activity within the gas space. The possible outcomes of electron collisions with an oxygen molecule are illustrated below.[40]

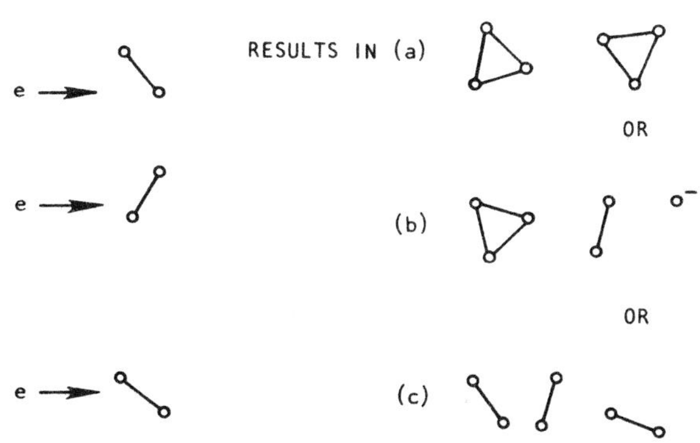

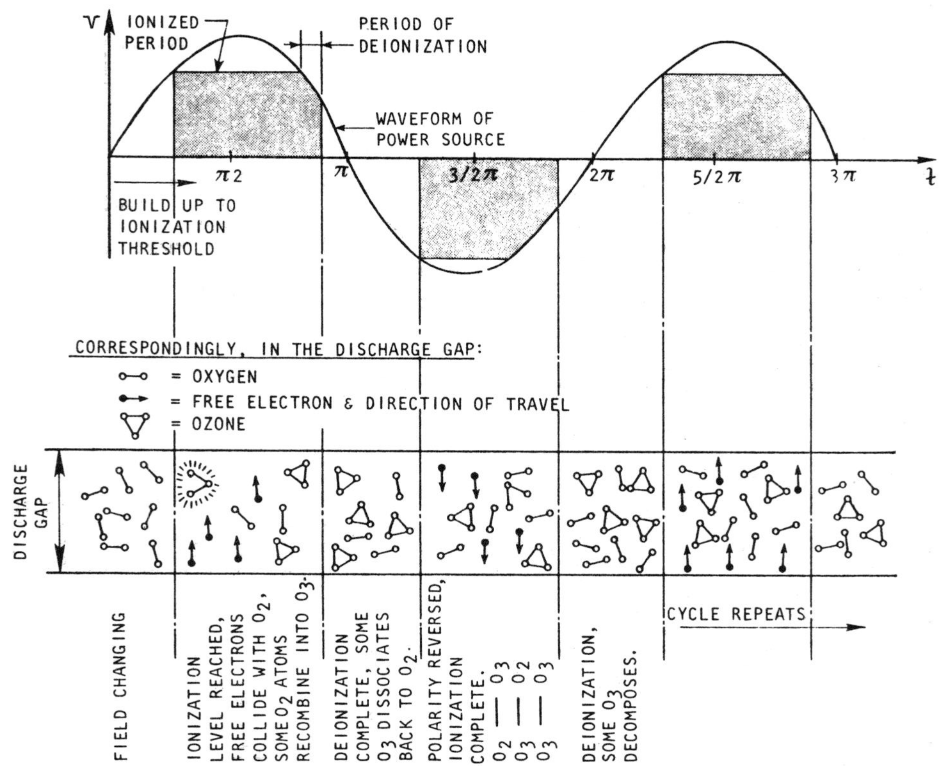

Figure 16. Formation of ozone during ionization/deionization[40].

Frequency can be increased to a maximum level as illustrated by Figure 17 which is a curve for an ozone generator with cylindrical dielectric tubes and an ionization gap of approximately 3mm. For this particular ozone generator the optimum frequency appears to be about 600 Hz.

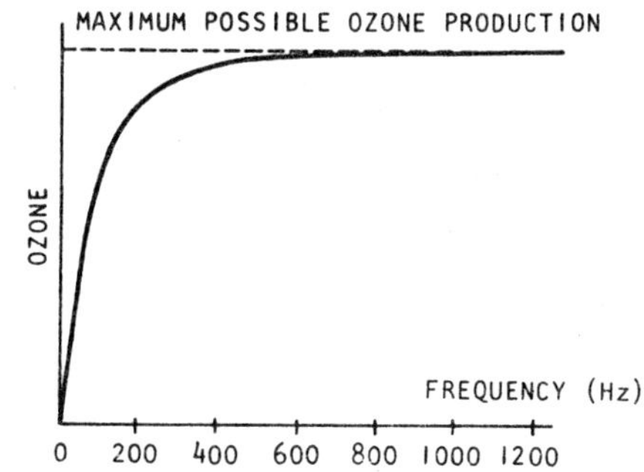

Figure 17. Relationship of ozone production and power supply frequency[42].

It would appear, therefore, that:

- since ozone production per unit area of electrode surface varies as the square of voltage, the higher the applied voltage the greater the production of ozone,
- the longer the period that the air is ionized, the greater the chance of ozone being generated, and,
- the higher the frequency, the greater the chance of generating ozone.

However, these statements each taken out of context can be misleading. As stated initially, the temperature within the corona discharge area is critical. The reverse reaction of ozone to produce oxygen dominates above 40°C. Above a maximum applied voltage, the amount of heat liberated can exceed the cooling capability. In addition, dielectric media are subject to deterioration at high voltages. Thus there is a practical maximum voltage which will produce the highest yield of ozone at the most efficient cooling cost and most efficient dielectric life.

Exposure of the generated ozone to ionization in the discharge gap will result in decomposition of ozone back to an oxygen molecule and a single oxygen moiety. Therefore, prolonged residence time of the air feed gas in the discharge area will expose the produced ozone to conditions of temperature and ionization which will reverse the formation reaction.

The production rate of ozone varies linearly with frequency, and higher frequencies cause less deterioration of dielectric media than do high voltages. On the other hand, power consumption efficiency at higher frequencies is less than at lower frequencies. Therefore, although more ozone can be produced from the same generator operated at frequencies of 600 Hz than at 50 Hz, the number of kilowatts required to generate a unit weight of ozone also is higher at the higher frequency.

Therefore, maximizing the yield of ozone for the lowest expenditure of electrical energy involves a balance between the many factors cited. The different types of commercially available ozone generators observed during this study are described below.

COMMERCIAL OZONE GENERATORS

There is a variety of types of commercial scale ozone generators available. These include the following types:

- horizontal tube, water cooled
- vertical tube, water cooled
- vertical tube, doubled cooled (oil and water)
- Otto plate type, water cooled
- Lowther plate type, air cooled

The current interest in ozone generators makes it likely that other types of ozone generators will be developed, but those listed above currently are the only types currently marketed actively.

Horizontal Tube, Water Cooled Ozone Generator

The horizontal tube, water cooled ozone generator currently is the most commonly used. Rated maximum capacities of typical ozone generators range from approximately 0.4 kg/hr (0.88 lb/hr) to 27.5 kg/hr (60.5 lb/hr) using air as the feed gas. Rated capacities would be approximately doubled if oxygen were used as the feed gas. Normally, however, plants operate ozone generators at 50% to 75% of rated capacity to minimize operation and maintenance costs and to minimize the amount of electrical energy required to generate a unit weight of ozone.

Figure 18 represents construction details of a typical tube ozonator. Figure 19 provides more detailed views of typical horizontal, tube type, water cooled ozone generators. These units are normally designed to operate

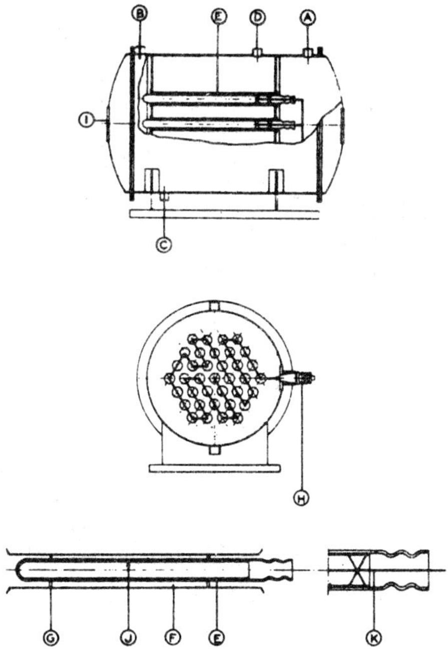

LEGEND

A-AIR INLET
B-OZONIZED AIR OUTLET
C-COOLANT INLET
D-COOLANT OUTLET
E-DIELECTRIC TUBE
F-DISCHARGE ZONE
G-TUBE SUPPORT
H-H.V. TERMINAL
I-PORT
J-METALLIC COATING
K-CONTACT

Figure 18. Construction details of a tube type ozone generator[43].

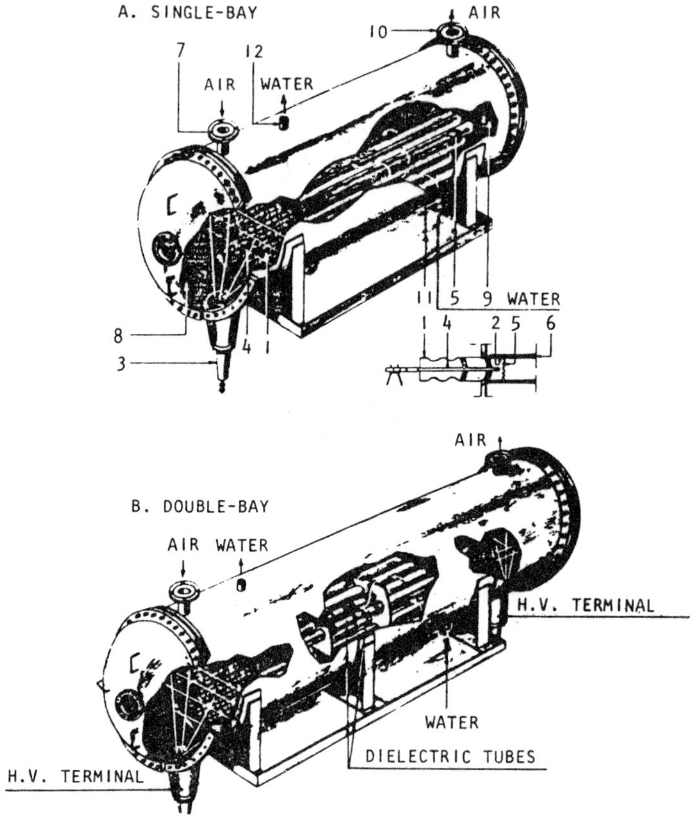

1. DIELECTRIC TUBE
2. METALLIC COATING
3. H.V. TERMINAL
4. CONTACT
5. CENTERING PIECE
6. IONIZATION GAP
7. AIR INLET
8. FRONT CHAMBER
9. REAR CHAMBER
10. AIR OUTLET
11. WATER INLET
12. WATER OUTLET

Figure 19. Typical details of tube type ozone generators[43].

at feed gas pressures of 8 to 10 psi and are, therefore, frequently used in pressurized porous diffuser contactor installations.

Ozone production rate typically is about 18 g/hr per generator tube at an ozone concentration ranging from 18 to 40 g/cu m at an internal pressure of 8 psi employing 60 Hz power for air dried to minus 60°C dew point and at a cooling water temperature of 15°C.

The majority of ozone generators utilize electrical supply at 50 or 60 Hz. However, when ozone production requirements exceed 10 kg/hr, it may be cost-effective to use frequencies up to 600 Hz, which will produce approximately double the ozone from the same equipment operating at 50 Hz.[41]

Vertical Tube, Water Cooled Ozone Generator

The vertical tube, water cooled ozone generator utilizes the cooling water both as the grounding electrode and the coolant. Input gas is pushed through the gas preparation system to the ozone generator by means of a low pressure blower. The ozone-rich air is drawn to the contactor under negative pressure for transfer into the liquid. The ozonator tubes are comparatively small in diameter.

The ozone generator consists of three chambers, or compartments, arranged vertically. As illustrated by Figure 20, dried gas is drawn into the upper compartment where it enters metal tubes which are the high tension (voltage) electrodes. The gas is drawn downward past the tubular metal electrode to emerge into the closed end of the glass dielectric tube.

The glass dielectric tube is almost entirely immersed in cooling water in the lower compartment. The cooling water bath acts as the ground electrode. The dried gas passes upward through the corona discharge area which is established in the annular space between the high tension electrode and the glass dielectric. The ozonized gas is discharged from the top of the dielectric tubes into the middle compartment, from which it is drawn to the contactor.

Otto Plate, Water Cooled Ozone Generator

The water cooled plate type (Otto) ozone generator consists of pairs of flat hollow blocks, separated by two glass plates and a gas space, as illustrated by Figure 21. One block serves as the low tension electrode and is water cooled, while the adjacent water spray-cooled block is at high potential.

Air enters the housing enclosing a number of generator elements and passes through the corona discharge area in the space between the glass plates to a central manifold passing through the glass plates and metal blocks. The ozone-rich gas is drawn through the manifold to the point of ozone application.

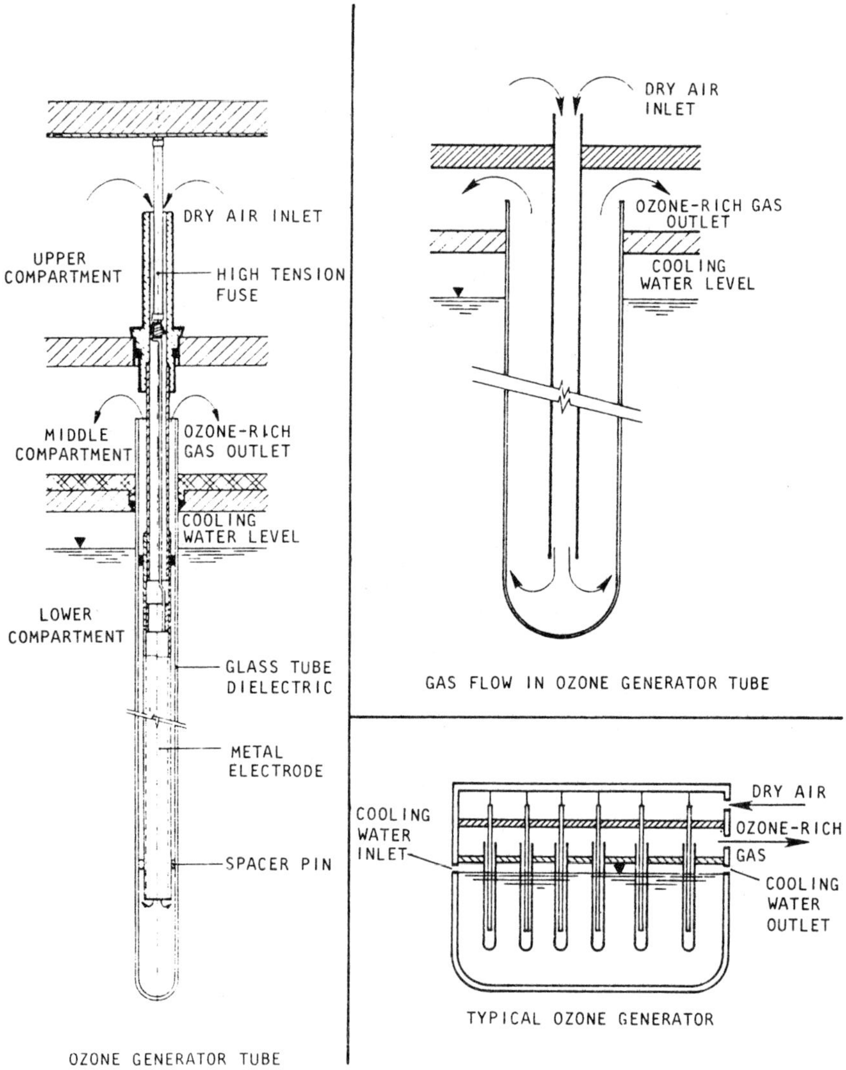

Figure 20. Details of vertical tube ozone generator.

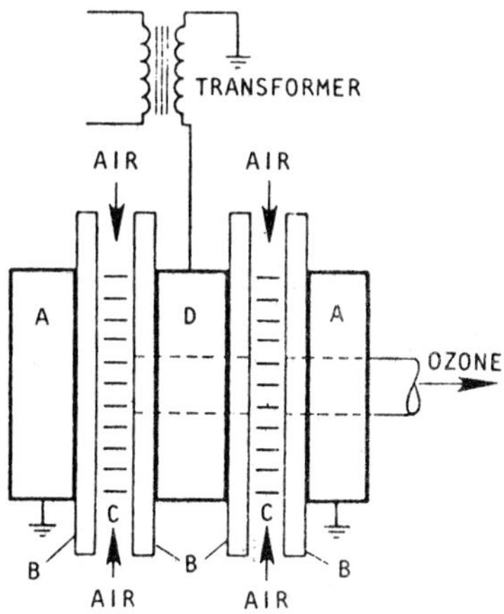

LEGEND: A. GROUND POTENTIAL WATER-COOLED BLOCK; B. GLASS DIELECTRIC; C. DISCHARGE GAP; D. HIGH TENSION WATER-COOLED BLOCK.

Figure 21. Otto Plate ozone generator module[44].

The Otto plate ozone generator operates at atmospheric or negative pressure. Its application typically is limited to negative pressure dissolution methods, such as by vacuum injection or inductive turbine applications. Alternatively, the ozonized gas emerging from the generator at atmospheric pressure can be pressurized by means of a water-seal compressor for dissolution by means of submerged porous diffusers.

Otto plate units are not as subject to damage due to high dew points as are the tube type units. Otto type units can operate with dew points as high as minus 30°C but at reduced production efficiency. One standard element produces up to 20 g of ozone/hr when operating on 450 watts of a 60 Hz service with air containing less than 0.1 g of water/cu m and at a cooling water temperature of 15°C. 20,000 Volts are applied across the electrodes at a maximum energy application of 250 watts/sq ft of single electrode area.

Vertical Tube, Double Cooled Ozone Generator

Vertical tube, double cooled ozone generators utilize both water and a non-conducting oil or other fluid for cooling. This unit incorporates a more complex tube design as illustrated by Figure 22. The high voltage

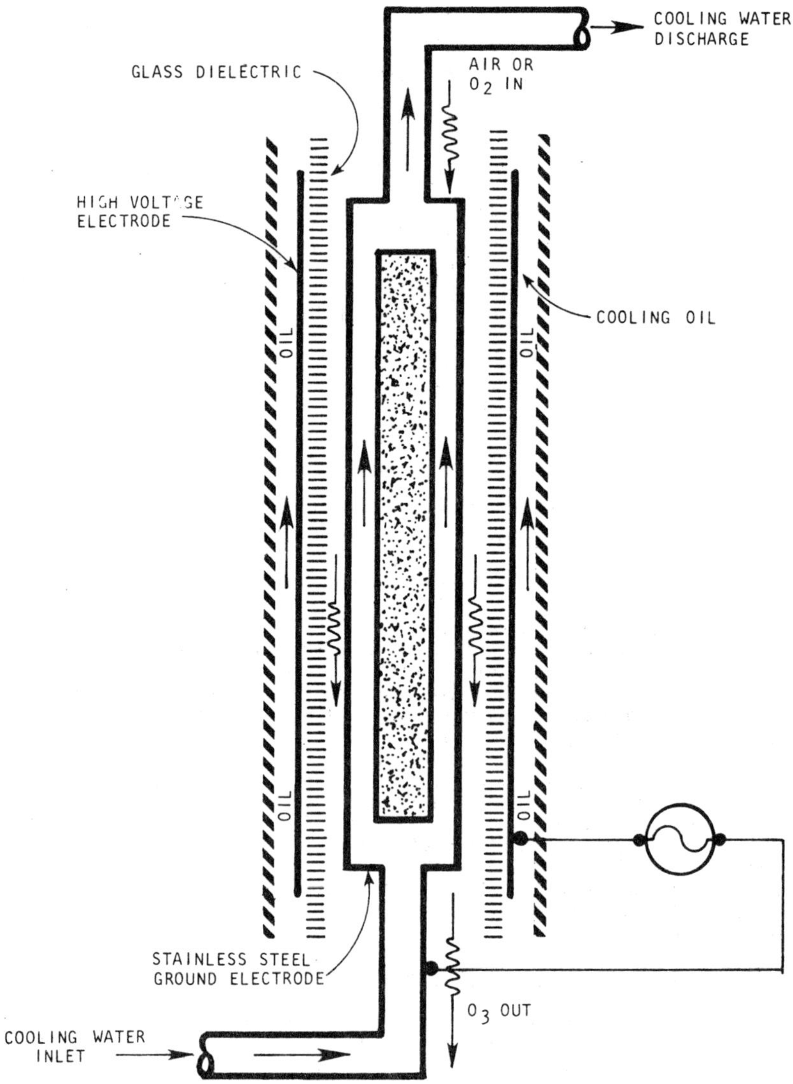

Figure 22. Vertical tube double-cooled ozone generator-tube detail [45].

electrode is cooled by a non-conducting oil or other fluid contained in a closed loop cooling system. The oil, in turn, is cooled by means of a heat exchanger. The low voltage electrode is cooled directly by water.

The current vertical tube, double cooled units operate at 10,000 volts and are controlled by varying the frequency from 60 to 2,000 Hertz by means of a frequency inverter. An apparent disadvantage of this unit is higher dielectric cost. However, ozone production rates, per ozone generator tube, are claimed to be more than 6 times those of conventional tube type ozone generators. A discharge ozone concentration of 2% in ozonized air and 3% in ozonized high purity oxygen is claimed. This type of unit was observed at St. Denis, Quebec, Canada, and is installed in the Rivière-du-Loup plant in Quebec. Units of this type also are installed in the Monroe, Michigan water treatment plant which began operation in early 1978.

Lowther Plate, Air Cooled Ozone Generator

The Lowther plate, air cooled ozone generator is a relatively new unit. The basic element or cell (illustrated by Figure 23) currently consists of the following components; an aluminum heat dissipator, a steel electrode coated with a ceramic dielectric, a silicone rubber spacer to establish the discharge gap, a second ceramic coated steel electrode with gas inlet and ozonized gas outlet and a second aluminum heat dissipator. Thirty or forty cells are manifolded to form a module. The unit, which uses ambient air for cooling, operates at an upper frequency range of 2,000 Hz at 9,000 volts at an upper gas pressure of approximately 15 psi. Figure 24 illustrates a typical Lowther type unit.

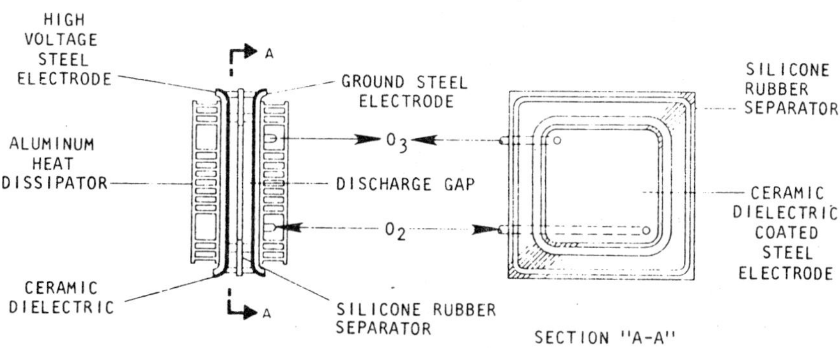

Figure 23. Lowther Plate ozone generator cell [46].

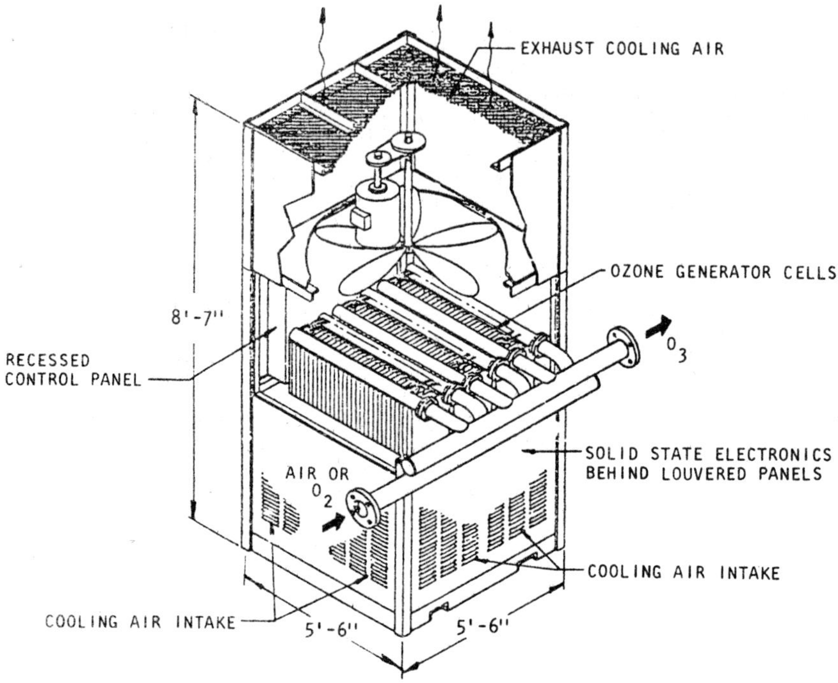

Figure 24. Air cooled Lowther plate type ozone generator.

There are presently no operational Lowther type ozone generators in municipal water treatment facilities, although such a facility is under construction in St. Johns, Newfoundland, Canada. The St. Johns plant will install six units, each capable of generating 115 pounds of ozone per hour. These units will use air as the feed gas to treat a nominal flow rate of 14,000 gpm (20.2 mgd peak flow) for color removal at an ozone dose rate of 2 mg/l. Two pairs of ozone generators will be in normal operation, with the third pair of generators being available for peak flow and backup. Full scale, municipal wastewater treatment applications of the Lowther ozone generator, to be in operation in 1978, will include the plants at Meander Creek, Ohio and Springfield, Missouri. Both of these plants will use the oxygen activated sludge sewage treatment process and, thus, will use oxygen as the feed gas for ozone generation.

OZONE GENERATOR COOLING

Cooling is required for the ozone generator to remove heat from the generator as rapidly as possible to prevent the decomposition of ozone back to oxygen, to assure maximum ozone production and to minimize dielectric heat stress. Water, oil and air are used as coolants. Figure 25 illustrates the effect of cooling water temperature on ozone production of a typical tubular type, water cooled ozone generator. The temperature of waters in the lakes of Switzerland and Austria remain relatively constant at about 4°C all year round and are used to provide cooling. However, the temperatures of many of the rivers in France and Germany can be as high as 20°C in the summer months, and supplementary cooling methods sometimes are required.

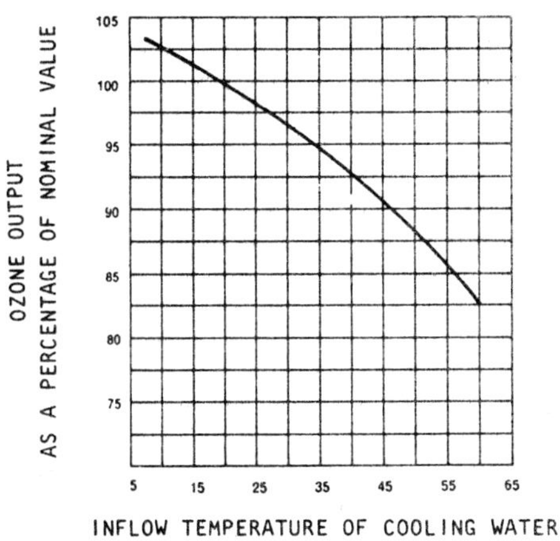

Figure 25. Ozone output in relation to inflow temperature of cooling water[43].

A maximum temperature rise of 6°C is considered allowable for the cooling water of tube type ozone generators. However, at the sites visited during May, 1977, a temperature rise of approximately 3°C was observed while the ozone generator was operating at less than maximum production capacity. Approximately 3 to 4 liters of cooling water at 20°C are required for each gram of ozone produced[43].

Finished water typically is used for cooling which, after having passed through the generator cooling system, is returned to the water treatment process ahead of disinfection. Closed loop cooling systems were observed at the three Düsseldorf plants in which boiler quality is used in the closed loop generator cooling system. The closed loop system is cooled by water from the treated water system. The large ozone generator facility at Neuilly-sur-Marne near Paris, France provides a separate refrigerant cooling system to cool the ozone generator cooling water in the summer months when both ozone demands and Marne River water temperatures are high.

The parameters of the ozone generator cooling system that should be measured include inlet water temperature, inlet water flow measurement, inlet water flow indicating and outlet water temperature (Figure 25). The size of the installation and degree of control sophistication determine the types of sensors used for measuring the various parameters. Standard thermometers inserted in pipe wells would be adequate in many installations, while more sophisticated temperature sensors could be used for remotely monitored facilities. In any case, high outlet temperatures should be sufficient reason to shut down the generator unit to protect the dielectric tubes.

Rotameters are sufficient for flow measurement for even large installations such as Neuilly-sur-Marne. Some means of sensing the flow of cooling water in the system is required.

Air Cooling

The Lowther plate-type, air cooled ozone generator does not require cooling water. Ambient air is claimed to provide sufficient cooling of the generator. A fan moves ambient air across aluminum heat dissipators removing heat from both electrode surfaces simultaneously. This generation is designed to operate at 85°F (29.5°C). Ozone generation efficiency drops 0.5% per degree F rise in temperature, so that if the generator is operated at 100°F (30.8°C) it will produce 7% less ozone than its rated capacity at 85°F.

USE OF OXYGEN-RICH FEED GAS

It is well recognized that the use of a feed gas with oxygen levels higher than the normal 20 to 21 percent of ambient air will result in higher rates of ozone production for a given power input. The use of high purity oxygen for wastewater ozonation has been widely promoted and a number of wastewater ozonation facilities using oxygen as feed gas will be placed in operation in the United States during 1978. However, in wastewater treatment, there is a potential use of oxygen-rich ozone contactor off-gases for application in the activated sludge process. In order to use oxygen-rich gas to generate ozone for drinking water applications, economics dictate that oxygen-rich contactor off-gases be recycled to the inlet of the gas preparation system so that the off-gases can be conditioned prior to their return to the ozone generator.

The Wasserwerk III Wittlaer of Duisburg, Federal Republic of Germany, is the only known municipal water treatment plant in the world in which high purity oxygen is used for generating ozone (Figure 26). The Duisburg Plant ozonation system schematic is provided in Figure 27. There are several industrial facilities using ozone generated from oxygen to treat industrial water, including the Hoechst Works near Frankfort, Federal Republic of Germany. The process diagram of the Hoechst Works industrial water ozonation system displayed at the May, 1977 Wasser Berlin meeting appears to be similar to the Duisburg System. It should be noted that the Duisburg Plant employs neither contactor off-gas ozone destruction nor purging of off-gas to maintain a certain level of oxygen in the recycle gas.

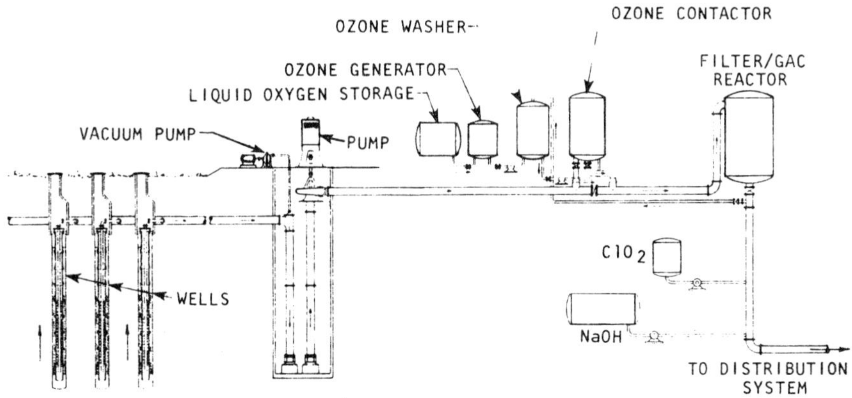

Figure 26. Duisburg Plant overall water treatment schematic.

The lack of operational data from municipal water treatment ozonation systems using recycled, high purity oxygen, despite the promise of the concept, necessitates taking guidance from industrial applications and other information. For example, ozone generated by a recycle oxygen system is used for the processing of oleic acid by Emery Industries, Inc., of Cincinnati, Ohio. Figure 28 is a composite Oxygen Recycle Ozonation System schematic incorporating the concepts promoted by Emery Industries, Inc. and Union Carbide Corporation. It should be noted that these systems incorporate destruction of contactor off-gas ozone and a purge system to maintain recycle off-gas oxygen levels.

The key to successful use of oxygen is an economical and reliable source of oxygen. This may take the form of liquid delivered from an off-site, cryogenic oxygen generating system, on-site oxygen generation by means of pressure swing adsorption units (PSA), or by an on-site cryogenic oxygen generation system. The use of liquid oxygen is practiced at the Duisburg, Germany plant and at the Hoechst (Germany) industrial water treatment plant. The Tailfer Plant in Brussels, Belgium is installing the capability to use liquid oxygen as the ozone generator feed gas in place of air for a period of one year. After this one year test period, a decision will be made either to continue using oxygen, return to air, or to use oxygen only during periods of high ozone requirements.

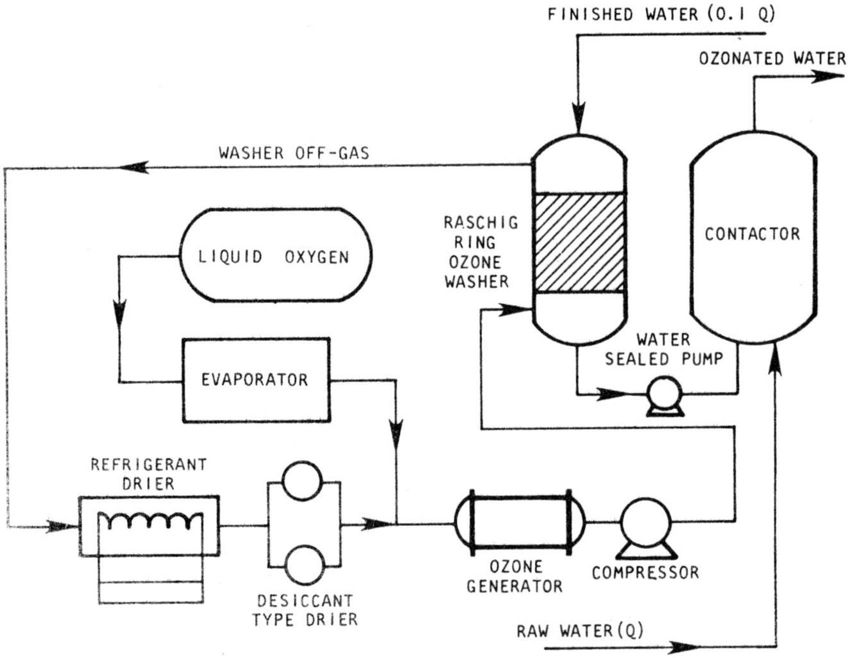

Figure 27. Duisburg water treatment plant ozonation system schematic.

Another mode of supplying the necessary oxygen is pressure swing adsorption (PSA) to generate gaseous oxygen on-site. This type of unit has not been applied in water treatment ozonation but will be utilized at the Meander Creek, Ohio, wastewater treatment plant for both the oxygen activated sludge and the ozone disinfection processes. The PSA unit has been demonstrated successfully in over 70 worldwide installations, more than half of which are in the U.S.A., and 50 of which are wastewater treatment facilities.

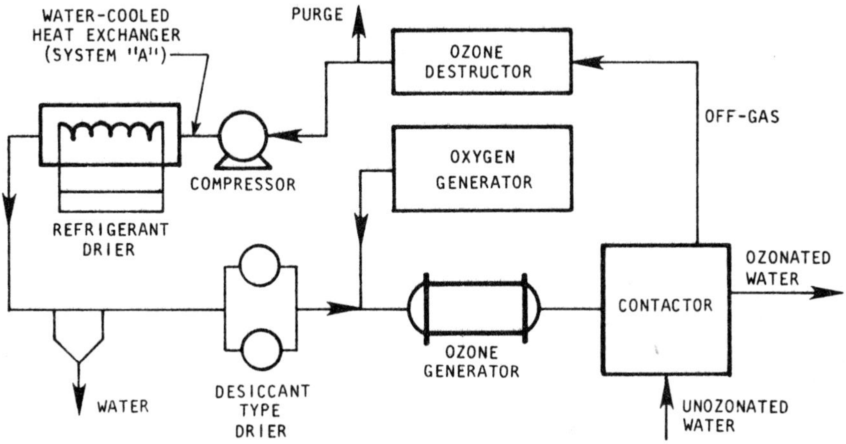

Figure 28. Composite oxygen-rich ozone generation system schematic. (proposed for installation in U.S.)

OZONE CONTACTING

Efficient diffusion of ozonized gas into the water is necessary from the standpoints of economics and achieving the intended process results. The ozonized gas and water generally are brought into close proximity in a contactor. The advantages and disadvantages of various types of contactors have been summarized in Table 3. The choice of an ozone contactor depends primarily upon the ozone process application. However, other factors include the pressure of the ozonized gas discharged from the ozone generator, local design practice, ozonation philosophy, regulatory agency requirements and allowable off-gas ozone concentrations. The various types of ozone contactors are well described in the ozone specialty literature.[47,48,49]

The primary applications of ozone observed during the site visits are listed below. Some of these applications are dependent on the mass transfer (M.T.) of ozone and others are chemical reaction rate (C.R.R.) dependent. Thus different types of contactors are used to optimize the efficient use of ozone in the various applications.

- Disinfection and Viral Inactivation (M.T., but regulated over 6 to 10 minutes).

- Iron and Manganese Oxidation (M.T.).
- Micropollutant Oxidation (M.T. or C.R.R., depending upon the nature of chemicals present).
- Microflocculation (M.T.).
- Taste and Odor Removal (M.T. or C.R.R. depending upon the nature of chemicals present).
- Pre-Conditioning of Biological Activated Carbon (M.T.).

If disinfection if the primary goal, the most prevalent technology is a two or three stage porous diffuser concept. Sufficient time and the proper amount of ozone must be provided initially to meet the initial ozone demand of the water. After this initial demand has been satisfied, additional ozone and contact time is provided to maintain a 0.4 mg/l ozone residual for a minimum four-minute period of time. These conditions have been adopted by French public health authorities for viral inactivation and disinfection by ozonation. The porous diffuser contactor satisfies these contacting requirements for this application of ozone.

A contactor for iron and manganese oxidation should be relatively unaffected by deposition of iron and manganese. Therefore, an injector or turbine contactor would be preferable to the use of the porous tube type.

In the interest of simplicity contactors have been divided into two classifications, negative pressure contactors and positive pressure contactors. Negative pressure contactors are those which draw ozone from the generator by negative pressure at the point of initial gas-water contact. Positive pressure contactors are those in which the ozonized gas is under positive pressure at the initial gas-water contact.

Negative Pressure (Low Pressure) Contactors

Contacting systems that require little or no pressurization of the ozonized gas include vacuum injection and inductive turbine types. Energy is required to provide the pressure head for vacuum injection or for turbine rotation.

Vacuum Injection--

Vacuum ozone injection is a negative pressure method by which the water to the contactor is passed through a Venturi under pressure. Ozonized gas is drawn into the water by the negative pressure at the Venturi throat. One variation of this method is the total vacuum injection method in which the entire plant water flow is passed through one or more Venturi sections, as illustrated by Figure 29.

Based on the observations of this survey, total vacuum injection and partial vacuum injection apparently have provided satisfactory service in all the required applications, but these applications were oxidation and for purposes other than viral inactivation at the visited plants. Overall system considerations are also important factors in contactor selection.

An example of this system was observed at Quebec City, Quebec, Canada. A variation on this method is partial vacuum injection in which only a portion of the water flow is passed through the Venturi section. The ozone-rich side stream subsequently is mixed with the remainder of the main flow in the contactor as illustrated in Figure 30.

Examples of this contacting method include the Konstanz plant, Germany (FRG). The three Düsseldorf, Germany plants (Holthausen, Flehe, and Am Staad) use a similar approach, with the exception that the side stream is finished water.

Turbine Diffusion--

Another means of low pressure contacting is the use of a motor driven inductive turbine which draws ozonized gas into water as it pumps water through the eye of the impeller and outward to the impeller tips. Figure 31 illustrates this type of mechanism while Figure 32 illustrates the multistage installation at the Lengg Plant, Zürich, Switzerland. The technique is used at many plants employing the Kerag ozonation system. Figure 33 illustrates a different impeller and ozonation chamber configuration applied at the Tailfer Plant of Brussels, Belgium. This unit is fabricated by Frings, but functions in the same manner as the Kerag impeller. Figure 34 shows details of the turbines used at Tailfer.

Positive Pressure Contactors

There is a wide range of positive pressure contactor systems which includes:

- Porous Diffusers
- Mechanical Turbines
- Packed Beds or Columns
- Spray Towers

They have been developed mainly in conjunction with the tube type ozone generators which discharge ozone-containing gas at pressures up to 10 psi. However, ozone-containing gas at atmospheric pressure from plate type ozone generators can be pressurized after generation for application in these units.

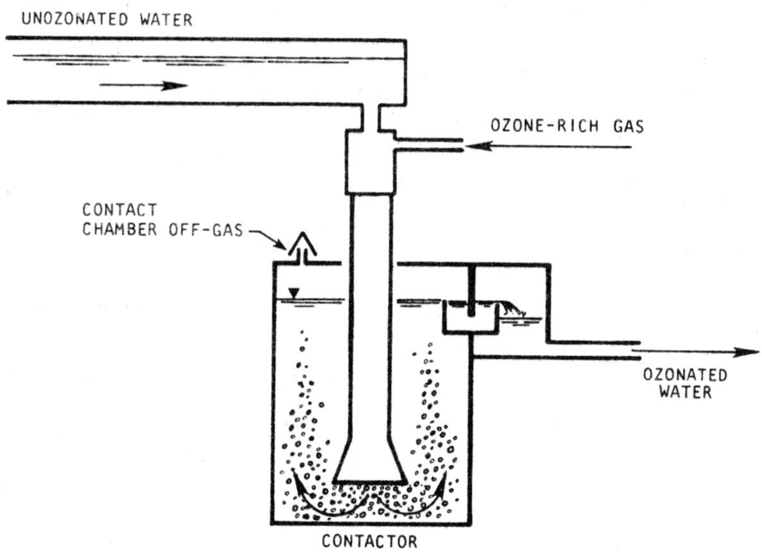

Figure 29. Total flow vacuum injector contactor for ozonation.

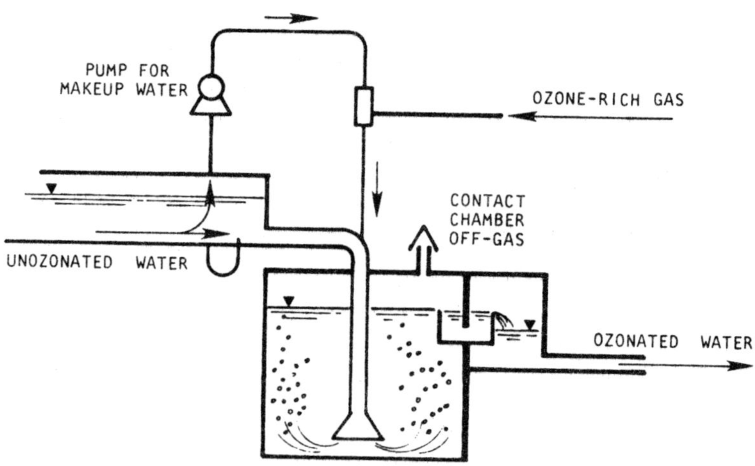

Figure 30. Partial flow vacuum injector contactor for ozonation.

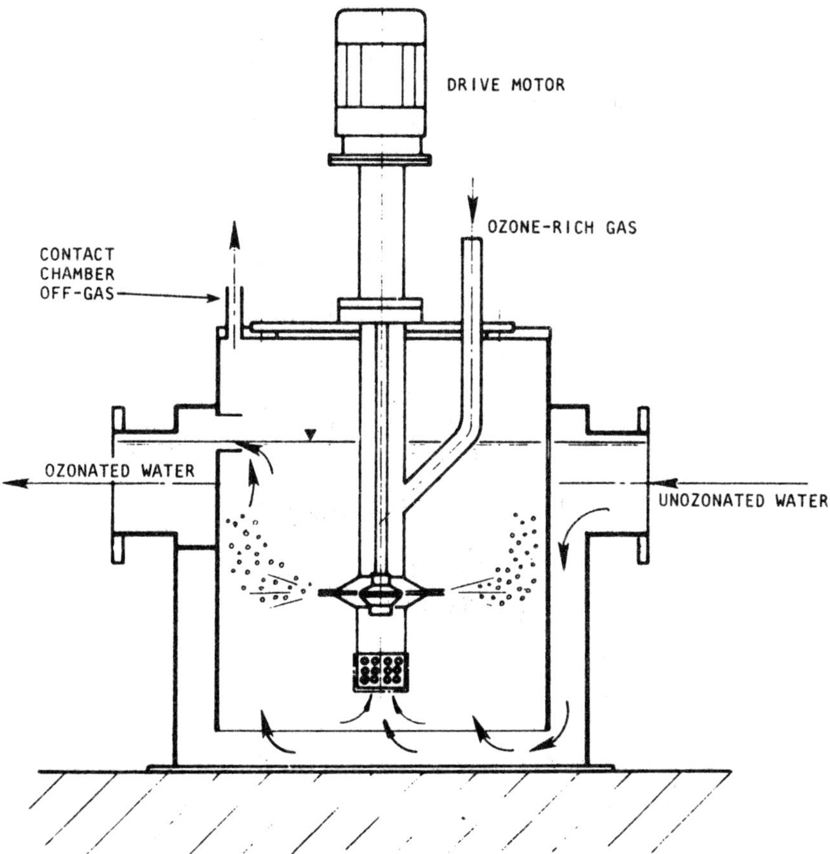

Figure 31. Turbine type ozone contactor [50]

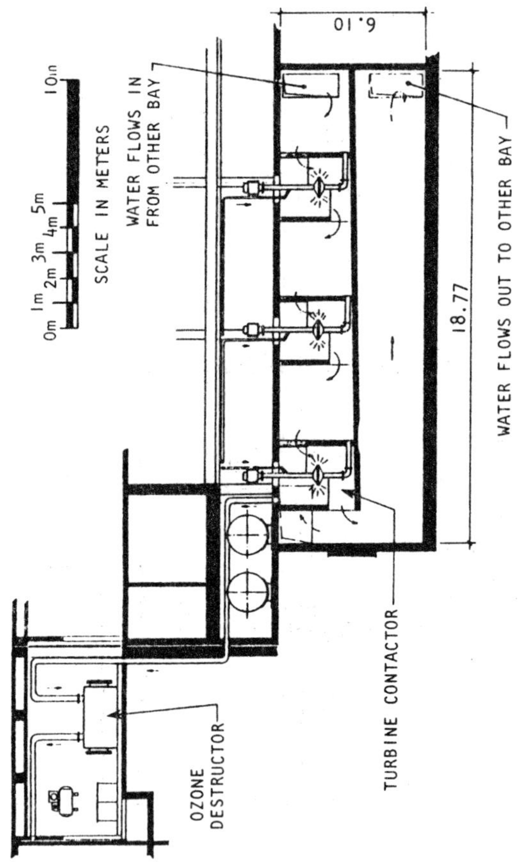

Figure 32. Multi-turbine contactor installation (Lengg Plant, Zürich)[30].

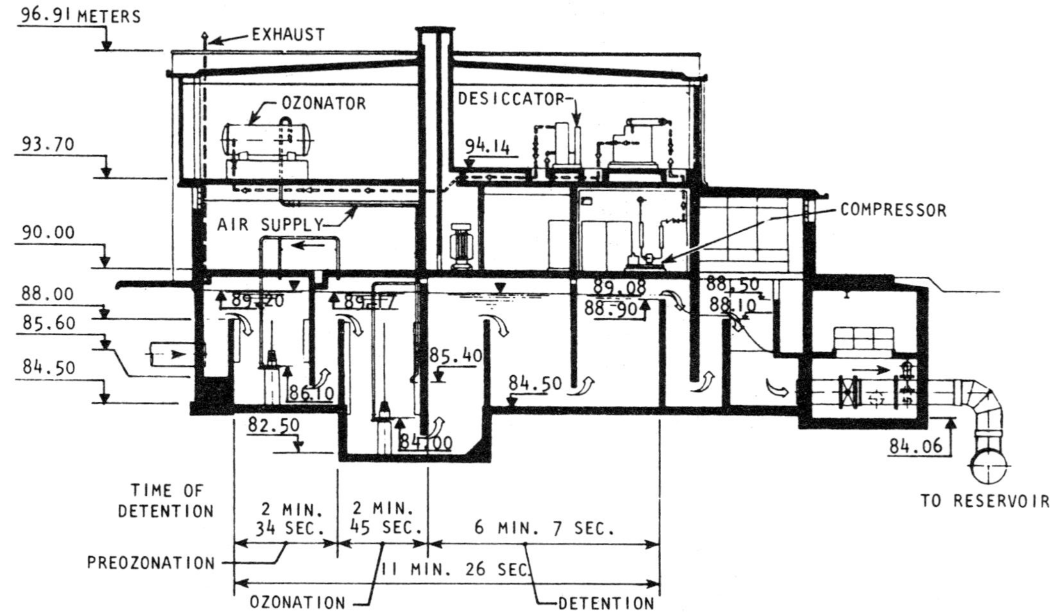

Figure 33. Tailfer water treatment plant, Brussels, ozone contact chamber[45].

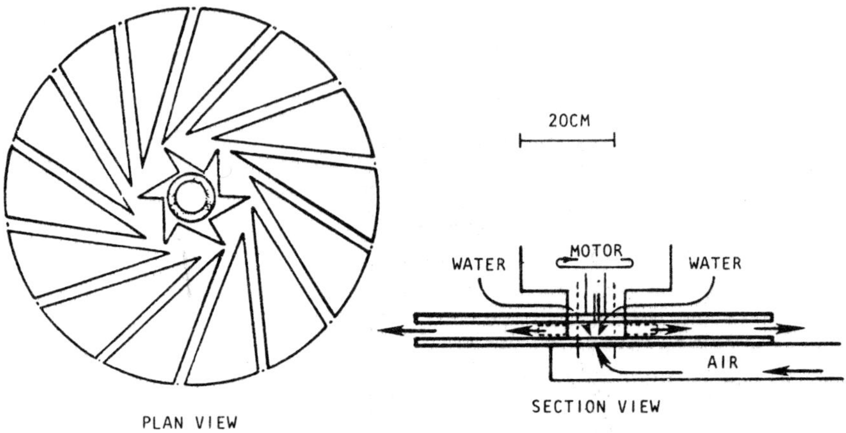

Figure 34. Turbine detail, Tailfer water treatment plant, Brussels [45].

Porous Diffusers--

Innumerable configurations of porous diffusers are possible, and many have been applied, some in combination with turbine mixing devices. Counter-current flow of gas and water appears to be a more efficient method of bringing ozone-rich air and water into intimate contact. Early applications of this mode of ozone contacting are illustrated in the following figures. Figure 35 illustrates the configuration of the Van der Made contactor which utilizes cocurrent contacting at a water depth of 5 meters (16.33 feet) in a covered tank.

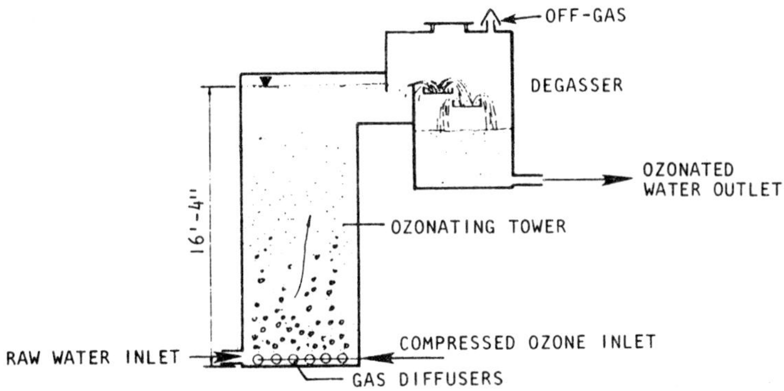

Figure 35. Van der Made ozone contactor[51].

Figure 36 illustrates the Welsbach contactor which utilizes counter-current contacting in an uncovered tank. This configuration is used at Whiting, Indiana.

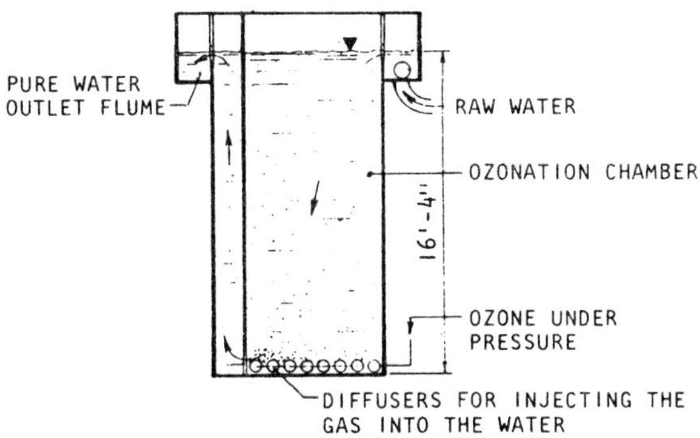

Figure 36. Welsbach ozone contactor[51]

Figure 37 illustrates the Torricelli contactor (a combination of injection and porous diffusion) which injects ozone at an equivalent water depth of 10 meters (32.75 feet) and recycles the covered contactor off-gas to the contactor inlet stream in a form of preozonation step.

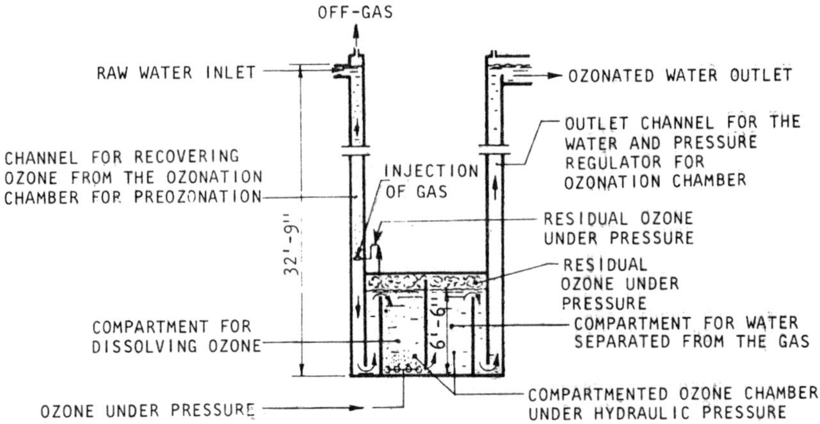

Figure 37. Torricelli ozone contactor[36].

Porous diffuser, multi-chambered contactors are more commonly used for disinfection purposes, particularly in France, and for viral inactivation where a specific ozone residual content in water must be maintained for at least four minutes. Multiple stage contactors (countercurrent - cocurrent - countercurrent - etc.) are common configurations as illustrated by Figure 38. Figure 39 represents the application of this concept at the Choisy-le-Roi plant near Paris. With multiple contact chambers, successively decreasing amounts of ozone can be fed into the chambers as the ozone demand of the water is satisfied.

Two stage countercurrent contactors are installed at the Morsang and Aubergenville plants in France and at the Sherbrooke plant in Quebec, (Canada) as illustrated by Figure 40. As with the multiple stage diffuser contactors, it is customary to feed less ozone into the second chamber (about 33% of the total ozone applied) because there is less ozone demand in the once-ozonated water.

Figure 41 illustrates an interesting two-level diffuser contactor system installed at the Kreuzlingen, Switzerland plant and in several other Swiss plants. Ozone-containing gas from the ozone generator is fed into the lower chamber which is configured so that the off-gases collect in a small air space above the liquid. When the pressure in this small air space exceeds that of the head created in the upper chamber by the influent chamber, the off-gases from the first (lower) chamber "escape" into the upper chamber. Since the upper chamber contains influent water which has the highest ozone demand, and since the off-gases from the lower chamber contain

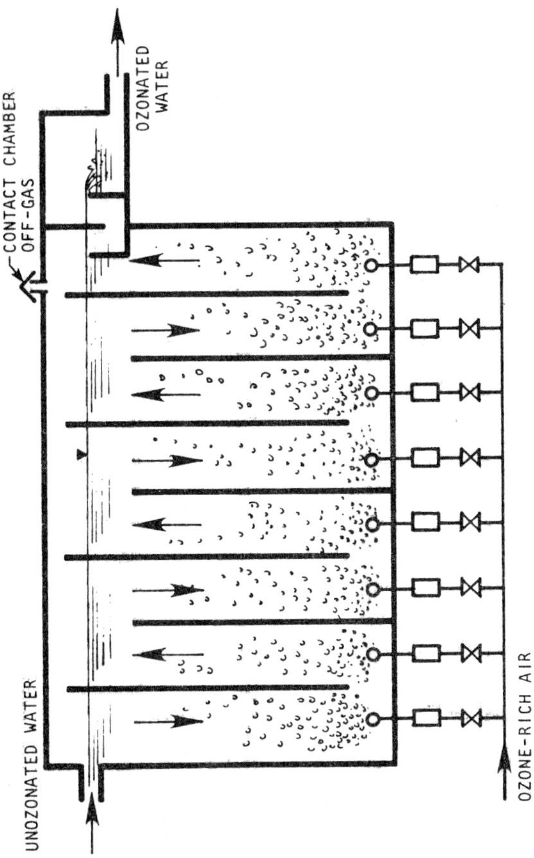

Figure 38. Multi-compartment, porous-tube contact chamber.

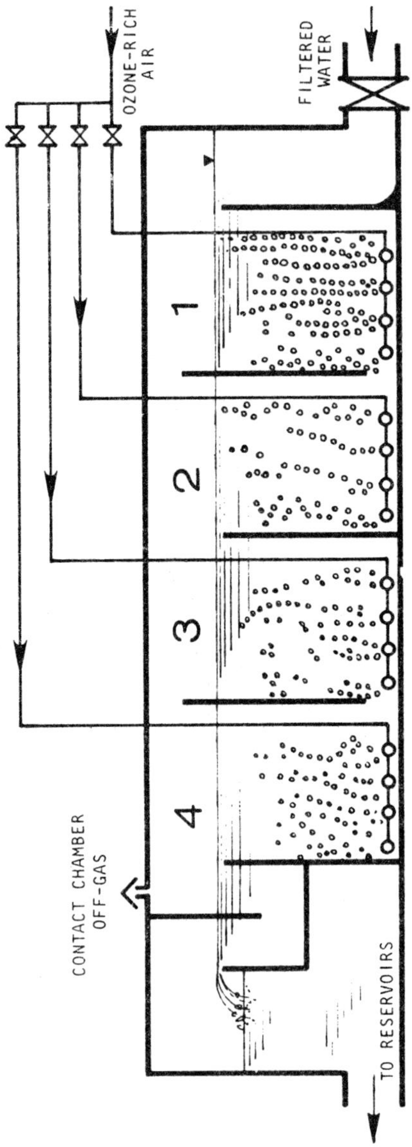

Figure 39. Choisy-le-Roi (Paris, France) water treatment plant ozonation chamber detail[52].

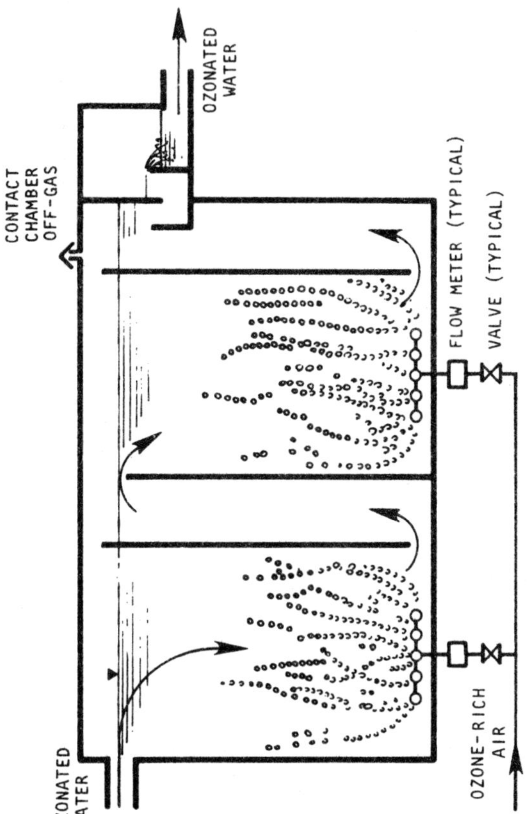

Figure 40. Two-compartment ozone contactor.

only 10 to 20% of the ozone feed, the technique effectively "recycles" the off-gases from the primary (lower contactor). Off-gases from the upper contactor at Kreuzlingen, which contain very small amounts of residual ozone, are discharged to the ambient atmosphere. Alternatively, they may be passed through an ozone destruction unit before discharge.

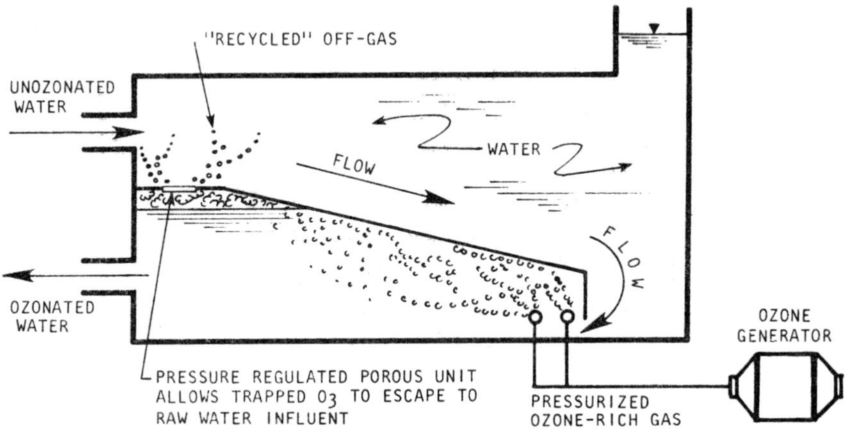

Figure 41. Two-level ozone contactor[47].

Turbine contactors are used in ozone disinfection applications but appear more frequently in ozone recycle systems or in processes where the primary purpose is destruction of waste ozone, at the same time accomplishing useful work (lowering ozone demand of the influent water, or assisting flocculation). Figure 42 illustrates two similar applications for Paris (Neuilly-sur-Marne) and Montreal, Quebec, Canada plants in which off-gases from the last porous tube contactor chamber are recycled to a preozonation chamber of the contactor.

Ozone-containing gas from the generators is fed to the porous tube contactors in different quantities (since once-ozonated water requires less ozone to main required residual). Ozone in the off-gas from these chambers is effectively "destroyed" by passing it into the influent water (which has the highest ozone demand).

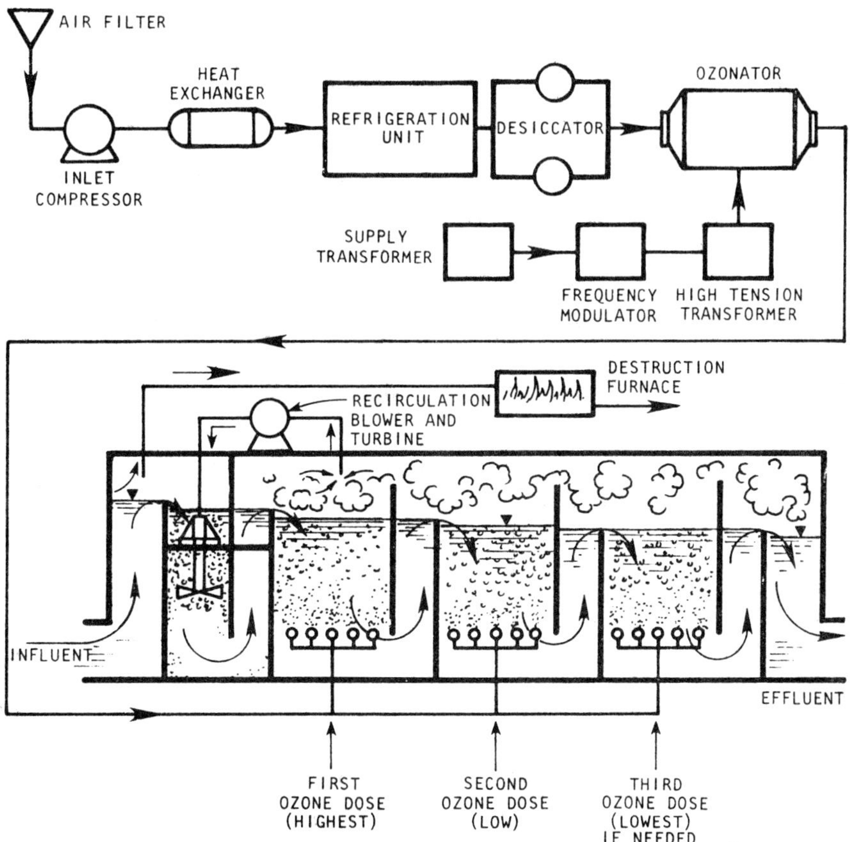

Figure 42. Porous tube contact chamber (Paris and Montreal)[53]

In the Mülheim (Figure 43), Rouen (Figure 44) and Langenau plants the ozone contactor off-gases are recompressed and recycled to a turbine ozone contactor in the influent to the plant. At Wuppertal, off-gases are fed to the influent plant water at the aeration step.

Packed Bed Contactor--

The packed bed or column is another means of promoting intimate contact between the water and ozone-rich gas. Water flows downward, countercurrently through ozone-containing gas through ceramic Raschig rings. Two examples of this are Duisburg (Figure 26) and Sipplinger Berg, Germany (FRG).

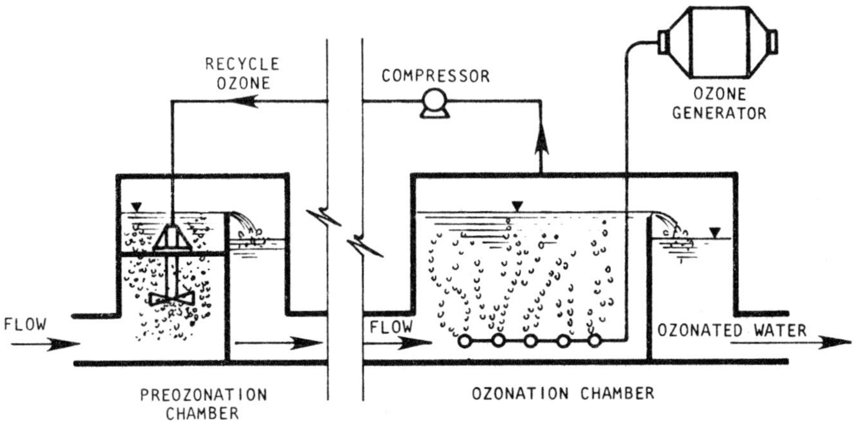

Figure 43. Dohne plant, Mülheim, Federal Republic of Germany.

Spray Nozzle Contactor--

The use of spray nozzles through which water to be treated is sprayed into a pressurized, ozone-rich atmosphere was observed at the Wuppertal plant, Benrath, Federal Republic of Germany. This application is used for iron and manganese oxidation, oxidation of organics and conditioning of granular activated carbon for biological removal of ammonia in a single contacting vessel. Figure 45 illustrates the contactor configuration.

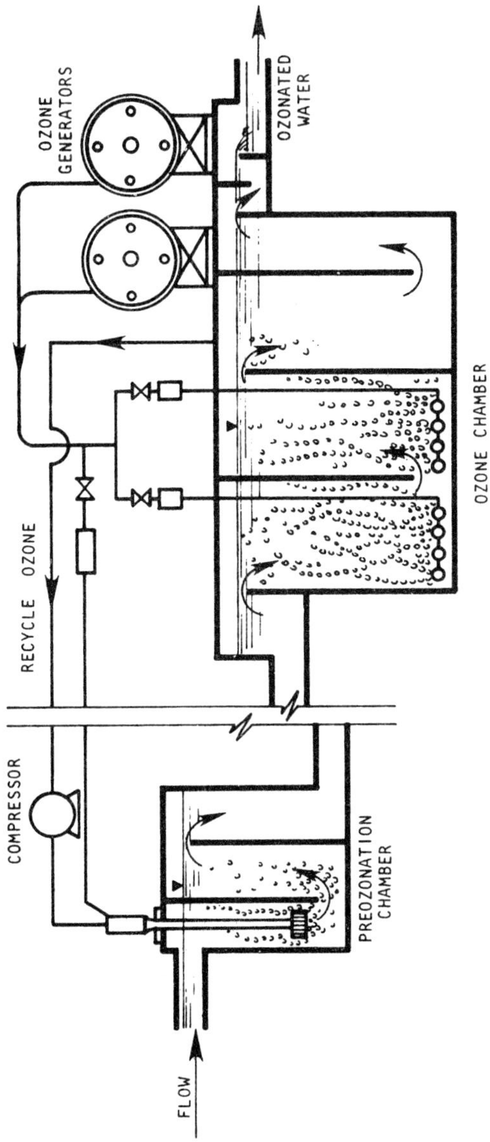

Figure 44. The ozonation system at the la Chapelle plant-Rouen, France.

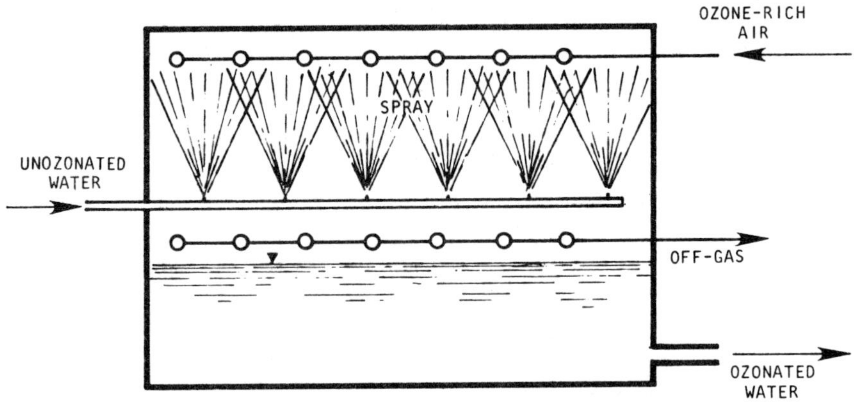

Figure 45 Liquid dispersion into ozone-rich gas.

OFF-GAS TREATMENT (OZONE DESTRUCTION)

Safety considerations require that efforts be made to reduce the ozone levels of contactor off-gases to 0.1 mg/l or less before discharge to the atmosphere. Methods of reducing ozone levels include the following:

- Dilution either by natural wind patterns in the areas of the treatment plant or by the provision of a blower which is designed to dilute the off-gas stream with air by either 10:1 or 20:1 ratios.

- Dry, granular activated carbon has been used for destruction of off-gases. However, there are reports of explosion problems with this form of ozone destruction, especially if the carbon is not kept wet. This method is in use at several plants visited, but should be approached cautiously, if at all.

- Wet, granular activated carbon is frequently effective. Examples of this type of installation include the Holthausen, Flehe, and Am Staad plants at Düsseldorf and Konstanz, Federal Republic of Germany. The effective life of the carbon beds is indicated to be approximately six months, although Konstanz used one charge of carbon for more than two years for this purpose.

- Thermal destruction of ozone contactor off-gas is commonly practiced. Ozone breaks down rapidly at temperatures in excess of 200°C. The applications of this concept appear to be satisfactory, but there are complaints about the fuel cost associated with raising the off-gases to the required temperature. A novel application of thermal ozone destruction was observed at the Clairfont plant in Toulouse, France where the exhaust pipes from the plant's diesel-driven electric generators supply the heat required for the ozone destruction. Contactor off-gases are sent directly to the diesel engine exhaust stacks.

- Several plants, including those of Düsseldorf, were making plans to convert to catalytic ozone destruction units. Plant operators expressed concern regarding the capital and operation cost of these units. However, the projected two year life of the catalyst in such units and the lower operating temperatures said to be required apparently justifies the initial investment. The proprietary nature of the catalysts supplied precludes further description of the material.

- Recycling of contactor off-gases to points in the water treatment system having a high ozone demand is another means of off-gas treatment. Advantage frequently is taken of this recycling for other treatment applications of ozone within the water treatment system. This concept was observed at several plants, such as the la Chapelle Plant of Rouen, France, and the Mülheim, Langenau and Wuppertal plants of the Federal Republic of Germany. On the other hand, contactor off-gases are recycled to the first compartment of the disinfection units at the following plants: Tailfer (Belgium), Neuilly-sur-Marne, Annet-sur-Marne (France) and Montreal (Canada).

The following four methods of recycling contactor off-gases were observed:

- Off-gas pressurization, submerged turbine diffusion in first compartment of disinfection contactor - Annet-sur-Marne, Neuilly-sur-Marne, Montreal (Figure 40).

- Off-gas pressurization, submerged turbine diffusion/flash mixing upstream of chemical clarification - Dohne, (Mülheim), Rouen-la-Chapelle (Figures 43 and 44).

- Vacuum induction, turbine mixing upstream of chemical clarification - Langenau.

- Inductive turbine recycling to first compartment of disinfection contactor - Tailfer (Brussels).

- Diffusion to plant influent tower - Wuppertal.

Provision is made for further treatment (destruction) of the recycled off-gas in some of the plants. In others, these gases are simply discharged.

The reuse of contactor off-gases is applicable when these gases contain quantities of ozone of 5 to 10% of that originally generated for the primary application. However, it should be realized that although reinjection will further reduce the concentration of ozone, the off-gases from this secondary contacting still contain ozone, and in amounts such that destruction or dilution should be considered before discharge to the atmosphere.

On the other hand, when the contactor off-gases contain smaller quantities of ozone (less than 5% of the original concentration) it may be more cost-effective to destroy or dilute and discharge these gases and utilize fresh ozone at a much higher concentration, for a second application.

CONTACTOR DESIGN CONSIDERATIONS

Process applications of ozone frequently will determine the type of contactor to be used. However, other factors such as flexibility, site constraints and plant size also must be considered. It is good practice to provide more than one contactor for many reasons, including maintenance. However, the question arises, how many contactors should be provided? A large number of contactors would result in flow distribution problems as well as high construction costs. It is understood that many of the Swiss plants are designed to operate on a multiple train basis. An ozonation train (ozone generator and contactor) is associated with a single raw water pump. If a second raw water pump were to be brought into operation, a second ozonation train would become operational as well. This independent train concept is likely to be inappropriate for larger plants where higher degrees of interdependence are economically justified.

The contactor configuration relates to the contactor selection and the layout of the overall facility. Economics dictate that common wall construction be used wherever possible. A depth of five meters is common to many porous diffuser contactor installations and is considered to be the minimum depth necessary to provide maximum mass transfer from porous diffusers. On the other hand, the effective water depth in vacuum injectors such as the three Düsseldorf water treatment plants and the Quebec City plant is ten meters. The inductive turbines are installed at water depths of 4 and 5 meters at Tailfer (Belgium) while those at Lengg (Switzerland) are at 3 meters water depth.

Measurement and control of the ozonized gas at each point of ozone application is necessary. Economics dictate that the ozone supplied not exceed the ozone required by more than the specified amount despite variable ozone demands due to flow rate changes and water quality variations. For example, a two compartment, countercurrent porous diffuser contactor used for disinfection should provide for ozonized gas flow measurement and control to each compartment. A means of measuring residual ozone levels should be provided at the discharge from both the first and second compartments. If virus inactivation is to be practiced, sufficient ozone should be supplied to satisfy the initial ozone demand and to enable a residual ozone level of 0.4 mg/l to be maintained in the second compartment. While this frequently means that two-thirds of the overall ozone dosage is added to the first compartment and one-third is added to the second compartment, there is no assurance that this situation will exist over the four seasons of a year or over the operational life of the facility. Ozonized gas measuring devices such as rotameters should be provided, as well as diaphragm or butterfly valves, and all should be fabricated from appropriate construction materials.

Access to the interior of the contactor should be provided through gasketed, low pressure hatchways. A method for dewatering the structures is required. The question of being able to observe the gas bubble pattern within the contactor appears to be one of process application and the desires of the operating agency.

The viewing ports of contactors used for iron and manganese oxidation appear to be generally useless, since they quickly become coated with opaque films of metal hydroxides. However, viewports on porous diffuser contactors appear to provide valuable visual access for the determination of diffuser deficiencies and the need for repairs or adjustments.

CONTROL AND INSTRUMENTATION

The degree of control incorporated into ozonation systems may be simple or sophisticated and may be localized or centralized. The trend in France and Switzerland appears to be toward highly sophisticated and centralized control, extreme examples being the small, remotely operated Kreuzlingen plant (Switzerland) and the large Neuilly-sur-Marne plant in France. At Kreuzlingen, the single plant operator merely checks plant equipment daily and prepares analytical solutions. His residence is alongside the plant, and he is immediately available in case of trouble, in which event alarms are signalled to his residence.

At the large Neuilly-sur-Marne plant, the ozonation system is highly automated, and there are only two operators for this section of the plant. They are present on a three-shift/day basis, and are trained in handling and maintaining automation equipment.

In contrast to this are the water treatment plants of the Federal Republic of Germany and the Province of Quebec, where higher degrees of local control and monitoring are practiced to ensure that the system is physically inspected.

Table 8 is a translation and summary of the ozone system record sheet used by the City of Pierrefonds (Quebec) water treatment plant to record data from their daily operations.

TABLE 8. CITY OF PIERREFONDS (QUEBEC) WATER TREATMENT PLANT

OZONE SYSTEM DATA RECORDED

REFRIGERANT COOLER	COOLING WATER
Units in operation Air discharge temperature (°C) Refrigerant flow indication	Pressure of ozonation, cooling water, and pressure at the cooling water inlet
DESICCANT DRIER	COMPRESSORS
Units in operation Inversion air pressure Regeneration temperature (°C) Pressure loss across the filter Gas discharge pressure (kg/sq cm)	Units in operation Discharge pressure (kg/sq cm) Air discharge temperature (°C) Pressure loss across the filter
OZONE GENERATOR	WATER COOLED HEAT EXCHANGER
Units in operation Inlet air temperature Inlet air flow rate (cu m/hr) Cooling water flow rate (cu m/hr) Cooling water inlet temperature (°C) Cooling water discharge temperature (°C) Discharge cooling water pressure (bar) Inlet air dew point (°C) Voltage (volts) Amperage (amperes) Power (KWAC) Power level selector	Air discharge temperature (°C) OZONE CONTACTOR Ozone residual, column 1 Ozone residual, column 2 GENERAL INFORMATION Ambient ozone level Plant water flow (cu m/hr)

While other ozonation systems may suggest other data to measure and record, Table 8 indicates the amount of information regarded as necessary to control this operational facility. These data may be obtained by remotely reporting sensors; however, in the case of Pierrefonds, personnel take direct readings of the sensing devices, thereby performing visual inspections of the system for process control and for detection of potential operational problems.

A number of parameters must be measured to provide a fully operable system; these include the following:

- There must be a means of providing a full temperature and pressure profile of the ozone generator feed-gas from the initial pressurization (by fan, blower, or compressor) to the ozone generator inlet. These measurements should be taken regularly, either in the form of simple mercury thermometers and pressure gauges or by more sophisticated sensors.

- There must be a means of measuring the moisture content of the feed-gas to the ozone generator. This procedure should be conducted with a continuously monitoring dew point meter or hygrometer. Without this device, the operation is unable to prevent ozone generator maintenance problems which arise from the use of uncertain feed-gas quality. Most dew point monitors and hygrometers were provided as direct reading instruments. However, it was understood that ozonation systems such as those installed at Quebec City and Kreuzlinger incorporate a lithium chloride sensing device which senses a high moisture content and sounds an alarm. It would appear that a direct reading instrument would be a more appropriate operational tool. Since either the dewpoint monitor or hygrometer are relatively expensive instruments, they must be specified or they would not be supplied on a competitive, low bid equipment supply basis.

- There must be a means of measuring the temperature, pressure, flow, and ozone concentration of the ozone-containing gas being discharged from an individual ozone generator and from the combined discharge from all the ozone generators. This is the only effective method by which ozone dosage and the ozone production capacity of the ozone generator can be determined. In view of the increasing reliability of gas phase ozone analyzers it should be possible to have continuous and accurate measurement of these parameters. At the very minimum, there should be frequent manual determination of these parameters.

- There must be a means of measuring the electrical power supply to the ozone generators. The parameters measured include amperage, voltage, power, and, if a controllable variable, frequency.

- There must be a means of measuring the flow and temperature of the cooling water to all water-cooled ozone generators. Reliable cooling is important to maintain constant ozone production and to protect the dielectrics. A drop in the rate of cooling water flow through the unit (or an excessive cooling water temperature) should be required to cause an automatic shut-down of the ozone generation system.

- There must be a means to monitor the several cycles of the desiccant drier, particularly the thermal swing unit. The plant superintendent of the Quebec City, Quebec water treatment plant found it necessary to add a high temperature shut-down switch on the discharge line from the heat regeneration system. This was to prevent the recurrence of a condition in which an excessively high regeneration temperature occurred with a resultant fire and desiccant destruction.

Figure 46 illustrates the required measurements for ozone generator operation. Ozonation control requires the measurement of ozone concentrations in:

- The ozone-rich gas from the ozone generator,
- The water after ozonation,
- The gas from the contactor, and,
- The ambient air in the rooms of the treatment plant using ozone, and possibly in the ambient atmosphere outside the treatment plant.

Means of ozone measurement include the following:

- Simple "sniff" test
- Dräger type detector tube
- Wet chemistry potassium iodide method
- Amperometric type instruments
- Gas phase chemiluminescence
- Ultraviolet radiation (254.7 nm) absorption

The smaller and older water treatment plants employ the more simple methods of ozone measurement. However, more sophisticated methods of ozone measurement are being incorporated into larger, newer plants as operational personnel become more confident of the accuracy and reliability of such methods.

The "sniff" test makes use of the fact that a worker is able to detect the odor of ozone at a concentration level of approximately 0.01 mg/l. This test is used for safety control in ozone hazard areas. This level of control is also used by the Düsseldorf plants to control the ozone dosage. The off-gas from the holding tank after ozonation is "sniff" tested to ensure that a trace of ozone remains in the off-gas. If no ozone can be detected, the generator output is increased.

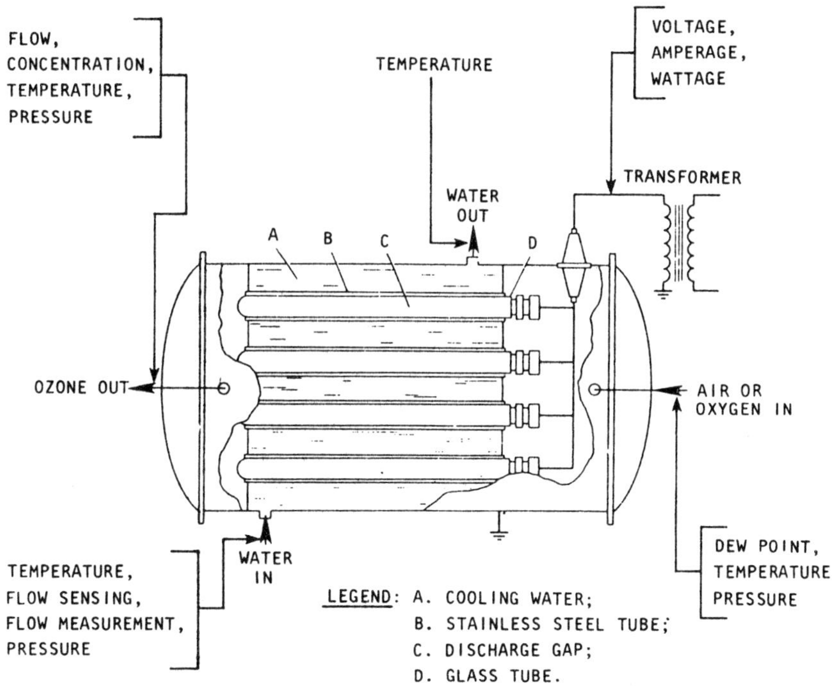

Figure 46. Tube type ozone generator - required measurements.

The Dräger type detector tube test utilizes portable equipment by which a specified volume of gas is drawn through a tube containing absorbing media. The absorbing media reacts immediately with the ozone in the gas and a constant colored stain is produced which varies in length according to the ozone concentration being measured. A calibration scale on the exterior of the tube enables direct reading of the ozone level in the gas. Literature indicates that ozone concentrations as low as 0.05 mg/l to as high as 300 mg/l by volume[54,55] can be measured by proper selection of detector tubes and volume of gas passed through the absorbent media.

The wet chemistry, potassium iodide (KI) method is the classical method for measuring ozone both in the gaseous phase and dissolved in water. The reaction involves the oxidation of iodide ion to iodine in water as shown by the following equation:

$$O_3 + 2I^- + H_2O \rightarrow I_2 + O_2 + 2(OH)^-$$

Starch is added as an indicator and the solution is titrated with sodium thiosulfate. The volume of thiosulfate consumed can be converted into an ozone value.

The KI procedure may be used as a day to day method of measuring ozone concentrations or as a means of calibrating instruments. The 14th Edition of Standard Methods for the Examination of Water & Wastewaters recommends the iodide method as the calibration standard for ozone instrumentation and provides a detailed description on how the test is to be performed for ozone in the gaseous form and dissolved in water.

It should be noted that in using the KI analytical method to determine the concentration of ozone in water, any oxidant present in solution that is capable of oxidizing iodide ion to iodine also will show up as "ozone". Thus, the KI procedure really measures total oxidants present. Examples of other oxidants that will be measured by this method are chlorine, chlorine dioxide, hydrogen peroxide, potassium permanganate, etc. At the Choisy-le-Roi (Paris) plant where chlorine dioxide is used in pretreatment to decompose organic complexes of iron and manganese and the subsequent ozonation process is controlled by monitoring an ozone residual of 0.4 mg/l, sodium bisulfite is added to destroy the excess chlorine dioxide before ozonation, so as not to produce a false "ozone residual" measurement.

Amperometric type instruments are familiar to the U.S. water treatment industry because of their common use to measure residual chlorine levels. This type of instrument utilizes a flow-through measurement cell containing two dissimilar metal electrodes. As the water sample flows past the electrodes, a current is generated which is proportional to ozone concentration in the water. The metals frequently used for electrodes in this instrument when measuring ozone are gold and copper. One requirement in the proper operation of such units is keeping the electrodes clean. Units of this type which have a successful record of application in closed loop control of ozonation are the Amperazur (Degrémont), Chlorosis and Degox instruments.

Gas phase chemiluminescence is used to measure ozone concentration in gases. The range is from 3.9 μg ozone/mV (0.002 mg/l) to more than 1962 μg ozone/mV (1 mg/l)[56]. Commercial gas phase chemiluminescence equipment is readily available.

Absorption of ultraviolet light radiation is a method which can be applied to measure ozone concentrations in both gas and water phases. In ambient air containing ozone, ozone is the only component that absorbs ultraviolet light in the 240 to 300 nm wave length range. The wave length used in these types of analyzers is 254.7 nm. The Dasibi ultraviolet ozone monitor for ambient air monitoring is used successfully at the Pierrefonds, Quebec water treatment plant and at the Union Carbide ozone generator test facility at Tonawanda, New York. The Sigrist single and double-beam spectrophotometric instruments measure ozone dissolved in water by the ultraviolet technique.

The use of control systems based on the measurements described above varies considerably in the water treatment plants that were inspected. The French plants using ozonation primarily for disinfection incorporate a closed loop control system by which the residual ozone level in the ozonized water is used to control the amount of ozone supplied to maintain that ozone residual. The key to successful operation of such a system is an accurate and reliable residual ozone analyzer. On the other hand, the older German ozonation systems which are used primarily for oxidation are manually controlled through a periodic "sniff" test of off-gas from the holding tanks after ozonation (ozone is used for iron, manganese and organics oxidation, and conditioning of biological activated carbon).

At present, it appears that continuous residual ozone monitoring equipment may be successfully applied to water that has already received a high level of treatment, as in the most common French application of ozone (disinfection and viral inactivation). However, a more cautious approach must be taken with the application of continuous residual ozone monitoring equipment for water that has received only chemical clarification. Continuous monitoring of ozone concentrations in gas phases appears to be reliable.

The Ozone Working Group of the German FIGAWA (Technischen Vereiningung der Firmen im Gas und Wasserfach eV) organization currently is evaluating methods for monitoring ozone concentrations in gas and in water. FIGAWA, which is roughly equivalent to the U.S. WWEMA organization (Water and Wastewater Equipment Manufacturers Association), already has published three "Technical Information" documents on various aspects of ozone technology which are included in this report as Appendix C. The FIGAWA Ozone Working Group reports on Ozone Analysis and Ozone Monitoring Equipment are expected to be published in 1978.

Various levels of controlling ozone generation are described below. Ozone production rates should be changed as water quality changes for several reasons. First, since excess ozone cannot be stored, power is wasted in generating excesses. Secondly, excess ozone must be destroyed before discharging, and this also requires energy. On the other hand, when water quality decreases, additional ozone is required immediately, since under-ozonation of certain organic contaminants present can produce partially oxidized intermediates which can, under certain circumstances, be more toxic than the initial materials (See Section 12, Oxidation Products of Organic Materials).

Manual Operation - Manual Sampling

The change in water characteristics brought about by ozonation is determined by visual inspection or chemical procedures. The required ozone production rate is controlled by manually changing either the voltage or the frequency of the electrical power supply.[40]

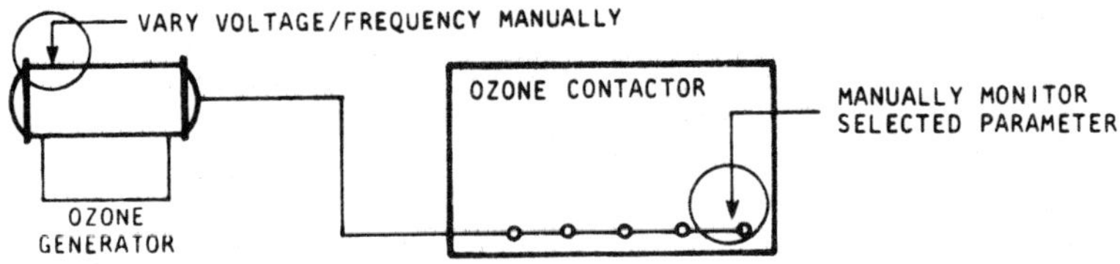

Manual Operation - Automatic Sampling

The ozone production rate is manually controlled through continuous monitoring of a selected parameter such as:

- the residual ozone level in water
- the residual ozone in off-gases
- the color of water (e.g., ozone blue, permanganate pink)

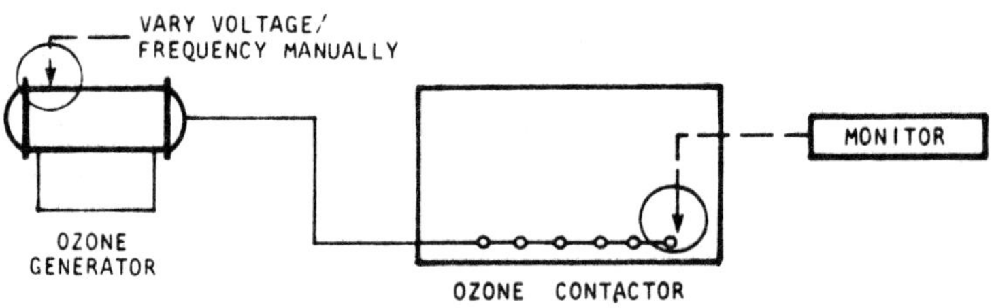

Closed Loop Control - Automatic Sampling

The ozone production rate is automatically controlled (voltage or frequency) by a signal generated by the continuous monitoring of a selected parameter.[40]

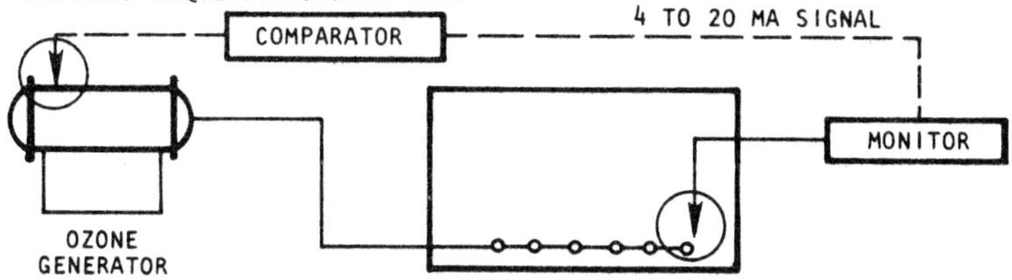

Closed Loop Control of Voltage/Frequency and Air Flow - Automatic Sampling

This system is fully automatic and is designed to operate at peak electrical efficiency with ozone-containing gas concentration controlled for optimum dissolution. Many system parameters are monitored and a pre-programmed controller adjusts both gas flow and the voltage or the frequency.[40]

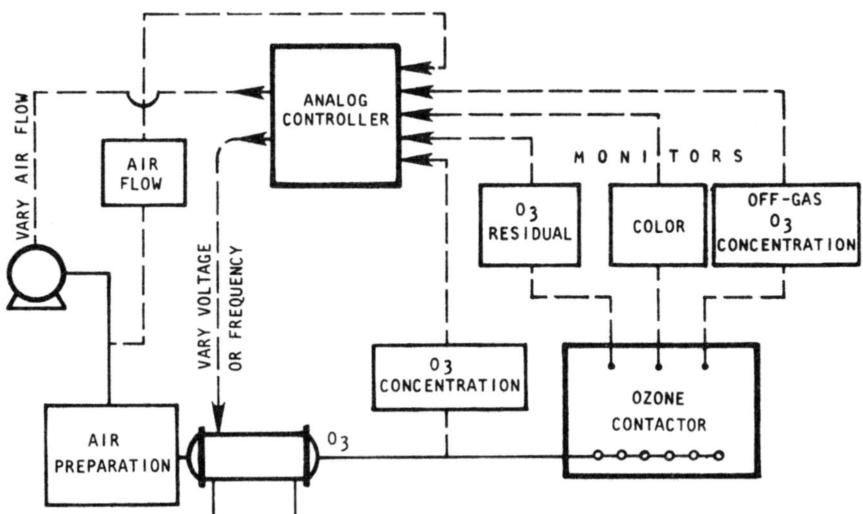

Full Computer Control - Automatic Sampling

The computer receives the same information as just above, and more from the control ozonation system, including diagnostics, pre-failure shutdowns, optimum energy balance, etc.[40]

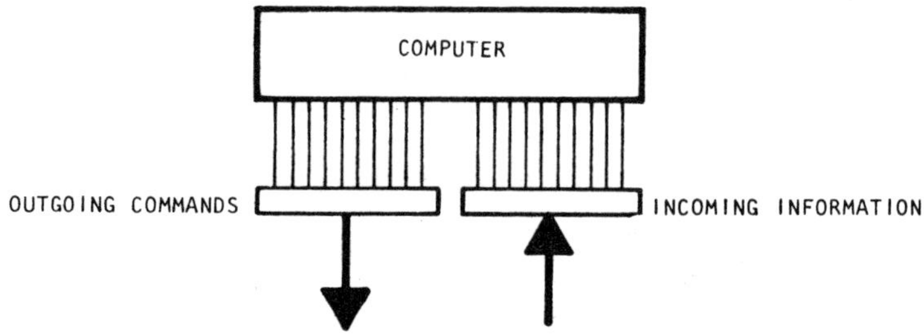

OCCUPATIONAL SAFETY

Ambient Air Quality

The plants visited varied in their approach to ozone safety. The ozone generator rooms of most facilities opened off control rooms, galleries or heavily used hallways. The ozonation processes of the Tailfer (Brussels, Belgium) and Neuilly-sur-Marne (Paris, France) plants are housed in structures separate from the remainder of the plant. However, it appears that the size of the ozonation facility dictated the separate structures rather than an intent to isolate the ozonation facility. In other plants, such as Lengg in Zürich and Annet-sur-Marne near Paris, the ozone generation facilities are placed in the heart of the treatment facilities. In general, while the European plants pay an extremely high degree of attention to safety with respect to chlorine storage and application, a far lower degree of concern appears to be paid to ozonation facilities.

A great deal of reliance appears to be placed in man's ability to detect the odor of ozone at levels far below the critical exposure level of 0.1 mg/l (by volume). Exceptions to this were the Tailfer plant of Brussels, Belgium and many plants in the Federal Republic of Germany. The Tailfer plant maintains negative pressure in the ozone generator room and the room over the ozone contactors. Continuous monitoring instruments are maintained to monitor levels of ozone in the rooms. Self-contained breathing apparatuses are located in hallways outside the rooms liable to ozone hazards. Many German (FRG) plants provide emergency high rate ventilation of ozone generator rooms as well as maintain self-contained breathing apparatuses. It was understood that it is standard practice to design ozonation systems to operate under negative pressure. Continuous monitoring instruments were observed in the ozonation rooms of the Pierrefonds, Quebec plant.

Ambient ozone exposure levels which have been proposed by appropriate U.S. organizations have been summarized in a recent paper.[45] The maximum recommended ozone levels are as follows:

- Occupational Safety and Health Administration (OSHA):

  Maximum permissible exposure to airborne concentrations of ozone not in excess of 0.1 mg/l (by volume) averaged over an eight hour work shift.

- American National Standards Institute/American Society for Testing and Materials (ANSI/ASTM):

  Control occupational exposure such that worker will not be exposed to ozone concentrations in excess of a time weighted average of 0.1 mg/l (by volume) for eight hours or more per workday, and that no worker be exposed to a ceiling concentration of ozone in excess of 0.3 mg/l (by volume) for more than ten minutes.

- American Conference of Government Industrial Hygienists (ACGIH):

  Maximum ozone level of 0.1 mg/l (by volume) for a normal eight hour work day or 40 hour work week, and a maximum concentration of 0.3 mg/l (by volume) for exposure of up to 15 minutes.

- American Industrial Hygiene Association:

  Maximum concentration for eight hour exposure of 0.1 mg/l (by volume).

Concern for safety, even at the risk of being overcautious, would be to follow practice that has been successfully applied to other oxidants over the years. This would be to generally isolate the ozonation system from the remainder of the plant. This should not be interpreted to mean a separate building but rather separate rooms, separate exterior entrances, separate heating and ventilation systems, noise control, etc. This method already is manifested in some of the European ozonation plants, but on a lesser scale.

There is a question whether prolonged exposure to ozone may impair a worker's ability to smell or be aware of ozone levels at less than critical levels.[45] Awareness of an odor of ozone should not be relied upon. Instrumentation and equipment should be provided to measure ambient ozone levels and perform the following safety functions:

- Initiate an alarm signal at an ambient ozone level of 0.1 mg/l (by volume). Alarms should include warning lights in the main control panel and at entrances to the ozonation facilities as well as audible alarms.

- Initiate a second alarm signal at ambient ozone levels of 0.3 mg/l (by volume). This signal would immediately shut down ozone generation equipment and would initiate a second set of visual and audible alarms at the control panel and at the ozone generation facility entrances. An emergency ventilation system capable of exhausting the room within a period of 2 to 3 minutes also would be interconnected to the 0.3 mg/l ozone level alarm.

Self-contained breathing apparatuses would be permanently stored at the entrance doors to the ozonation facilities. Controls for the high rate room ventilation system also should be provided at the exterior of entrance doors. Hand operated gas detector devices, such as the units of National Dräger of Bendix/Gastec, consisting of a volumetric pump and direct reading ozone detector tubes should be permanently stored in the vicinity of the self-contained breathing apparatuses. Use of the Dräger tube apparatus was demonstrated several times at Federal Republic of Germany water treatment plants as part of their safety program.

NOISE CONTROL

The control of noise is of obvious concern for installations in the United States. Noise from the water cooled ozone generation system (aside from the air preparation system) is very low. Therefore, most attention is paid to the air preparation system, particularly to those involving high pressure compressors or medium pressure blowers.

The most common solution to the noise resulting from medium pressure, positive displacement blowers, is isolation of the unit. These units commonly are located in a separate room at the lowest level of the building and air is brought to the intake filter by means of a vertical duct from the roof. The walls, ceiling, and floor generally are made of plain concrete having no special noise treatment properties.

There are special noise treatment rooms at the Dohne, Mülheim (Germany) and the Tailfer, Brussels (Belgium) plants. Perforated metal, false wall attachments are used at the Dohne plant while the Tailfer plant uses special sound-proofed doors for the blower room.

Facilities using low pressure fans in advance of refrigerant cooler and desiccant drier frequently were noisy. There seemed to be little concern with this problem, although the ozone generators often were in the same room. An example of a facility having this noise problem is the Holthausen plant at Düsseldorf, Germany.

Close attention should be paid to noise control of these facilities. Reference to current experience in the wastewater treatment field can provide significant guidance. For example, careful selection of positive displacement blower rotational speed and tip speed will minimize problems with this potentially noisy unit.

Isolation of noisy components of the system is a partial solution; however, the use of special construction materials would further reduce the problem and increase the value of preventive maintenance. The requirements of the Occupational Safety and Health Administration offer sufficient design guidance, provided the designer is aware of the major sources of noise generation within the system.

SELECTION OF CONSTRUCTION MATERIALS

Considerable effort was made during the study to identify and evaluate materials used in operational ozonation systems. Although special care must be given to selection of materials in contact with "dry" and "wet" ozonized gas, very little other concern in construction material selection appears necessary.

The materials of construction used in the air preparation systems appear to be those which are common to air preparation systems for other applications. The piping material used throughout the air preparation systems observed were common steel or galvanized steel of an appropriate pressure rating. No special coatings of reinforced concrete ozone contact structures were observed. Normal piping materials were used to convey the ozonized waters from the contactors. Ozone concentrations of less than 5 mg/l do not appear to pose threats to the normal cast iron or concrete pipes used in such service. It does not appear that special coatings of these types of piping, in this service, are necessary.

Selection of materials for surfaces in contact with "dry" or "wet" ozonized gas requires care. Further information is required to enable determination of the most cost-effective material. For example, there is considerable difference between the United States and European manufacturer's choice of material for ozone generator fabrication. Problems have been reported with the use of plastic piping systems in ozone service both in the United States and Europe.

European ozone generator manufacturers appear to use 316 or 316L stainless steel for those portions of the generator in contact with ozonized gas. In contrast, United States manufacturers use 304, 304L and 321 stainless steel and aluminum. Table 9 provides information on the various grades of stainless steel.

TABLE 9

COMPARISON OF STAINLESS STEELS

DESCRIPTION	AISI STAINLESS STEEL GRADES 304	304L	316	316L	321
	ANALYSIS (PERCENT)				
Chromium	18 to 20	18 to 20	16 to 18	16 to 18	17 to 19
Nickel	8 to 12	8 to 12	10 to 14	10 to 14	9 to 12
Carbon	0.08 max	0.03 max	0.08 max	0.03 max	0.08
Manganese	2.00 max	2.00 max	2.00 max	2.00 max	2.00 max
Silicon	1.00 max	1.00 max	1.00 max	1.00 max	1.00 max
Other Elements		-	Mo 2 to 3	Mo 2 to 3	Ti=5x%C min
RELATIVE COST*	1.00	1.06	1.33	1.38	1.34

* Relative cost based on prices provided by Midwest stainless steel supplier for 3/16" x 96" steel plate in 10,000 pound lots in January 1978.

Extra low carbon grades (304L and 316L) and the stabilized grade (321) are used in welding applications. Tungsten Inert Gas (TIG) arc welding is recommended. The relatively high cost of 316L stainless steel may be one of the reasons that United States ozone generator manufacturers prefer to use 304L and 321.

Selection of system piping material entails consideration of the gas being conveyed and its pressure. The gas piping material commonly used for the gas preparation system is carbon steel or galvanized steel of an appropriate pressure rating. Stainless steel piping generally is used for "dry" or "wet" ozonized gas service. As previously discussed, European plants would use 316L stainless steel for welded piping systems and 316 pipe for non-welded systems. On the other hand, United States manufacturers would use 304L and 304 stainless steel piping for similar applications. If a flanged piping system is used, Viton and silicone rubber are appropriate gasket materials.

The use of plastic pipe in ozonized gas service must be considered with caution, as illustrated by problems in European applications and a recent United States wastewater installation. Contacts with representatives of the plastic pipe industry indicate that little if any research has been expended on studies of the effect of ozone on plastic piping materials. So-called "hard" or unplasticized polyvinyl chloride (UPVC) is the only plastic piping material which is considered acceptable by those with experience in conveying ozone in plastic pipe. The authors have cautiously interpreted this to mean materials as described under ASTM Designation D1784-75 to be rigid poly(vinyl chloride) compounds, Class 12454-B. Class 12454-B is that material previously described as Type 1, Grade 1 under the previous ASTM D1784-65T. Unplasticized PVC may include ASTM D1784-75 Class 12454-C (Type 1, Grade 2). Pipe designations for these two materials are covered in ASTM Specification D1785-74 as PVC 1120 and PVC 1220, respectively.

Solvent welded joints should not be used. Only flanged couplings or threaded couplings should be used. Viton or silicone rubber gaskets or restrained Teflon gaskets should be used with flanged couplings. Threaded couplings should be prepared on schedule 80 or 120 pipe with a sharp, clean threading machine. Teflon tape should be used to wrap the threads. Ozone has been found to embrittle PVC pipe and to destroy the beam strength of this material. "Hard" PVC pipe was observed to provide satisfactory service under negative pressure ozonized gas applications. It was observed at only one positive pressure ozonized gas application, Langenau, Federal Republic of Germany.

Plastic pipe technical personnel have suggested the consideration of polyethylene (PE) plastic pipe as covered by ASTM specification D2239-74. The recommendation would be for pipe material described as PE 2306, 3306, 3406. This pipe material is fusion welded. Another suggestion is for a newly marketed piping system using FRP (Fiberglass Reinforced Plastic) lined with Teflon. The caution must be added that there is no known published information on ozone testing or application of these materials.

"Wet", pressurized, ozone-rich gas from the contactors would be expected to require the highest quality material, such as 316L stainless steel. However, it has been stated that the blowers used to repressurize the contactor off-gases from the Rouen-la-Chapelle plant in France and the Montreal, Quebec, Canada plant are constructed of aluminum. Selection of aluminum for the blowers has been explained on the basis that the ozone concentrations in the contactor off-gas are very low.

Another word of caution must be added; these comments should not be extended to include systems using oxygen-rich gases such as high purity oxygen. Time constraints precluded a detailed evaluation of materials used in construction of the Duisburg, Federal Republic of Germany plant. It is well recognized that special attention must be taken in the design of systems handling high purity oxygen.

OPERATION AND MAINTENANCE

The cost of operating and maintaining an ozonation system is of obvious concern to those evaluating the total cost of applying ozone as part of a water treatment facility. These costs may be assigned to the four components of the ozonation system (i.e., air preparation, electrical power supply, ozone generator, and ozone contactor).

Operation and maintenance of the air preparation subunit generally is predictable in terms of difficulty and cost. The frequency with which the filter medium is changed is a function of the ambient air quality. However, the inspected plants indicated that filter media cleaning or replacement initially should be scheduled twice per year. The power cost and maintenance requirements of the gas compresser and refrigerant units (such as oil changes) can be derived from manufacturers' literature. There are reportedly few operational problems with these units. Experience indicates a minimum absorber media life in excess of ten years for the desiccant driers. The energy costs to regenerate the desiccants depend upon the moisture loadings on the unit, but 16 hours is a reasonable estimate for an exhaustion-regeneration cycle. The desiccant should be checked regularly as should the regeneration cycle controls.

The operation and maintenance of the electrical power supply subunit is difficult to estimate since there is little available information. Apparently few (or no) problems have been experienced with this subsystem. Guidance may be taken from industrial experience with auto transformers. Transformer dielectrics and casings should be checked regularly.

Two factors that affect operation and maintenance of the ozone generators observed include the effectiveness of the air preparation system and the period at which the ozone generator is required to operate at maximum capacity. Maintenance of the ozone generators commonly is scheduled once per year, or twice a year in some plants. Approximately one man-day is necessary to service a 240 tube unit. Dielectric replacement due to failure as well as breakage during maintenance may be as low as 1 to 2%. However, it appears reasonable to predict an average tube life of ten years if a feed gas dew point of minus 60°C is maintained and if the ozone generator is not required to operate for prolonged periods at its rated capacity. An exception to this generalization is the major ozonation system of the Tailfer water treatment plant that serves Brussels, Belgium. The Tailfer plant currently maintains its ozone generators every six months and replaces the ozone generator tubes every two years, due to the deterioration of glass dielectric subjected constantly to high voltages.

The ozone generator production efficiency should be checked frequently in terms of ozone output per kilowatt hour of electrical energy input.

Operation and maintenance of the contactor also must be considered. Turbine diffusers require electricity to power the drive motors, while porous diffusers require regular inspection and maintenance to ensure a uniform distribution of ozone-rich gas in the contact chambers. Experience with maintenance of the ozone contact chambers in the Morsang plant (France) indicates that even after air purging of the contact chambers, maintenance personnel entering the chambers should be equipped with self-contained breathing apparatuses, because of the presence of ozone.

The following maintenance schedule is based on operational experience in Quebec City (Canada):

Daily

Monitor or measure major operating parameters including the following:

- Temperatures
- Pressures
- Dew points
- Production rates (lbs/KW)

Every Six Months

- Check oil level in compressors
- Check desiccant unit switchovers
- Check transformers

Every Year

- Complete overhaul of system
- Change parts where required
- Clean ozone generator thoroughly

Several of the inspected plants maintain a service contract with the ozonation system supplier for regular maintenance of the system. Both the Annet-sur-Marne plant near Paris and the Lengg plant in Zürich use this method to provide the expertise to maintain the various components of the system. Trailigaz suggests this approach to their customers, especially those new to the operation of ozonation systems.

SECTION 9

COSTS OF OZONATION

INTRODUCTION

No assessment of a given technology is complete without a review of its capital and operating costs. These costs are important both in terms of total magnitude and in terms of their relationship to each other. In the case of equipment or engineered systems for municipal use, the tradeoff between capital and operating costs, along with total cost magnitude, is of principal interest to both the municipal government and its consulting engineers. Thus, a principal objective of this investigation was to gather and document cost data for ozonation systems.

There is very little U.S. experience in ozonation systems; thus, the data presented herein are drawn largely from European and Canadian experiences. Further, the information presented in this section is not an in-depth economic analysis based on a rigorous analysis of a large data base. Rather, this section is a report on the content of various conversations and inquiries of knowledgeable and respected persons and firms in the ozone technology field. Most of the cost information contained herein was obtained from European and U.S. equipment manufacturers, European water treatment plant personnel, European research institute officials, and European and Canadian equipment suppliers.

Since most of the data presented were obtained from European operating experiences, the use of these data without defining precise conditions could lead to inaccurate conclusions. Costs vary greatly from country to country. Economic conditions within a country vary, personnel costs differ, and the price of energy is not constant. Moreover the cost of materials of construction may vary. With regard to capital investments, amortization periods are rarely the same and depreciation and interest data are difficult to obtain.

Ozone can be used for many purposes as evidenced by the multiple applications in Europe. Ozone cost data presented should not be viewed as the "cost of a disinfectant". In fact, ozone can be applied in such a way as to aid in microflocculation of previously coagulated water, improve organics removal, bacterial disinfection and viral inactivation. These beneficial side effects can be achieved as part of the disinfection and oxidation process and thus allow the designer to "spread" the cost of the ozonation system over other functions besides disinfection/oxidation. Ozonation was also found to be highly beneficial to the operation of activated carbon facilities (Section 13), another advantageous side effect. Thus, in evaluating the capital and operating costs of ozonation relative to the entire plant process, it may be misleading to consider it only as a disinfectant/oxidant.

CAPITAL COSTS

Capital costs* of an ozonation system generally include the costs of ozone generators, air preparation equipment, associated instrumentation, costs of needed ancillary facilities including the contactors, housing and power supplies, and installation. During the course of this investigation, the following key variables were identified as the principal determinants of installed system cost:

- Raw Water Quality and Ozone Requirements
- System Size and Cost of Housing
- Cost of On-Site Power
- Energy Demand of System Components
- Degree of Automation

These are discussed in turn below, followed by several illustrative case histories.

RAW WATER QUALITY AND OZONE REQUIREMENTS

Sizing of ozonation systems is critical, both from the standpoint of cost and process performance. For this reason, the common practice in Europe is to conduct pilot studies of 18 to 36 months in duration. Through the pilot studies, in which ozone is tested in conjunction with other unit treatment processes, the capacity required for accomplishing the desired objective can be determined. The duration of the pilot study allows evaluation of variations in temperature and water quality over several seasons. The pilot plant approach is well suited to the common European practice of turnkey construction and/or private ownership of water treatment plants. In the U.S. and Canada, the more common practice is separate design and purchase of equipment and facilities through competitive bidding. Thus, it is incumbent upon the municipality and its consulting engineers to conduct pilot studies.

---

* Capital costs typically are defined as equipment depreciation plus interest. Amortization of capital investment (the European terminology) is the terminology used throughout this section. Amortization (in European terminology) is the same as depreciation plus interest (North American terminology).

For smaller systems where pilot studies may be too costly, a more limited approach to ozone demand determination has been reported recently. Legeron[57] has defined a method whereby the ozone demand of a raw water supply can be closely estimated in the laboratory. The method was compared with actual operating plant experiences and found to be approximately equivalent, except where processing included physical-chemical treatment. In these cases, the procedure greatly over-predicted the amounts of ozone required. A full pilot plant study would be needed in this case.

The yearly average ambient air temperature, the peak temperature and water temperature are key factors in providing refrigeration and desiccation equipment. The air fed to an ozonator must be cooled and dried. The less it must be cooled, the lesser the capacity of refrigerant equipment that must be provided. In the case of water temperature, the colder the water is initially, the less expense that must be incurred for cooling. Most tube type generators are water cooled. If, as in the case of the Lake of Constance in southern Germany, the water temperature is a constant 4° to 6°C, very little refrigeration is ever required. In a much warmer climate, refrigeration equipment must accomplish much more work and thus must be designed and sized accordingly.

Thus the design engineer should work to: 1) define the application for which ozone is intended; 2) look at the ozone application in the context of the other unit processes; and 3) determine present and future ozone demands by analyzing raw water supply and providing standby capacity. As many manufacturers of ozonation equipment are well versed in its application, the designer should gather information from this source also.

SYSTEM SIZE

The capital costs of ozonation systems demonstrate significant economics of scale (decreasing unit cost versus output). The equipment is quite costly in small sizes, less so in larger sizes. System costs in various size ranges are given in this section. These general costs were obtained through conversations with equipment suppliers and manufacturers, both in North America and Europe.

Ozonation systems normally are considered to be in one of four size ranges: laboratory units (less than 0.9 kg/day, 2 lbs/day); very small units (0.9 to 45 kg/day, 100 lbs/day); medium size units (45 to 225 kg/day, 100 to 500 lbs/day); and large systems (greater than 225 kg/day, 500 lbs/day). Table 10 shows various system sizes and related costs. Costs given are for equipment only and do not include housing and installation. Costs are expressed in dollars/pound of ozone generation capacity/day using air as the feed gas.

TABLE 10. COST RANGE OF OZONATION SYSTEMS (USING AIR)

System Size	Approximate Cost $/lb/day	$/kg/day
Laboratory Unit (up to 2 lbs)	$3,000 - 10,000	$6,667 - 22,222
Very small systems (2-100 lbs)	$2,000 - 4,000	$4,444 - 8,888
Medium size systems (100-500 lbs)	$1,000	$2,222
Large systems (>500 lbs)	$600 - 800	$1,333 - 1,778

The figures shown in Table 11 were supplied by a large vertically integrated* water company located in Paris. The costs were converted to dollars using the exchange rate of 4.8 French francs/dollar.

It can be noted that the costs given here correlate well with the more generalized costs in Table 10. The added cost increments are due to the substantial costs for housing and installation which amount to 22.2 to 33% of the total. All costs given are for systems which generate ozone using air as the oxygen source material.

TABLE 11. CAPITAL COSTS OF SMALL TO MEDIUM OZONATION SYSTEMS (INCLUDING HOUSING AND INSTALLATION)

Daily Production Capacity		Cost vs capacity		Costs for Housing and Installation ($)	Total	Cost of Housing as % of Total
lbs	kg	lbs/day	kg/day			
3.3	1.78	$15,783	35,073	$16,667	$ 52,083	32
6.6	3.0	9,470	21,044	20,833	62,500	33
13.2	5.9	7,102	15,792	25,000	93,750	26.7
30.8	13.9	4,058	9,018	31,250	125,000	25
66.0	29.7	2,841	6,313	41,667	187,500	22.2
132.0	59.4	2,052	4,560	62,500	270,833	23

* The company referred to designs, constructs, manufactures equipment and installs and operates waterworks, hence the term "vertically integrated".

Overall, this investigation has shown that housing and installation can constitute 20 to 33% of total costs (European and Canadian experience) depending on the size of installation. Thus, one should attempt to minimize housing needs by purchasing larger ozone generator unit modules, if feasible. For example, tube type ozonators are available in unit sizes which can produce much more ozone than an Otto plate type generator of roughly the same size, offering potential for major space savings in larger systems. The Lowther plate type unit is said to be competitive with tube type ozonators in terms of output per unit space.

Table 12 shows capital costs for a wider range of systems. These figures were supplied by a major U.S. manufacturer. The costs given include the following system components and services including installation:

- ozone generators
- dissolution equipment
- gas preparation equipment including compressors, dryers and residual ozone decomposition equipment
- instrumentation and valving
- oxygen generation including process and equipment design support
- engineering assistance to enable the buyer's consultant to accomplish plant layout and design, specification, purchase and procurement of equipment
- safety review
- ozonation system engineering data package which includes process design drawings, instrument and control system drawings and installation specifications
- field engineering support for operator training, plant checkout and start-up.

The cost data presented do not include costs for:

- liquid pumping
- inter-process piping
- ozone contact chamber
- structure (housing)

TABLE 12. CAPITAL COSTS OF INSTALLED OZONATION SYSTEMS

Production Capacity of Ozone		Total Cost	Cost ($/lb capacity/day)	
lbs/day	kg/day		$/lb-day	$/kg-day
10	4.5	$ 125,000	12,500	27,778
100	45	325,000	3,250	7,222
500	225	800,000	1,600	3,555
1,000	450	1,000,000	1,000	2,222
2,000	900	1,550,000	775	1,722
3,500	1,575	2,150,000	614	819

Again, it can be noted that generalized figures given in Table 10 are consistent with these figures and that significant economics of scale are realized in larger systems.

Several comments should be made here, in order that the reader not be misled. The wide range of services furnished for the above costs will not necessarily be provided by every manufacturer or equipment supplier. Each manufacturer or equipment supplier provides levels of equipment and service depending on: 1) the company's objectives and policies; 2) the relationship of the supplier to the consulting engineer designing the plant; and, 3) the degree of system sophistication desired by the customer.

In the laboratory size units, for instance, the manufacturer/supplier may view a sale as a one time transaction and ask a very high price. On the other hand, the small unit may be a test unit which will, if the test proves successful, lead to the sale of a much larger system. In this case, the manufacturer/supplier may choose to offer the laboratory size unit at a greatly reduced price or on a rental basis.

Also, the market for ozonation systems in the U.S. is not yet well established; thus, many U.S. and European firms are competing for future shares of this market when and if it develops. For these reasons, the list of system components and services listed above should not be viewed as "typical". The list does provide a good checklist of services which are available and should be supplied by every manufacturer and/or supplier.

One further qualification should be made. In smaller size units, the manufacturer and/or supplier normally is asked to supply a unit capable of producing a given amount of ozone per day. Engineering or other services may not be requested by the design engineering firm and thus are not provided.

Finally, the reader is referred to another report prepared recently for the EPA's Office of Drinking Water which documents capital and operating costs of ozonation systems as well as adsorbents and other oxidants. The report, prepared by Temple, Barker and Sloane, Inc. and dated August 1977 is entitled "Economic Impact Analysis of A Trihalomethane Regulation for Drinking Water". Capital costs for ozonation systems cited in the report on page C-39 are shown in Table 13.

TABLE 13. OZONE CAPITAL COSTS

Design Capacity (MGD)	Cost of Complete Unit	Cost per MGD of Capacity
0.015	$ 4,550	$303,000
0.075	$ 7,530	$100,400
0.184	$ 18,280	$ 99,350
0.457	$ 35,515	$ 77,700
1.058	$ 65,080	$ 61,500
1.457	$ 82,780	$ 56,800
12.665	$ 401,345	$ 31,700
79.790	$1,538,290	$ 19,300
728.013	$7,726,415	$ 10,000

*Assumes unit sized for maximum day to deliver 2 mg/l. Cost includes all capital expenditures.

Note that data in Table 13 are not stated in terms of dollars per unit of ozone production. Computation of costs on this basis, assuming 2 mg/l ozone dosage and design flow, yields the data shown in Table 14.

TABLE 14. OZONE CAPITAL COSTS

Design Capacity (MGD)	Production Capacity lbs/day	Cost $/lb O_3/day	$/kg-$O_3$ day
0.015	0.25	18,186	40,713
0.075	1.25	6,092	13,538
0.184	3.07	5,956	13,235
0.457	7.62	4,659	10,353
1.058	17.65	3,688	8,195
1.457	24.30	3,406	7,569
12.665	211.25	1,900	4,222
79.79	1,330.90	1,156	2,569
728.013	12,143.26	636	1,413

In this report, ozone was viewed as a substitute disinfectant for chlorine and not as an oxidant capable of performing other functions. The capital costs are expressed in terms of plant design capacity.

In reporting on total capital and operating costs of ozonation systems, the Temple, Barker and Sloane study cites costs in terms of ozone followed by a chloramine residual. Again, this assumes use of ozone only as a disinfectant and further assumes use of chloramine as a terminal treatment step for residual. Combining these two costs can be misleading if not studied carefully. Under certain conditions (low TOC, cold finished water temperature, absence of ammonia,), ozone can be used as a terminal treatment step. Each treatment situation must, therefore, be reviewed on an individual basis. A plant owner or operator should not assume that chloramine or other oxidant is needed.

COST OF ON-SITE POWER

An important variable in total capital cost is the availability, reliability, and cost of on-site power. Ozonation systems are highly energy intensive, typically requiring about 22 kwh/kg (from air) for production in the 8 to 20 kilovolt range needed for ozone production. Because of the energy intensive nature of the production, it is important to optimize the use of incoming electrical power, i.e., use incoming energy as efficiently as possible to do real work. Thus, power factor, the ratio of power used to do "real work" versus the total amount needed to induce magnetic induction in motors plus accomplish work, becomes extremely important. Ozone generators and auxiliary equipment provided usually have a power factor rating of 0.80 to 0.97. A higher power factor represents more efficient use of input power.

ENERGY DEMAND OF SYSTEM COMPONENTS

Since the production of ozone is highly energy intensive, the various equipment components of an ozonation system will be discussed briefly in order to show relative energy demands and equipment tradeoffs.

Much of the information presented here is excerpted from a paper by Maurice Paré of Degrémont Corporation in France and is entitled "Operational Cost of Ozone and Energetic Balance". For a more rigorous discussion of system component alternatives and tradeoffs, the reader is referred to that paper[58].

The reader should be aware that conclusions of Paré's paper are based on equipment costs, energy costs, and amortization practices in France, and are valid and applicable in Europe. In North America, energy and equipment costs may be quite different. In some cases, for example, a high pressure air supply and preparation system consisting of compressors, water cooled heat exchangers, air filters and pressure swing type driers can be significantly lower in first cost than a comparable low pressure system. The cost effectiveness considerations therefore must take into account capacity and the redundancy of the components of the air preparation subsystem.

Paré's paper examines ozonation systems between 1 kg/hr and 30 kg/hr production capacity. It is a fairly rigorous theoretical examination of a "typical" water cooled, tube type ozone generation system using diffusers for contacting. As noted above, equipment costs, amortization rates, and other factors indigenous to North American practices may tend to reduce the value of the data presented. However, the reader can learn about the relative energy costs of various ozonation subsystem components and the comparison of capital and operating costs of the equipment itself.

The production of ozone and its utilization in water treatment can be characterized by the following sequence:

- supplying and drying of air
- production of ozone
- contacting with the water to be treated
- recovery or eventual destruction of excess ozone

Each of these system components has been discussed in detail in the earlier section dealing with engineering. The purpose of discussing them briefly again is to permit an insight into equipment selection with energy demand as the principal concern.

The largest portion of energy required for ozonation processing is in its generation and contacting. Lesser percentages are required for air preparation and destruction or recovery of excess ozone.

Supplying and Drying Air

Paré examines two methods of supplying air:

- supplying at high pressure starting with compressors (relative pressure between 5 and 7 bars);
- supplying at low pressure starting with a blower (pressure lower than 0.7 bar).

These possibilities are dependent upon the procedure adopted for air drying. Paré considers drying by means of refrigerant cooling and high pressure, heatless (pressure swing) desiccant drying and two stage drying by means of refrigerant cooling followed by low pressure, thermally regenerated desiccant drying.

With regard to these two system components, the following conclusions were reached: Energy demand for the system using water cooled heat exchangers for cooling at high pressure, and heatless, desiccant drying amounts to 0.1 kwh/cu m of air; energy demand for the system using blowers, refrigerant cooling, and thermally regenerated desiccant drying amounts to 0.06 kwh/cu m.

Paré's study indicated that, as a comparison of the total cost (amortized capital investment plus operation and maintenance) of the air preparation subsystem, the high pressure air preparation system is 50% to 85% more costly than the low pressure air preparation system. However, taking the entire ozone generation system costs into account, Paré notes that the total cost difference between a system using a high pressure air preparation system and a system using a low pressure air preparation system would be approximately 10% for a system in the 1 to 2 kg/hr range. Paré concludes that this small difference explains the great number of small installations which are equipped with high pressure systems. The additional 10% in total cost is compensated for through ease of operation and maintenance, largely due to compactness of the installation[58].

Although not discussed by Paré, the lower capital investment would make the low pressure air preparation system more attractive in the "low bid" U.S. method of equipment supply.

Ozone Production

Paré considers only one method of ozone generation, that being a horizontal, dielectric tubular type generator operated at a frequency of 50 Hz, using air as the starting material. Four concentrations of ozone (by weight) in air are considered: 0.5, 1.0, 1.5, and 2.0%. Energies consumed for ozone production only at these concentrations are, respectively, 16, 17, 19 and 21.2 kwh/kg.

The conclusion is drawn that, using the various system combinations considered, the optimal working concentration of ozone made from air lies between 1.0 and 1.25% by weight (approximately 1.15%).

Ozone Contacting

The energy demand of the method of contacting considered by Paré is taken into account in the air supply system. The contacting is accomplished by means of porous tubes located at the base of the contact chamber which is 4 to 5 meters in depth. The water depth of the contact chamber determines the operating pressure of the ozone generator. Thus, the pressure supplied by the air supplying/drying system also furnishes the energy for contacting in this example. This is not always the case. It will be shown in a specific case study later in this section that energy for contacting using turbines can amount to as much as 75% of the amount of energy required for generation.

Ozone Destruction

Paré considers four methods of destruction of the excess ozone: 1) thermal destruction at 150°C, 2) thermal destruction with partial recovery of energy using a gas/gas cross flux exchanger, 3) destruction by catalytic action and 4) recycling of excess ozone.

Thermal destruction is calculated to demand energy at 0.05 kwh/N cu m of air. It is estimated that up to 50% savings of energy can be realized using thermal destruction with partial energy recovery. This is accomplished by using flux exchangers to preheat the air leaving the contact chambers. No energy demands are cited for catalytic destruction, although it is noted that air must be preheated to 60° to 80°C in order to enhance the catalytic action. Experiences with catalysts have been recent and the working lives can only be estimated. The estimate is that the catalytic material must be replaced every two years.

It is concluded that the most favorable solution for ozone destruction in the first three cases is thermal destruction with partial energy recovery. Paré determines that the cost of thermal destruction at a production capacity of 15 kg/hr is optimized at 1.3 weight % concentration of ozone. At this concentration, the cost of destruction represents 6% of total system cost and is not cost-effective relative to thermal destruction with partial energy recovery.

Other methods of destruction not discussed are 10:1 dilution with air, destruction on wet activated carbon, chemical destruction, and combined thermal-catalytic destruction. The latter uses heat at much lower levels (60 to 80°C) plus the catalyst to effectively degrade the excess ozone. Because of the lower temperature requirements of the combined process, the needed heat often can be provided by using waste heat from other plant processes.

Using the various treatment trains shown in Figure 47, Paré estimates the ranges and distribution of costs for air supply/drying systems and ozone generation. The cost distributions are for production at the determined optimal concentration of ozone in air of 1.15%. Amortization of equipment is over a 20 year period. Breakdown of annual equipment amortization and operating costs is as follows:*

Air production	
amortized equipment cost	4 - 7%
energy cost of operation	15 - 18%
Ozone generation	
amortized equipment cost	18 - 28%
energy cost of operation	50 - 60%

---

* This breakdown assumes that ozone contacting is accomplished using diffusers for which gas pressure is developed in the air production train, and appears in the cost as an air preparation cost. It does not include other components of annual cost, such as the cost of personnel and maintenance.

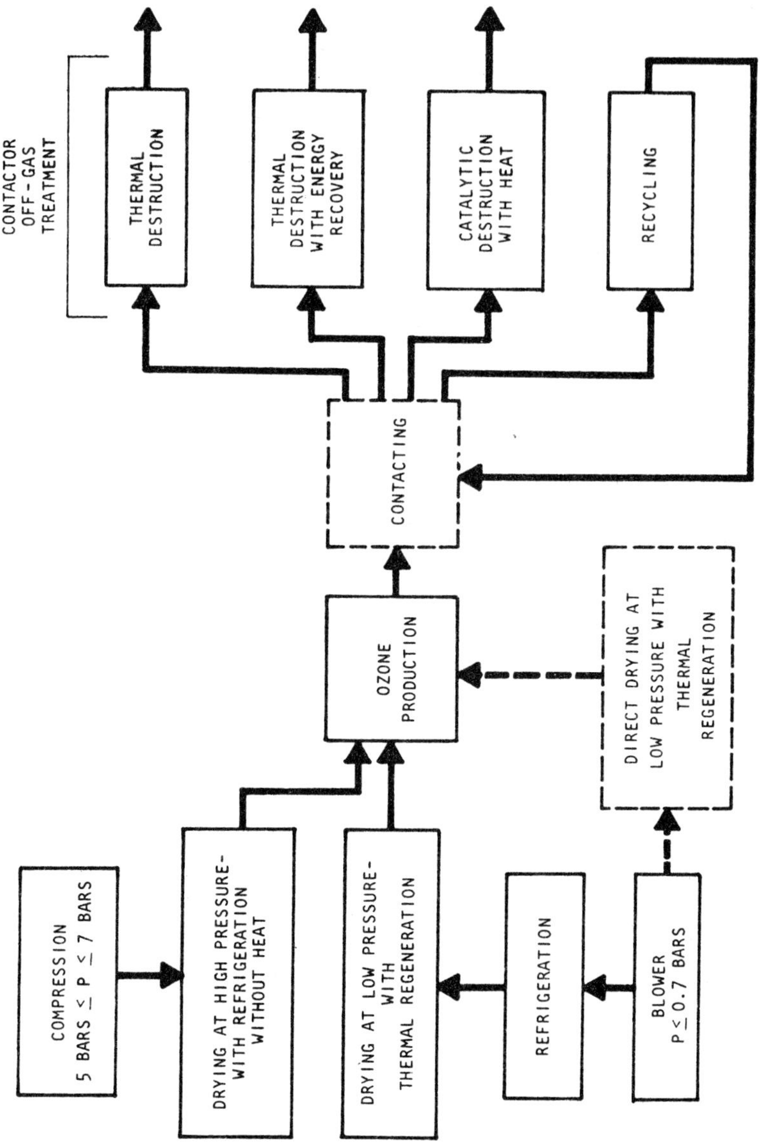

Figure 47. The different possibilities of energy consumption in ozonation systems[42]

One can readily conclude that the energy demand for ozone production constitutes the greatest cost of ozonation system costs, including amortization of capital costs of the equipment.

INSTRUMENTATION AND AUTOMATION

One of the most important variables in capital costs of ozone generation and contacting systems is the desired sophistication of the control system. The most inexpensive system is a manual system with some type of feedback to the operator which allows a necessary adjustment in treatment. An example of this type of control system was seen by the site visit team at the Quebec City, Canada, plant. See Appendix B for a detailed description of this plant.

The second level of sophistication is a semi-automatic system. Here, several key control variables are monitored. The most important variables are: 1) the residual ozone concentration in the treated water; 2) the concentration of ozone being produced; and 3) the concentration of excess ozone coming off the contact chamber which must be recycled or destroyed. A semi-automatic system would provide automatic control for these variables and manual control with feedback for other variables. The semi-automatic system requires interruption by the operator for control. An example of a plant exhibiting this type of control is the Pierrefonds plant in the Montreal, Canada, suburbs. A description of this plant is included in Appendix B.

The most sophisticated system, and the one requiring the largest initial capital outlay, is a fully automatic system. This system requires no operator interruption for control and it provides measurement of many key variables of operation in addition to the ones mentioned above. An example of this type of system is the 600,000 cu m/day Neuilly-sur-Marne plant in the Paris suburbs.

To summarize, choice of a control system can affect cost greatly. One should not assume that purchase of a manual system, since the initial capital investment is much less, is indeed more inexpensive. An automatic system, complete with sensors, a software package and a computer or microprocessor can optimize control, thus making system operation much more efficient. The added investment for a more sophisticated control system could be recovered through savings in personnel, energy and chemical costs.

OPERATING COSTS

Operating costs of an ozonation system are defined as including energy, maintenance, and spare parts. Personnel costs are not included except in maintenance costs. During site visits to 27 plants using ozonation processes, questions were asked regarding personnel needed for ozonation system operation. The site visit team was told repeatedly that additional personnel were not needed for operation of the systems over those required for operation of the remainder of the treatment plant.

Costs included in this section were obtained largely through site visits to 20 European plants in May 1977 and 7 Canadian plants during August 1977, all of which use ozone for some purpose. Additional cost data were obtained from recent literature.

Categories of Operating Costs

Each category of operating costs merits a brief description. Energy demand of an ozonation system depends on the 1) method of ozone generation; 2) method of air preparation; 3) amount of refrigeration that is required for generator cooling water; 4) method of contacting; 5) the dosage required; 6) ozone destruction requirements; 7) coolant pumping; and 8) cost of generating or purchasing oxygen.

The subject of energy required for different methods of contacting is complex and cannot be covered in detail here (the different methods are discussed in detail in the engineering section). There are several methods ranging from the diffusion sparger which requires little or no energy expenditure to the submerged turbines which generally are recognized as being the most efficient mass transfer devices but highest energy demanding methods of contacting.

A brief example of tradeoffs serves to place energy demand of contactors in the proper perspective. At Rouen-la-Chapelle, France (described in Appendix B), porous diffuser contactors are used for disinfection. In France, the standard contacting practice for disinfection is that ozone must be supplied in sufficient quantity over a minimum period of time so that a residual of 0.4 mg/l of dissolved ozone can be measured after 4 minutes of contact time once the initial ozone demand is satisfied. The rationale behind this theory can be traced to the work of Coin, Gomella, and Hannoun in 1964 and 1967 which showed that for viral inactivation to be assured, this period of contacting is necessary[59,60]. If ozone must be applied for this period of time in a concentration great enough to satisfy initial demand and leave the desired residual after an additional 4 minutes, one should strive to acccomplish this objective in the least expensive way. Thus, at Rouen and other French plants, porous diffusers are commonly used when ozone is applied at or near the end of the treatment process for disinfection.

In Belgium and Germany, there exist no disinfection standards of this kind. At Tailfer (Brussels), for example, submerged turbines are used for contacting because ozone can be used more efficiently (i.e., the mass transfer of gas to liquid is greater than with porous diffusers. But, in order to gain this efficiency, more energy must be used.

When the cases presented later in this section are studied, one can see that contacting at Tailfer demands 75% as much energy as does ozone production. At Lengg (Zürich) the figures are more difficult to analyze; total energy demand for both production and contacting is about 35 kwh/kg of ozone generated and applied. If one accepts 22 kwh/kg as an average amount of energy required to produce ozone, this means that 13 kwh/kg is required for contacting at Lengg which also uses submerged turbines for contacting. The point to be made is that energy demand of the total ozonation system varies greatly, depending on the system components.

Maintenance and spare parts sometimes are grouped together as one cost item. Ozone generators of the tubular type normally are shut down once per year, sometimes every six months, for cleaning of the tubes and other general maintenance. This requires several man-weeks of time, depending on the number of ozone generators in the system. Spare parts normally consist of replacement tubes, broken during cleaning or which have deteriorated from years of operation at high voltages. A "rule of thumb" for maintenance costs, related to the study team by a large manufacturer, is that maintenance cost should be about 15 to 20% of the production cost. If ozone production cost is $0.35/lb ($0.78/kg), then maintenance costs, in order to operate at a high efficiency, should be about $0.07/lb (0.16/kg) of ozone generated.

Actual Operating Costs

A number of cost figures were obtained during site visits to Europe and Canadian plants. Few have been defined precisely enough, however, to report them with complete confidence. Lest the reader be misinformed, only those costs for which operating conditions have been defined and verified will be reported. Two of these sets of operating costs, those for Tailfer (Brussels, Belgium) and Lengg (Zürich, Switzerland) are summarized in the case histories which follow later in this section. Additional data are given in the following paragraphs.

A large, fully automated plant in the Paris, France area, which doses ozone at an average rate of 2.5 mg/l, has an ozone operating cost of 0.035 French francs/cu m of water. The plant is amortized at an annual interest rate of 11% for a period of 20 years. Costs include all aspects of operation and maintenance plus amortization of capital investment, and an assumed inflation rate of 10%. Costs are based on annual production at half of rated capacity. The price paid for electrical energy is 0.10 French franc/kwh.

Converted to $/1000 gallons, using a currency exchange rate of 4.8 francs/dollar (the exchange rate in mid-1977), costs amount to $0.0276/1000 gallons. Electrical energy costs at this plant are 2.1¢/kwh (U.S. cents/kwh).

Other operating costs were reported by a large French water production company located in Paris for several plants which use ozone and which are operated by this company. Ozone treatment costs, including energy, capital cost and maintenance were given as 0.05 French franc/cu m, which converts to $0.0395/1000 gallons. Amortization of equipment is over a 10 year period. Cost of purchased electrical power is 0.15 franc/kwh or 3.12 U.S. cents. Water production costs for these same plants range from $0.079 to $0.39 per 1000 gallons. Average ozone dosage rates were not given, but can be assumed to be between 1.5 and 3.0 mg/l.

These operating costs are summarized in Table 15. The highest costs, 3.95¢/1000 gallons, occur in plants which are paying the highest rate for electricity and amortizing the capital investment over the shortest period.

The second highest cost reflects a relatively high average dosage of 2.5 mg/l. The Lengg plant, which has the lowest operating cost, pays less for electricity and has a lower average ozone dosage than do the other plants.

TABLE 15. OZONE OPERATING COSTS

Plant(s)	Ozone Operating Costs ¢/1000 gal	¢/cu m	Electrical Cost (¢/kwh)	Avg Dosage (mg/l)	Amortization Period (yrs)
Tailfer (Belgium)	2.52	0.66	3.01	1.7	20
Lengg (Switzerland)	1.75	0.46	2.0	1.5	20
Large Automated Paris Plant	2.76	0.73	2.1	2.5	20
Several French Plants	3.95	1.04	3.12	1.5-3.0	10

It appears from this admittedly small amount of data, that ozone can be produced and utilized for 1.75 to 4.0¢/1000 gallons (0.46 to 1.06¢/cu m) of water treated.

OTHER FACTORS IMPACTING COSTS

Other factors can impact capital and operating costs; one of these is standby capacity. In every system that operates continuously or almost continuously, a certain degree of redundancy is necessary. The question is, how much is enough? A consensus emerged during the course of this study among the European systems that one (1) extra ozone generator is adequate to provide standby capacity. A number of water plant superintendents voiced strong support for 100% backup redundancy. Manufacturers will provide as much ozone capacity as is desired by the customer, but normally recommend 10 to 25% over projected needs. The question of how much standby capacity to install must be addressed in each individual case by the customer and the design engineer.

A current trend in the ozone equipment marketplace of some importance is the availability of "pre-engineered" or "packaged" units. The trend is emerging because of the costs of the site-specific engineering services provided by many manufacturers. If some of these services (listed earlier under capital costs), such as providing detailed system drawings, can be standardized, significant reductions in the price of smaller systems can be achieved. Most manufacturers provide packaged units.

Another development in the marketplace that can affect overall water plant operating costs is the use of waste heat from the ozone generators to heat plant buildings. This approach is to be used in the St. John's, Newfoundland plant, currently under construction and which will use air cooled ozone generators. Estimated savings using this approach, assuming an annual ambient temperature of 60°F, are about $3600/yr. In colder temperatures, more savings would accrue while warmer temperatures would reduce savings.

CASE HISTORIES ON CAPITAL COSTS

Two cases are given here, one for the Brussels, Belgium waterworks and the other for the Lengg waterworks serving Zürich, Switzerland.

Tailfer (Brussels)

The Tailfer plant, described in detail in Appendix B, is a 260,000 cu m/day plant (68.7 mgd) located about 75 km northeast of Brussels. The water is treated at Tailfer and pumped 75 km to the city of Brussels.

The ozone production capacity at Tailfer, as actually installed, is 24 kg/hr (1267.2 lbs/day). Six horizontal, tube type dielectric units, each capable of producing 4 kg/hr, are installed. Currently, air is used as the feed gas, although the plant will be studying the use of pure oxygen feed to allow production of more ozone from the same number of ozone generators during periods of peak ozone demands in the summer. If oxygen is used in the future on a part-time (peak load) basis, this will make Tailfer a truly unique ozonation plant, and only the second municipal water treatment plant known to be generating ozone from oxygen.

The building which houses the ozone generation system plus contact chambers is large enough to enable the installation of additional ozone generators to increase production to 32 kg/hr (1690 lbs/day) when air is used as the feed gas.

The complete cost of the current system including the building plus equipment was 148,000,000 Belgian francs in 1976. Using 35.2 Belgian francs/dollar as the exchange rate (as of mid 1977) this equates to $4.024 million.

The cost for the building, including the contacting chambers, was approximately 50 million Belgian francs or $1.42 million.

Equipment cost, which includes the transformer, the ozone production chain, thermal destruction, venting, safety analyzers and process analyzers with transmitters, amounted to 98 million francs or $2.784 million.

Building costs thus comprise 33.8% of total costs, which is roughly consistent with figures given in Tables 10, 11 and 12. Cost of ozonation expressed in $/lb capacity/day amounts to approximately $2200, which is high for system costs at that capacity. Much of this cost is due to the method of contacting chosen and the standby housing capacity.

Tailfer amortizes buildings and equipment over a 20 year period at an annual fixed cost of 15,000,000 Belgian francs, or $426,000.

The energy required for ozone generation varies between 13.5 and 19 kwh/kg of ozone produced. Energy for air compression and desiccation is about 3 to 4 kwh/kg of ozone. Consequently, total energy costs for ozone production vary between 17 and 22 kwh/kg. The cost of electricity is 1.06 BF/kwh (3.01¢/kwh).

Operating Costs at Tailfer--

Operating costs for an ozonation system have been defined as including the following items: energy, maintenance and spare parts.

Total energy costs of ozonation at Tailfer amount to 3.34 million Belgian francs/year ($94,886) for ozone production and 2.487 million BF ($70,653) for ozone injection. Total energy cost of ozonation is 0.03 BF/cu m of water treated. Total amount of water treated/year amounts to 94,900,000 cu m.

Converted to U.S. dollars, energy for ozone production at 1.7 mg/l average yearly dosage, amounts to $94,886. Energy for injection costs $70,693, for a total energy bill of $165,579.

Manpower costs for maintenance, calculated for 6 ozone generators, is approximately $28,400 year. This is based on twice per year cleaning of the generators.

Spare parts, mostly tubes and fuses for the ozone generators, amount to about $11,600 per year. Uniquely, Tailfer replaces all dielectric tubes once every two years. Other plants visited never replace all dielectric tubes, only the ones that fail or are broken at the time of routine cleaning.

Summing, total annual ozonation system costs amount to $631,450 or 2.52 cents/1000 gallons (0.66¢/cu m) of water produced.

Cost of water production is 32¢/1000 gal (8.44¢/cu m). Thus, ozonation costs are 7.9% of total operating costs.

Capital costs of ozonation at Tailfer are quite high and are not considered to be truly representative of ozonation system costs. Operating costs are likewise somewhat high, the twice annual maintenance, high rate of tube replacement and energy for turbine contacting contributing to that factor.

The following Tables 16, 17 and 18 summarize capital and operating costs at Tailfer.

TABLE 16. CAPITAL COSTS AT TAILFER

Item	Cost ($)
Building & Injection Chambers	$1,420,000
Equipment	2,784,000
TOTAL	$4,204,000

TABLE 17. ANNUAL OZONATION SYSTEM COST AT TAILFER

Item	Cost ($)
Amortization	$426,136
Maintenance	28,409
Spare Parts	11,364
Energy: Ozone Production	94,886
Energy: Ozone Injection	70,653
	$631,448

TABLE 18. OTHER PERTINENT COST DATA FROM TAILFER

Item	Datum
Cost of Electricity	3.01¢/kwh
Yearly Average Ozone Dosage	1.7 mg/l
Energy Demand of Ozonation System	up to 10 kwh/lb
Current Ozone Production Capacity	1267.2 lbs/day (472.97 kg/day)
Unit Cost of Ozonation/1000 gal water	2.52¢/1000 gal (0.66¢/cu m)

Lengg

The City of Zürich, Switzerland currently is served by three waterworks: Lengg, Hardhof and Moos. The largest and most modern of these waterworks is the Lengg plant (described in Appendix B).

The Lengg plant is a 250,000 cu m/day (66 mgd) plant equipped with an ozonation system capable of producing 33 kg/hr (1742 lbs/day). Capital costs for the 6 vertical, tube type Kerag ozonators, including the ozonator building section in the Lengg lake waterworks, were 2.8 million Swiss francs. Using an exchange rate of 2.49 Swiss francs/dollar (prevalent in mid-1977), this amounts to $1.12 million.

Capital costs at Lengg are amortized at a 12% interest rate, which amounts to 336,000 Swiss francs per year. Maintenance costs are 2% of total costs or 56,000 Swiss francs. Service costs (assumed to be parts and labor for replacement thereof) are 16,000 francs. Energy for ozone generation and refrigerant water amount to 0.13 Swiss cent (Rp)/cu m. Summing, total operating costs amount to 1.15 Rp/cu m of water treated. These figures are based on an annual average ozone dosage of 1.5 mg/l and water production of 48 million cu m/year.

Converted to dollars, operating costs are 1.75¢/1000 gallons of water produced. Costs are summarized in Table 19.

TABLE 19. OZONATION COSTS AT LENGG

Energy and Water Costs			
Ozone Production and Contacting (0.016 kw/cu m)			= 0.08 Rp/cu m
Auxiliary Services (0.0062 kw/cu m)			= 0.031 Rp/cu m
Refrigeration Water (5 cu m/hr)			= 0.021 Rp/cu m
	Total		= 0.132 Rp/cu m (Rp=Swiss cent)
Amortization (12% of 2,800,000 Fr)		= 336,000 Fr	
Maintenance Cost (2% of 2,800,000 Fr)		= 56,000 Fr	
Maintenance Service Cost (by Manufacturer)		= 16,000 Fr	
	Total	= 408,000 Fr	= 1.02 Rp/cu m
Total Cost			= 1.15 Rp/cu m
Cost in U.S. Dollars			= 1.75 ¢/1000 gallons

The cost of electricity in Zürich is 0.05 Swiss francs/kwh or 2.0¢/kwh (U.S.). Energy demand is given in terms of kilowatt hours per cubic meter of water produced (53 kwh/cu m). Using the average ozone dosage rate of 1.5 mg/l, an energy demand of 35.2 kwh/kg of ozone produced can be calculated. This figure is the combined figure for ozone production and contacting.

Compared with Tailfer costs, Lengg capital costs are much lower, \$643/lb of ozonation capacity/day versus \$2200/lb. Operating costs are somewhat lower, 1.75¢/1000 gallons versus 2.52¢/1000 gallons at Tailfer.

Direct comparisons of Lengg with Tailfer can lead to inaccurate conclusions unless specific conditions are studied carefully. Some of the obvious differences in the two systems are as follows:

- Tailfer's ozonation system is housed in a separate building which also includes the ozone contacting chambers and capacity for expansion. Lengg's system was retrofitted into existing building space. Contact chamber costs may not be included in capital costs of the ozonation system, but rather in plant capital costs.
- The amount of safety equipment at Lengg is not as great as in Tailfer.
- Ozone production is accomplished at 50 Hz frequency in Tailfer, 350 Hz at Lengg.
- Amortization periods and interest rates are different for the two systems.
- The ozone generation equipment was supplied by different manufacturers.
- Annual maintenance costs at Lengg are \$22,490 and \$28,409 at Tailfer. This could mean that personnel costs are different or that ozone generators are cleaned once/yr (which is normal practice) versus twice/yr at Tailfer or both. Lengg also has a service contract with the manufacturer for ozone generator maintenance. Tailfer has a high rate of maintenance and more frequent replacement of tubes.

SUMMARY

The capital costs of ozonation systems can range from a low of \$600/lb capacity/day for a large system to a high of \$4000/lb capacity/day for a relatively small system. An added cost is the housing for the system which can range from approximately 20 to 33% of equipment costs.

Total capital and operating costs depend on energy demand, amount of maintenance required, amortization period, interest rates and cost of energy. Based on limited data, total costs for ozonation process steps were found to range from 1.75¢ to about 4¢/1000 gallons of water treated.

When one considers the many applications of ozone, including its effectiveness in breaking down synthetic organics, these costs can be justified. Also, ozone combined with granular activated carbon has been demonstrated in Europe to be an effective combination for the simultaneous, biological removal of dissolved organic carbon compounds and ammonia. Under ideal circumstances (see Section 13), the biological activated carbon created by the ozone-granular activated carbon combination also can result in longer operating lifetimes of the carbon and much longer periods between regeneration. Cost savings of the biological activated carbon (BAC) approach have not yet been investigated, but they are believed to be substantial.

SECTION 10

PUBLIC HEALTH ASPECTS OF OZONE USAGE

Ozone (O_3) is an allotrope of oxygen (O_2). It is 1.5 times as dense as oxygen and 13 times more soluble in water. It can be manufactured from dry air or from oxygen by passing these gases through an electric field of high potential sufficient to generate a "corona" discharge between the electrodes. Up to 3% by weight of ozone in air or 6% by weight in oxygen are about the maximum yields[61] of ozone for optimum use of electrical energy.

Ozone is highly unstable and must be generated on site. Its oxidation potential (-2.07V) is greater than that of hypochlorous acid (-1.49 V) or chlorine (-1.36 V), the latter agents being widely used in water treatment practice. It is thought to decompose according to[62]:

$$O_3 + H_2O \longrightarrow HO_3^+ + OH^-$$

$$HO_3^+ + OH^- \rightleftharpoons 2HO_2^\cdot$$

$$O_3 + HO_2^\cdot \longrightarrow HO^\cdot + 2O_2$$

$$HO^\cdot + HO_2^\cdot \longrightarrow H_2O + O_2$$

The free radicals ($HO_2^\cdot$ and $HO^\cdot$) may react with a variety of impurities such as metal salts, organic matter including microorganisms, hydrogen and hydroxide ions. They are more potent germicides than hypochlorous acid -- by factors of 10- to 100-fold. Experimental evidence indicates Gram-negative bacteria to be on the order of 10- fold more susceptible than viruses to ozone[63].

Morris[63] observes that the criteria of Coin et al.[59,60] for ozone disinfection appear to comprise a valid standard. These criteria are based upon poliovirus inactivation rates which showed 0.3 mg/l ozone residual after a 3 minute contact time to be the theoretical concentration-contact time relationship[64]. A safety factor is added, however, so that the criteria of Coin et al. are 0.4 mg/l of residual ozone after a 4-minute contact time. Gomella[65] observes these relationships to be as effective for the several river waters tested as for distilled water. Further, he notes that the 0.4 mg/l ozone residual corresponds to an oxidation potential of 600 to 650 mv, whereas an equivalent application of chlorine compounds corresponds to 550 to 650 mv.

Several plants in European cities that use ozone were visited: Zürich, Kreuzlingen, Konstanz, Sipplingen, Langenau, Paris, Rouen, and Toulouse. Microbiological data were requested at each plant, but were only available at Zürich, Rouen and Toulouse. By law, all German plants sample for microbiological analyses. Often, the samples are analyzed at other laboratories. By German law, all total plate counts/ml and coliforms/100 ml must be zero.

Finished water data for plants in Zürich, Rouen and Toulouse show:

Location	Record Examined	avg (min-max) Total Plate Count/ml	Total Coliforms/100 ml	
Zürich	1 year	5 (0-624)	0 (0-3)	ClO_2 added at end
Rouen	6 months 18 samples	116 (0-600)	0 (0-0)	ClO_2 added at end
Toulouse	5 months 24 samples	15 (0-125)	0 (0-0)	ClO_2 added at flash mix Ozone is terminal disinfectant

It is important to note that chlorine in some form is used at all three of the above plants. In Zürich, chlorine dioxide (ClO_2) is added terminally, and at Rouen Cl_2 is added terminally. At Toulouse, ClO_2 is added at the flash mix. In all cases, chlorine compounds added at all of the European plants visited are used minimally -- below taste levels. Thus, any discussion of the public health aspects of ozone usage based upon data from these plants cannot be divorced from chlorine usage. As the two processes (ozone and chlorine) are used in these European plants, it appears that microbiological quality control is as good as that in the U.S. where higher concentrations of free or combined chlorine alone are utilized.

An important additional aspect of ozonation that relates to microbiological control is its ability to convert some biorefractory organic compounds to biodegradable forms. The discovery of this effect occurred in Kreuzlingen, Switzerland when ozonation was substituted for terminal chlorination in the late 1950s. In Europe this is referred to as the "Kreuzlingen experience".

The Kreuzlingen plant obtains its water from Lake of Constance (the Bodensee) through which the Rhine River flows. Ozone, applied as a terminal treatment about 20 years ago, resulted in massive slime growths in the distribution system and water meters. Interestingly, Dr. Cyril Gomella observed that Paris had similar experiences after World War II -- but never before that time (when half the city's water was obtained from wells and the other half from surface sources which were treated with slow sand filters, a process which could be expected to decrease the biodegradable component).

Dr. H. Sontheimer from the Engler-Bunte Institut, University of Karlsruhe has developed some criteria for drinking water as a result of his studies of the experiences of a number of plants. These criteria utilize the parameter

DOC (dissolved organic carbon) which is the same as the dissolved TOC (total organic carbon) method familiar in the U.S.

- Water with a DOC $\geq$ 2 mg/l is not suitable for distribution.
- Water with a DOC $\leq$ 0.2 mg/l can be ozonated terminally without any real problem of aftergrowths.
- Application of 1 mg/l O_3:1 mg/l DOC results in 75% conversion of DOC to biodegradable materials.
- Application of 2 mg/l O_3:1 mg/l DOC results in 90% conversion of DOC to biodegradable materials.
- Ozonated water applied to a normal granular activated carbon filter will result in 30% DOC removal across the carbon unit process. (The sometimes considerable froth which accumulates on the water surface after ozonation accounts for 5% of the DOC loss.) Usual data at different plants range between 100 and 200 g DOC removed per cubic meter of GAC (BAC) per day.

Dr. Sontheimer and his colleagues recommend the practice of providing carbon adsorption beds following ozonation. The carbon removes the residual ozone and allows bacterial growth to develop which removes some of the biodegradable organics and ammonia. The resulting water is then chlorinated minimally, and the total plate count procedure is used to regulate chlorine application thusly:

- If the total plate count (TPC) is $\geq$100/ml, at least 0.1 mg/l free chlorine must be used by German law. This requires that 0.1 mg/l free chlorine residual must be present after a reaction time of 15 to 30 minutes.
- If the TPC is <100/ml, less chlorine can be used.

Hence, aftergrowth problems in distribution systems are well recognized in Europe and the ozone users of Europe resort to the same basic control technique as does the American art -- namely the use of chlorine. In Europe, however, ozone contact is relied upon for the main disinfection process and chlorine is added for distribution system control, whereas in the U.S., chlorine is relied upon for both functions. The main difference is a much lower chlorine residual in Europe than in the U.S. along with the potential for a lower total trihalomethane content. For example, the Sipplinger Berg plant of West Germany uses microstrainers, ozonation, and rapid sand filtration and delivers the water to a 150 km-long distribution system. The water is chlorinated to the maximum allowed by law - 0.6 mg/l - at the entrance to the system.

Dr. Dietrich Maier[66] stated that total haloforms in treated Sipplinger Berg waters are 5 micrograms/l (μg/l) with chloroform accounting for 2.5 to 3 μg/l. He also said that Lake Zürich water treated at the Lengg Plant results in a similar haloform distribution. Bromine in the Bodensee water

(water source for Sipplinger Berg) is less than 10 μg/l, and after ozonation the brominated portion of the haloform concentration is higher than before ozonation (13 mg/l).[67]

The effects of ozone upon some organic compounds have been studied. Dr. Cyril Gomella supplied the data of Table 20 which shows the effects of varying concentrations of ozone at three different contact times upon anionic detergents and COD. Ozone has much more effect on detergents than on the gross organic parameter, COD.

The data of Table 21 showing the effects of ozone upon a number of specific organics also were supplied by Dr. Gomella. There is a very broad range of effects by ozone from compound to compound. Two were not touched at all -- n-tetradecane and n-pentadecane and only one slightly -- dihexylphthalate. (di-2-ethylhexylphthalate). Two phthalates present in trace amounts appear also to have been little affected -- diisobutylphthalate and butylisobutylphthalate. However, n-octane and trichlorobenzene show greater than 90% removals.

It is necessary to be cautious in interpreting such data. First the study team did not determine the ozone content or contact time that effected these changes. But more importantly, for almost each and every compound an oxidized product is produced and the health effects of these products presently are unknown. Also, Dr. Gomella is quite aware of this problem and hopes to initiate such research this coming year.

The only data available which reflect the performance of an activated carbon bed acting as a biological filter after an ozonation process was supplied by Mr. Daniel Versanne, Table 22. These are based upon the performance of the Rouen Plant in northwestern France two months after it commenced operation. Mr. Versanne stated that more recent results show even better performance.

Summing the chemicals present (except for DBP and DEHP which -- judging by their high concentration probably were been added as internal standards) gives an estimate of unit process efficiencies in oxidizing or decreasing the concentrations of these materials. The first ozonation step (O_3 concentration = 1 mg/l, time of contact = 3 minutes) results in a 63% quantitative decrease, and the filtration through sand followed by carbon yields a 34% decrease. Both processes together effect a 76% decrease. The post ozonation step (O_3 concentraton = 1 mg/l, time of contact = 12 minutes) yields a 39% decrease, and all processes together yield an 85% efficiency in removing impurities.

Only one class of chemicals shows an increase, this being equal to or greater than C 20 isoalkane, but the amounts are so small it is likely not significant. The 34% decrease across the filtration step using sand followed by carbon is quite close to the 30% DOC removal noted by Dr. Sontheimer for a biological carbon bed.

The chemical evidence available does not indicate any untoward health hazard to be associated with the use of ozone in water treatment. The potential exists as indicated by an increase in concentration of one group

TABLE 20 EFFECTS OF OZONE ON ANIONIC DETERGENTS AND COD

Ozone, mg/l		Time of Contact (min)	After Ozonation			
			Detergent*		COD*	
Dose	Residual		mg/l	% removal	mg/l	% removal
1.35	0.19	4	0.052	58	8	16
1.66	0.34	4	0.048	61	7.5	21
0.70	0.45	4	0.083	27	8	20
2.05	0.48	4	0.043	65	8.5	11
2.03	0.68	4	0.048	58	7.5	25
2.40	0.80	4	0.031	73	7.5	25
2.50	0.93	4	0.035	72	8	16
3.85	1.13	4	0.015	87	6	40
3.31	1.30	4	0.016	87	6	37
2.95	0.34	10	0.025	80	7	26
3.80	0.90	10	0.012	90	6	40
3.85	1.18	10	0.003	98	6	40
3.17	0.27	20	0.013	90	7.5	25
4.10	0.75	20	0.003	98	6.5	32
4.83	0.90	20	0.002	99	6	40
6.90	2.95	20	t	100	6	40

* Initial sample concentrations:

Detergent: 0.114 or 0.124 mg/l

COD: 10.0 or 9.5 mg/l

Data courtesy Dr. Cyril Gomella

TABLE 21 EFFECTS OF OZONE UPON ORGANICS IN WATER

Compound	Filtered Water µg/l	Ozonated Water µg/l	Percent Removal
Chloroform	102	60	41
Carbon tetrachloride	6	4	33
Trichloroethylene	15	13	13
n-Octane	1	0	>90
1,2-Hexanediol	10	2	80
n-Undecane	8	3	60
Trichlorobenzene	3	0.2	>90
Trichlorobenzene isomer	2	0.3	85
n-Dodecane	0.5	0.2	60
n-Tridecane	1	0.7	30
Diphenylether	1.5	0.3	80
n-Tetradecane	-	0.3	0
n-Pentadecane	0.2	0.2	0
Dibutylmaleate	1.5	0.2	85
n-Hexadecane	0.5	0.2	60
n-Heptadecane	<0.1	<0.1	-
n-Octadecane	0.4	0.3	25
Diisobutylphthalate	t	t	-
Butylisobutylphthalate	t	t	-
n-Nonadecane	0.8	0.3	60
Dibutylphthalate*	138	25	82
n-Eicosane	0.3	-	>66
Dihexylphthalate	7.1	6.9	3
Di-(2-ethylhexyl)phthalate	127	84	34
Dioctylphthalate	42	38	10
Octyldecylphthalate	41	37	10
Didecylphthalate	18	16	10

Data courtesy Dr. Cyril Gomella, SETUDE, Paris, France

* High vapor pressure

TABLE 22 ROUEN PLANT PERFORMANCE, ORGANICS

Retention Time Relative to DBP	Compound	Approximate Concentration, µg/l			
		Raw	After Pre-Ozone	After Filt+AC	After Post-Ozone
0.08	isoalkane >C 8	80	17	13	2
0.09	methylcyclohexane	14	10	5	5
0.10	isoalkane >C8	16	8	6	t
0.13	isoalkane >C 8	72	40	17	5
0.16	isoalkane >C 8	4	4	3	t
0.18	n-nonane	4	t	t	t
0.20	isoalkane >C 8	4	t	t	t
0.27	n-decane	2	t	t	t
0.31	isoalkane >C 10	2	t	t	t
0.33	isoalkane >C 10	0.4	t	t	t
0.36	n-undecane	1	0.5	0.5	t
0.38		18	2.5	1	1
0.44	isomers of 4-tert-	22	2.5	1	0.5
0.47	butylcresol or of	6.5	1	0.5	0.4
0.50	tert-butylanisole	6	1	0.5	0.5
0.51		0.8	0.8	0.8	0.8
0.54	bicyclohexyl	70	33	30	30
0.59	not identified	10	t	t	t
0.63	n-tetradecane	0.7	0.3	0.3	0.3
0.67	not identified	0.4	0.4	0.4	0.4
0.71	n-pentadecane	0.5	0.2	0.2	0.2
0.75	isoalkane >C 15	0.2	t	t	t
0.78	n-hexadecane	1.5	0.6	0.6	0.6
0.85	n-heptadecane	1	0.3	0.3	0.3
0.88	not identified	0.4	t	t	t
0.92	n-octadecane	1.5	0.5	0.5	0.5
0.94	DIBP	7	3.5	3	3
0.97	DIBP	4	2.5	2	2
1.00	DBP	650	390	320	240
1.05	n-eicosane	1.2	t	t	0
1.07	isoalkane ≥C 20	0.4	0.4	t	0
1.10	isoalkane ≥C 20	0.4	t	t	0
1.15	isoalkane ≥C 20	0.4	t	t	0
1.16	not identified	3	2	1	0.1
1.22	not identified	t	t	t	t
1.27	isoalkane ≥C 20	0	t	t	0.2
1.32	DEHP	330	220	130	117
Σ(less DBP + DEHP)		355.3	131.0	86.6	52.8
% Removal by process			63%	34%	39%
by train				76%	85%

Data courtesy Mr. Daniel Versanne, two months after start-up of Rouen Plant. More recent results show even better performance.

of materials. But even if that indication is real, it is likely for most compounds to be of much less potential hazard than is the use of chlorine and chloro-organics that result from chlorine's use. And there appears to be no escaping the need for continued chlorination of most public water supplies. It must be remembered, however, that many new oxidized moieties and biodegradation products are being produced and we do not know what these are, but in general, these are compounds of less -- not greater -- toxicity.[62]

U.S. water treatment professionals have been somewhat reluctant to include a biological process in water treatment. Two very real possibilities must be studied before such a treatment can be recommended. First, such a unit process would most certainly result in an increase in *Pseudomonas* bacteria content. These organisms are ubiquitous, have a propensity to grow in minimal media, and some strains are opportunistic-pathogens. They are genuinely hazardous to infants and the elderly -- population groups at high risk due to their lesser resistance.

Secondly, a biological bed intuitively can be expected to produce an increased endotoxin content in the treated waters. Endotoxins are lipopolysaccharides which are synthesized in the outer cells of Gram-negative bacteria. They have the capability of producing fever (pyrogenic effect) when injected into the blood stream of some animals. Many endotoxins when in the intestines have the capacity to produce effects similar to gastrointestinal infections. However, endotoxins in the intestines -- and in water wherever Gram-negative bacteria are present -- are a part of our natural daily lives, and hence are not ordinarily hazardous agents. Nonetheless, it would be desirable not to increase the concentrations of these materials by virtue of a water treatment unit process.

The endotoxin content can now be quantified down to μg/l levels by use of the *Limulus* (horse shoe crab) lysate assay. Analyses of renovated wastewaters showed values ranging from less than 0.313 μg/l (which employed a reverse osmosis unit in the treatment sequence) to 1250 g/l (which samples had been delayed in shipment). Analyses of drinking water samples showed values ranging from less than 0.625 μg/l (lime softening plant, no carbon used) to 125 μg/l (granular activated carbon in the complete treatment sequence) to 500 μg/l (only chlorination practiced).[68] However, this work did not examine any of the specific unit processes in a before and after mode.

Recent unpublished work[69] on filtered activated sludge effluent in Dallas, Texas, in which two carbon columns were used in series and from which samples were taken before, in between, and after the carbon columns is shown in Table 23. Looking at both total (bound + free) and filtered (free) endotoxins, there is a 73 to 83% decrease across the first carbon column, but a small increase in endotoxin levels occurs through the second carbon column, the entire process resulting in a 67 to 69% decrease for total and filtered endotoxins, respectively.

TABLE 23. ENDOTOXIN VALUES AND CARBON ADSORPTION

Sample		Endotoxin Equivalents, ng/ml		
		Before Carbon	Midpoint	After Carbon
12-14	Total	600	192	192
	Filt	480	48	96
12-21	Total	500	100	200
	Filt	250	50	100
1-25 (SA)	Total	250	100	100
	Filt	125	50	75
1-25 (CS)	Total	300	60	60
	Filt	120	30	30
$\bar{X}$	Total	412.5	113	138
$\bar{X}$	Filt	243.8	44.5	75.25
% Change, Total			73	67
% Change, Filt			82	69

SA = San Antonio data

CS = College Station Data

Total = not filtered

filt = filtered through 0.45 μm Millipore

Additional information of public health interest can be obtained from the data of Table 24. These data compare the standard plate count and total coliform values of samples (activated sludge + filtration + carbon adsorption) subjected to four different disinfection procedures. The bacteriological tests were run on samples before shipment from Dallas to College Station, Texas (Texas A&M University) and after their arrival. According to both tests, the ozonated samples performed just as well as the chlorinated samples. The endotoxin values of these samples are shown in Table 25.

TABLE 24. EFFECTS OF SHIPMENT TIME ON BACTERIOLOGICAL QUALITIES OF WATER SAMPLES-PYROGENIC PROJECT

Date of Sample (1977)	Standard Plate Count/ml				Total Coliforms/100 ml			
	UV	Cl_2	pH	O_3	UV	Cl_2	pH	O_3
2/8 before	112	43	1	0	0	0	0	0
after	1,820	475	500	310	23	0	0	0
2/28	24	4	0	0	2	0	0	0
"	125	35,000	44,000	213	49	23	5	0
3/14	81	8	27	3	7	0	0	0
"	50	10	25	-	23	0	0	2
3/28	280	3	13	3	4	1	0	0
"	235	11	59	379	23	0	0	0
4/4	140	26	26	3	14	0	0	0
"	5,540	3,800	9,340	11,800	23	0	0	0
4/11	240	5	17	8	6	1	0	0
"	63	14	13	0	2	0	0	0
4/18 "	79	65	60	50	15	1	0	0
	7	128	96	8	2	2	2	0
$\bar{X}$ Before Shipment	137	22	21	10	7	<1	0	0
$\bar{X}$ After Receipt	1,120	5,634	7,719	2,118	21	4	1	<1

0 = <1/ml for SPC, <2/100 ml for TC

TABLE 25. SAMPLE ENDOTOXIN ASSAYS

Sample Date	Endotoxin Equivalents, μg/l Total (Free)			
	UV	Cl_2	pH	O_3
2/8	320 (320)	160 (80)	32 (16)	32 (8)
2/28	-	-	-	-
3/14	80 (48)	80 (8)	80 (8)	80 (8)
3/20	30 (22)	30 (22)	9 (2)	30 (3)
4/4	90 (30)	30 (30)	12 (6)	15 (12)
4/11	30 (22)	47 (25)	5 (3)	30 (6)
4/18	47 (25)	47 (25)	47 (25)	30 (6)

The impact of biological activated carbon treatment on drinking water supplies presently is being studied at Texas A&M University. At the time of this report, data from two water samples were available:

Sample	Endotoxins Before GAC μg/l	After GAC μg/l
#22 Filt.	1.56	1.56
Total	3.12	1.56
#77 Filt.	3.12	3.12
Total	3.12	3.12

The GAC (granular activated carbon) for #22 had been in use since 1968 without regeneration and for #77 had been in use only 5 weeks. In neither case is there an increase in endotoxins through the GAC.

Additional data is necessary, but the following presumption is possible: Biologically treated wastewater has a high level of endotoxin (~6000 μg/l) which is reduced by both filtration and adsorption mechanisms in GAC columns. Since there is still a biodegradable carbonaceous component remaining in the wastewater, additional growth of Gram-negative bacteria can occur in subsequent GAC. In the case of drinking water supplies, higher quality waters (in terms of a biodegradable carbonaceous component) are utilized and thus less substrate is available for the growth of Gram-negative bacteria. If ozone is applied prior to the GAC, the biodegradable carbonaceous component will be increased, but the amount of increase will be relatively small and consequently should impact on the endotoxin levels in only a minor manner.

In order to answer the above presumption (or hypothesis) as it pertains to drinking water, it is necessary to expand current research efforts to include:

- pseudomonas assays on samples from water treatment plants taken before and after GAC beds;

- apply pseudomonas, limulus, and biodegradable component assays to GAC beds functioning as Biological Activated Carbon (BAC) beds which are immediately preceeded by ozonation.

Of course, it is also necessary to implement rather strenuous efforts to identify oxidized organics specifically that result from the O_3 - BAC unit process.

SECTION 11

CHLORINE DIOXIDE

HISTORY OF USE

Chlorine dioxide was first discovered in 1811 in the form of a greenish yellow gas by Sir Humphrey Davy[70]. Davy prepared the compound by reacting potassium chlorate with hydrochloric acid and named the resulting compound "euchlorine". In 1834, Watt and Burgess identified the bleaching properties of the compound. Around 1900, it was discovered that ClO_2 could be used in a dilute acetic acid solution for the bleaching of paper pulp. Other experimentation with ClO_2 was carried out during this period, but it had little practical application due to the lack of a safe and economical way of synthesizing the chemical. In the 1930s, the Mathieson Alkali Works developed the first commercial process for making ClO_2 from sodium chlorate[70]. By 1939, Mathieson Chemical Corporation had succeeded in making sodium chlorite a commercial product.

With the advent of this commercial feed stock for ClO_2 generation, ClO_2 processes quickly found practical application. The No. 2 Niagara Falls, New York, Water Treatment Plant was the first U.S. water plant to use ClO_2 (1944). The oxidant was used to treat a potable water supply for taste and odors arising from phenolic compounds. Other plants that soon adopted the ClO_2 process for water treatment were Greenwood, South Carolina and Tonawanda, North Tonawanda, Lockport and Port Colbourne, New York[2]. The use of ClO_2 was not only gaining favor by the water treatment plants, but in 1946, three pulp mills, one in Canada, and two in Sweden, began using ClO_2 for bleaching wood pulp. This use grew rapidly thereafter; today 60 or more U.S. pulp mills are using ClO_2 for pulp bleaching.

The use of ClO_2 has continued to expand in the treatment of potable water. A 1958 survey of 150 municipal water works believed to be using ClO_2 reported that 56 were using the oxidant predominantly for taste and odor control[71]. In addition to taste and odor (T/O) control, these plants listed other uses for ClO_2: disinfection (15 plants), algae control (7 plants) and iron and manganese removal (3 plants).

The use of ClO_2 in U.S. municipal water supplies has leveled off in the last 15 years. Our current survey of ClO_2 usage found 84 U.S. plants using this oxidant and most of these plants were older plants. The results of this survey are summarized later in this Section.

CURRENT USES OF ClO_2

Considerable quantities of chlorine dioxide are used daily for bleaching in the pulp and paper industry. Some 500+ tons of chlorine dioxide per day are produced in the 60 U.S. pulp mills for this purpose. It is also used in large amounts in the textile industry for bleaching and dye stripping, as well as in the bleaching of flour, fats, oils and waxes[70].

In treating drinking water, chlorine dioxide is used for taste & odor control, color removal, iron and manganese oxidation, oxidation of organics, disinfection and for providing a lasting residual in distribution systems.

Chlorine dioxide currently is used in 84 U.S. water treatment plants. Of this number, 63 responded to the ClO_2 questionnaire and the results are summarized in Appendix F. The Hamilton, Ohio water plant is the only U.S. plant adding ClO_2 to its process water to serve as the final disinfectant. Other U.S. plants supplement ClO_2 dosages with surplus chlorine in order to achieve the prescribed chlorine residual in the finished plant water. Besides providing disinfection to plant water, ClO_2 is most frequently used to control taste and odor problems as well as undesirable concentrations of iron, manganese and algae. Average dosages of ClO_2 can range from 0.10 to 1.5 mg/l, depending on whether the oxidant is used for final treatment (disinfection) or for pretreatment (removal of algae, Fe, Mn, etc).

In Europe, at least 495 water treatment plants are known to be using chlorine dioxide for water treatment. ClO_2 is more commonly used in Europe as a final disinfectant or for residual for finished water than for pretreatment purposes as frequently practiced in United States. European plants report that ClO_2 provides a longer oxidant residual than chlorine and accordingly, add lesser concentrations of ClO_2 than the U.S. plants. The average dosage of ClO_2 in Europe ranges from 0.1 to 0.5 mg/l. Other than providing an oxidant residual for disinfection, ClO_2 also is used in Europe for controlling taste and odor problems and to a lesser degree for destroying organic complexes of iron and manganese early in the treatment process.

There are 10 water treatment plants in Canada, all in the Province of Ontario, that are reportedly using ClO_2 for water purification.

Before continuing the discussion of ClO_2 practices it is important for the reader to understand that plant water in the United States and Europe is not routinely analyzed solely for ClO_2 residual. Rather, plant water is analyzed by techniques normally used to determine residual chlorine. These techniques include the colorimetric OTA and DPD methods as well as amperometric titration methods. Each of these methods determines free chlorine, hypochlorite, chlorine dioxide, chlorite, chlorate and ozone. When chlorine dioxide is synthesized using excess chlorine or hypochlorite, these analytical procedures applied to measure "chlorine dioxide" also measure other oxidants present.

If care is taken to control the stoichiometry of reactants to produce chlorine dioxide, the amount of chlorine dioxide measured by the OTA, DPD or amperometric titration methods can be calculated from the total oxidant concentration actually measured.

CHARACTERISTICS OF CHLORINE DIOXIDE

Chlorine dioxide (whose color in water changes from yellow-green to orange-red as the concentration increases) is soluble in water at room temperature to about 2.9 g ClO_2/l at 30 mm Hg partial pressure, or more than 10 grams in chilled water (see Figure 48). The boiling point of liquid ClO_2 is 11°C and the melting point is minus 59°C. ClO_2 has a density of 2.4 (air = 1) and its molecular weight is 67.5. The oxidant normally is used in aqueous solution, is 5 times more soluble in water than chlorine gas, and does not react with water or ammonia as does chlorine. It should also be noted that ClO_2 is quite volatile and therefore can be removed easily from solution with minimum aeration[2].

The compound has a disagreeable odor, similar to chlorine gas, and is detectable by humans at 17 mg/l. Chlorine dioxide is distinctly irritating to the respiratory tract at a concentration of 45 mg/l in air. Concentrations of ClO_2 in air above 11% can be mildly explosive[71].

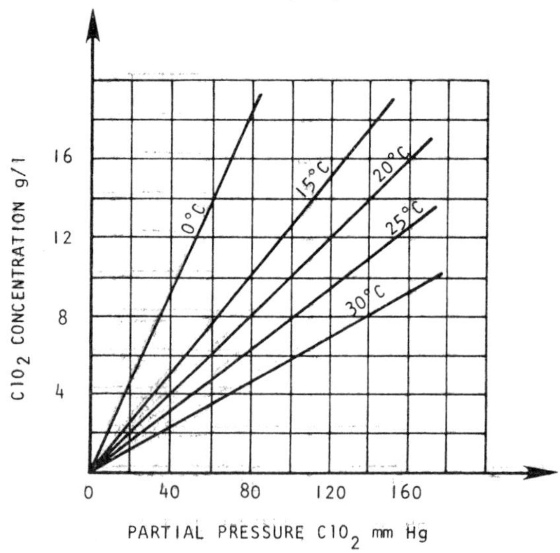

Figure 48. Solubility of ClO_2 in water.

Chlorine dioxide, as a gas or liquid, may be readily decomposed upon exposure to ultraviolet light. It is also sensitive to temperature and pressure, which are two reasons why ClO_2 is generally not shipped in bulk concentrated quantities. Industrially, ClO_2 gas is mixed with inert gases to about 10% concentration of ClO_2. This greatly reduces the hazard of explosion.

Although ClO_2 has about 1.15 times the oxidation potential of chlorine[70], this capacity to oxidize is not usually realized at the narrow pH ranges normally encountered in water or wastewater treatment practices[2]. Chlorine dioxide does not belong to that family of "available chlorine" compounds (those compounds which hydrolyze to form hypochlorous acid), nevertheless the oxidizing power of ClO_2 has been referred to as having an "available chlorine" content of 263%, calculated as follows[2]:

The amount of chlorine in chlorine dioxide is 52.6% by weight. Since the chlorine atom undergoes five valence changes in the process of oxidation to the chloride ion:

$$ClO_2 + 5e^- = Cl^- + 2O^{-2}$$

the equivalent available chlorine content is 52.6 x 5 = 263%. In effect, this indicates that chlorine dioxide <u>theoretically</u> has about 2.5 times the oxidizing power of chlorine. This can be substantiated by the reactions liberating iodine from iodide in the acid/starch-iodide analytical procedure:

Chlorine dioxide:

$$ClO_2 + 5I^- + 4H^+ \longrightarrow Cl^- + 2.5I_2 + 2H_2O$$

Chlorine:

$$HOCl + 2I^- \longrightarrow OH^- + Cl^- + I_2$$

However, the oxidizing capacity of chlorine dioxide is not all used in water treatment practice because the majority of its reactions with substances in water only cause the chlorine dioxide reduction to chlorite:

$$ClO_2 + e^- = ClO_2^-$$

The redox potential of this couple is 1.15 V. Now compare this to the action of chlorine using the hypochlorous acid to chloride couple:

$$HOCl + H^+ = 2e^- = Cl^- + H_2O \ (E_o = 1.49V)$$

It is common knowledge that the last reaction shown does occur. In waterworks practice therefore, it is safe to say that chlorine is a better oxidant than chlorine dioxide except under special conditions. One of these special conditions is the reaction of chlorine dioxide with phenols where six valence changes take place. In this instance chlorine dioxide utilizes its full oxidizing potential of 263% "available chlorine" and a little more[2].

SYNTHESIS OF CHLORINE DIOXIDE FROM SODIUM CHLORITE

Three processes for the synthesis of ClO_2 from sodium chlorite have been developed and are commercially applied to water treatment operations. The optimum efficiency of ClO_2 is a function of chemical feed rates, pH and retention time in the ClO_2 reactor. The processes, which are designed to operate with commercially available feedstock chemicals are presented as follows:

Preparation of ClO_2 From Acid and Sodium Chlorite:

$$5NaClO_2 + 4HCl \longrightarrow 4ClO_2 + 5NaCl + 2H_2O$$

(Sodium Chlorite) (Hydrochloric Acid) (Chlorine Dioxide) (Salt)

or

$$10NaClO_2 + 5H_2SO_4 \longrightarrow 8ClO_2 + 5Na_2SO_4 + 4H_2O$$

(Sulfuric Acid) (Sodium Sulfate)

This process can use either hydrochloric acid or sulfuric acid as the acid reagent for ClO_2 generation. The HCl reagent is more commonly used in water treatment operations. It should be noted that although the HCl method above requires 25% more $NaClO_2$ than the $NaClO_2/Cl_2$ method (as discussed in the next section), the $NaClO_2/HCl$ method theoretically does not result in free chlorine in the product water as does the $NaClO_2/Cl_2$ method. The latter method uses excess chlorine in the reagent mix to lower the pH for a more complete generation of ClO_2 with free chlorine carrying over into the product water. Analytical results gathered over a three year period at the Zürich, Switzerland Lengg plant[76], however, have shown that the so produced ClO_2 solutions in fact contain 4 to 7.5% free chlorine, 0 to 4.5% $NaClO_2$ and 90 to 96% ClO_2.

Extra care must be exercised when mixing $NaClO_2$ and acids. Sulfuric acid must never come into contact with solid $NaClO_2$ as the reaction is explosive. Therefore, solutions of $NaClO_2$ in water are employed exclusively in water treatment.

Preparation of ClO_2 From Gaseous Chlorine and Sodium Chlorite:

This process is a two step reaction, first beginning with the formation of an aqueous chlorine compound -- hypochlorous acid.

$$Cl_2 + H_2O \longrightarrow HOCl + HCl$$

(Chlorine Gas) (Water) (Hypochlorous Acid) (Hydrochloric Acid)

These intermediate products in turn react with sodium chlorite to form ClO_2.

$$HOCl + HCl + 2NaClO_2 \longrightarrow 2ClO_2 + 2NaCl + H_2O$$

The summation of these two equations is:

$$Cl_2 + 2NaClO_2 \longrightarrow 2ClO_2 + 2NaCl$$

and the stoichiometry of reactants is 1 mole Cl_2/2 moles $NaClO_2$, or 0.5 molar.

In practice, however, it is recommended that the chlorine to chlorite ratio be 1:1 by weight (1.28 molar). This provides an excess of chlorine which increases the reaction rate necessary to attain a high yield of ClO_2. At molar ratios of less than 1:2 of chlorine to sodium chlorite, the activation is incomplete and unreacted chlorite carries over into the product water[71].

Stoichiometrically, 1.68 lbs (0.762 kg) of 80% pure $NaClO_2$ (commercially referred to as technical grade) are mixed with 0.5 lb (0.227 kg) of chlorine to produce 1.0 pound of ClO_2. Thus water treatment plants maintain at least a 1:1 ratio of these reagents for ClO_2 production. The chlorine solution is normally at a minimum concentration of 500 mg/l. When a 1:1 weight ratio of chlorine to technical grade $NaClO_2$ is used to prepare ClO_2 and this solution is analyzed by OTA, DPD or amperometric techniques, the total oxidant measured will consist of approximately 50% ClO_2 and 50% free residual chlorine.

It is also important to note another reaction that takes place during the $NaClO_2/Cl_2$ production of ClO_2. The excess chlorine that is added causes the following reaction:

$$\underset{\text{(as excess chlorine)}}{HOCl} + ClO_2 \rightleftharpoons \underset{\text{chlorate ion}}{ClO_3^-} + HCl$$

The true amount of free available chlorine residual therefore can be reduced because of the oxidation of the ClO_2 by the free chlorine to the chlorate ion (ClO_3^-)[72].

ClO_2 Preparation From Sodium Hypochlorite and Sodium Chlorite:

Smaller localities that use hypochlorite feed systems find it advantageous to generate ClO_2 by acidifying the chlorite solutions with HCl or H_2SO_4. The stepwise reactions of hydrochloric acid and sodium hypochlorite forming ClO_2 are as follows:

$$NaOCl + HCl \longrightarrow NaCl + HOCl$$

$$HCl + HOCl + 2NaClO_2 \longrightarrow 2ClO_2 + 2NaCl + H_2O$$

Although the three reagents are stored separately, the $NaClO_2$ and NaOCl can be fed into the ClO_2 generator using one pump. The system also can be arranged to combine streams of $NaClO_2$ and acid proportionally into a single stream of hypochlorite solution.

For much larger uses of ClO_2, such as for bleaching paper pulps, generally greater than the demand of most water treatment operations, chlorine dioxide can be synthesized from sodium chlorate ($NaClO_3$). For large scale operations, this approach has been found to be more economical, since sodium chlorite actually is manufactured by reacting sodium chlorate with hydrogen peroxide in a basic solution of sodium hydroxide. The intermediate step which produces sodium chlorite from sodium chlorate is circumvented in large scale production of ClO_2.

ADVANTAGES AND DISADVANTAGES OF CHLORINE DIOXIDE

Chlorine dioxide offers many advantages for water treatment operations. One feature is that its oxidative efficiency is not impaired over a wide range of pH. A second desirous feature is that it does not react with such chlorine-demanding materials as ammonia and nitrogenous compounds. In addition, EPA has noted that when ClO_2 is synthesized with no free chlorine present as an end product and added to the water being processed, no trihalomethanes (THMs) are formed[73].

There are other advantages to using ClO_2. Chlorine dioxide destroys taste-producing phenolic compounds in the water supply. Of particular interest, this oxidant also can destroy the chlorophenol compounds which have objectionable tastes and odors and which result from chlorination.

ClO_2 also serves as an algaecide. Chlorine dioxide has been shown to be successful in controlling musty and fishy tastes that are characteristic of the pre2.sence of such algae as <u>Anabena</u>, <u>Asterionella</u>, <u>Synura</u> and <u>Vorticella</u>[2].

Manganese can cause offensive black waters when the metal is present in minute concentrations. Chlorine dioxide quickly oxidizes the manganese compound according to the overall reaction:

$$2ClO_2 + MnSO_4 + 4NaOH \longrightarrow MnO_2\downarrow + 2NaClO_2 + Na_2SO_4 + 2H_2O$$

The rate of reaction is much faster than that of chlorine. Manganese is slowly oxidized by chlorine and may require 24 hours before complete oxidation has occurred. By this time, the water may be well into the distribution system with the black MnO_2 compound precipitating inside the pipes. Because of its high cost, ClO_2 use is generally limited to water operations containing manganese in concentrations of 1 mg/l or less.

Chlorine dioxide also can be used to remove iron from the water supply, although the reaction rate is not as rapid as it is for manganese. Both metal compounds are oxidized more efficiently by ClO_2 at pH 7 or above. The overall reaction of iron with ClO_2 is as follows:

$$ClO_2 + Fe^{++} + NaOH + H_2O \longrightarrow Fe(OH)_3 + NaClO_2$$

The ClO_2 is not consumed by other chlorine-demanding materials and oxidizes soluble ferrous iron to insoluble ferric hydroxide. This heavy, brown floc typically is removed by sedimentation and filtration.

Chlorine dioxide is as effective as chlorine as a disinfectant and, as previously mentioned, is less sensitive to changes in pH. At a typical operating range of pH 6 to 10, chlorine partially loses its effectiveness for disinfecting; ClO_2 does not. Also, because ClO_2 does not react with ammonia and other nitrogenous compounds, it can achieve a higher residual and bacteriological kill than chlorine[6].

Chlorine dioxide can be generated at any water plant that uses chlorine or hypochlorite. The oxidant is produced either seasonally or year-round depending on the need. Most plants in the United States operate the ClO_2 generation system on a manual basis, whereas most European countries have automated systems. The manual system is applicable to plants of constant flow and water quality. For variable flows or variable water quality, the manual system can be adjusted to proportion the chemical reagent feed rates to meet the ClO_2 demand.

One concern of using chlorine dioxide in a water treatment process is the excess chlorine that typically carries over into the product water. Excess chlorine in the product water is common to ClO_2 production in U.S. water plants and in many European plants in Germany, France and England. Excess free chlorine that is discharged from the ClO_2 reactor vessel may react with organics and form trihalomethanes - several of which are suspected carcinogens. These products are not formed when ClO_2 is produced in the absence of free chlorine residual.

Chlorine dioxide does not react chemically with water as does chlorine, and therefore is expelled from solution by aeration. The strength of ClO_2 deteriorates under these conditions and the remaining solution will contain neglible amounts of chloric acid ($HClO_3$)[2]. Aqueous solutions of ClO_2 are also sensitive to photodecomposition, the rate of degeneration being a function of both exposure time and intensity of ultraviolet light. The end products from such a reaction are chloric and hydrochloric acids[2].

There are current uncertainties regarding the health effects of chlorine dioxide in drinking water -- particularly the effects of the chlorite ion (ClO_2^-). When ClO_2 oxidizes organics (or any other reducing agent), the following reaction can take place:

$$\text{organic} + ClO_2 \longrightarrow \text{oxidized organic} + ClO_2^-$$

The ClO_2 reverts to the chlorite ion which then can remain in solution. Incomplete chemical reactions within the ClO_2 reaction vessel also can produce chlorite ions from the $NaClO_2$ reagent. Although there are reportedly no epidemiological or toxicological studies that identify adverse health risks due to the chlorite ion, the EPA, HEW and many European organizations currently are examining, in closer detail, the potential health effects of the chlorite ion. More definitive data are expected in 1978.

The chemical reagent $NaClO_2$ is safe to use for water treatment operations, provided that the operator understands proper handling and storage of the chemical. Rubber gloves, apron and safety goggles are recommended when the chemical is handled. $NaClO_2$ is readily combustible in the presence of organic material (i.e., straw, leather, rags, etc.) and therefore should never be allowed to splash on clothing or remain or evaporate on the floor. Sodium chlorite also can explode if reacted directly with an acid because of the rapid evolution of ClO_2 gas. Stored $NaClO_2$ should be protected from heat. When the temperature of the chemical reaches 347°F, $NaClO_2$ decomposes rapidly with sufficient heat liberated to make the decomposition self-sustaining. An explosion will result if the chemical is contained in closed drums or tanks and subjected to such temperatures.

U.S.A - ClO_2 QUESTIONNAIRE AND PLANT VISITATION SUMMARY

General

A total of 105 questionnaires was forwarded to U.S. water treatment plants which were reported by state health departments to operate ClO_2 systems for water treatment operations. Based on the questionnaires returned, it was learned that 63 plants have ClO_2 systems in use. A follow-up telephone survey to those plants not responding verified that 21 more water plants operate ClO_2 systems. Thus, a total of 84 U.S. water treatment plants use ClO_2 for water treatment processes. The following is a brief summary of those 63 plants that responded to the questionnaires. Pertinent data are presented in Table 26.

Distribution of Plants Using ClO_2 Based on Plant Flows

Almost 48% of those respondents (30 plants) confirming the use at ClO_2 on their questionnaires, have plant flows of less than 5 mgd. Most of these 30 water plants are located in EPA Regions 4 and 5. There are 18 plants (29%) using ClO_2 that have plant flows between 5 and 10 mgd. Region 5 has 8 of these plants. Of the remaining plants, there are 6 water plants using ClO_2 having plant flows between 10 and 20 mgd and 9 plants with flows greater than 20 mgd. Only 4 of the 63 U.S. plants are privately owned and operated.

Uses of ClO_2

The water treatment plants responding to the questionnaires indicated that the major reason for using ClO_2 is to control taste and odor problems in the water supplies. Other than controlling T/O problems, ClO_2 is used frequently to reduce levels of iron, and manganese, to disinfect and to provide a chlorine residual in the plant water (this latter objective is accomplished by adding excess chlorine to the ClO_2 reactor to insure that free chlorine passes through the reactor and into the plant water). Most plants list multiple uses for adding ClO_2 at their plants.

TABLE 26. ClO_2 QUESTIONNAIRE RESULTS BY EPA REGIONS

EPA Regions	1	2	3	4	5	6	7	Total
No. Plants Surveyed	3	10	21	30	35	2	4	105
No. Plants Responding	1	3	9	21	26	2	1	63
Public	1	2	8	20	25	2	1	59
Private		1	1	1	1			4
Process:								
Acid-$NaClO_2$					1			1 (b)
Cl_2-$NaClO_2$		2	7	21	23	1	1	55 (b)
HCl-Hypochlorite/$NaClO_2$		1	1		1			3 (b)
Plant Size (mgd)								
<5 mgd		1	2	14	11	1	1	30
5-10 mgd	1		4	4	8	1		18
10-20 mgd			1	2	3			6
>20 mgd		2	2	1	4			9
Purpose for ClO_2 addition (c)								
Taste and Odor	1	3	7	1	24		1	37
Organics		2	4	3	2		1	12
Disinfection			3	7	11		1	21
Fe & Mn				17	1			18
Color				3	2			5
Cl_2 residual (d)		1	2	7	10	1	1	12

(continued)

TABLE 26. (continued)

EPA Regions	1	2	3	4	5	6	7	Total
ClO_2 Dosage								
Range (mg/l)		0.1-0.2	0.1-0.4	0.25-7.0	0.1-2.0	0.6		(b)
Average (mg/l)		0.15	0.17	1.23	0.63	0.6		(b)
Measured Residuals (e)								
Disinfection								
Range (mg/l)		0.5-1.0	0.5-3.0	1.0-2.0	0.6-2.6	0.3		(b)
Average (mg/l)		0.75	0.9	1.5	1.36	0.3		(b)
Taste & Ordor								
Range (mg/l)		1.0	0.1-0.9	0.3-10	0.1-2.0		0.12	(b)
Average (mg/l)		1.0	0.2	3.0	0.62		0.12	(b)
Number of Plants Using								
Constant Dosage			5	9	9		1	(b)
Intermittant Dosage		1	3		6			(b)

(a) 105 U.S. water treatment plants were surveyed but, based on questionnaire results and follow-up telephone surveys, the best estimate is that 84 of these plants use ClO_2. The above data summarize only 63 plants which responded to the questionnaire.
(b) For the line items, not all plants provided all the information requested. Information presented is for those plants responding.
(c) Many plants list multiple purposes for ClO_2 addition.
(d) Excess Cl_2 added to ClO_2 generation process to provide chlorine carry over to plant water for achieving chlorine residual.
(e) U.S. water treatment plants generally do not measure specifically for ClO_2 residual but rather for the summation of oxidants that are anlayzed collectively by the test used: DPD, amperometric, etc.

ClO_2 Dosages

The dosage of ClO_2 in U.S. water treatment operations depends on whether the oxidant is used to pretreat the influent water, treat the plant water, or to treat the effluent water. Higher dosages of ClO_2 generally are used to pretreat raw water supplies for controlling algae, iron and manganese. Taste and odor problems also can require higher dosages of ClO_2. Adding ClO_2 as a disinfectant requires much lower concentrations. For U.S. water plants, the approximate range of ClO_2 added is 0.1 to 1.5 mg/l. Many plants seasonally add higher dosages of ClO_2, depending on the raw water quality.

Generation of ClO_2

The most common method of generating ClO_2 at U.S. water treatment plants is by the chlorine/sodium chlorite reaction. Of the 63 plants responding, 59 use this process. Many plants using the chlorine/sodium chlorite technique add chlorine in excess of the recommended 1:1 weight ratio of chlorine to sodium chlorite (technical grade). U.S. manufacturers recommend this 1:1 ratio to insure a maximum yield of chlorine dioxide. Chlorine is fed to the ClO_2 reactor using single pass chlorination. There are 3 water treatment plants using the acid/hypochlorite/sodium chlorite technique for ClO_2 production. One plant (Ann Arbor, Michigan) uses the hydrochloric acid/sodium chlorite method for generating ClO_2. A second plant in Contra Costa County, California is using the sulfuric acid/sodium chlorite method of generation.

Most of the responding U.S. plants have been in service for at least 15 years. As a result many of the older ClO_2 systems do not reflect the more recent improvements in ClO_2 generation.

Methods of Analysis and Monitoring

Only a limited amount of information was provided by the plant respondents on this subject. From the data available, however, it is concluded that U.S. water treatment plants generally do not analyze plant waters specifically for ClO_2 residual. Rather, the plant water is analyzed for total oxidant or "chlorine residual", as this is the standard practice of U.S. plants. A portion of the "chlorine residual" measured, however, can be attributed to ClO_2 residual by knowing the stoichiometry of reactants employed and the chemical reactions involved. Therefore, although several U.S. water treatment plants reported that "ClO_2" is measured at the plant, their incomplete explanations of the test procedures used indicates that ClO_2 is measured conjunctively with chlorine, and together these oxidant residuals are registered as the "chlorine residual" or "chlorine dioxide" reported.

The common methods used to measure ClO_2 and/or chlorine in the plant water are DPD, amperometric titration, spectrophotometry and OTA (which has been deleted from the 14th edition of Standard Methods). Of these, only the spectrophotometric procedure can be conducted so as to determine ClO_2 in the presence of free residual chlorine.

Other

Responding U.S. plants often noted the high cost of sodium chlorite as a disadvantage of using ClO_2 for water treatment. A few plants reported their concern for the potential explosiveness of sodium chlorite if not properly handled and stored.

Most plants operating a ClO_2 generating system use polyvinyl chloride piping and appurtenances.

Limited operational problems are reported with chemical feed pumps, particularly with the suction tubing from the pump to the $NaClO_2$ solution tank. Thin walled rubber tubing becomes limp and weak after several months of service. Pump suction causes the walls to collapse and the chemical feed stops. Heavy walled tubing or PVC pipe does not prevent this problem.

Several plants expressed concern about the public health risks regarding the addition of chlorite to potable water. No plants reported any knowledge of the organic oxidation products formed as a result of ClO_2 treatment.

SUMMARY OF U.S. PLANT VISITS - OVERVIEW

During September and October 1977, 13 water treatment plants in the United States were visited. The purpose of these visits was to learn first hand information about the engineering and operational aspects of ClO_2 in water treatment operations. Of the thirteen plants visited, five were located in Georgia, five in Ohio, and one each in West Virginia, Kentucky and Michigan. These plants offered a broad picture of the use of ClO_2 in United States water treatment works. The reasons for choosing these 13 plants for visitation include the following:

- purpose of ClO_2 application
- type of ClO_2 generation systems
- type of water treatment process
- source of raw water supply
- size of plant
- type of oxidant analysis

Most of the plants visited use ClO_2 for the control of taste and odor problems. The sources of these T/O contaminants are attributed to phenols, algae and organics in the raw water supplies. Most of these taste and odor bodies are natural in origin. Those plants that are supplied water (directly or indirectly) from surface reservoirs typically use ClO_2 to reduce the concentration of manganese and iron in the raw water. The occurrence of these metals is generally seasonal. Of the 84 water plants that are reportedly using ClO_2, Hamilton, Ohio, and Fayetteville, Georgia are the only plants using the chemical solely for disinfection. Fayetteville adds ClO_2 with excess chlorine to the raw water. Hamilton, however produces ClO_2 stoichiometrically and adds it as a final disinfectant. No other oxidant is added at this plant, whereas other plants supplement ClO_2 addition with a prechlorination and/or post-chlorination step.

Seven of the thirteen plants visited use ClO_2 for the removal of manganese and iron. The point of application is typically ahead of the filters in order that the insoluble compounds can be filtered from the plant water. The other six plants that are troubled with phenolic compounds or other taste producing agents add the ClO_2 either before or after the filtration process. The Hamilton, Ohio plant adds ClO_2 in the clearwell for disinfection.

The method of ClO_2 generation is predominantly the sodium chlorite/chlorine system. Olin Corporation manufactures the $NaClO_2$ supplied to the water plants although the chemical generally is delivered to the plant by a local retailer. Several plants purchase liquid sodium chlorite either in 150 to 200 pound drums (37% concentration) or by tank truck (approximately 28 to 30% concentration). The other form of $NaClO_2$ is the anhydrous technical grade which is 80% pure $NaClO_2$. The larger plants such as Columbus, Ohio (35 mgd) and Atlanta, Georgia (70 mgd) prefer the liquid $NaClO_2$ because there is no need to bother with storing the drums, mixing the dry chemical or working with a potentially explosive chemical. (Powdered $NaClO_2$ has been known to "spark" when walked upon with shoes.) Production of ClO_2 from liquid $NaClO_2$ is easily controlled because the exact concentration of the liquid $NaClO_2$ feed can be controlled. With batch mixes, the operator must add the correct proportions of $NaClO_2$ to water in order to prepare the desired concentration of $NaClO_2$ feed. Smaller plants, such as Bethesda, Ohio (0.079 mgd) and Covington, Kentucky (10 mgd), have elected to use anhydrous $NaClO_2$ or liquid $NaClO_2$ which are both delivered in drums. These plants do not have the facilities or money to operate large scale liquid feed systems as discussed above. Water treatment plants also use ClO_2 on a seasonal basis and thereby can easily accommodate the storage of several drums of $NaClO_2$ in the plant in anticipation of using the chemical for a few months each year.

Only one of the thirteen plants uses acid as a reagent for ClO_2 production -- Ann Arbor, Michigan. The plant formerly used chlorine with $NaClO_2$ to generate ClO_2 but the plant manager could not satisfactorily control the pH inside the ClO_2 reactor vessel. Chlorine costs were escalating and at the time and the chlorinators were requiring frequent service. The plant therefore switched to hydrochloric acid which is delivered to the plant in 150 pound drums. The ClO_2 production reportedly works much more efficiently.

The ClO_2 reactor vessels at the plants visited were made by one of two manufacturers-- Wallace & Tiernan or Fischer & Porter. The only apparent difference between the two brands of equipment is that the Fischer & Porter unit has an effluent sampling port on the reactor vessel. Both reactors are upflow units. Those plants that have the Fischer & Porter type of unit do not sample routinely and analyze the chemistry of the ClO_2 product in the reaction chamber. The Ann Arbor plant has installed a sampling port on its ClO_2 generator and reportedly analyzes the sample periodically. Hamilton, Ohio and Toledo, Ohio fabricated their plant ClO_2 reactors. The implicit reasoning for doing this is that the design and construction of the reactor is simple enough for most plant personnel to master. The reactors reportedly are performing satisfactorily and the construction appeared acceptable. Most of the ClO_2 reactors observed are transparent for purposes visual inspection of the ClO_2 color produced. A few reactors were opaque (Hamilton, Ohio and Columbus, Ohio) but viewing tubes were installed on the discharge piping for visual inspection of the ClO_2 product.

Chemical feed pumps for $NaClO_2$ typically are of the diaphragm type. Manufacturers of this equipment include Wallace & Tiernan, Fischer & Porter and BIF. Only one plant reported problems with the diaphragm pump. For some undetermined reason, the plastic diaphragm cracks and causes the feed system to malfunction. The pumps can be metered to provide variable flow rates. Less than half of the plants visited have back up $NaClO_2$ feed pumps.

The piping for the ClO_2 system generally is consistent from plant to plant. Most piping from the $NaClO_2$ pumps and from the gas chlorinators to the ClO_2 reactor is polyvinyl chloride--presumably schedule 80, solvent welded. The effluent piping from the ClO_2 reactor is PVC (schedule 80) or a heavy walled rubber tubing approximately 1.5 to 2 inches in diameter. This latter material reportedly can be furnished by the manufacturer of the ClO_2 reactor vessel. The piping from the chlorine cylinders to the gas chlorinators is welded black steel. There is some variance from plant to plant as to the tubing material that is for the suction line of the $NaClO_2$ feed pumps. Some plants use Tygon tubing and others use PVC (schedule 40 or 80). Tygon tubing has a tendency to lose its rigidity after several months of use and is subsequently replaced. If the tubing walls are weakened too much, the pump suction partially collapses the tubing and thereby restricts the flow of $NaClO_2$ liquid.

Most of the ClO_2 systems lack chlorine gas leak detectors, although such units are found in the chlorine storage rooms of many of the larger plants. About half of the plants have floor drains around the $NaClO_2$ feed system. These drains generally empty into either a municipal sewer or into a gravel trap. At least half of the plants visited have an air exhaust system in the chlorine storage room. The switch to activate this system typically is next to the entrance door (some inside and some outside the entrance door). Each plant has a glass panelled entrance door for visual inspection of the chlorine room interior. Nearly every plant has one or more gas masks for emergencies in the chlorine room.

During each plant visit, the plant operator was asked about the operational controls on the ClO_2 system. For example, when was the last time the dials and gauges were calibrated on the chlorinators, what is the proper operating water pressure for the injection of chlorine gas into the water feed system, what is the correct flow of water into the ClO_2 reactor vessel, etc. Most operators did not know this information and had no records of what the answers might be. It appears that most of the plants are content to operate the various pressure and vacuum settings of the ClO_2 system that have been used at the plant in the past. Several of the plants are operating the ClO_2 reactor vessels at too high a water pressure. This can be detected by the initial formation of the red-brown color of ClO_2 at the very top edge of the reaction vessel. One plant is producing ClO_2 in a plug flow manner as evidenced by the cyclic formation of the red-brown color in the reactor. In this case, much of the chlorine solution passes through the vessel without reacting with chlorite ions. A few of the plants are producing very pale lime-yellow ClO_2 solutions, whereas others are producing rich red-brown ClO_2 solutions. The intensity of the color that is formed indicates the different concentrations of chlorine dioxide produced, and is dependent upon proper mixing action inside the ClO_2 reactor vessel. Equally important are the feed ratios of the chemical reagents.

Manufacturers of ClO_2 generation systems recommend a 1:1 feed ratio by weight of chlorine and sodium chlorite. Most plants that were visited achieved this ratio and even higher. Columbus, Ohio and Hamilton, Ohio operate the ClO_2 generation system using a 1:1 ratio of Cl_2 and $NaClO_2$. Other plants such as Bethesda, Ohio and Toledo, Ohio operate the $Cl_2/NaClO_2$ feed systems at 1:5.6 and 6.5:1 weight ratios, respectively. The reasons for this wide range of ratios differ from plant to plant. Several plants add excess chlorine to provide the necessary chlorine residual in the finished water. Other plants are surcharging the ClO_2 reactor with chlorine in order to lower the pH in the reactor for a better rate of ClO_2 production. There is evidence that a few other plants may not have a complete understanding of how to feed the chemical reagents properly. The $NaClO_2$ concentrations in the process water range from 0.2 to 1.0 mg/l for most plants visited. The concentration of $NaClO_2$ in Fayetteville, Georgia water is 4 mg/l $NaClO_2$ where the ClO_2 is used to control manganese problems. Bethesda, Ohio is estimated to have a concentration of 17.4 mg/l $NaClO_2$ in the plant water added for the abatement of problems due to manganese and algae. This is an unusually high concentration to be added in any case, and it is of compounded concern when knowing that is is being added to the finished water and not the raw water.

In addition to being fed into the ClO_2 reactors, chlorine also is added to the plant water in other treatment processes. It is interesting to note that free chlorine residuals in the finished water of the plants visited reportedly range from approximately 0.2 to 1.5 mg/l. The free residuals in the distribution systems are about 0.1 to 0.5 mg/l. One plant is reportedly chlorinating the finished water far above what would appear to be an adequate dosage: 2.5 to 4.0 mg/l free chlorine residual with an upper limit of 6 mg/l.

Only one plant specifically monitors the ClO_2 residual in the finished water -- Hamilton, Ohio. The ClO_2 is measured spectrophotometrically to levels of less than 0.2 mg/l as ClO_2. The other plants measure chlorine residual (free and/or combined) to assure that an oxidant level is achieved in the treated water. The OTA and DPD analytical methods are the common tests to measure chlorine residual. One plant uses the H-acid technique. These methods are further discussed in Standard Methods[75] and the Handbook of Chlorination[2]. At a few plants it is difficult to determine how well the ClO_2 is performing because of the abundance of other chemicals that are added to the plant water; i.e, alum, lime, soda ash, polyphosphates, powdered activated carbon, potassium permanganate and chlorine.

EUROPEAN ClO_2 PLANT VISITATIONS AND QUESTIONNAIRE RESULTS

In Appendix F are listed the pertinent features of 74 European drinking water treatment plants which responded to the questionnaire and which employ chlorine dioxide in one or more places in their treatment processes. It was not possible to develop an accurate count as to the total number of European water treatment plants employing chlorine dioxide for several reasons. First the chlorine dioxide market is served by a great many more companies smaller than those who serve the ozone generation market. Secondly, there is no association representing the interests of chlorine dioxide suppliers, similar to the International Ozone Institute, through which we could work to develop such data.

Nevertheless, we have been able to determine that there are at least 495 European water treatment plants employing chlorine dioxide. This number was derived from discussions with suppliers of chlorine dioxide generation equipment who were exhibiting at the Third International Symposium on Ozone Technology, Paris, France, May, 1977, and at Wasser Berlin, Berlin, Germany, May 1977,

Degrémont, Reuil Malmaison, France	100
CIFEC, Paris, France	35
Wallace & Tiernan, Germany	60
CFG (Chemie und Filter GmbH), Heidelberg, Germany	250
ALLDOS	20
Fischer & Porter, USA (all in Italy)	30
Total	495

In Canada, only 10 water treatment plants are known to be using chlorine dioxide, mainly for taste and odor control, and these are all located in the Province of Ontario, in the vicinity of Toronto. Seven plants are known in Africa, 3 in Asia, 3 in South America, one in Iceland, one in New Zealand and 3 in the Indonesian area. Therefore, at least 523 plants throughout the world are known to be using chlorine dioxide for some purpose.

During May, 1977, the site visitation team visited 15 European water treatment plants which employ chlorine dioxide. Four of these plants are in France, one in Belgium, eight are in Germany and two in Switzerland. Only two of the chlorine dioxide plants visited did not also employ ozonation, either before or after use of chlorine dioxide.

As a general statement, Europeans use chlorine dioxide when chlorine cannot be used, mainly because of higher costs. In Europe, chlorine dioxide is said to cost 3 to 3.5 times more than does chlorine. The two primary purposes for which chlorine dioxide is utilized are for pretreatment and post-treatment. In pretreatment, chlorine dioxide is utilized at the Paris suburb plants of Choisy-le-Roi (on the Seine), Neuilly-sur-Marne (on the Marne), and at Annet-sur-Marne (upstream of Paris on the Marne) for breaking up organically bound manganese and iron and for predisinfection of water before filtration. At Toulouse, in southwestern France on the Garonne River, chlorine dioxide is used in pretreatment for predisinfection, taste and odor control, and for color removal.

At the Brussels, Belgium Tailfer plant, both chlorine and chlorine dioxide are used in pretreating Meuse River water. Chlorine performs the predisinfection function, while chlorine dioxide decomposes the organic complexes of iron and manganese and also predisinfects.

Dosages of chlorine dioxide employed for pretreatment generally range from 1 to 1.5 mg/l in Europe. These and other dosage or residual levels must be considered in terms of the method of preparing ClO_2 and the method of analysis employed, as discussed earlier in this section.

It is generally agreed by Europeans that chlorine dioxide, having a stronger oxidation potential than chlorine, is capable of oxidizing organic complexes of iron and manganese, whereas chlorine is not. Once converted to their ionic states, both iron and manganese will be oxidized to stages at which they easily hydrolyze and can be removed by settling and filtration.

In post-treatment (for residual), chlorine dioxide was being used only at Choisy-le-Roi and Annet-sur-Marne of the plants visited in France and Belgium.

In Germany and Switzerland, chlorine dioxide was being used only for post-treatment at all 10 water treatment plants visited in these countries. Germans and Swiss use a maximum of 0.3 mg/l of chlorine <u>or</u> chlorine dioxide (or a combination of the two) to provide residual for distribution systems. This dosage level provides trace amounts of chlorine or chlorine dioxide in the distribution systems. The tacit rule in these countries is that if the finished water leaving the plant requires greater than 0.3 mg/l chlorine or chlorine dioxide to provide stable residuals in the distribution system, then the water treatment process must be modified to reduce oxidant demand until this level of dosage is attained.

There are four methods of synthesizing chlorine dioxide on-site at European drinking water treatment plants that the survey team observed:

- Technique #1. Addition of chlorine gas to water, followed by addition of excess chlorine water solution to aqueous solutions of sodium chlorite.

- Technique #2. Addition of chlorine gas to water under pressure, then addition of this solution under pressure to aqueous solutions of sodium chlorite under pressure.

- Technique #3. Addition of chlorine gas to water and recirculation of this aqueous solution in a closed loop with continued addition of chlorine until a pH below 2.7 has been attained. Then addition of this pH 2.7 chlorine solution to aqueous solutions of sodium chlorite.

- Technique #4. Addition of dilute HCl solution to aqueous solutions of sodium chlorite.

Technique #1 is the most common and widespread method in Europe, and is the basis for the older Wallace & Tiernan generators and for those units marketed by Chlorator (Karlsruhe, Germany). In technique #1, excess chlorine deliberately is added to sodium chlorite solution so as to maximize conversion of chlorite ion to chlorine dioxide. However, this results in free residual chlorine also being present.

Technique #2 is the newer Wallace & Tiernan procedure, which is in use at the Brussels, Belgium Tailfer plant and is to be installed at the Berlin, Germany Jungfernheide plant.

Technique #3 is the basis for the patented CIFEC system, introduced during the past 2 to 3 years and which has been installed in some 35 European water treatment plants. The site survey team observed CIFEC equipment in operation in France at Choisy-le-Roi, Annet-sur-Marne and at the Toulouse Clairfont plant.

Techniques #2 and #3 are claimed by their manufacturers to provide 98 to 100% conversion of chlorite ion to chlorine dioxide with no excess free chlorine being present.

Technique #4 uses mineral acid to convert chlorite ion to chlorine dioxide. Since chlorine gas is not used in this process, there should be no "excess" chlorine present with the desired chlorine dioxide. However, Valenta & Gahler[76] have shown over a period of three years at the Zürich Lengg plant, that there is an average of 4 to 7.5% free chlorine present in the so-produced ClO_2.

Details of these methods of generating chlorine dioxide are described later in this Section under "Engineering Aspects of Chlorine Dioxide Generation Systems".

European plants analyze for ClO_2 colorimetrically (OTA, DPD, H-acid) or amperometrically. As noted under the discussion of USA plants, these procedures actually determine the "total oxidant" present, ClO_2, chlorine, OCl^-, ClO^-, ClO_2^- and ozone. However, when controlled excesses of chlorine are added to sodium chlorite solutions the amounts of free chlorine dioxide present are known from the stoichiometry used.

Pertinent Features of Plant Visitations

France--

At Choisy-le-Roi, on the Seine River in the Paris suburbs, chlorine dioxide is used in pretreatment, as necessary, for oxidation of organics (the plant personnel suspect the presence of geosmin) and for oxidizing manganese. Raw and settled waters (before and after pretreatment with chlorine dioxide) are analyzed daily for manganese and the dosage of chlorine dioxide is adjusted accordingly. Chlorine dioxide for pretreatment is dosed at about 1.5 mg/l.

Later in the process and before ozonation, any residual chlorine dioxide is destroyed by treatment with bisulfite. This step is necessary so that extraneous concentrations of chlorine dioxide oxidant will not be measured during ozone monitoring. The Choisy ozone disinfection step is controlled by monitoring residual ozone.

There are currently 10 chlorine dioxide generators used at Choisy for pretreatment at the 10 chemical mixing sites. All use the older generation process involving addition of excess chlorine in water to sodium chlorite solution. Equipment for chlorine dioxide generation for pretreatment is manufactured by Trailigaz of Garges-lès-Gonesse. The plant is being modified to combine these 10 chlorine dioxide addition points into a single operation. At that time, a single CIFEC generator (see Engineering Design subsection) will replace the 10 existing ClO_2 generators.

For post-treatment with chlorine dioxide, a CIFEC generator was installed in mid-1976. Before this, post-treatment after ozonation had been provided with chlorine. However, it has been found that residuals of chlorine dioxide last longer in the Choisy-le-Roi distribution system than do residuals of free chlorine.

Post-treatment dosage of chlorine dioxide at Choisy-le-Roi is 0.5 mg/l for 6 months of the year, with lower dosages being applied in winter. Plant personnel are concerned with this "high" level of chlorine dioxide being dosed, because of the probable re-formation of chlorite ion. As a result, there is a pilot program being conducted to "super-ozonize" the water so as to lower the amount of chlorine dioxide to be added. Super-ozonation involves dosing with as much as 10 mg/l of ozone, to obtain a residual ozone concentration of 0.6 to 0.7 mg/l after 10 minutes of contact time.

Controls for addition of chlorine dioxide in pretreatment at Choisy are manual, but are automatic in the CIFEC system used for post-treatment.

At Neuilly-sur-Marne, on the Marne River in the Paris suburbs, chlorine dioxide is used in pretreatment only for oxidation of organics and predisinfection. Dosage is 0.5 mg/l.

At Annet-sur-Marne, on the Marne River upstream of Paris, chlorine dioxide is used both for pre- and post-treatment. In pretreatment, 1 mg/l dosage is used for predisinfection. For post-treatment, 0.5 mg/l of chlorine dioxide is added after ozonation to provide a residual of at least 0.05 mg/l at the extremities of the distribution system. Pretreatment with chlorine dioxide is practiced the year around, but post-treatment with chlorine dioxide is practiced only in summer. During winter months, chlorine alone is adequate and is less costly.

Generation of chlorine dioxide at Annet originally was conducted with Trailigaz equipment (excess chlorine water added to sodium chlorite), but since late 1976 the material has been generated in a CIFEC apparatus, which has given satisfactory performance.

The Annet-sur-Marne plant is being constructed in stages, with the first 25,000 cu m/day capacity train having been placed in operation in 1973. Chlorine dioxide was installed for post-treatment as a security measure, since it already was present for predisinfection. However, because of the high quality of treated water, Annet is planning to install chlorine as the only post-treatment step in the second 25,000 cu m/day treatment stage. Chlorine dioxide will be used as a predisinfectant in the second train, and will be available for post-treatment if necessary.

At the Clairfont plant in Toulouse (southwestern France on the Garonne River), chlorine dioxide is used in 1.5 mg/l dosages for pretreatment only (predisinfection, tastes, odors and color). Chlorine dioxide is prepared in a CIFEC apparatus, and its performance has been satisfactory since it was installed in 1975.

Although not installed today at Aubergenville and at Rouen-la-Chapelle, chlorine dioxide was used at these plants before ozonation was incorporated. At Aubergenville, chlorine dioxide was used when phenols were present in the raw water. Today the phenol content of the plant raw water is much lower than it was, and ozonation can handle any surges.

Before the new ozone/activated carbon treatment system was installed at Rouen-la-Chapelle, the treatment process from 1964 to 1975 involved drawing deep well water and treating with chlorine dioxide. Since the newly installed ozonation process removes phenols, there is no longer a need to use chlorine dioxide. The cheaper chlorine is used at Aubergenville and at Rouen to provide residuals for the distribution systems.

Belgium--

At the Brussels, Belgium Tailfer plant, chlorine dioxide is used in pretreatment for decomposing organic complexes of iron and manganese, for color removal and for bacterial disinfection. It is added early in the treatment process along with chlorine, sulfuric acid, alum and silica. Production of chlorine dioxide is monitored in the analytical laboratory by analyzing for the amounts of reactants added (chlorine, water and sodium chlorite solution). Residual chlorine dioxide is monitored two minutes after addition.

The Tailfer chlorine dioxide reactor is a glass column 5 feet high and 1 foot in diameter, packed with Raschig rings. Chlorine addition to water is conducted under pressure with a pulsator, and both water and chlorine flows are monitored. Plant personnel consider it necessary to achieve a chlorine concentration in water of 2 g/l in order to attain 98 to 99% conversions of chlorite ion to chlorine dioxide and to use stoichiometric amounts of both reactants (chlorine and chlorite). If this concentration of 2 g/l chlorine in water is not attained, then excess chlorine is required to convert all chlorite ion to chlorine dioxide.

Tailfer plant personnel also advise that chlorine dioxide is much less stable in the presence of excess chlorine than when the stoichiometric amount of chlorine is employed.

Reaction time in the Tailfer chlorine dioxide reactor is about 1 minute. After passage through the reactor (both reactants under pressure), the product stream is diluted with water to the concentration at which addition is made to the process water stream. Dilution of reactants is not practiced before the reactor, since the necessary 2 g/l concentration of chlorine in water then would not be obtained.

Safety features of the Tailfer chlorine dioxide reactor include monitors for the chlorine flow, water flow and sodium chlorite flow. If any of these flows drop below a preset rate, the chlorine pump will shut off.

Germany--

German water treatment plants are not usually allowed to generate chlorine dioxide from acid and sodium chlorite (SONTHEIMER, 1977, private communication). This is because the pH must be controlled accurately, and many of the smaller German water treatment plants do not have this control capability. On the other hand, by adding excess chlorine water to sodium chlorite, complete conversion of chlorite ion to chlorine dioxide is assured, and some free residual chlorine is provided as well. Finally, since German water treatment techniques normally require that a maximum of only 0.3 mg/l of chlorine, chlorine dioxide or a mixture of the two can be dosed to product water to provide residuals, the combination of chlorine dioxide containing excess chlorine has been selected as the "standard" for German water treatment plants using ClO_2.

None of the German water treatment plants visited use chlorine dioxide for pretreatment -- all of these plants use chlorine dioxide as a post-treatment to provide residual for their distribution systems.

At all three City of Düsseldorf plants (Holthausen, Flehe and Am Staad), 0.1 mg/l of chlorine dioxide (containing excess chlorine) is added to the plant outlet water for residual. At the Holthausen plant, chlorine dioxide is generated by means of an Alldos International system. In this generator, 2,000 g of chlorine are mixed with 2,000 liters of water per hour, and this solution is passed through the Raschig ring packed reactor with a solution containing 2,000 g of sodium chlorite per hour to produce chlorine dioxide solution. There are two reactors at this plant, each 3 feet high by 4 inches in diameter. Since the plants in Düsseldorf operate on demand (there are only two small reservoirs to handle product water from all three Düsseldorf plants), chlorine dioxide production operates intermittently. The Alldos unit at Holthausen is automated, and safety features are programmed with the plant water flow. If this drops, the unit is programmed to shut down.

At the Flehe plant, a Chlorator chlorine dioxide generator system is used. In this unit 3 parts by weight of chlorine are added to 1 part of sodium chlorite. Thus the 0.1 mg/l dosed to the plant outlet actually contains 0.07 mg/l of chlorine plus 0.03 mg/l of chlorine dioxide.

During winter, Flehe uses chlorine for residual, because it is cheaper. Flehe also uses chlorine whenever the granular activated carbon is changed.

Use of chlorine dioxide at the Düsseldorf Am Staad plant is similar to its use at Flehe.

At the Duisburg plant (Rhine River near Düsseldorf), chlorine dioxide is generated in a Chlorator system. Dosages of 0.20 to 0.24 mg/l are added to provide residuals in the distribution system. The stoichiometry at Duisburg is 300 g of chlorine in water per hour added to solution containing 150 g of sodium chlorite per hour (a 400 molar % excess of chlorine). After addition of chlorine dioxide plus chlorine, the total oxidant concentration is monitored 15 minutes later, to be sure to attain a residual concentration of 0.1 mg/l before the water enters the distribution system for its 20 km travel to the city.

The Duisburg plant has two chlorine dioxide generators on line, one of which is for standby, but which will be used in the coming plant expansion. At that time, a third chlorine/water system will be installed, but not a third chlorine dioxide generator.

At the Wuppertal plant (Rhine River near Düsseldorf), 0.1 to 0.15 mg/l of chlorine dioxide plus residual chlorine is added at the reservoir just before product water enters the distribution system. A Chlorator system is used at Wuppertal. The distribution system is 1100 km long, has a 12 hour average residence time, and consists of two main pipes. The older pipe is more than 100 years old. Residual chlorine dioxide is less stable in the older distribution line than in the newer one.

At the city of Konstanz (southern Germany on the Lake of Constance), chlorine dioxide is prepared in a Chlorator system, then added at the outlet of the plant reservoir before water enters the distribution system. Normal dosage of chlorine dioxide plus free chlorine at Konstanz is 0.25 mg/l for the 7 km distribution system. Whenever Konstanz water is sent longer distances, however, the dosage is raised to 0.35 mg/l.

The Styrum West plant of Mülheim, Germany treats drinking water by breakpoint chlorination, flocculation, sedimentation, rapid sand filtration, ground water storage in wells, then chlorine dioxide when pumping water from the storage wells into the distribution system. Chlorine dioxide is generated in a 22 year old Bran & Lübbe system (excess chlorine water is added to sodium chlorite solution). The stoichiometry here is 300 g of chlorine (in water) to 300 g/l of sodium chlorite solution. This plant was one of two visited in Germany that does not use ozone in the water treatment process (an ozone system is planned for addition within the next year).

West Berlin has 7 water treatment plants, but only one, Jungfernheide, uses chlorine dioxide. This plant draws river Spree water, then filters it through sand in the ground which has bullrushes growing on it. Passage through this medium adds oxygen to the water and at the same time destroys a large portion of the bacteria. Filtration rate in the sand/bullrushes beds is 2.5 to 3 m/day. Aeration of the filtered water follows (for iron & manganese oxidation), then filtration and chlorine residual.

During summer, taste and odor causing components are present in the Spree, and chlorine dioxide is employed rather than chlorine. A Chlorator generation apparatus was at the plant in May, 1977, but a new Wallace & Tiernan unit was to be installed shortly.

During winter when chlorine alone is used, the sodium chlorite portion of the chlorine dioxide generation system is operated once each month to keep it in operational condition. Residual chlorine concentration in water leaving the plant is less than 0.1 mg/l, with no traces being detected at distribution system sampling taps throughout the city. Water from Jungfernheide is mixed with waters from the other six Berlin water treatment plants in the city distribution system.

The water works at Langenau in southern Germany used to treat groundwater and add chlorine dioxide before distribution. In recent years, however, Danube River water is treated along with groundwater, especially during periods of high water demand in summer. The advanced water treatment process includes ozonation, followed by addition of ferrous sulfate, then filtration through sand, then granular activated carbon. When chlorine dioxide was added to this water, the color changed from blue to green (apparently caused by higher dissolved iron contents). Since it is desired to maintain the blue color of ozonized water at Langenau, chlorine dioxide has been replaced by chlorine. At the small dosages of chlorine applied (0.3 mg/l maximum), the blue color of the plant water is unaffected.

Switzerland--

The use of chlorine dioxide in Zürich water treatment plants began in 1971, when it replaced chlorine for residual in the Moos and Lengg plants serving the city of Zürich. Raw water drawn from the Lake of Zürich is of high quality. When first installed, chlorine dioxide was prepared from excess chlorine/water and sodium chlorite. In 1974, the preparative method was changed at the Lengg plant to the technique involving acid. This change may have been caused by complaints of chlorine tastes in the treated waters. An excellent discussion of the use of chlorine dioxide at Zürich is given by Valenta and Gahler[76].

HCl (33%) and 24.2% sodium chlorite solution are stored in individual 9 cu m, reinforced polyester tanks. Each ingredient is stored in a separate room beneath the plant in an area surrounded by a concrete lip, which will contain the total content of each storage vessel in the event of leakage or a spill. These catchment basins are not drained, to prevent the acid or chlorite from passing into the sewage treatment plant. If a spill occurs, the material would be pumped to ground level, then into a tank for disposal.

Valenta and Gahler[76] report that chlorine dioxide concentrations above 10% in air are explosive. Aqueous solutions greater than 30 g/l in concentration also pose this danger. Generation of an equivalent amount of chlorine dioxide from acid requires 25% more sodium chlorite than by using chlorine, but acid eliminates the presence of excess chlorine in the product.

From the 9 cu m storage tanks, HCl and sodium chlorite solutions are pumped to 100 liter tanks in which they are diluted automatically with water to 9% (HCl) and 7.5% (sodium chlorite). Both of these diluted solutions now are pumped at equal volumes and rates through a double metering pump to the Raschig ring packed reactor column, where chlorine dioxide solution is produced. Following production, the chlorine dioxide solution is further diluted to 10 to 20 mg/l concentration for addition to the product water. Because chlorine dioxide mixtures with air can be explosive, air is continuously drawn out of the reactor product by means of an injector and diluted with the vacuum-causing water flow.

Dilution water at Zürich is softened to avoid precipitation of calcium compounds in the chlorine dioxide generation equipment. The Lengg chlorine dioxide reactor is a Raschig ring packed column 5 feet tall and 6 inches outside diameter. The pumping rate of both feed solutions through the double metering pump is 40 l/hr. Conduits are made of PVC piping.

There are 3 metering pumps in each chlorine dioxide line at Lengg, and these are coupled with the pump which regulates plant water flow. When the plant flow slows, the chlorine dioxide unit will shut off. If any of the chemical storage tanks become cloudy due to precipitation of salts, an alarm sounds, notifying plant personnel that there is a problem with the chlorine dioxide unit. All pumping operations of the chlorine dioxide generators are automated, controlled at the plant control panel, and are paced by the water flow from the plant reservoir, where dosage occurs.

The Lengg chlorine dioxide generation and addition system is manufactured by Wallrabenstein, Remchingen, Germany, and all operations are automatically controlled.

Analytical control of the chlorine dioxide system at Lengg is accomplished by monitoring the stream of combined HCl and sodium chlorite solutions and determining residual chlorine dioxide, chlorine and sodium chlorite in the treated water. Control of the starting chemicals is by density measurements.

For the analysis of chlorine dioxide and related materials in product water, the following procedure is used: Two parallel 10 ml samples of water are subjected to 3 or 4 step iodometric titration with 0.05N sodium thiosulfate in various pH ranges using starch indicator. Each sample is diluted with 90 ml of double distilled water, and the iodine produced is titrated as follows:

- The first titration is conducted in neutral solution, and determines 20% of the chlorine dioxide and all of the chlorine.

- The solution from the step above is acidified, then titrated again. This determines 80% of the chlorine dioxide plus sodium chlorite.

- The second 10 ml sample is acidified and titrated as the control. This determines the total of chlorine, sodium chlorite and chlorine dioxide present, and the number should total the sum of values obtained from titrations in the first two steps above.

- The titrated solution from the three steps above is mixed with NaOH, allowed to stand in indirect light until the solution is colorless (the chlorine dioxide disproportionates), phosphoric acid is added and the solution again is titrated. This determines 40% of the chlorine dioxide plus the chlorine and sodium chlorite present.

When the acid system was first installed at Lengg, the above analytical titrations were conducted several times daily. It has been found that the amount of chlorine dioxide present is always in the 90% to 96% range, sodium chlorite ranges from zero to 4.5%, and chlorine ranges from 4.0% to 7.5%. Since there has been a three year period of these analyses, indicating the water to contain these ingredients in the ranges presented, analyses currently are conducted weekly. These data show that when chlorine dioxide synthesized from acid and $NaClO_2$ is added to water, 90 to 96% of the oxidant present is ClO_2.

The Zürich distribution system is fed by all local treatment plants and groundwater pumping stations. The network surrounds the city and is 1430 km in total length, but the average residence time of water in this system is only 1 day. Electricity costs are lower at night, so most of the plant operations involving electrical power are performed at night.

At the present time, the total distribution system of Zürich and each of the 20 reservoirs is being interconnected (see plant description in Appendix B). When this is completed, all water treatment plant operations at the Lengg and new Moos plants and all reservoir operations will be controlled by a computerized system housed in a new facility being constructed at the

Hardhof pumping station site. This facility is being constructed to withstand bombings, with the subterranean levels being protected against nuclear explosions. These new facilities will house the analytical laboratories and all water supply services for the city.

Separate chlorine dioxide generation facilities are being installed in the Hardhof subterranean level. The pumping station currently can draw 75,000 cu m/day of groundwater and treat it with chlorine dioxide for addition to the Zürich water supply system. This capability is being extended to allow an additional 75,000 cu m/day to be drawn from the Limmat River, treated with chlorine dioxide, then recharged to the ground for emergency use.

The Kreuzlingen plant is located in northern Switzerland, on the Lake of Constance, and treats this lake water by chemical addition, coagulation, filtration, ozonation, final filtration (sand then granular activated carbon) then chlorine dioxide for residual. Dosages of chlorine dioxide added at the exit of the plant reservoirs are 0.2 mg/l to maintain a residual of 0.05 mg/l in the Kreuzlingen distribution system. HCl and sodium chlorite solutions are used to generate the chlorine dioxide, by a system very close in design to that used in the Lengg plant, but manufactured and installed by Schnellenbraun & Co., Winterthur, Switzerland.

There are two separate chlorine dioxide generating units installed at Kreuzlingen. Concentrated solutions of acid and sodium chlorite are prediluted to the proper concentrations, then pumped through a double metering pump at equal rates into the reactor, taking care to exclude air from the system. Product chlorine dioxide solution is stored until use in an airtight container, being careful to prevent entry of atmospheric air, which could produce an explosive mixture. The off-gases from the reactor and storage tank are passed through a granular activated carbon column for destruction.

All chlorine dioxide operations at Kreuzlingen are designed to operate automatically and unattended. When more chlorine dioxide solution is required, the double metering pump is signalled to begin operation. When the first dilution tanks become low in acid or chlorite, the dilution pumps automatically turn on to draw the starting materials from the main storage tanks. Levels in the main acid and sodium chlorite tanks are checked daily.

The older plant at Kreuzlingen also used chlorine dioxide as a posttreatment for the past twenty years or so. When ozonation was first installed at the old Kreuzlingen plant, no post-treatment was practiced after ozonation. This led to the growth of slimes in the distribution system. When chlorine dioxide post-treatment was installed at Kreuzlingen, however, the slimes problem disappeared. In the ozone literature, this situation has come to be referred to as "the Kreuzlingen experience".

Summary of European Use of Chlorine Dioxide

Chlorine dioxide is used mostly for providing a residual for distribution systems. High quality groundwater can be treated just with small amounts of chlorine or chlorine dioxide and sent directly to distribution systems. Surface waters which require additional processing steps are treated so that the dosage of chlorine dioxide employed as post-treatment remains less than 0.6 mg/l. Only in France are levels as high as 0.5 to 0.6 mg/l of chlorine dioxide being added as post-treatment, and efforts are being made to lower this level. Choisy-le-Roi personnel are studying superozonation to reduce the oxidant demand of treated Seine River water, thus allowing less chlorine dioxide to be used.

In European distribution systems, the residual of chlorine dioxide is more stable than chlorine residuals.

German water treatment practice for post-treatment with chlorine or chlorine dioxide limits dosage to no more than 0.3 mg/l (calculated as free chlorine or total oxidant). If the oxidant demand is higher, German water works may add as much as 0.6 mg/l (calculated as free chlorine) of chlorine or chlorine dioxide, but then the circumstances must be presented to the proper health authorities and special permission obtained.

In pretreatment, chlorine dioxide is used primarily to break down organic complexes of iron and manganese. Prechlorination will oxidize free iron and manganese, but not the organically complexed heavy metals. Pretreatment with chlorine dioxide also will destroy phenolic tastes and odors and remove color in some cases. The addition of chlorine dioxide early in the process also can provide a degree of predisinfection, as will pretreatment with chlorine, ozone, potassium permanganate, and other strong oxidants.

Europeans agree that the unit cost of generating chlorine dioxide on-site is about 3 to 3.5 times the cost of using gaseous chlorine. Thus, where both oxidants can be employed, chlorine generally is the choice. During summer months, the oxidant demands at some plants are sufficiently high (and residual life therefore is sufficiently short) so that chlorine dioxide is used rather than chlorine. In the colder winter months, these plants then switch back to chlorine.

European systems generate chlorine dioxide by adding:

1) excess chlorine in water to sodium chlorite solutions at ambient pressure;
2) stoichiometric chlorine in water to sodium chlorite solutions under pressure;
3) stoichiometric chlorine in water to sodium chlorite solutions at low pH; or,
4) HCl solution to sodium chlorite solutions.

Chlorine dioxide generation systems generally are instrumented and the process is controlled at the plant control panel. Failure of any component in the system triggers alarms at the control panel. In all water treatment plants visited, chlorine dioxide is generated either in a separate building away from the main plant processing, or in a separate enclosed room having access and exit only from the outside of the building. The only exception to this was the Berlin Jungfernheide plant, which has chlorine dioxide generation in a room accessible only from the inside of the main plant building. At Jungfernheide, however, access to the chlorine dioxide room is by means of double doors, which can only be opened one at a time.

European drinking water plant personnel are aware that sodium chlorite in the dry state can spark if spills are allowed to dry out and are walked upon. Thus, spills are cleaned up quickly. Similarly, plant personnel are aware that chlorine dioxide gas in air can form explosive mixtures under certain conditions, and chlorine dioxide generators are designed to avoid building up gaseous concentrations in air.

Most of the treatment plants visited in Europe using chlorine dioxide regard its generation and application as quite simple and routine. If any problem occurs with the chlorine dioxide generation or application systems, a call to the supplier or his local representative brings quick response. Plant personnel generally were not as knowledgeable about the chemistries involved with generating and using chlorine dioxide as about the technology of ozone and activated carbon, for example. They advise that chlorine dioxide generation and application is so simple and gives such few problems that there is little need to be concerned about its scientific intricacies.

The usually employed colorimetric and amperometric analytical procedures determine total oxidant, when oxidants other than ClO_2 are present in solution

Miscellaneous Comments on European Chlorine Dioxide Systems

CFG (Chemie und Filter GmbH), Heidelberg, Germany--

CFG manufactures dosing pumps, and such a pump is designed into their chlorine dioxide generator, which uses HCl and sodium chlorite solution. A monitor reads the redox potential of the HCl solution and automatically adjusts its feed rate so as to guarantee the proper stoichiometry of HCl being added for the amount of sodium chlorite being pumped.

During operation of this generator, the product chamber is deaerated every five minutes automatically so as to prevent buildup of gaseous chlorine dioxide in air which may produce an explosive mixture.

CFG claims to have some 250 of these chlorine dioxide generators installed in water treatment plants throughout Europe, although the acid-chlorite generation system is not allowed in Germany. CFG is represented in the United States by Fluid Controls Ltd., Clearwater, Florida.

Wallace & Tiernan, Germany--

W & T representatives state that some 60 European drinking water treatment plants have been using W & T chlorine dioxide generators since about 1960. All are based upon addition of excess chlorine in water to solutions of sodium chlorite.

A new generator has been developed recently, and has been installed in about 20 water treatment plants as of May, 1977. The maximum production rate of this new generator is 5 kg/hr. Chlorine/water solution at a chlorine concentration of 5 g/l produces a pH below 2.0. Normally the solubility of chlorine in water at ambient pressure is 0.5 to 3 g/l, but at 7 bars, the 5.0 g/l concentration can be obtained. The new W & T generator thus operates at 7 bars pressure. Adding 1 part of chlorine in water under 7 bars pressure to 1 part of sodium chlorite (in solution, also at 7 bars pressure) normally produces a solution containing 7.5 g/l of chlorine dioxide.

However, at 7.5 g/l, chlorine dioxide solutions can produce explosive mixtures more easily, thus W & T has designed their reactor to produce maximum ClO_2 concentrations of 5.0 g/l in solution. In addition, the chlorine/chlorine dioxide gas phase is drawn through an injector where excess gases are dissolved in water, and the danger of explosion is eliminated. At the same time, the newly formed chlorine dioxide solution is diluted to 10 to 20 mg/l, at which concentration it is added to water at the treatment plant.

All operations in the new W & T system are controlled automatically.

Fischer & Porter, USA--

Representatives of F & P advise that there are 30 operational ClO_2 plants in Italy using F & P generators. All of these plants have been installed within the past three years, and all generate ClO_2 by adding acid to $NaClO_2$ solution.

European ClO_2 Questionnaire Results Summary

In mid-1977, questionnaires were mailed to 835 water plants in the Federal Republic of Germany (F.R.G.), 150 plants in France, the key plants in Great Britain which use ClO_2, 20 plants in Switzerland, and plants in Austria. As with ozone, more than 1000 European plants received the ClO_2 questionnaire, a blank copy of which is included in Appendix A.

Seventy-four questionnaires have been returned. Most of these are from the Federal Republic of Germany -- 44, while 23 were received from French plants, 5 from Great Britain, and 2 from Austria.

A brief summary of the information derived from the questionnaires is given by country in the following pages.

Federal Republic of Germany--

Of the 44 German plants responding, 19 are between 1,000 and 10,000 cu m/day in size, 17 are 10,000 to 50,000 cu m/day capacity, 4 are less than 1,000 cu m/day capacity, and 6 are larger than 50,000 cu m/day. Eighteen of the plants began using ClO_2 between 1960 and 1970, 15 have initiated use since 1970, and only 6 initiated its use before 1960.

Chlorine dioxide use in Germany is primarily for bacterial disinfection (33 plants) and to provide a chlorine residual (22 plants). Other uses include taste removal (9 plants), and iron and manganese oxidation (2 plants each).

Dosages of ClO_2 are very low in Germany, averaging 0.1 to 0.3 mg/l for disinfection. The highest dosage reported was 0.6 mg/l. The residual in the distribution system is likewise very low, usually less than 0.1 mg/l. Dosages of ClO_2 for taste and odor removal are slightly higher than for disinfection, usually 0.3 to 0.5 mg/l.

Chlorine dioxide is prepared by the same method in all plants responding to the questionnaires except one, by reacting chlorine gas plus water with sodium chlorite, usually a 30% solution. The only plant which prepares ClO_2 by the acid method is the Vallendar waterworks, which uses both the chlorine plus sodium chlorite and the hydrochloric acid plus sodium chlorite methods of preparation. Therefore most German plants add both ClO_2 and chlorine to their treated water.

Ten different manufacturers supplied chlorine dioxide equipment to the forty-four plants.

Several methods are used to analyze for ClO_2 or total oxidants, which is what the plants actually measure. Many plants use the DPD method, 3 measure electrochemically, and several use the orthotolidine method and iodometric titration. Each of these methods measures total oxidants, rather than ClO_2 specifically.

Ten of 44 plants reporting data indicated that ClO_2 is their only water treatment processing step.

Austria--

Questionnaires were mailed to the 11 largest water companies which serve nearly half of the Austrian population. Only two of the eleven use ClO_2, but they are by far the two largest ones, producing at least 80% of the total water treated by the 11 waterworks, more than 100,000 cu m/day. Chlorine dioxide has been used since 1967 and 1973, respectively, at their two plants (Nüssdorf Groundwater Works and Übergangskammer d. II).

Both plants use ClO_2 for bacterial disinfection and to provide a residual in the distribution system. The average dosage at both plants is 0.25 mg/l. The total oxidant residual in the distribution system is 0.08 mg/l from one plant and is not measured by the other. Total oxidant is measured using the DPD method.

Chlorine dioxide is prepared by reacting sodium chlorite and chlorine gas plus water in a Wallace and Tiernan reactor.

Great Britain--

Five questionnaires were received from British plants out of the approximately 20 that were mailed. The plants reporting data have an average design capacity of 97,000 cu m/day. The longest period that ClO_2 has been in use is 16 years in the Sutton Hall Treatment Works. Overall, ClO_2 has been used an average of 9 years in the five plants.

The purposes for which ClO_2 is used are taste and odor control (3 plants), bacterial disinfection (5 plants), to provide a chlorine residual

(4 plants), prevention of nuisance growths (1 plant), and to remove phenolic tastes (1 plant). Dosages range from 0.05 mg/l to 0.5 mg/l, averaging 0.28 mg/l of total oxidant.

As in Germany and Austria, ClO_2 is prepared in all reporting plants by reacting $NaClO_2$ solution and chlorine gas, using excess chlorine.

In Great Britain, ClO_2 is used not only as a disinfectant, but quite often is used as a distinct treatment step or added in a series of unit processes. The British believe that by adding it in multiple steps, a better residual will be maintained. In many cases, a reaction tank or basin is provided immediately after ClO_2 dosing.

Chlorine dioxide and ozone are not both used in any reporting plant in Great Britain.

France--

Questionnaires were mailed to 150 plants in France. Twenty-three plants completed and returned the questionnaire. The plants range in size from 1,000 cu m/day to 600,000 cu m/day. The average size is 57,300 cu m/day. Many of the plants have been using ClO_2 for 15 years or more. The plant that has used ClO_2 for the longest period is the Carcassone Plant which has used it since 1946. The average period of usage for 18 plants is 10.3 years.

Each of the French plants reporting prepare ClO_2 in the same way, by reacting chlorine gas with water, then with sodium chlorite solution.

Chlorine dioxide is used for 11 different purposes in France, more than in any other country. The primary purpose is for bacterial disinfection, reported by 22 plants. Other uses include organics removal (10 plants), taste removal (12 plants), odor removal (10 plants), color removal (10 plants), viral inactivation (6 plants), iron oxidation (5 plants), turbidity reduction (3 plants), and manganese oxidation (2 plants). Only one plant reported the use of ClO_2 for the purpose of providing a chlorine residual. Also, one plant notes that it uses ClO_2 to prevent the formation of chlorophenols.

Dosages are somewhat higher in France than in other countries because of the multiple applications. The highest dosage reported was 2.3 mg/l at the LeMans plant which adds ClO_2 after filtration and before ozonation, then adds more ClO_2 as a terminal treatment step. The average dosage for 17 plants reporting data is about 0.66 mg/l. It should be noted that when ClO_2 is used for taste and odor control, the dosage ranges from 0.6 to 1.3 mg/l.

The residual chlorine dioxide concentration measured in the distribution system is very low, usually reported as zero. The highest residual concentration reported was 0.25 mg/l. The actual amount of ClO_2 measured depends upon how the Cl_2 + $NaClO_2$ reaction was conducted as well as upon the stoichiometry of reactants used.

ENGINEERING DESIGN FOR CHLORINE DIOXIDE GENERATION SYSTEMS

Introduction

As discussed, there are two basic types of chlorine dioxide generation systems used in water treatment plants: the $NaClO_2/Cl_2$ system and the $NaClO_2/HCl$ system. Both methods employ the same type of ClO_2 reactor; only the feed systems for the Cl_2/HCl chemicals are different. The following text will discuss engineering aspects of the design and operation of both ClO_2 systems. Much of the discussion is typical of United States operations.

Chlorine Dioxide Generator

ClO_2 reactor vessels are designed as mixing chambers, providing for the continuous production of chlorine dioxide. The aqueous reagents are merged at the bottom of the vessel and flow upward through the vessel where chlorine dioxide is produced. The solution exits from the top of the reactor and travels through the discharge piping to a dilution or storage tank and/or to the point of application in the treatment process.

The typical ClO_2 reactor vessel is made from Pyrex glass or polyvinyl chloride (PVC). The PVC material is used for higher operating pressures. Dimensions of the overall vessels may vary, but typically are from 36 to 42 inches high and 8 inches in diameter. The approximate weight of the unit is 50 to 60 pounds. Most ClO_2 reactors are wall mounted and positioned at eye level for easy viewing. The glass generating units allow the operator to inspect the color of ClO_2 visually that is being produced. If the chamber is made of opaque PVC, a sight glass is installed on the discharge tubing.

The mixing action inside the ClO_2 vessel is caused by the solution flowing around and through Raschig rings. These porcelain rings are about 1 inch in size and are placed inside the vessel in a random manner. Some plant-fabricated units use PVC rings (cut from PVC pipe) instead of porcelain rings. Both materials are inert.

ClO_2 reactor vessels are sized according to maximum ClO_2 production rates. At least three sizes are available based on the following approximate rates: 140 pounds ClO_2/day; 370 pounds/day; and 1500 pounds/day. Other reactor sizes are available.

A sampling valve is a desired feature to have on a ClO_2 generator. Mounted at the discharge end of the vessel, this port can be made of PVC, tygon tubing, or glass. Samples can be drawn off periodically in order to check the composition of the reactor product. Chemical feed inlets on the reactor vessel are made of PVC, glass, or certain types of rubber tubing. Teflon, glass, and Lucite are suitable but not practical. Rubber-lined pipe should be avoided .

$NaClO_2$ Feed System

The $NaClO_2$ reagent is delivered to the plant either as an anhydrous technical grade solid (80% pure $NaClO_2$) or as a liquid (20% to 37% concentration). The powdered $NaClO_2$ (readily soluble in water) is mixed with water in open drums. Batch mixes normally are 2 to 3 pounds of $NaClO_2$ per gallon of water. $NaClO_2$ solutions are fed directly to the ClO_2 generator.

Diaphragm metering pumps are used commonly to feed $NaClO_2$ solutions to the ClO_2 reactor. These positive displacement pumps are simple to operate and can be metered over a wide range of flow rates. The diaphragm or piston compartment is made from PVC material. The pump and motor (120 volt, single phase) are usually small enough to operate on top of a drum or a bench.

The piping for aqueous $NaClO_2$ is either PVC, rubber or Tygon. Tygon tubing looses its rigidity after several months in service and must be replaced. The softened walls occasionally collapse due to the pump suction and subsequently restrict the flow of $NaClO_2$ to the pump.

The $NaClO_2$ is relatively stable in either the dry flake or fluid phase but is very combustible in the presence of organic material (rags, gloves, brooms, etc.). Therefore, the solution never should be allowed to evaporate on the floor. Any spills should be washed down the floor drain or mopped with a technical sulfite solution. Solid $NaClO_2$ is also explosive when contacted with acids. The oxidant decomposes rapidly when heated above 347°F (175°C). If the heat is confined, such as in a building on fire, the chemical drums will explode. Thus, the chemical should be handled with appropriate caution.

HCl Feed System

This is similar to the $NaClO_2$ feed system. The same types of pump and piping that are recommended for $NaClO_2$ solutions are also recommended for HCl.

The HCl solution can be stored either in inert plastic tanks (such as polyester, polyethylene, etc.) or in fiberglass tanks. $NaClO_2$ solutions also can be stored in these types of containers. Some plastic tanks become increasingly opaque after months of storing these chemicals. When the transparency is reduced to the point where the operator cannot see the liquid level of the chemical through the tank wall, the tank is replaced with a new tank. The tanks, however, remain structurally safe. Storage tanks are readily available on the market.

In the event of an accident, the floor area around the liquid chemical storage tanks and pump equipment should be confined by a continuous concrete step 6 to 8 inches high and 6 to 8 inches wide. This barrier prevents the liquid from escaping to the surrounding floor areas if a line ruptures or a tank fractures. It is recommended to have floor drains inside the barrier which empty to a sanitary sewer system. Some European systems do not have floor drains; instead the spilled chemicals are pumped to trucks for transporting to hazardous waste disposal sites.

Piping for all chemical feed systems is exposed and mounted along the walls and ceiling. This allows the plant personnel to inspect and repair the lines if a leak should occur.

The Cl_2 Feed System

Most chlorination systems can be modified to produce ClO_2. The standard chlorinator regulates the rate at which chlorine gas is injected into the water feed system. Chlorine solution then is piped to the ClO_2 reactor vessel for ClO_2 production. The chlorinator should have at least a vacuum regulator, a flow valve and a flow meter. The flow meter indicates the rate in pounds/day at which chlorine is injected into the feed water. The chlorinator can be controlled manually or automatically, the latter system responding to a water flow transmitter and/or a chlorine residual analyzer controller.

The chlorine ejector is sized according to the maximum chlorine feed rate and the hydraulic conditions of the water supply line and chlorine solution line.

Liquid chlorine is supplied in either one-ton cylinders (2000 pounds of chlorine) or 150 pound cylinders (150 pounds of chlorine).

Different materials are used for transporting chlorine liquid, chlorine gas and aqueous chlorine. Liquid chlorine is always packaged in steel containers. The chemical will not react with iron and steel in the absence of free moisture. Schedule 80 seamless steel pipe (with at least 300 pound malleable iron, and preferably 2000 pound forged steel fittings) is used for liquid chlorine. The liquid oxidant readily attacks and dissolves PVC. Chlorine gas that is dry is also inert to ferrous materials. When chlorine gas is contained above atmospheric pressure, PVC should never be used, as it becomes porous. If the chlorine gas should reliquify, the PVC will deteriorate more rapidly. Chlorine gas under vacuum can be piped using PVC, reinforced fiberglass and steel[2]. Aqueous chlorine solutions generally are piped with PVC, fiberglass, Saran or rubber-lined steel once beyond the injector. Neoprene-lined hose will decompose if used with chlorine solution.

Production of ClO_2 From Chlorine-Sodium Chlorite System

This approach uses aqueous chlorine (commonly as HOCl) and aqueous $NaClO_2$ to produce ClO_2. Figure 49 is a schematic of such a system, which consists of a ClO_2 generator, a gas chlorinator, a storage reservoir for liquid $NaClO_2$, and a chemical metering pump. Sodium chlorite solution can be prepared from the dry chemical by adding it to water. The minimum recommended feed ratio of Cl_2 to $NaClO_2$ is 1:1 by weight. Further, additional chlorine can be injected into the reactor vessel without changing the overall production of chlorine dioxide.

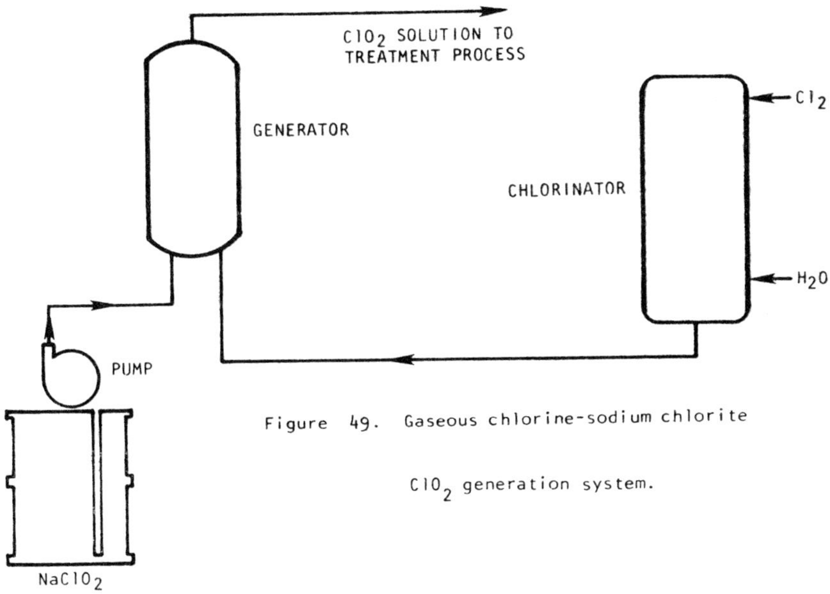

Figure 49. Gaseous chlorine-sodium chlorite ClO_2 generation system.

A disadvantage of this process is the limitation of the "single pass" gas chlorination. Unless using increased pressure, this equipment is not able to achieve high concentrations of chlorine which provide for a more complete and controllable reaction with the chlorite ion. A French firm, CIFEC, recently has developed a variation of this process using a multiple pass enrichment loop on the chlorinator to achieve a higher concentration of chlorine in water and thereby maintain better yields of chlorine dioxide without having to use excess chlorine (Figure 50).

The purpose of the multiple pass recirculation system is to allow enrichment of the chlorine solution to a level of 5 to 6 g/l. At this concentration, the pH of the solution will drop to 3.0 and below. A "single pass" results in a chlorine concentration in water of about 1 g/l, which produces a pH of 4 to 5. If the $NaClO_2$ solution is mixed with the aqueous chlorine at this pH, only a 60% yield of chlorine dioxide reportedly is obtained. The remainder is unreacted chlorine (in solution) and chlorite ion. When quantitative yields of ClO_2 are achieved, there is virtually no free chlorite or free chlorine carrying over into the product water.

Using the CIFEC recirculating loop approach with the injector, some 46 kg/hr (2434 pounds/day) of chlorine can be dissolved in the recirculating water. Makeup water to the loop is added at the rate of 1 cu m/hr (4.4 gal/min) through an aspirator in which the water is mixed with chlorine. Thus a chlorine/water solution of 6,000 mg/l is made at the rate of 1 cu m/hr (4.4 gal/min).

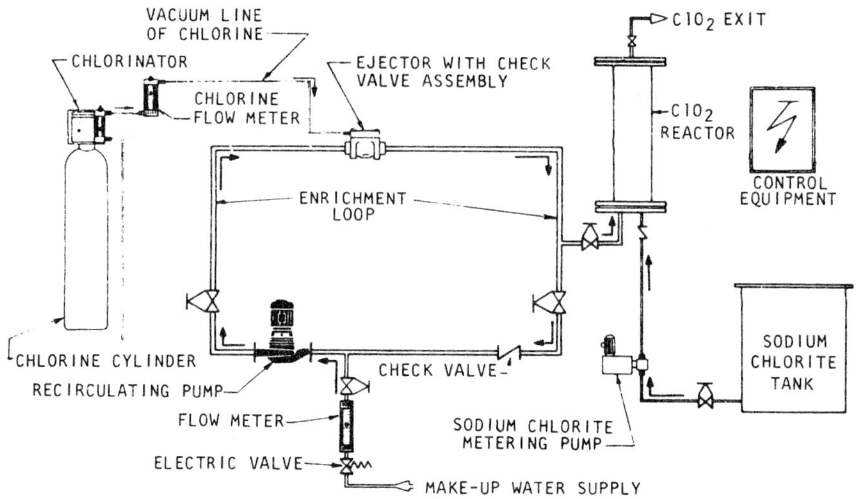

Figure 50. The CIFEC System.

Both the aqueous chlorine and the aqueous $NaClO_2$ solutions are pumped to the ClO_2 reactor. A control system monitors the equipment. When the chlorine cylinder empties, the unit shuts down automatically. If the recirculating pump fails, or if the water pressure drops, the unit also will shut down and alert the operator that the system has malfunctioned.

The chlorine dioxide solution is injected directly into the process water by a remote controlled injector, which is also connected to or is a part of the control panel. The aqueous stream containing chlorine dioxide is under pressure to avoid explosive hazard from entrapped air. Pressurizing the system prior to the ClO_2 reactor avoids the need to pump the chlorine dioxide solution after it is formed. The injector is set to a predetermined feed rate, and the system operates automatically.

If variable feed rates are desired on the CIFEC system, the injector can be connected to an analytical monitor. As the residual chlorine dioxide values go up or down, the meter will automatically increase or decrease its output. All of these operations can be recorded, so that the system requires only periodic inspection by plant personnel. The entire CIFEC unit operation also can be connected to the central control panel of the water treatment plant.

The CIFEC system thus claims to address three problems inherent with generation of chlorine dioxide:

- It automatically adjusts the chlorine/water solution to the proper pH for synthesizing chlorine dioxide in maximum yield, and minimizes the presence of either excess chlorite ion or chlorine;
- It eliminates the necessity for a chlorine dioxide reservoir or for a pump which must pass chlorine dioxide solution;
- The flow of chlorine dioxide solution can be readily adjusted downward from the maximum production rate by a factor of 20. That is, if the maximum production rate is 10 kg/hr (530 pounds/day), the CIFEC equipment can be programmed to produce as little as 0.5 kg/hr (26.5 pounds/day) at a constant rate. Competitive units are reported (by CIFEC) to be adjustable over more narrow ranges.

Wallace and Tiernan offers a single pass system which is pressurized to produce a higher concentration of aqueous chlorine for ClO_2 production. Gaseous chlorine is injected into feed water at about 7 bars (102 psi) to produce a chlorine concentration of 5 g/l which is added to the $NaClO_2$ solution also under 7 bars pressure, and ClO_2 is produced. The ClO_2 solution then is diluted to 10 to 20 mg/l (as ClO_2) for water treatment operations. This system is described in greater detail in Appendix B under the Tailfer (Brussels, Belgium) plant description.

Production of ClO_2 From Hypochlorite-Hydrochloric Acid-Sodium Chlorite System

The production of ClO_2 using the chemical reagents NaOCl, HCl and $NaClO_2$ is shown schematically in Figure 51. NaOCl solution is reacted with $NaClO_2$ solution with the pH adjusted to a maximum of 4 using hydrochloric acid (or sulfuric acid). The reactions are as follows:

$$NaOCl + HCl \longrightarrow NaCl + HOCl$$

$$HOCl + 2NaClO_2 + HCl \longrightarrow 2ClO_2 + 2NaCl + H_2O$$

The flow scheme is arranged whereby the acid for both reactions is added simultaneously, with the excess from one reaction carrying over to react with the HOCl. The $NaClO_2$ is added last, just before the combined solution enters the packed bed.

Production of ClO_2 From the Acid - Sodium Chlorite System

The combination of acid and sodium chlorite is claimed to produce an aqueous solution of ClO_2 with no free chlorine being present. It is for this reason that many plants prefer this approach for ClO_2 production. The acid based process minimizes the difficulty of differentiating between chlorine and chlorine dioxide for establishing an oxidant residual. Fischer and Porter is introducing an automatic ClO_2 generator which uses chlorite and hydrochloric acid as reagents.

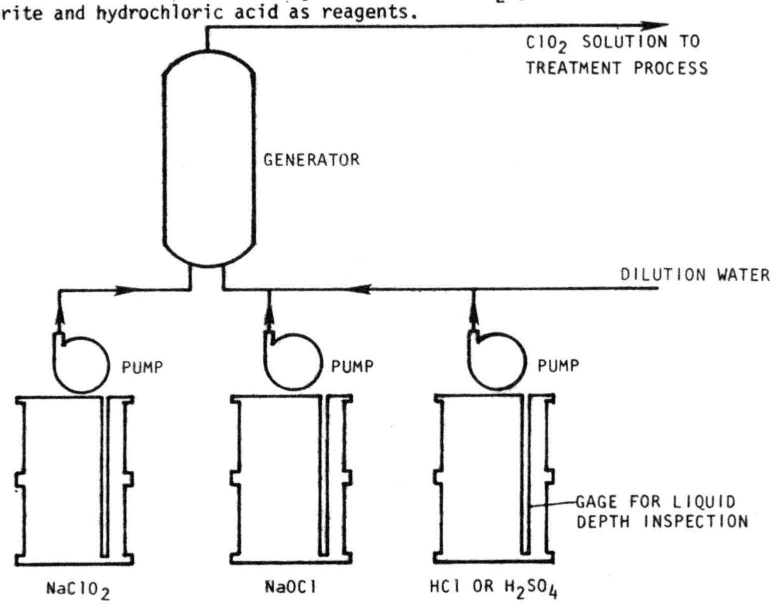

Figure 51. Sodium chlorite-hypochlorite-acid ClO_2 generation system.

To illustrate a large scale operation of ClO_2 production, Figure 52 presents a schematic of the Lengg Waterworks of Zürich, Switzerland (acid-sodium chlorite layout). This system uses liquid chemicals delivered by tank trucks as feed stocks. Each storage tank has a level sensor to avoid overfilling. The tanks are installed below ground in concrete bunkers which are capable of withstanding an explosion. There are no floor drains in these bunkers (to avoid contamination of local sewers), and any spillage must be pumped with non-corrosive pumps. Primary and backup sensors with alarms warn of any spillage. The extreme safety precautions are required because of the explosiveness of the chemicals when combined in these concentrations (32%-HCl; 24% $NaClO_2$).

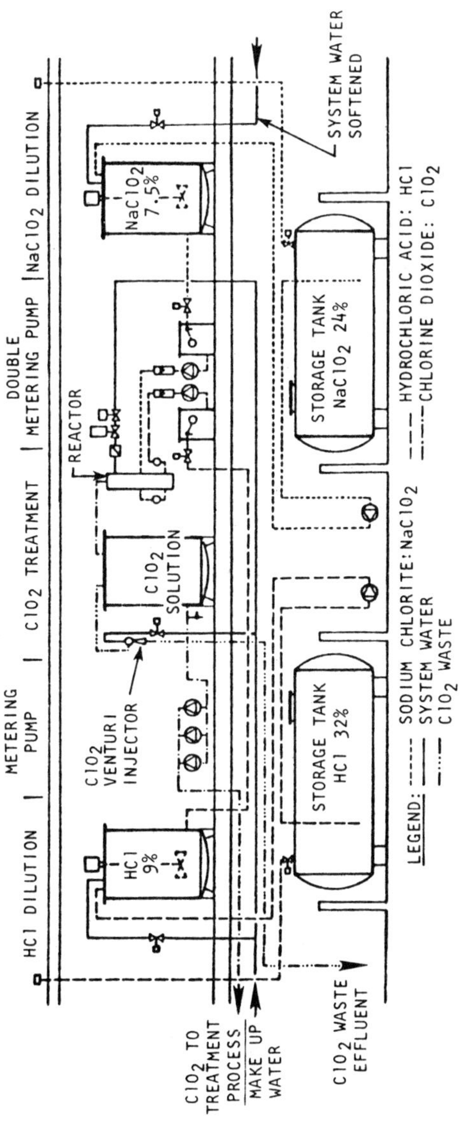

Figure 52. Lengg Waterworks, Zürich, Switzerland. Chlorine Dioxide System.

Because of the potential explosiveness, the reagent chemicals are diluted prior to the production of ClO_2. The dilution is carried out on a batch basis which is controlled by level monitors. Proportionate quantities of softened dilution water, along with the chemical reagents, are pumped to mixing vessels. After the reactor is properly filled, an agitator within the container mixes the solution for 15 minutes. Solutions of 9% HCl and 7.5% $NaClO_2$ are produced in the chemical preparation process. Pumping equal volumes of these reactant concentrations into the reactor provides the proper stoichiometry.

ClO_2 subsequently is manufactured on a batch basis and the rate of production is controlled by the liquid level in the final ClO_2 reservoir. A level switch activates a calibrated double metering pump which feeds the diluted reagent chemicals to the reactor at an identical rate. The reactor is a small cylindrical vessel packed with Raschig rings in which the exothermic reaction of the two chemicals occurs. The ClO_2 product is diluted again prior to passing into the ClO_2 storage tank.

The metering pumps operate in conjunction with the main plant pumps. The ClO_2 solution is subjected to a final 1:100 dilution and mixed with process water in an hydraulically mixed chamber. The ClO_2 solution is metered into the treated water stream at the end of the treatment process as well as before injection into the plant reservoir.

The process is monitored in the laboratory to accomplish three objectives:

- Control of stock concentrations of HCl and $NaClO_2$
- Control of chemical mixing and dilutions
- Control of Cl_2 and ClO_2 residuals in the treated water.

The control of the initial concentrations of chemical reagents is made at the time of delivery by a simple determination of density. Weight per unit volume (density) is checked against known values from tables.

Control of process chemical concentrations and final residuals is accomplished by using a three or four step iodometric titration. More detail on the Lengg ClO_2 system is provided in the plant description in Appendix B.

Other Design Precautions

A ClO_2 system should be well designed for efficient and safe production of ClO_2. In addition to the preceding discussion regarding the operation and maintenance of a ClO_2 system, the following items also should be considered for any plant producing this oxidant:

- Use of inert materials where contact is made with oxidants;

- Provision of separate rooms for each of the chemical components with monitors and alarms for failure warning. The rooms should be designed to isolate and contain spilled or leaked chemicals and facilitate their cleanup;
- Provision of any off-gas collection and neutralization from the ClO_2 production;
- Adequate ventilation and air sensing/alarm systems in working areas;
- Provision of gas masks and first aid kits outside of entrance doors to chemical storage room(s);
- Provision for wash down in chemical storage and mixing areas;
- Clear glass housing on the reactor for visual check of the reaction;
- Flow monitoring equipment on influent and effluent lines;
- Provision of softened process water to avoid calcium buildup in the system;
- Management program for frequently checking solution strengths of chemicals. Testing equipment should be on-site;
- Management programs for routine shutdown and careful inspection of system components;
- Avoidance of air contacting ClO_2 solutions after generation so as to avoid forming explosive concentrations of ClO_2 in air.

Costs for Producing Chlorine Dioxide

The cost for generating ClO_2 on-site is dependent on what method is used to generate the oxidant: chlorine/sodium chlorite, acid/sodium chlorite, etc., and particularly the costs for chlorine, acid and sodium chlorite. There are other factors that affect capital and operating costs for on-site ClO_2 production. These include:

- rate of ClO_2 production
- level of automation
- back up equipment
- availability of existing equipment to be incorporated into a ClO_2 system
- chemical storage facilities

- reagent chemicals ordered--liquid, powdered, drums, tank car, rail car, etc.
- size and frequency of chemical shipments
- level of training of plant operators

Although it is difficult to assign cost figures to ClO_2 production because it is site-specific, ClO_2 production costs generally follow economics of scale. Larger plants that generate ClO_2 on-site normally can produce it at a lower cost per pound than can smaller water treatment plants. However, the larger plants typically will incur higher capital costs for ClO_2 production because of higher levels of system automation, chemical storage facilities, stand-by equipment, etc.

It is most important to realize that when the acid/sodium chlorite method is used for generation more than 90% of the oxidant contained in solution is ClO_2. However, when the chlorine/sodium chlorite method is used in 1:1 weight ratio, particularly when the $NaClO_2$ is technical grade, the total oxidant in solution contains about 50% ClO_2 and 50% other oxidants, primarily free residual chlorine. Therefore, when a treatment plant synthesizes ClO_2 by the chlorine/sodium chlorite methods, analyzes for ClO_2 by a method which determines total oxidants and reports the results as "ClO_2", the true concentration of ClO_2 is about half that reported.

Estimated Costs in the United States

With this brief introduction, cost analyses for ClO_2 should be based on the stoichiometry of producing ClO_2. Examining first the $HCl/NaClO_2$ method in the United States, ClO_2 costs approximately $1.80 to $2.60 per pound produced using hydrochloric acid and sodium chlorite as the chemical reagents (1977). If, however, the $Cl_2/NaClO_2$ method is used, ClO_2 costs approximately $1.35 to $2.00 per pound in the United States. These estimates are based on the following chemical costs in the United States furnished by several chemical distributors (1977): $0.25/lb of hydrochloric acid, $0.0675 to $0.15/lb of chlorine and $0.78 to $1.15/lb of technical grade sodium chlorite.

Capital costs for a ClO_2 generation system are largely dependent on the method of ClO_2 production, production rate and chemical feed systems. The major equipment required for a ClO_2 generation system includes a ClO_2 reactor, chemical feed pump(s), chemical storage tank(s), chlorinator(s) (only for chlorine/sodium chlorite method), weighing scales and piping/electrical appurtenances. The following costs for this equipment are quoted as 1977 prices, FOB.

The price of a ClO_2 reactor ranges from $650 to $1200, the price variance being based on reactor capacity, operating pressure, and valve/piping appurtenances.

Feed pumps that are capable of handling such chemical oxidants generally cost between $400 and $800. These are plunger pumps and proportion the chemical feed rates by a manual metering system. Automatic feed pumps, which proportion the chemical feed rates according to the plant flow or the chlorine residual in a process tank, typically cost more than manual feed pumps.

Chemical storage tanks with lids are made either from fiberglass or PVC. Steel tanks are not recommended. The smaller storage tanks cost from $75 to $500 for capacities up to 250 gallons. Larger tanks must be special ordered. Chemical mixers are optional depending on daily requirements of chemical reagents.

Chlorination systems generally are available at most U.S. water treatment plants. Several U.S. water plants have used an on-site chlorinator(s) to furnish chlorine to a newly installed ClO_2 reactor for ClO_2 production ($Cl_2/NaClO_2$ method). The chlorine feed rate then is adjusted so that the desired chlorine residual in the plant water continues to be maintained. Using this strategy, a plant will not necessarily require a second chlorinator to be used for a ClO_2 generating system. If such an option is not feasible at a water treatment plant, however, then a second chlorinator (or more) is needed for the new ClO_2 generation system.

Most plants, however, use a separate chlorinator for their ClO_2 generation system. Weighing scale(s) for measuring chlorine contents in the cylinders also should be provided. Chlorinators for ClO_2 reactors range from $1700 (wall mounted, 500 lbs chlorine per day) to $4000 (floor mounted, 2000 lbs chlorine per day). Evaporating chlorinators are more expensive. Weighing scales for 150 pound chlorine cylinders cost from $300 to $500.

If a water treatment plant contracts the work for furnishing and installing a ClO_2 generation system, installation charges can be expected to cost roughly 100% to 200% of the material costs. For example, if a 10 to 15 mgd water treatment plant determines that it needs a ClO_2 generating system to abate a seasonal taste and odor problem in its raw water, the plant then requires (using the chlorine/sodium chlorite method) a ClO_2 reactor, a chlorinator (wall mounted), a chlorine cyclinder weighing scale, a storage tank for aqueous $NaClO_2$, a chemical feed pump for $NaClO_2$, and piping/electrical appurtenances.

For purposes of this example, it is assumed that there is adequate room for chemical storage, chemical tank, ClO_2 reactor and proper ventilation and safety equipment for the ClO_2 generating system. A rough estimate for the material cost for this simple system is $3000 to $4000. The final cost for furnishing and installing such a ClO_2 system may range from $6000 to $9000. Nominal maintenance expenses for a system like this may range from $50 to $100 per year. These figures were reported from several U.S. plants visited and are said to be conservative. Larger ClO_2 installations for 50 to 100 mgd plants where chemicals may be delivered by tank car or rail car will experience higher operating costs.

This quick estimating exercise should not in any way be considered as a substitute for the thorough and detailed analysis that is required for estimating capital costs for a ClO_2 generating system. Each water treatment plant will be different and therefore the cost for implementing a ClO_2 system will be site-specific.

There are water treatment plants that purchase ClO_2 equipment directly and install the system without procuring outside services. This approach requires experienced professionals within the plant labor force who fully understand the work involved. If this approach is taken, it is highly recommended to have the equipment manufacturer inspect the overall ClO_2 system with the plant personnel prior to startup. The manufacturer's service person also can assist with fine adjustments of the system for increasing efficiency. (In the United States, the major manufacturers of chlorinators also manufacture ClO_2 reactors and some chemical feed pumps.) Services of the equipment manufacturer's representative cost around $300 to $400 per day, with one day normally being sufficient time to complete the work.

Estimated Costs in Europe

During the May, 1977 visitations to European drinking water treatment plants, the survey team was advised repeatedly that the costs for using ClO_2 in Europe are about 3 to 3.5 times greater than those for chlorine -- on an equivalent dosage basis.

Costs for chlorine and sodium chlorite as of late 1977 are given in Table 26.

TABLE 26. EUROPEAN CHEMICAL COSTS FOR ClO_2

Country	Cl_2 $/kg	Cl_2 $/lb	$NaClO_2$ $/kg	$NaClO_2$ $/lb	Cost ratio $NaClO_2/Cl_2$	Chemical costs for 1 mg/l dosage ClO_2 prepd. from 1:1 ratio of Cl_2 + $NaClO_2$, in ¢/1000 gal
Italy	0.40	0.18	1.82	0.83	4.55	0.84
Netherlands	0.62	0.28	1.50	0.68	2.42	0.80
England	0.43	0.20	1.96	0.89	4.56	0.91
USA	0.16	0.07	2.96- 2.49	1.34- 1.13	18.5- 15.6	1.18- 1.00

From the above it can be seen that the ratio of the cost of $NaClO_2$ to chlorine in Italy, the Netherlands and England ranges from 2.4:1 to 4.6:1, whereas in the United States the ratio is 15:1 to 18:1. Thus, comparing costs for ClO_2 generated by the chlorine/sodium chlorite method using 1:1 weight ratios of the two reactants on the basis of this chemical cost alone, the cost of ClO_2 should be from 3 to 8 times higher in the USA than in Europe. However, the cost of chlorine in the USA is 25% to 40% lower than the cost of chlorine in Europe. Therefore, the relative cost of chemicals required to prepare chlorine dioxide (from a 1:1 weight ratio of chlorine to sodium chlorite) is 110% to 150% the costs in Europe.

Costs of Chlorine Dioxide at Tailfer--

Tailfer plant personnel made available a series of four quarterly expense reports on water production costs for the year 1976. During 1976, the Tailfer plant produced 21,671,385 cubic meters of water at a total chemical cost (including costs for producing ozone) of 16,355,039 Belgian francs (Bf, or $464,632, or $0.0214/cu m). Of this, the total costs for sodium chlorite were 2,760,463 Bf ($78,422). Therefore, costs for sodium chlorite at Tailfer during 1976 were $0.00362/cubic meter of water processed. To this should be added the cost of associated chlorine, plus some electrical energy, to arrive at the cost for chlorine dioxide (35.2 Bf = $1 U.S. in mid-1977).

Schalekamp[2] cites the operating cost for chlorine dioxide in Zürich, Switzerland at an average dosage of 0.5 mg/l as 0.4 Swiss Roppen/cu m (0.85¢/1000 gal). The cost of using sodium chlorite in Zürich to generate chlorine dioxide at this average dosage is given by Schalekamp as 0.22 Swiss Roppen/cu m (0.42¢/1000 gal).

The city of Karlsruhe, Germany installed a new 104,500 cu m/day (27.6 mgd) water treatment plant in 1977 which employs aeration, coagulation, filtration and ClO_2 addition (average dosage 0.20 mg/l). Capital costs for the chlorine dioxide generation equipment (4 kg/day, Cl_2 + $NaClO_2$) were 15,000 DM (7,000 at 2.0 DM/$ exchange rate). The cost of chlorine is 0.6 DM/kg (18¢/lb at 2.0 DM/$ exchange rate) (W. Kühn, Private Communication, 1978).

The CIFEC system is a new method for generating ClO_2 at water treatment plants from Cl_2 and $NaClO_2$. High yields of ClO_2 are claimed to be generated with little or no free residual chlorine being present. Although cost data on this type of operation are not readily available, the manufacturer reports that the capital cost for a 20 kg ClO_2/hr (1058 pounds ClO_2/day) CIFEC system is about $100,000 (1977). Installation charges are not included. A 200 g ClO_2/hr (11 pounds/day) CIFEC system costs about $24,000 (1977). Production rates of the CIFEC system are reported in terms of ClO_2 produced (kg or pounds) rather than Cl_2 (kg or pounds) added as is generally the case with ClO_2 systems in the United States.

SECTION 12

OXIDATION PRODUCTS OF ORGANIC MATERIALS

In this section we will discuss the results of an extensive survey of the literature dealing with current knowledge of the nature of specific organic products formed when aqueous solutions are treated with ozone or with chlorine dioxide. For a more complete understanding and comparison we are also including a brief discussion of reactions of organic materials with chlorine.

INTRODUCTION

There does not exist a great deal of literature in which the investigators deliberately set out to isolate and identify the oxidation products of organic compounds from actual drinking water supplies or wastewaters. Most of the research has been pointed at determining the dosage of oxidant necessary to reduce the concentration of a particular organic compound to below a limit detectable by means of an analytical technique specific for that compound. It has not been generally recognized that oxidation products formed from these materials may still be present, may not be detectable by the analytical technique used to determine the original compound, and may pose a public health concern. It has only been relatively recently that Rook[77] has shown chloroform to be a product of the chlorination of drinking water supplies containing humic materials. Since then, research into the formation and nature of organic oxidation products under drinking water treatment conditions has accelerated.

Most of the experiments dealing with organic oxidation products from ozonation and from treatment with chlorine dioxide have not been conducted under controlled water or wastewater treatment plant conditions, i.e., dilute aqueous solutions of compounds, low dosages of oxidant, short contact times, no pH control, etc. At the rather low concentrations of organic materials normally encountered in water and wastewater treatment plants, oxidation of one organic compound usually produces other compounds in even lower concentrations. Therefore, recent research studies have utilized the approach of starting with relatively concentrated solutions of organic compounds, and using high dosages of oxidants for prolonged periods of time, in order to force the oxidations to proceed to later stages. Others have started with high concentrations of organic compounds, but have deliberately underdosed with oxidant in order to be sure of producing partially oxidized, early intermediate materials for isolation and study.

We have reviewed the available literature on the use of ozone and chlorine dioxide with the intent of determining specific oxidation products which have been isolated and identified without regard to the relationship of their experimental synthetic conditions (mostly laboratory studies) to actual drinking water treatment plant conditions under which ozone and chlorine dioxide normally are employed. A list of specific organic chemicals known to be formed upon oxidation by ozone and/or chlorine dioxide will be useful to toxicologists and will allow incorporation of these materials into toxicological testing programs, if desired.

Similarly, once it is known with certainty which specific organic compounds can be formed upon oxidation with ozone, reaction conditions can be designed for drinking water treatment processes which will minimize their formation.

Background

The capability of one substance to oxidize another is measured by its "Oxidation Potential", normally expressed in volts of electrical energy (referenced to the hydrogen electrode). The Oxidation Potential is a measure of the relative ease by which an atom, ion, molecule or compound is able to lose electrons, thereby being converted to a higher state of oxidation. If the oxidation potential of substance A is higher than that of substance B, then substance B will be oxidized in the presence of substance A. Oxidation potentials of representative oxidants encountered in drinking water treatment are listed in Table 27.

Although the relative position of an oxidant in this table is indicative of its ability to oxidize other materials, it does not indicate how fast one material will be oxidized by another, nor how far toward completion the oxidation reaction will proceed. One cannot tell from oxidation potentials alone whether a specific organic compound will be oxidized completely (to carbon dioxide and water) or only to the first of several intermediate stages.

One significant fact can be learned from this table, however, at this point. As will be discussed in detail later in this section, it is rare that organic compounds treated with even as powerful an oxidant as ozone will be converted totally to carbon dioxide and water, under conditions normally encountered in drinking water treatment plants.

Therefore, no other commonly employed and less powerful water treatment oxidant (such as chlorine, bromine, chlorine dioxide, etc.), all of which have lower oxidation potentials than ozone, will oxidize an organic material completely to carbon dioxide and water if ozone will not.

All oxidants weaker than ozone will be less effective than ozone in converting organic compounds to carbon dioxide and water, and thus may produce higher quantities of partially oxidized organic materials under water treatment plant conditions. This fact has not yet been fully appreciated by the water supply industry. Once understood, a logical conclusion

TABLE 27. OXIDATION-REDUCTION POTENTIALS OF WATER TREATMENT AGENTS*

Reactions	Potential In Volts (E°) 25 °C
$F_2 + 2e = 2F^-$	2.87
$O_3 + 2H^+ + 2e = O_2 + H_2O$	2.07
$H_2O_2 + 2H^+ + 2e = 2H_2O$ (acid)	1.76
$MnO_4^- + 4H^+ + 3e = MnO_2 + 2H_2O$	1.68
$HClO_2 + 3H^+ + 4e = Cl^- + 2H_2O$	1.57
$MnO_4^- + 8H^+ 5e = Mn^{2+} + 4H_2O$	1.49
$HOCl + H^+ 2e = Cl^- + H_2O$	1.49
$Cl_2 + 2e = 2Cl^-$	1.36
$HOBr + H^+ + 2e = Br^- + H_2O$	1.33
$O_3 + H_2O + 2e = O_2 + 2OH^-$	1.24
ClO_2 (gas) $+ e = ClO_2^-$	1.15
$Br_2 + 2e = 2Br^-$	1.07
$HOI + H^+ + 2e = I^- + H_2O$	0.99
ClO_2 (aq) $+ e = ClO_2^-$	0.95
$ClO^- + 2H_2O + 2e = Cl^- + 2OH^-$	0.9
$H_2O_2 + 2H_3O^+ + 2e = 4H_2O$ (basic)	0.87
$ClO_2^- + 2H_2O + 4e = Cl^- + 4OH^-$	0.78
$OBr^- + H_2O + 2e = Br^- + 4OH^-$	0.70
$I_2 + 2e = 2I^-$	0.54
$I_3 + 2e = 3I^-$	0.53
$OI^- + H_2O + 2e = I^- + 2OH^-$	0.49
$O_2 + 2H_2O + 4e = 4OH^-$	0.40

*Handbook of Chemistry & Physics, 56th Edition, 1975-76. CRC Press Inc., Cleveland, Ohio, p. D-141-143.

to be drawn is that when oxidants are used as the terminal step in treating water supplies, the water should contain minimal amounts of dissolved organic materials.

Fundamental Principles

All oxidants react with organic materials by one or more of three different mechanisms:

- Addition
- Substitution
- Oxidation

In some cases, oxidants will react with organic compounds by all three mechanisms, although in sequential steps.

Addition--

Occurs with organic compounds containing aliphatic unsaturation, such as olefins. Chlorine can add across an olefinic double bond to produce a dichloride:

$$RR'C = CRR' \;+\; Cl_2 \longrightarrow \begin{array}{cc} RR'C & - \; CRR' \\ | & | \\ Cl & Cl \end{array}$$

Hypochlorous acid can add across a double bond to form a chlorohydrin:

$$RR'C = CRR' \;+\; HOCl \longrightarrow \begin{array}{cc} RR'C & - \; CRR' \\ | & | \\ Cl & OH \end{array}$$

Ozone can add across a double bond to form an ozonide:

$$RR'C = CRR' \;+\; O_3 \longrightarrow \begin{array}{c} O\text{-}O\text{-}O \\ |\quad\;\; | \\ RR'C - CRR' \end{array}$$

This last reaction occurs readily in non-aqueous solvents, but as soon as water is added, the ozonide hydrolyzes to other products, with cleavage of the former double bond:

$$\underset{\text{ozonide}}{\begin{array}{c} O\text{-}O\text{-}O \\ |\quad\;\; | \\ RR'C - CRR' \end{array}} \longrightarrow \underset{\text{alcohol}}{RR'CH_2OH} \;+\; \underset{\text{ketone}}{O=CRR'}$$

Substitution--

Involves replacement of one atom or functional group with another. For example, chlorine can react with phenol to produce o-chlorophenol. In this reaction the ortho-hydrogen atom is replaced by chlorine:

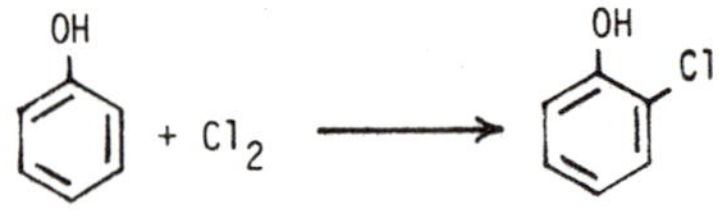

phenol o-chlorophenol

Oxidation--

Involves the introduction of oxygen into the organic molecule, with or without degradation of the organic compound. For example, oxidation of phenol with either chlorine, chlorine dioxide or ozone can produce catechol as a first oxidation product:

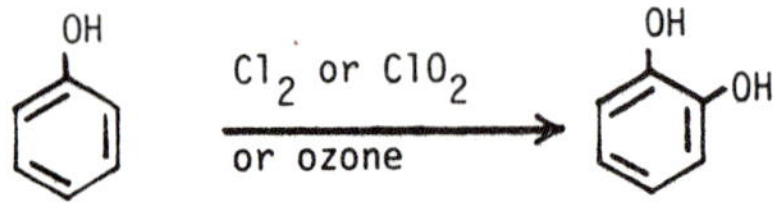

catechol

This specific reaction also can be viewed as an insertion reaction, whereby oxygen is inserted between the ring carbon and hydrogen to form the hydroxy group on the ring.

Oxidation also can involve cleavage of carbon-carbon bonds to produce fragmented organic compounds. For example, ozonation of styrene produces formaldehyde, benzaldehyde, and benzoic acid[30]:

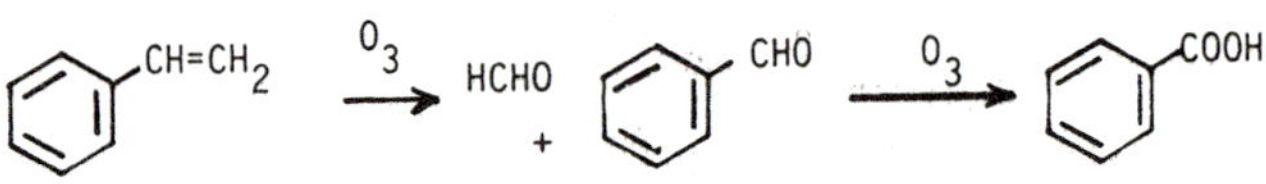

styrene benzaldehyde benzoic acid

At the last stage in treatment of organic compounds with oxidants, oxidation also can involve production of carbon dioxide and water:

$$HCOOH + O_3 \longrightarrow CO_2 + H_2O$$

formic acid

CHLORINE

There is considerable literature dealing with the reactions of chlorine with organic compounds in aqueous solution. This will not be reviewed in detail, but representative reactions will be cited to illustrate the significant points for drinking water treatment scientists. Before discussing specific organic oxidation products, however, a brief synopsis of the chemistry of chlorine in water is in order, since many of the aqueous reactions of "chlorine" do not involve free chlorine, but rather hypochlorite ion or hypochlorous acid.

Reactions of Chlorine With Water

When chlorine gas is dissolved in water the following reaction occurs[78]:

$$Cl_2 + H_2O \rightleftharpoons HOCl + H^+ + Cl^-$$

The HOCl (hypochlorous acid) can ionize in water as follows:

$$HOCl \rightleftharpoons H^+ + OCl^-$$

At pH below 6.5, the solution contains almost 100% of the chlorine in the form of HOCl[79], but at pH 9.0 or higher, nearly all of the chlorine is in the form of hypochlorite ion, OCl^-. In the pH range of 6.5-9.0, therefore, dissolved chlorine will be present as both HOCl and hypochlorite ion, in varying amounts depending upon the specific pH.

As indicated in the table of Oxidation Potentials, HOCl is a stronger oxidizing agent (1.49 v) than is free chlorine (1.36 v), so that HOCl is actually more to be desired when using chlorine as an oxidant in aqueous solution. Hypochlorite ion is the weakest of the three species (0.9 v).

Reactions of Chlorine With Organic Compounds

The direct reaction of chlorine with unsaturated compounds proceeds mostly by addition. For example, maleic acid treated with chlorine will produce dichloromaleic acid:

$$HOOC\text{-}CH = CH\text{-}COOH + Cl_2 \longrightarrow HOOC\text{-}CH(Cl) - CH(Cl)\text{-}COOH$$

However, since chlorine forms hypochlorite ion or hypochlorous acid when added to water, reactions of HOCl with organic compounds are more germane to this discussion.

Ethylene reacts with an alkaline solution of chlorine to give ethylene chlorohydrin. The reaction was once thought to involve the simple addition of HOCl across the ethylenic double bond. However, the reaction mechanism is now understood to be a two-step process in which chlorine cation attacks one of the unsaturated carbon atoms, followed by combination with an hydroxide ion. The overall effect, however, still can be viewed as addition of HOCl across the unsaturation[80]:

$$H_2C = CH_2 + HOCl \longrightarrow H_2C(Cl) - CH_2(OH)$$

Treatment of ethylene chlorohydrin with strong alkali results in elimination of the elements of hydrogen chloride with formation of ethylene oxide, which is an epoxide containing no chlorine:

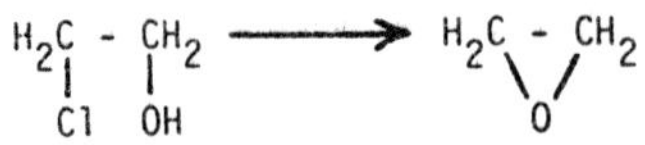

Thus there is the possibility of forming epoxides by chlorination of drinking water supplies if aliphatic double bonds are present in dissolved organic materials, and if alkaline conditions are encountered during treatment -- such as lime treatment after prechlorination.

Hypobromous acid adds similarly across aliphatic double bonds. In both cases, the organic chlorohydrin compounds, which may be isolated as intermediates, contain halogen.

The Haloform Reaction--

The uni-halogen containing trihalomethanes, chloroform, bromoform and iodoform, are formed by a process known as the haloform reaction. The reaction seqence occurs when acetone is treated with hypochlorite in alkaline solution:

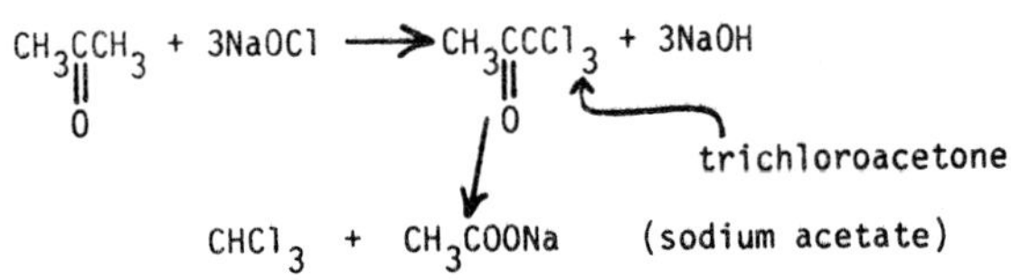

In the first phase of the reaction, one of the methyl groups in acetone becomes fully substituted by chlorine. Since sufficient alkali is produced in this first step to cause cleavage in the second step, both reactions proceed together[80]. Notice that in addition to chloroform, a non-halogen-containing, oxidized organic compound (sodium acetate) also is produced.

Recently, Suffet, et al.[81] have identified 1,1,1-trichloroacetone in chlorinated drinking water supplies in two Philadelphia treatment plants. One of these plants employs post-ammoniation for disinfection, but the trichloro intermediate in the haloform reaction was produced by chlorination and not upon chlorammoniation.

Chloroform also is produced by the action of hypochlorite upon ethanol, which is first oxidized to acetaldehyde, followed by chlorination[80]:

$$CH_3CH_2OH + NaOCl \longrightarrow CH_3CHO + NaCl + H_2O$$

$$CH_3CHO \xrightarrow{3NaOCl} CCl_3CHO + 3NaOH$$

$$CCl_3CHO \longrightarrow CHCl_3 + HCOONa$$

(sodium formate)

Similarly, reaction of HOCl with acyl substituted aromatic compounds proceeds by substitution of chlorine for alkyl hydrogen atoms, followed by oxidation with formation of chloroform. Thus methyl β-naphthyl ketone undergoes hypochlorite oxidation readily[80], to produce chloroform:

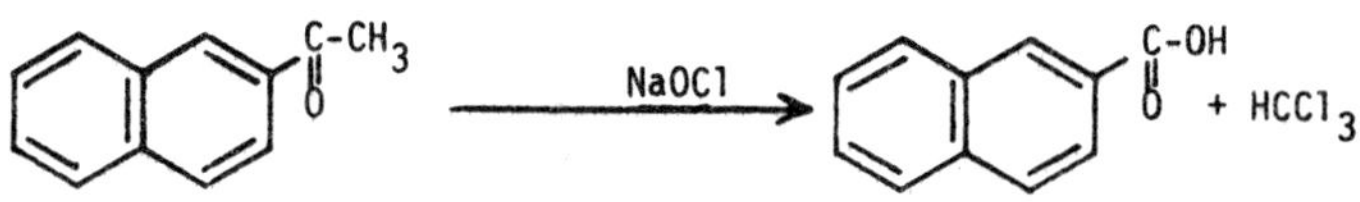

β-naphthoic acid

Similarly, trihalo derivatives of the type $C_6H_5COCX_3$ are intermediates in the conversion of acetophenone into benzoic acid and chloroform by the action of NaOCl or NaOBr[80]:

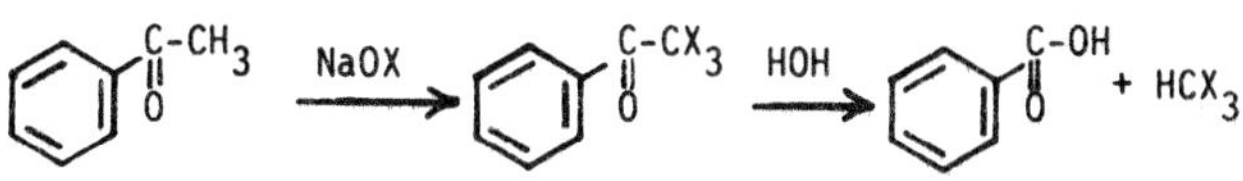

acetophenone

3-Acetophenanthrene treated with hypochlorite produces 75% yields of 3-phenanthroic acid and chloroform[80]:

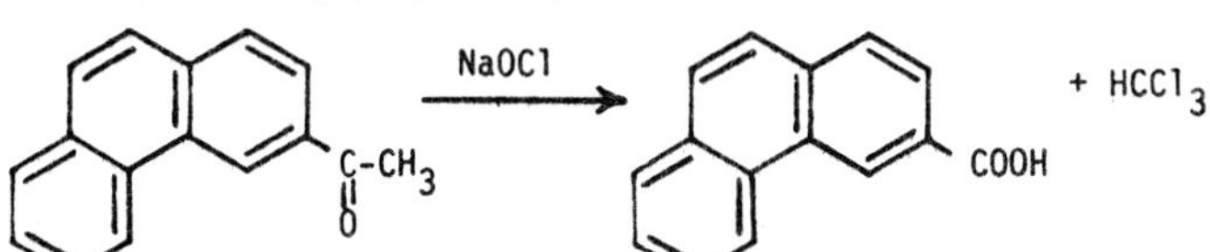

3-phenanthroic acid

Rook[82] has determined the chloroform-producing potentials of a series of aromatic compounds that served as models for the more chemically complex fulvic acids. Resorcinol (1,3-dihydroxybenzene), 1,3-dihydroxynaphthalene, 3,5-dihydroxybenzoic acid and phloroglucinol (1,3,5-trihydroxybenzene) all produced high yields of chloroform when 0.001 M solutions in water were treated with 0.012M chlorine 2 hours at 15°C at pH 7 and pH 11. Rook considers the mechanisms to involve ring chlorination, followed by ring rupture to produce fragments which then form chloroform.

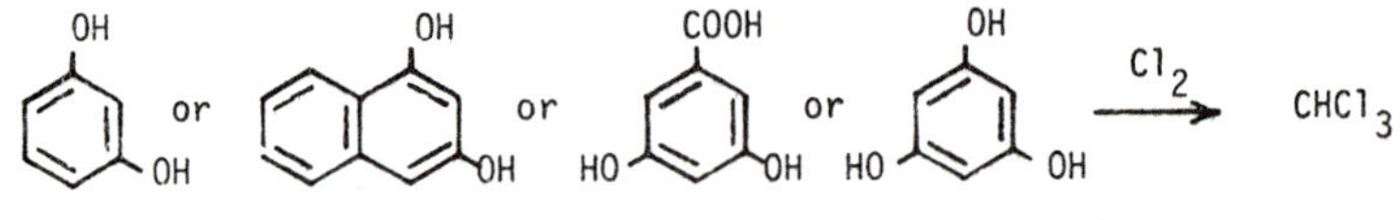

Reactions of Chlorine With Phenol--

Burttschell, et al.[83] followed the chlorination of phenol in laboratory studies simulating drinking water plant conditions. Aqueous solutions of 20 ppm of phenol were treated with 40 ppm solutions of chlorine (2/1 ratio of chlorine/phenol, previously dissolved in water, at pH 8. Products were isolated and identified by paper chromatography, infrared and ultraviolet spectrophotometry. The course of reaction is indicated in Figure 53 (parenthetical percentages are those of products found after 18 hours of reaction). 2,6-Dichlorophenol was shown to be the major product responsible for taste and odors in chlorinated drinking waters.

These same chlorinated phenolic materials were produced when phenol was treated with chloramine, but much more slowly. After 18 hours of reaction with chloramine, nearly all phenol was recovered unreacted. However, after 5.5 days of reaction, significant amounts of 2,6- and 2,4-dichlorophenols were isolated.

Miscellaneous Reactions--

Reaction of NaOBr with nicotinamide produces 67% yields of 3-aminopyridine[80]:

Finally, Mills[84] cites the work of Jolley[85] who analyzed chlorinated municipal wastewaters. The yield of chlorinated organic compounds isolated at any location totalled less than 1% of the chlorine dosed. Furthermore, nearly all the chlorinated compounds identified contained only a single chlorine atom. Thus, most of the chlorine added reacted by oxidation, forming chloride ion and non-chlorinated organic compounds.

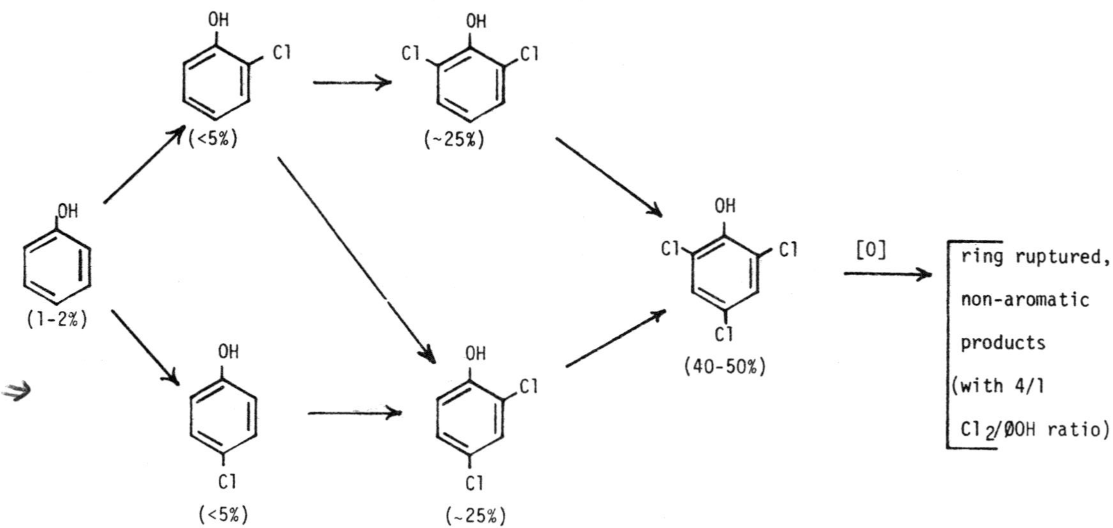

- parenthetical numbers = amounts isolated after 18 hours -- chlorination of a 20 mg/l phenol solution with 40 mg/l chlorine at pH = 8.

Figure 53. Reactions of chlorine with phenol[83].

Summary of Chlorination Reactions

The major points involving chlorine reactions with organic materials are:

- Chlorine in water forms HOCl, over pH ranges normally employed in water treatment plants. HOCl is both an actual chlorinating and oxidizing material in aqueous solution.

- HOCl adds to isolated double bonds present in dissolved organic compounds to form chlorohydrins. In alkaline solution, chlorohydrins eliminate HCl, producing epoxides.

- The haloform reaction is unique to chlorine, bromine and iodine and involves both halogenation and oxidation by hypohalite. With chlorine, an organic methyl group adjacent to an activating carbonyl group first is perchlorinated, then is cleaved oxidatively. This cleavage produces a molecule of chloroform <u>and</u> a molecule of an oxidized organic material which does not contain halogen -- normally an organic acid.

- Chlorination of phenol proceeds stepwise, producing mono-, di- and tri-chlorophenols. Ring chlorination is maximum at 2/1 chlorine/-phenol molar ratios.

- Chlorination of phenol using 4/1 molar ratios of chlorine/phenol or higher proceeds past the ring chlorination stage by oxidative ring rupture, producing aliphatic compounds.

- Many aromatic di- and poly-hydroxy compounds produce chloroform when treated with chlorine.

- Chlorination of organic materials can produce a wide variety of organic oxidation products, which do not necessarily contain chlorine. Non-chlorinated aldehydes and acids have been isolated from alcohols. Non-chlorinated organic acids (and chloroform) are produced upon chlorination of methyl ketones. If all chloroform measured in a given water supply was produced by the haloform reaction involving carbonyl-activated methyl groups, there will also be present an equivalent molar amount of non-halogen-containing organic acids, or their degration products.

REACTIONS WITH OZONE

Reactions With Phenol

Eisenhauer[86] ozonized aqueous solutions of phenol for 30 minutes (until phenol was destroyed) and isolated catechol, p-quinone, <u>cis</u>-muconic acid, oxalic acid and fumaric acid:

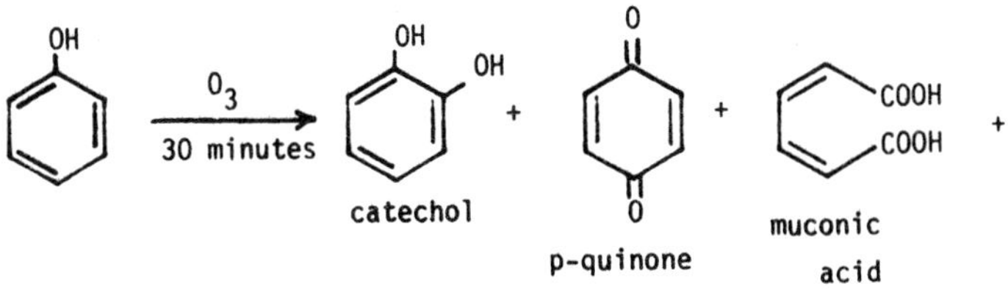

HOOC-CH=CH-COOH (fumaric acid)

+ HOOC-COOH (oxalic acid)

When ozonation was conducted for only 10 minutes, Eisenhauer isolated a 20% yield of catechol, but only 70% of the phenol was destroyed. This indicates that upon oxidation of phenol, other organic compounds (10%) or CO_2 are produced along with the catechol.

Gabovich, et al.[87] treated 10 mg/l aqueous solutions of phenol with ozone dosages of 0.7 to 2.3 mg/l. To attain 90% destruction of phenol (to 1 mg/l concentration) required 1.85 mg of ozone/mg of phenol; complete destruction of phenol required 2.3 mg ozone/mg phenol.

Bauch, et. al.[88] found monobasic and polybasic (aliphatic) acids upon ozonation of water solutions of phenol. They concluded that oxidation of phenol by ozone proceeds via the ozonide and produces hydrogen peroxide. Initial phenolic oxidation products themselves consume additional ozone.

Bauch & Burchard[89] ozonized aqueous solutions of phenol and isolated and identified maleic acid, tartaric acid, glyoxylic acid, oxalic acid and CO_2:

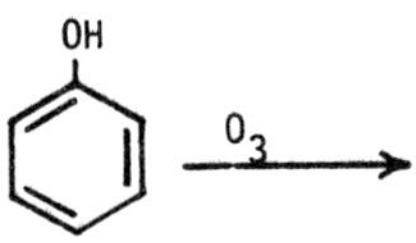

HOOC-CH=CH-COO + HOOC-CH(OH)-CH(OH)-COOH +

maleic acid tartaric acid

HOOC-CHO + HOOC-COOH + CO_2

glyoxylic acid oxalic acid

Smith, et al.[90] found that the rate of disappearance of phenol upon ozonation in water is increased by combining ultrasonics and/or Raney nickel with ozonation. After 2.5 hours of ozonation, an initial solution of 500 mg/l phenol showed the complete absence of phenol. On the other hand, the COD of the original solution decreased only slowly, and was still fairly high after the phenol had disappeared, indicating the presence of intermediate oxidation products. Considerable carbon loss indicated the formation of some CO_2.

Mallevialle[91] ozonized 100 to 200 mg/l aqueous solutions of phenol with 25 mg/l ozone doses and identified catechol, o-quinone, hydroquinone and p-quinone as oxidation products:

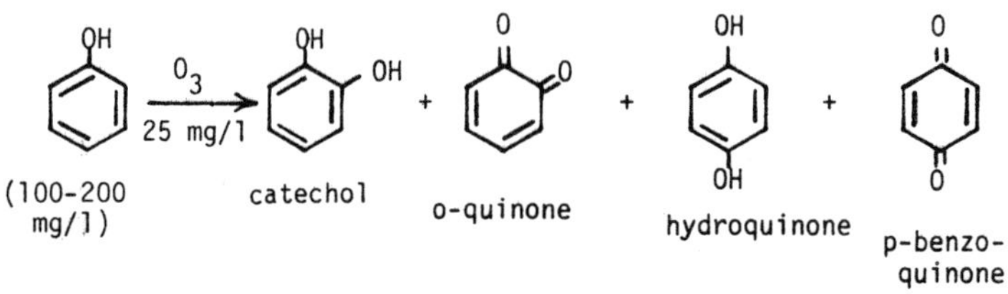

Spanggord & McClurg[92] were the first investigators to identify resorcinol as an initial oxidation product, along with catechol, upon ozonation of phenol in water:

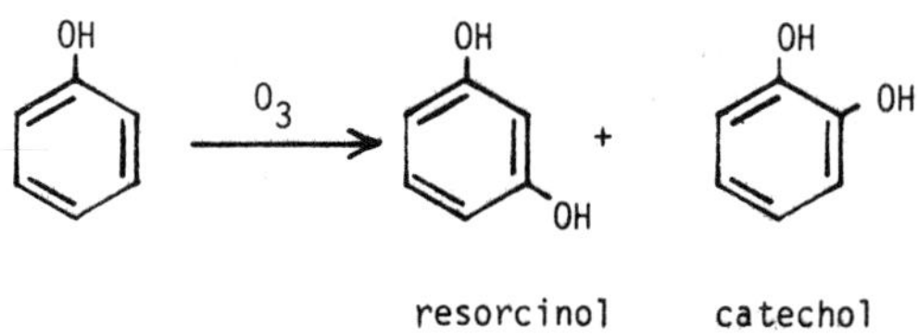

Gould & Weber[93] have made the most complete study to date on the ozonation of phenol. They found that the early oxidation products (catechol and hydroquinone) are further oxidized as ozonation continues, and fall to relatively insignificant concentrations as the reactions proceed. Glyoxal is formed by ring rupture, but itself decreases to a low concentration level as the reactions proceed. Glyoxylic acid is the main oxidation product isolated after 30 minutes of treatment with ozone, together with smaller amounts of oxalic acid.

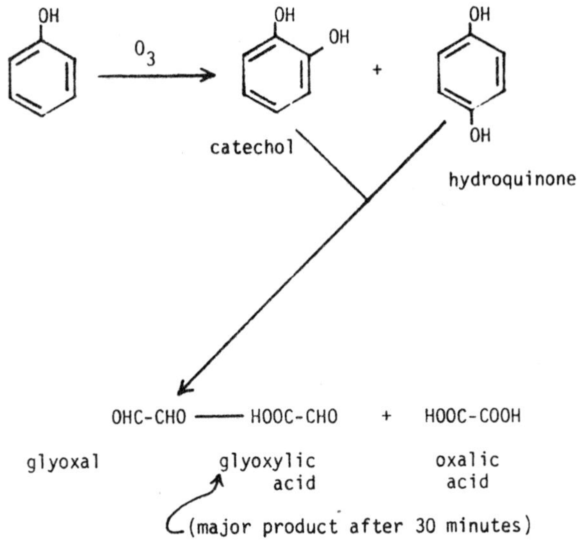

Throop[94] showed that ozone dosages of 5.32 mg/l produced non-detectable quantities of phenol, starting with concentrations of 110 ppb of phenol in water. This dosage is equivalent to 48 parts ozone/part of phenol. However, ozone dosages of 25.5 mg/l (200 parts ozone/part phenol) were required to produce a measureable (trace) amount of residual ozone in the solution. This confirms that although ozone rapidly destroys phenol itself, significant amounts of ozone-demanding oxidation products are formed.

Reactions With Other Phenols

Hillis[95] studied the oxidation of 14 phenols with ozone over the pH range 4 to 10, but did not identify oxidation products. With 30 mg/l concentrations of phenols and ozonation conducted 4 to 12 minutes [except for pentachlorophenol (PCP), which required 35 minutes], residual concentrations of phenols of less than 0.10 mg/l were obtained. However, COD values were reduced only about 50%, indicating that organic carbon-containing oxidation products still present.

With solutions of phenol, phenolsulfonic acid, hydroquinone and pyrogallol, COD was destroyed steadily upon ozonation. However, with pentachlorophenol and β-naphthol, there was a steady rise in COD values during the first few minutes of ozonation, followed by a steady reduction in COD values.

The amount of ozone required to lower the concentration of these phenols from 30 mg/l to below 0.5 mg/l in 10 minutes (except for PCP) was 2.0 to 3.0 g/g phenol (4 to 6 moles ozone/mole phenol).

Bauch, Burchard & Arsovic[88] compared the rates of ozonation of phenol with cresols and xylenols, and also isolated and identified ring ruptured oxidation products. Cresols decomposed more rapidly than phenol, and m-cresol decomposed faster than the o- and p-isomers. Cresols reacted faster with ozone in acid solution than in basic solution. 80% Decomposition of cresols was accomplished with 2 moles of ozone/mole of cresol (85 g ozone/-100 g cresol).

Upon initial ozonation, the methyl group in cresols oxidized to the carboxylic acid. o-Cresol, for example, produced salicylic acid:

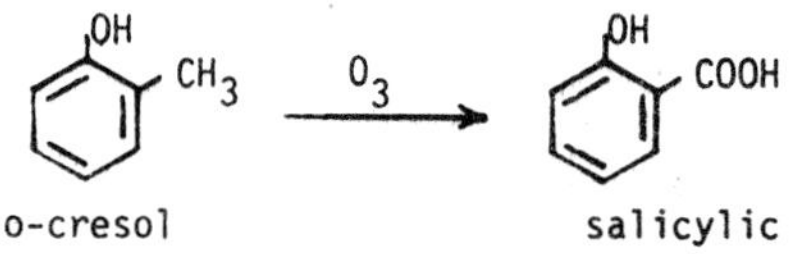

Continued ozonation of cresols ruptured the aromatic ring and produced maleic acid (which further oxidized to mesotartaric acid), acetic acid, propionic acid, glycolic acid, glyoxylic acid, oxalic acid and CO_2:

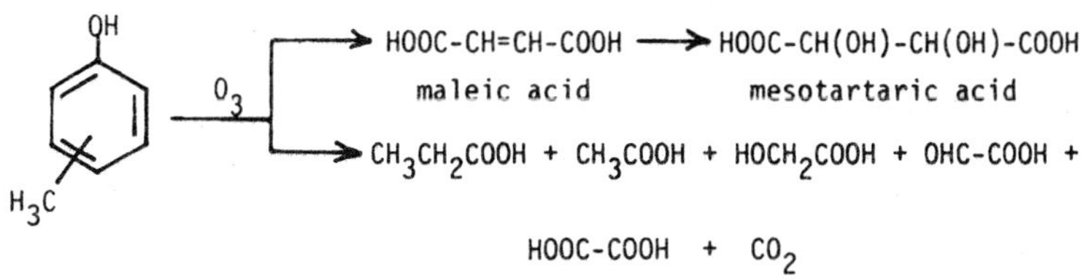

All 3 cresols (o-, m- and p-) formed the same oxidation products upon ozonation. Only the rates of oxidation varied.

Xylenols with ortho or para hydroxy groups reacted fastest with ozone and produced the same oxidation products as the cresols[88]. In addition, 1,2,3- and 1,2,4-xylenols produced diacetyl, glyoxal (which disproportionates to glyoxylic acid), hydroxyphthalic acid and ketoaldehydes:

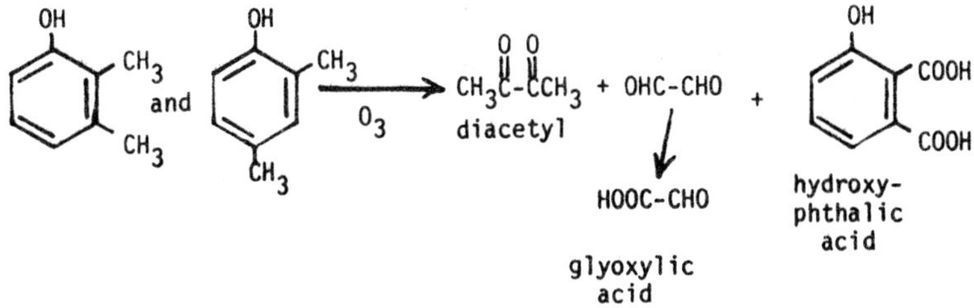

Gilbert[96] ozonized 1 mmole/l of 2-nitro-p-cresol with 10 mg of ozone/-minute until the cresol was destroyed. He found that 90% of the original nitrogen was converted to nitrate ion, indicating rupture of the aromatic ring.

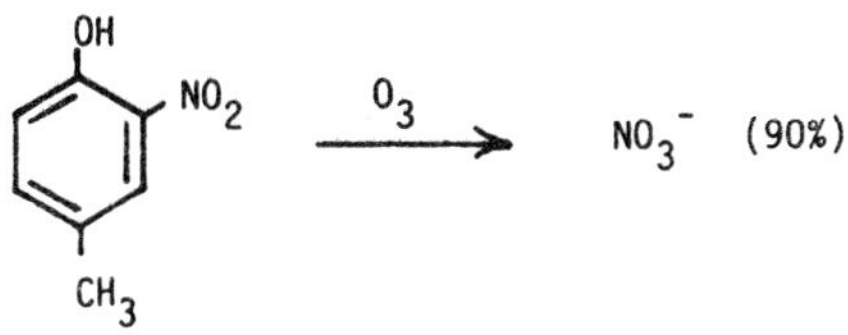

Reactions With Chlorinated Phenols

Shuval & Peleg[97] compared the ozonation of phenol with o-chlorophenol. At the same initial pH, the rates of oxidation with ozone are the same for both compounds, but the rate is fastest at higher pH. Starting at pH 10 the reactions are faster when the pH is maintained at 10, rather than letting it drop (to 2.5) as ozonation proceeds.

In addition, the rate of formation of chloride ion was followed during ozonation of chlorophenol. In all experiments, about 80% of the aromatic chlorine was converted to chloride ion upon ozonation, indicating that covalent C-Cl bonds are broken with ozone. There was an induction period during which the concentration of o-chlorophenol decreased but without formation of chloride ion. On the other hand, after all o-chlorophenol had disappeared, chloride ion still was being produced upon continued ozonation. This indicates that the active oxidation species attacks the aromatic ring at a site or sites other than the chlorine site, producing chlorinated aliphatic compounds as intermediates.

Gilbert[96] ozonized aqueous solutions of 2-chloro-, 4-chloro-, 2,3-dichloro-, 3,5-dichloro- and 2,4,6-trichlorophenols until the phenols could not be detected by gas chromatography and no phenolic functionality could be detected by 4-aminoantipyrine. This required 3.2 to 5 mmoles of ozone/-mmole of phenol. The rate of oxidation increased from mono- to trichlorophenol.

After ozonation, 60 to 95% of the chlorine was found as chloride ion. Ozonation of 4-chlorophenol produced chloride ion at the start of ozonation; chloride ion was found only after 40% of the 2-chlorophenol had been degraded. The different rates of dechlorination are explained in terms of different electron density distributions on the aromatic rings.

2,4-Dichlorophenol produced formic and oxalic acids upon ozonation, in addition to chloride ion:

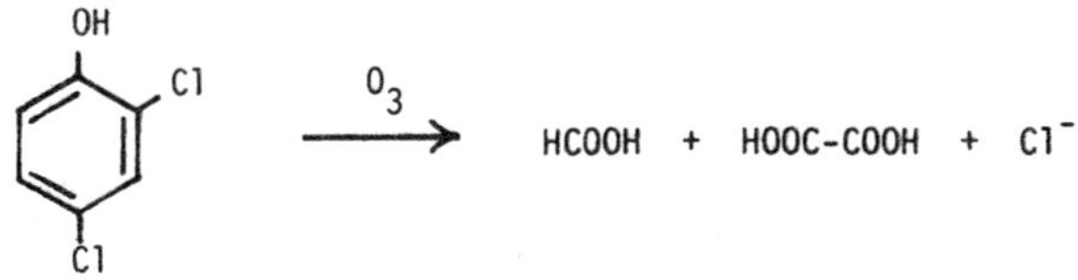

Biodegradability of the ozonized products was higher with increasing degree of oxidation and with decreasing chlorophenol concentration. After total oxidation of the phenols, the COD had been reduced from 200 to 100 mg/l and TOC had been reduced from 72 to 59 mg/l.

After chlorophenols had disappeared, thin layer chromatography techniques indicated the presence of carbonyl or carbonyl/carboxylic acid functionalities. In instances of incomplete dechlorination, chlorinated aliphatic moieties were isolated but not identified.

Gilbert[96] ozonized 4-chloro-o-cresol and identified 67% of its oxidation products. After 80 minutes of treatment with ozone (800 mg of ozone total dosage added to 1 mmole of chlorocresol) none of the starting cresol was present and 100% of the chlorine was found as chloride ion. In the ozonate, methylglyoxal, pyruvic acid, acetic acid, formic acid and oxalic acid were isolated and identified, along with CO_2. The course of reaction is as follows:

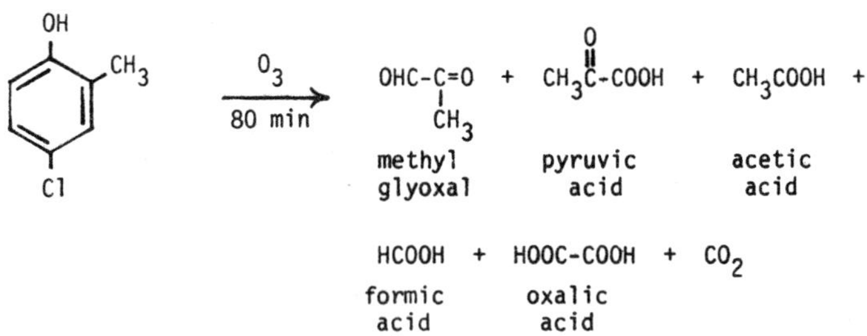

Methylglyoxal was produced from the beginning of the reaction, its concentration reaching a maximum after 60 minutes of reaction, then slowly decreasing. This means that its rate of formation from the cresol is faster than its rate of oxidation.

Pyruvic acid and acetic acid concentrations increased steadily during ozonation, even after complete elimination of the cresol, indicating that these two acids are produced from the initial oxidation products of the cresol.

All TOC was accounted for by these organic compounds at various times during ozonation. Therefore, the above compounds, plus CO_2, water and chloride ion are the only oxidation products of this cresol.

Figure 54 summarizes the reactions of phenols with ozone.

Reactions With Other Aromatics

Ahmed & Kinney[99] ozonized 0.8693 g of 3,8-pyrenequinone 33 hours in water and isolated 0.56 g of 1,2,3,4-benzenetetracarboxylic acid plus acetic acid:

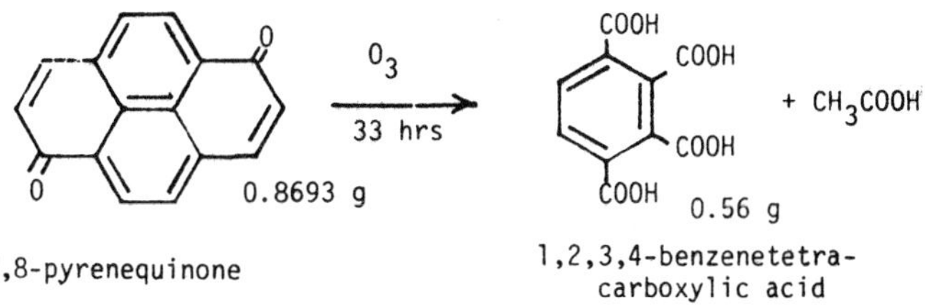

Kinney & Friedman[100] ozonized an alkaline solution of phthalic acid 24 hours and isolated 28% of the carbon as oxalic acid, 34% as CO_2, 3% as acetic acid and 35% as other water soluble acids:

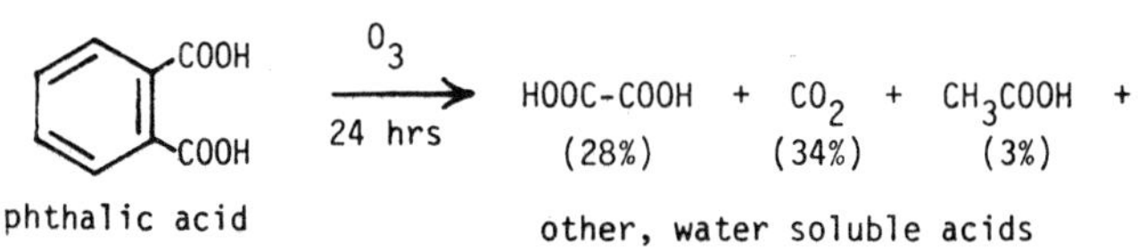

Kinney & Friedman[100] also ozonized a solution of pyrene 24 hours and isolated 2.4% of the carbon as acetic acid, 0.6% as CO_2, 0.1% as oxalic acid and 19.9% as water soluble acids.

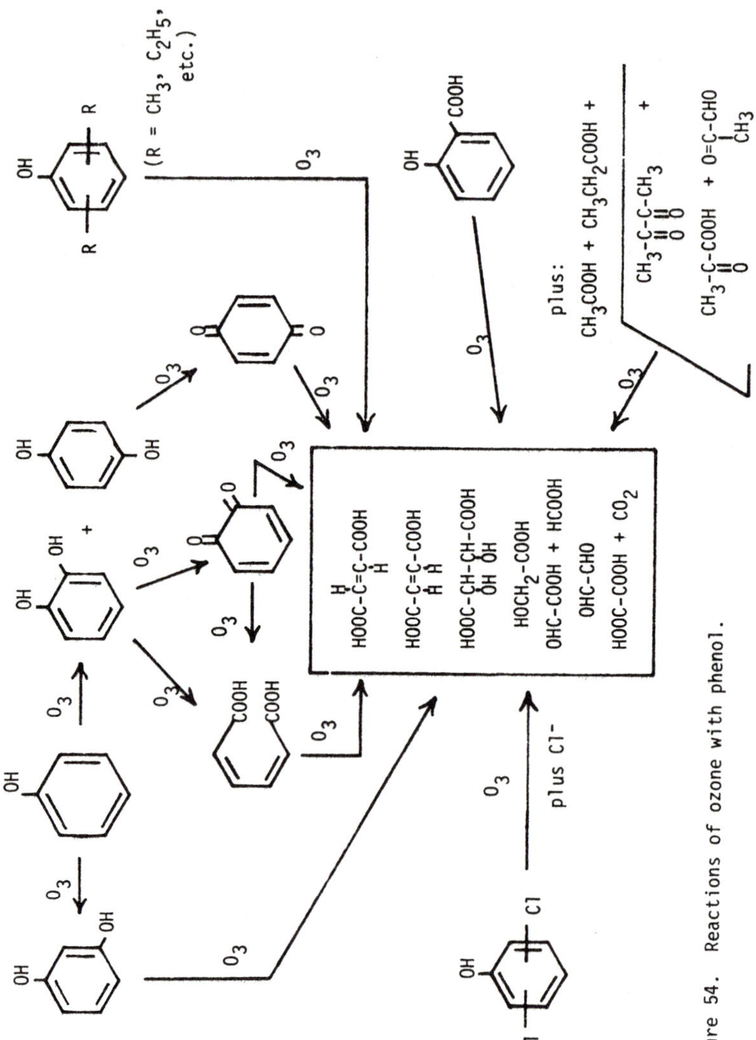

Figure 54. Reactions of ozone with phenol.

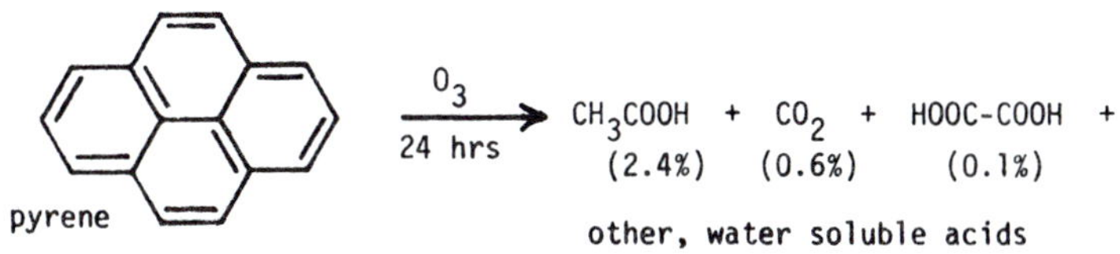

Sturrock, et al.[101] ozonized phenanthrene in a 1/1 water/methanol solution (methanol is resistant to ozone). Water was used to avoid the formation of explosive ozonides or peroxides, which are formed under neat or non-aqueous solvent conditions. After ozonation of 15 g of phenanthrene, 1.3 g of unreacted phenanthrene was recovered, along with 16.4 g of oxidation products. These were identified as 2'-formylbiphenylcarboxylic acid, 2'-hydroxymethyl-2-biphenylcarboxylic acid, diphenide and diphenic acid.

It is of significance that only one ring in phenanthrene was opened by treatment with ozone. This indicates that the material is fundamentally resistant to ozone.

Il'nitskii, et al.[102] found that 3,4-benzopyrene in distilled water was destroyed much more rapidly than when the pyrene was added to raw drinking water containing added soil particles. In 1 minute, ozone destroyed 61% of the pyrene in distilled water, but only 33% in naturally occurring raw water. In 2.5 minutes of ozonation, 100% of the pyrene was destroyed in distilled water, but only 60% was destroyed in this time (same concentration) in raw water containing added soil particles.

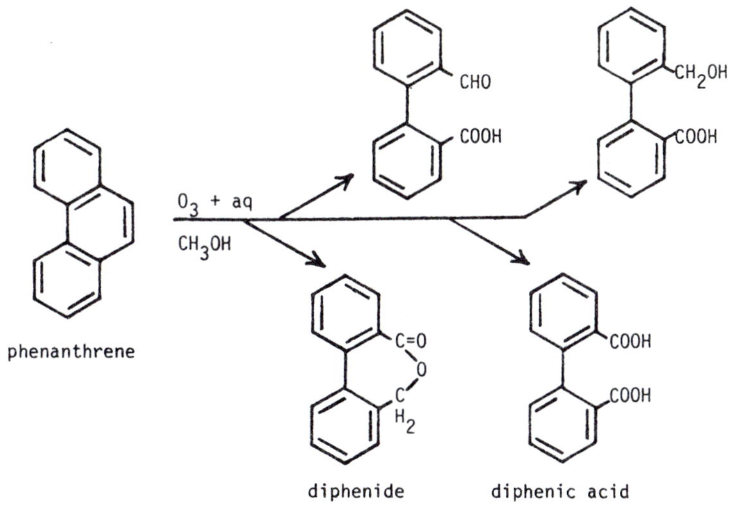

These investigators concluded that 3,4-benzopyrene is adsorbed onto fine soil particles and thus is "protected" from oxidation. In designing the ozonation treatment for waters containing 3,4-benzopyrene, an effective filtration step should precede ozonation. Protection of organic compounds from oxidation by naturally occurring colloidal materials also was noted by Mallevialle[48] in studies with aldrin (see-- "Reactions With Pesticides").

Il'Nitskii[103] ozonoized 0.6 to 1.2 mg/l concentrations of 3,4-benzopyrene in water maintaining a 0.4 mg/l residual of ozone. In 7.5 minutes the pyrene had been oxidized to below the detectable limit.

Gabovich, et al.[104] compared the oxidation rate of 3,4-benzopyrene when treated with chlorine and with ozone. Chlorine reduced the concentration of the pyrene 5 to 10 times in 0.5 to 2 hours. Ozonation for 3 to 5 minutes reduced the concentration 10 to 50 times.

3,4-Benzopyrene (4 μg/l) treated 3 minutes with 2.5 mg/l of ozone was reduced in concentration to 0.06 μg/l; treating the same concentration 3 minutes with 4.5 mg/l of ozone lowered the pyrene concentration to 0.04 μg/l[4].

Reichert[139] dissolved 3,4-benzopyrene in 1 ml of acetone, then diluted this to 1000 ml with water. Samples containing 1 to 100 microg/l of pyrene were ozonized with 0.5-1.5 mg/l doses. In 30 minutes of ozonation, 99% decomposition of the pyrene was observed.

Gabovich, et al.[87] also studied the rates at which ozone would reduce the concentrations of other aromatic compounds in water. Diethylbenzene in concentrations of 10 to 100 mg/l upon treatment 7 to 10 minutes with quantities of ozone similar to those used in drinking water treatment (1 to 5 mg/l) was reduced in concentration to 0.5 to 0.8 mg/l. 2,4-Dinitrophenol at 50 mg/l was reduced in concentration to 0.35 mg/l using 2 mg ozone/mg of phenol; using 5 mg ozone/mg phenol, the final concentration of 2,4-dinitrophenol was lowered to 0.05 mg/l.

Chlorobenzene reacts with ozone slower than does phenol, probably because of its lower solubility in water, but gives the same oxidation products as does phenol. In addition, HCl, chlorotartaric acid and o-, m- and p-chlorophenols are formed[88]:

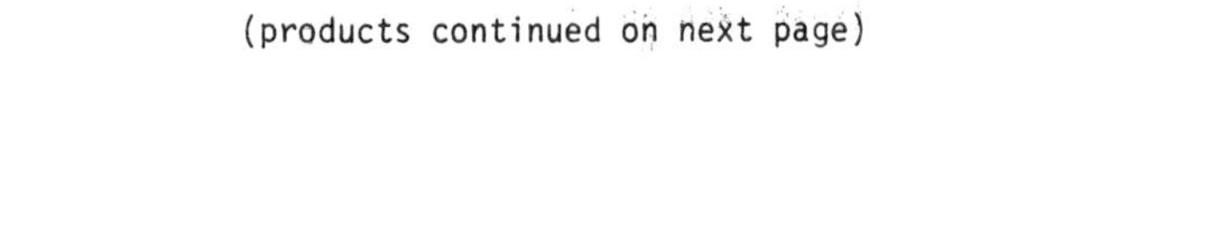

(o-, m- and p-chlorophenols)

(products continued on next page)

+ HOOC-CH(OH)-C(Cl)-(OH)-COOH + HCl +

same ring-ruptured, aliphatic
oxidation products as from
ozonation of phenol

Chlorocresols, chlorophenols, naphthols, thiophenols and polyhydroxyphenols give similar oxidation products as do phenols upon ozonation[88].

Hoigné[106] showed that ozonation of benzoic acid caused 10% decarboxylation to produce CO_2.

Mallevialle[91] ozonized 100 to 200 mg/l aqueous solutions of salicyclic acid. The TOC remained constant during the first 10 minutes of ozonation (25 mg/l total dosages), but then dropped steadily. Phenol, catechol and three unidentified phenols were isolated from the ozonate. 2,3-Dihydroxybenzoic acid was shown to be absent. Three moles of ozone/mole of salicylic acid were required to destroy all of the starting acid. Infrared analyzes showed the products to have strong -OH and -COOH absorptions, indicating that a mixture of carboxylic acids was formed.

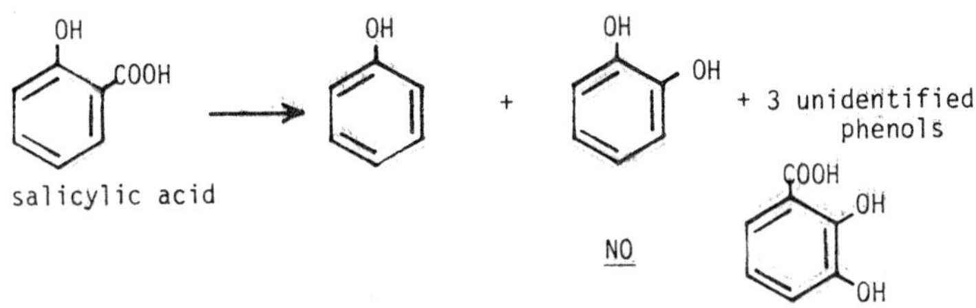

Spanggord & McClurg[92] ozonized aqueous solutions of N,N-diphenylhydrazine hydrochloride at pH 7 and identified ring- and N-hydroxylated derivatives, plus free uns-diphenylhydrazine:

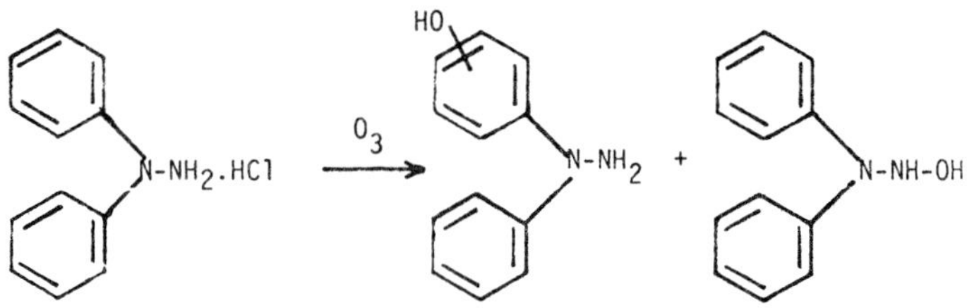

Jürs[107] reviewed the then current literature and concluded that ozonation of benzene (in benzene) produced a triozonide which, when treated with water, rapidly decomposes to form glyoxal, glyoxylic acid and oxalic acid:

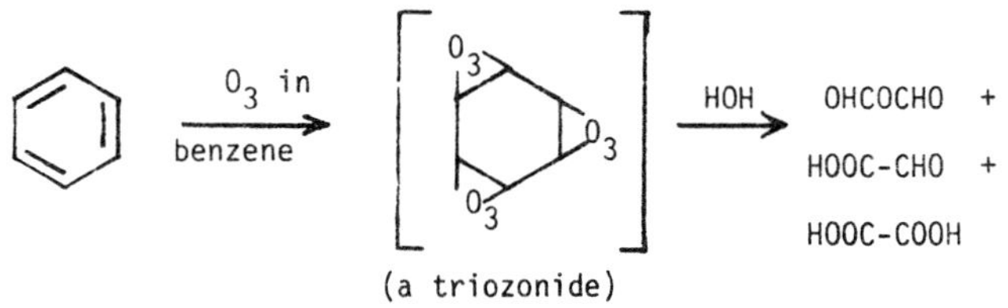

Jürs also concluded that ozonation of phenol, even in water, proceeds through a triozonide, which is transient, decomposing into aldehydes, oxalic acid, glyoxal and hydroperoxides:

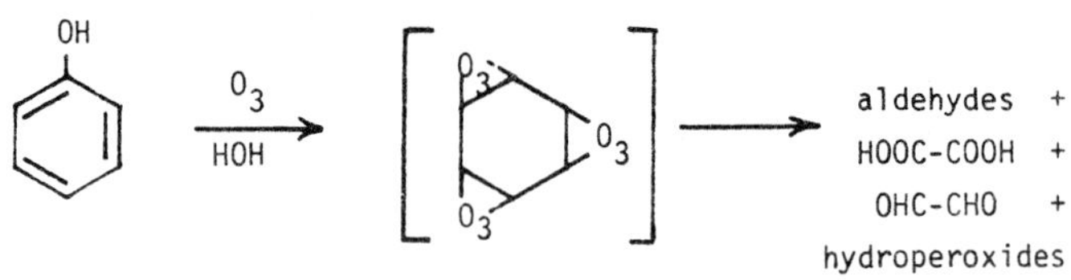

Ozonation of naphthalene in water produced salicylic acid[107].

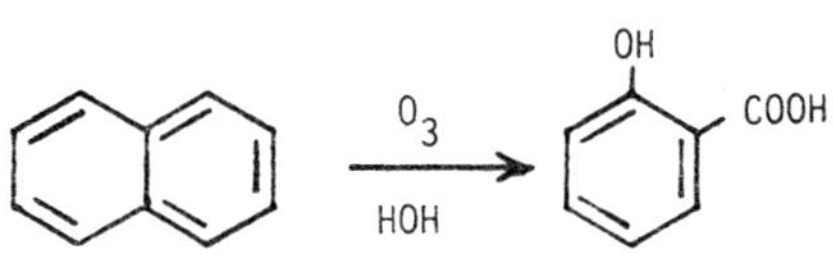

Ozonation of indole and skatole proceeds through ozonides which, when treated with water, form o-aminobenzaldehyde and/or o-aminobenzoic acid[107].

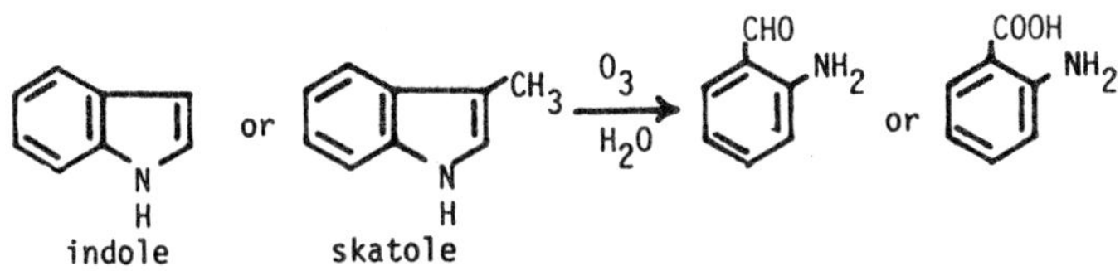

Yocum[108] studied the ozonation of styrene under aqueous conditions, but only to the benzoic acid stage of oxidation. He found that initial cleavage of the exocyclic double bond occurred rapidly, producing benzaldehyde and formaldehyde. Further oxidation of formaldehyde to CO_2 and water occurred rapidly, as did the oxidation of benzaldehyde to benzoic acid. This last step (benzaldehyde to benzoic acid) required 1.47 moles of ozone/mole of aldehyde.

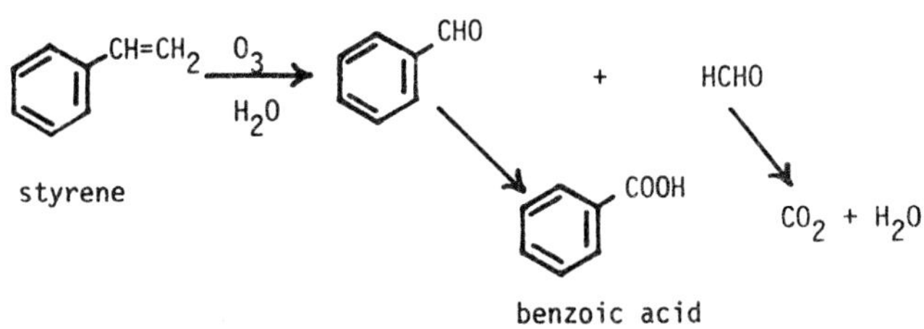

Further ozonation of benzoic acid solutions caused attack at the o-, m- and p- positions on the aromatic ring. After 30 minutes of ozonation of a 300 mg/l solution of acid with 4 moles ozone/mole ØCOOH, the benzoic acid was 85% oxidized to other products.

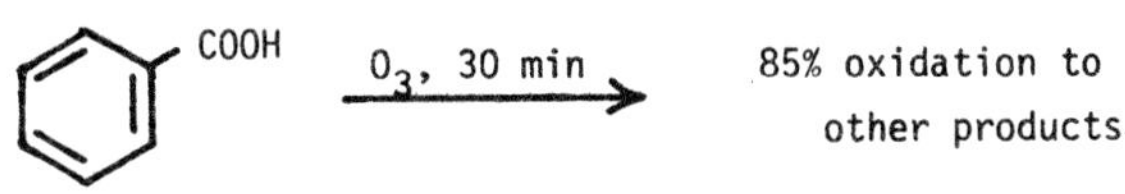

Finally, Yocum[108] determined that the biodegradability of ozonized styrene is much higher than that of styrene itself. The starting BOD_5/TOC ratio was 0.47, but after 150 minutes of ozonation it had increased to 2.69. This shows that ozonation in aqueous solution can convert relatively non-biodegradable compounds into compounds which are biodegradable.

Ozonation of naphthalene-2,7-disulfonic acid[96] for 120 minutes produced formic, oxalic and mesoxalic acids, (accounting for 25% of the TOC) sulfate ion, plus organic carbonyl compounds and organic sulfonic acids. After 300 minutes of ozonation, nearly complete desulfonation was achieved (nearly quantitative yield of sulfate ion). Glyoxal and mesoxalic acid semialdehyde were also identified. The BOD_5/COD ratio increased from 0 to 0.8, indicating that the oxidation products are biodegradable.

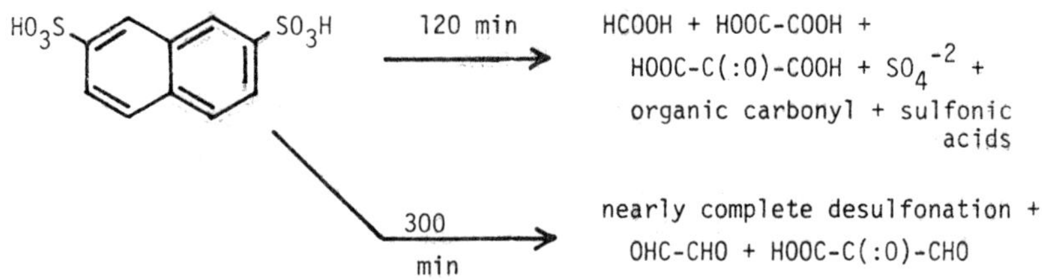

Ozonation of 4-aminobenzoic acid[96] gave formic and oxalic acids, ammonia and nitrate. After 80 minutes of ozonation, only 70% of the organically-bound nitrogen was measured as ammonia and nitrate. This indicates that organic compounds containing nitro or amino groups still were present. The BOD_5/COD ratio increased from 0 to 0.4.

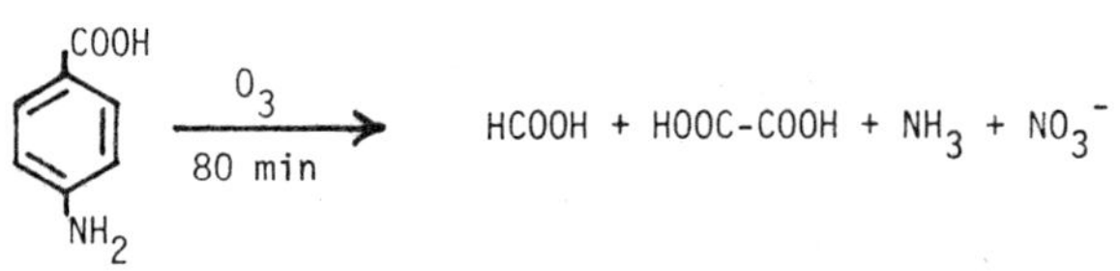

Chian & Kuo[109] ozonized aqueous solutions of o-toluidine and found acetic and oxalic acids:

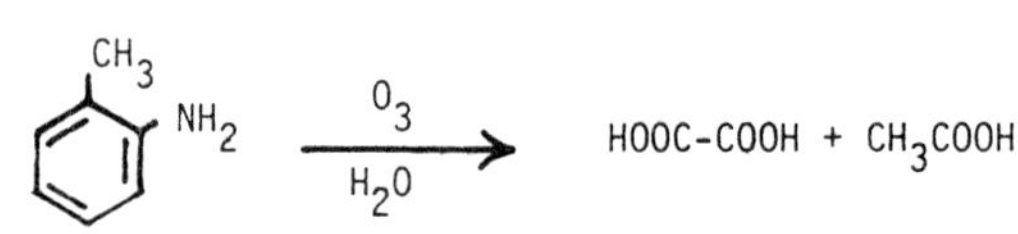

o-toluidine

Ozonation of aqueous solutions of N,N-diethyl-m-toluamide produced formic, acetic and oxalic acids[109].

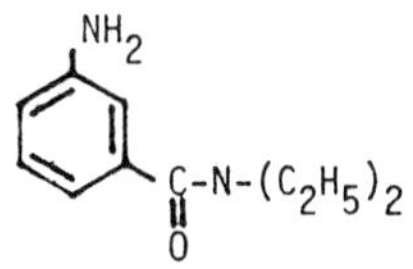

$$\xrightarrow{O_3} \text{HCOOH} + \text{CH}_3\text{COOH} + \text{HOOC-COOH}$$

Gilbert[110] found that 3 to 8 kg of ozone are required to remove 1 kg of aromatic sulfonic acids from aqueous solutions. Ozonation of p-toluene sulfonic acid in water formed organic peroxides, the concentrations of which decreased to zero in 120 minutes, plus H_2O_2.

Many papers by Prengle and his co-workers[111,112,113], which have appeared since 1973 describe the combination of ozone with ultraviolet radiation. The combination oxidizes compounds at a faster rate than does ozone alone. Prengle, et al.[112] followed the UV/ozonation of pentachlorophenol (PCP) by gas chromatography. After 1 hour of treating a 7 mg/l aqueous solution at pH 9.6 with ozone/UV, the PCP concentration had dropped and the amount of chlorine found in solution as chloride ion was more than 50% of that available. In addition, the TOC value had dropped. This indicates that the aromatic ring had been ruptured, forming CO_2, in addition to chloride ion.

Similarly, a 100 mg/l solution of o-dichlorobenzene treated 30 minutes by UV/ozonation showed 100% destruction of the aromatic compound, but only 50% of the available chlorine was recovered as chloride ion.

Reactions With Aliphatic Compounds

Dobinson[114] ozonized aqueous solutions of malonic acid and identified hydroxymalonic acid and ketomalonic acid as products:

$$\text{HOOC-CH}_2\text{-COOH} \longrightarrow \text{HOOC-CH(OH)-COOH} +$$

$$\text{HOOC-C(=O)-COOH} \quad \text{(ketomalonic acid)}$$

Jürs[107] in reviewing the literature, concluded that in non-aqueous, non-polar solvents (carbon tetrachloride, hexane, etc.) ozonation of aliphatic double bonds proceeds through polymeric ozonides. In non-aqueous polar solvents (such as acetic acid), monomeric ozonides are favored.

In water, however, Jürs concluded that ozonides hydrolyze to peroxydiesters, which further hydrolyze to dialcohols and aldehydes[107]. Peroxyacids decompose in water to aldehydes or acids and H_2O_2. Acids and aldehydes can recombine to form dialcohols.

Pryde, et al.[115] studied the ozonolysis of aliphatic unsaturated materials in water. Methyl oleate (5 g in 25 g of distilled water) ozonized 20 to 30 minutes with 2 to 3% ozone in oxygen produced an 82% yield of compounds containing new carbonyl groups.

$$\underset{\text{methyl oleate}}{CH_3(CH_2)_7\text{-}CH = CH\text{-}(CH_2)_7COOCH_3} \xrightarrow{O_3} \text{compounds containing } \gt C = O \text{ groups}$$

1-Decene (0.2 to 0.5 g) ozonized 16 hours in water produced an aldehyde, a dimethyl acetal, a methyl ester and a hydrocarbon, all from cleavage of the double bond:

$$\underset{\text{decene}}{CH_3(CH_2)_7CH = CH_2} \quad \underset{\text{decene}}{(\text{or } R\text{-}CH{=}CH_2)} \xrightarrow{O_3} RCHO + RCH(OCH_3)_2 + RCOOCH_3 + RCH_3$$

These authors concluded that ozonolysis of dispersions of slightly soluble (in water) organics in water proceeds through hydroperoxides and/or dihydroxyperoxides, both of which rearrange readily in water to produce acids, hydrocarbons + CO_2 or aldehydes + CO.

$$RR'CH = CH_2 \xrightarrow{O_3} RR'CH(OOH)\text{-}CH_2(OH) \;\text{ or }\; RR'CH(OOH)\text{-}CH_2(OOH)$$

$$RR'CH(OOH)\text{-}CH_2(OH) \rightarrow \text{acids} + CO_2 \qquad RR'CH(OOH)\text{-}CH_2(OOH) \rightarrow \text{aldehydes} + CO$$

Dorfmann[116] ozonized cysteine in water and isolated cystine as an initial oxidation product. Cystine then formed unidentified degradation products upon continued ozonation:

$$\underset{\text{cysteine}}{HSCH_2CH(NH_2)COOH} \rightarrow \underset{\text{cystine}}{\text{-}[SCH_2CH(NH_2)COOH]_2} \rightarrow \text{degradation products}$$

Kraznov, <u>et al.</u>[116] ozonized aqueous solutions of aliphatic alcohols and aldehydes. Ethanol, butanol and octanol in dilute aqueous solution with 33 mg/l ozone dosages produced aldehydes, then acids, but <u>no</u> CO_2. The rate of oxidation increased with increasing pH:

$$\left.\begin{array}{l} CH_3CH_2OH \\ CH_3CH_2CH_2CH_2OH \\ CH_3(CH_2)_6CH_2OH \end{array}\right] \xrightarrow[33\ mg/l]{O_3} \text{aldehydes} \longrightarrow \text{acids (no } CO_2)$$

Secondary alcohols readily produced ketones upon ozonation, which oxidized further to organic acids + H_2O[116]. Ketonic intermediates were lower boiling than the alcohols, and were more readily stripped into the gas phase:

$$RR'CH_2\text{-}OH \xrightarrow{O_3} RR'C{=}O \xrightarrow{O_3} R\text{-}COOH + R'\text{-}COOH + H_2O_2$$

Aldehydes formed peroxy acids, which produced organic acids + H_2O_2 on continued ozonation in dilute aqueous media[116].

$$RCH = O \xrightarrow{O_3} R\text{-}\overset{O}{\overset{\|}{C}}\text{-}O\text{-}O\text{-}H \xrightarrow[H_2O]{O_3} RCOOH + H_2O_2$$

Gilbert[110] ozonized 1 liter solutions (0.001 mole/l) of maleic acid at an initial pH of 5. After 50 minutes of ozonation, during which time 195 mg of ozone (39%) was utilized, oxidation of maleic acid was complete. Degradation of maleic acid was rapid, and glyoxylic and formic acids were formed simultaneously as maleic acid disappeared. After complete disappearance of maleic acid, formic acid was ozonized to CO_2 and water, and glyoxylic acid was ozonized to oxalic acid:

$$HOOC\text{-}\overset{H}{C}{=}\overset{H}{C}\text{-}COOH \longrightarrow HOOC\text{-}CHO + HCOOH$$

$$HCOOH \longrightarrow CO_2 + H_2O$$

$$HOOC\text{-}CHO \longrightarrow HOOC\text{-}COOH$$

The total amounts of maleic, glyoxylic, oxalic and formic acids and CO_2 determined on each sample analyzed was equal to the amounts calculated on the basis of this mechanism. Therefore, these specific materials are the only oxidation products of maleic acid. Gilbert also concluded that under his experimental conditions, little oxidation of oxalic acid occurs upon ozonation.

Prengle, et al.[112] studied the UV/ozonation of 1,4-dichlorobutane and of chloroform in water. With ozone alone, 1,4-dichlorobutane was 50%

destroyed in 1 hour of ozonation, whereas UV/ozonation destroyed 100% of the material in 1 hour. The stoichiometric amount of chloride ion was found after 1 hour of UV/ozonation.

UV/ozonation for 2 hours resulted in 80% destruction of chloroform. However, only 25% of the theoretical amount of chloride was recovered, therefore, much of the "destroyed" chloroform must have been stripped out of solution. The rate of decrease of TOC was almost as rapid as the rate of decrease of chloroform concentration, therefore there are few non-volatile, carbon-containing organic oxidation products of chloroform.

Chian & Kuo[109] studied the oxidation of several refractory aliphatic compounds in aqueous solution with ozone and ozone/UV combinations, and with the pH contolled by addition of HCl or NaOH solutions as needed during the reactions. 1-Propanol (408 mg/l) ozonized 1 hour at pH 9 (1980 mg total ozone dose) produced a solution with zero ozone concentration. Propionaldehyde (85 mg/l) was produced, and carbon balance analyses showed that this compound and 1-propanol (145 mg/l remaining) were the only two organic materials present. TOC was reduced from 225 to 210 mg/l, indicating that some CO_2 had been formed.

$$CH_3CH_2CH_2OH \longrightarrow CH_3CH_2CHO + CO_2$$

UV/ozonation of 1-propanol (410 mg/l) under the same conditions produced 95 mg/l of propionaldehyde; 120 mg/l of propanol remained, and TOC was reduced to 205 mg/l.

In both cases of ozonation of 1-propanol, TOC levels did not begin to fall significantly until after aldehyde production had peaked (45 minutes with ozone alone, 40 minutes with UV/ozone).

Propionic acid (490 mg/l) was ozonized at pH 9 for 2 hours (2960 mg total ozone dose). Again, dissolved ozone concentration was zero, indicating that the ozone demand had not been satisfied. Unreacted acid concentration was 235 mg/l and TOC had dropped from 235 to 200 mg/l. Oxidation products were not identified.

For the balance of this work of Chian & Kuo[109] lower concentrations of organic compounds were chosen so that ozone residual concentrations of 4 mg/l or higher were present. Ozonation of propionic acid at pH 7 resulted in 17% TOC reduction and formation of acetone. No monocarboxylic acids were detected and acetone accounted for over 85% of the TOC of the reaction mixture:

$$CH_3CH_2COOH \longrightarrow CH_3\text{-}\underset{\underset{O}{\|}}{C}\text{-}CH_3 \quad \text{(85\% yield)}$$

UV/ozonation of 2-propanol substantially increased the rate of TOC removal. An initial TOC of 110 mg/l was reduced to 20 mg/l (85%) after 135 minutes of UV/ozonation. 2-Propanol disappeared after 30 minutes and the concentration of acetone formed reached its maximum value at this time. On continued UV/ozonation, the acetone level decreased to zero after 75 minutes.

C_2 to C_6 monocarboxylic acids were not detected in the ozonate, but NaOH addition was required throughout the reaction in order to maintain the pH constant. This indicates that formic and/or oxalic acids were produced from the ozonation of acetone.

$$CH_3CH(OH)CH_3 \longrightarrow CH_3\text{-}\underset{\underset{O}{\|}}{C}\text{-}CH_3 \longrightarrow HCOOH \text{ and/or } HOOCCOOH$$

Ozonation of methyl ethyl ketone (MEK) 2 hours at pH 7 produced a small amount of acetate ion. MEK accounted for over 85% of the ozonate TOC. Much of the loss of MEK was attributed to air stripping (40% TOC removed after 2 hours). By contrast, after 2 hours of UV/ozonation, MEK was completely eliminated. Acetate ion reached its maximum concentration after 35 minutes, then decreased to zero at 105 minutes. Trace amounts of acetone and ethanol were detected.

$$CH_3CH_2C(:O)CH_3 \longrightarrow CH_3COO^-$$

Ozonation of acetic acid 2 hours at pH 7 (adjusted at the beginning, but allowed to rise during treatment) resulted in 14% removal of TOC. Glyoxylate anion concentration increased with time, but was always low:

$$CH_3COOH \xrightarrow[UV]{O_3} OHC\text{-}COO^-$$

UV/ozonation of acetic acid was much more rapid. After 75 minutes, no acetate was detected. Glyoxylate ion (at lower concentrations than with ozone alone) remained level in concentration during the first hour, but then decreased to zero. No alcohols were found, nor formic acid, but oxalate ion was found in the ozonate:

$$CH_3COOH \xrightarrow[UV]{O_3} OHC\text{-}COO^- \longrightarrow {}^-OOC\text{-}COO^-$$

Ozonation of diethyl ether 2 hours at pH 9 produced ethyl acetate and acetate ion as the major oxidation products, but small amounts of acetaldehyde, methyl formate, ethanol, acetone and ethyl formate also were isolated. During the later stages of ozonation, TOC reduction rates became slower, when ethyl acetate and acetate ion were the major constituents. Good agreement between the calculated and experimentally determined TOC indicated that all the oxidation products were accounted for and that CO_2, therefore, is <u>not</u> an oxidation product:

$$CH_3CH_2OCH_2CH_3 \xrightarrow{O_3} CH_3COOC_2H_5 + CH_3COO^- \text{ (major) } +$$

$$CH_3CHO + HCOOCH_3 + CH_3CH_2OH + CH_3C(:O)CH_3 + HCOOC_2H_5$$

UV/ozonation of diethyl ether resulted in 94% removal of TOC under the same conditions, because of further oxidation of ethyl acetate and acetate ion to CO_2. Formation of acetate ion occurs by 2 routes: from the ether and from ethyl acetate. Small amounts of the same other organics as obtained with ozone alone also were isolated after UV/ozonation.

Quantitative determination of oxalate and glyoxylate showed that acetate concentration reached zero when oxalate concentration reached its maximum. This indicates that the degradation route of acetate by UV/ozonation is through glyoxylate and oxalate to CO_2. Since the amount of oxalate formed was not equal to the amount of acetate present, it is likely that glyoxylate (the precursor of oxalate) can be oxidized directly to CO_2.

$$CH_3COO^- \longrightarrow OHC\text{-}COO^- + {}^-OOC\text{-}COO^- \longrightarrow CO_2$$

Spanggord & McClurg[92] ozonized aqueous solutions of oleic acid and isolated three organic acids. During ozonation, the pH dropped to 3.8:

$$CH_3(CH_2)_7CH{=}CH(CH_2)_7COOH + O_3 \rightarrow C_8H_{17}COOH + OHC(CH_2)_7COOH + HOOC(CH_2)_7COOH$$

Ozonation of diethylamine produced acetaldoxime plus an unidentified nitrogen-containing compound, not a nitrosamine[92]:

$$(C_2H_5)_2NH \xrightarrow{O_3} CH_3CH{=}NOH + \text{unidentified compound}$$

Spanggord & McClurg[92] ozonized concentrated solutions of ethanol in water (several percent) with very large doses of ozone (several thousand mg) for 1 to 2 hours. Acetaldehyde and acetic acid were identified along with a dihydroperoxide, which was shown to exhibit mutagenic activity:

$$CH_3CH_2OH + O_3 \rightarrow CH_3CHO + CH_3COOH +$$

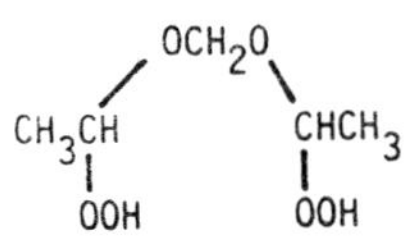

These authors do not believe that the dihydroperoxide will form in drinking water supplies containing low concentrations of ethanol and when ozonized under conditions normally employed in drinking water treatment (low ozone dosages, short contact times).

Gilbert[96] ozonized 1 liter samples of 1 mmole/l aqueous solutions of aliphatic compounds with 10 mg ozone/ minute until the initial compound became undetectable. Oxalacetic acid consumed 1.8 mmole of ozone/mmole of acid, and a 60% yield of oxalic acid was isolated along with mesoxalic acid (both formed by oxidation of glyoxylic acid intermediate) and formic acid:

$$HOOCC(:O)CH_2COOH + O_3 \longrightarrow OHCCOOH + HOOC\text{-}COOH\ (60\%) + HOOC\text{-}C(:O)\text{-}COOH + HCOOH$$

Dihydroxyfumaric acid consumed 1.4 mmoles ozone/mmole of acid and rapidly produced oxalic acid as the major product, plus traces of dihydroxytartaric acid, mesoxalic acid and the semialdehyde of this acid:

$$HOOC\text{-}C(OH)\text{=}C(OH)COOH + O_3 \longrightarrow HOOC\text{-}COOH\ (major) + HOOC\text{-}C(OH)_2C(OH)_2COOH + HOOC\text{-}C(:O)\text{-}COOH + HOOC\text{-}C(:O)\text{-}CHO$$

Malonic acid consumed 4 mmoles ozone/mmole acid. Oxalic acid and mesoxalic acids were the major oxidation products isolated. The concentration of tartronic acid increased from the start of ozonation, then decreased, forming mesoxalic acid and H_2O_2:

$$HOOCCH_2COOH + O_3 \longrightarrow HOOC\text{-}COOH + HOOCCH(OH)COOH \swarrow HOOC\text{-}C(:O)\text{-}COOH + H_2O_2$$

Tartronic acid was converted totally to mesoxalic acid and in 40 minutes of ozonation had used 1 mmole ozone/mmole acid. There was no decrease in TOC during this time nor formation of H_2O_2:

$$HOOC\text{-}CH(OH)\text{-}COOH + O_3 \longrightarrow HOOC\text{-}C(:O)\text{-}COOH$$

Glyoxal disappeared after 50 minutes of ozonation, producing glyoxylic acid which further oxidized to oxalic acid, then disappeared after 60 minutes of total ozonation. No H_2O_2 was formed:

$$OHC\text{-}CHO + O_3 \longrightarrow HOOC\text{-}CHO \longrightarrow HOOC\text{-}COOH$$

Ozonation of the 6-carbon containing muconic acid, HOOCCH=CHCH=CHCOOH, produced 2-carbon containing fragments and only traces of 3-carbon containing compounds. The 4-carbon containing fumaric acid, HOOCCH=CHCOOH, behaved similarly upon ozonation.

Kuo, Chian & Chang[118] treated 2-propanol and acetic acid with ozone and UV/ozone. 2-Propanol formed acetone, which then formed acetic acid and oxalic acid plus traces of formaldehyde and formic acid upon continued ozonation:

$$CH_3CHOHCH_3 + O_3 \longrightarrow CH_3C(:O)CH_3 \longrightarrow CH_3COOH + HOOC\text{-}COOH + HCHO + HCOOH$$

Acetic acid produced glyoxylic acid initially, which rapidly formed oxalic acid, which slowly formed CO_2 upon continued ozonation or UV/ozonation:

$$CH_3COOH + O_3 \rightarrow HOOC\text{-}CHO \rightarrow HOOC\text{-}COOH \rightarrow CO_2$$

Schalekamp[119] reported that the Lake of Zürich water contains various organic aldehydes (heptanal through tetradecanal) in concentrations of 8 to 40 nanog/l. After ozonation (1 to 1.5 mg/l dosages) these aldehydes plus hexanal were present in higher concentrations (up to 920 nanog/l). Passage of ozonized Lake Zürich water through activated carbon reduced the concentrations of these aldehydes to below the levels originally present.

Schalekamp concludes that ozonation should be used only before an adsorption step in processing drinking waters.

Gilbert[110] ozonized aqueous solutions of several aliphatic compounds. In pure solutions at pH 3 to 7, 1 kg of COD is removed from solution with 1.2 kg of ozone. In wastewaters, 2 to 5 kg of ozone are required to remove 1 kg of COD. Ethanol ozonized 350 minutes produced acetaldehyde, acetic acid, formic acid, CO_2 and H_2O_2:

$$CH_3CH_2OH + O_3 \rightarrow CH_3CHO + CH_3COOH + HCOOH + CO_2 + H_2O_2$$

Tartaric acid ozonized 80 minutes at pH 3 and pH 7 produced dihydroxytartaric acid, glyoxal, oxalic acid, mesoxalic acid and H_2O_2:

$$HOOC\underset{OH}{C}H\text{-}\underset{OH}{C}HCOOH + O_3 \rightarrow HOOCC(OH)_2C(OH)_2COOH + OHC\text{-}CHO + HOOC\text{-}COOH + HOOC\text{-}C(:O)\text{-}COOH + H_2O_2$$

Malonic acid ozonized 90 minutes at pH 4 produced tartronic acid, mesoxalic acid, oxalic acid and CO_2 plus H_2O_2:

$$HOOCCH_2COOH + O_3 \rightarrow HOOC\text{-}CH(OH)\text{-}COOH + HOOC\text{-}C(:O)\text{-}COOH + HOOC\text{-}COOH + CO_2 + H_2O_2$$

Finally, Gilbert[110] showed that the presence of H_2O_2 has a remarkable catalytic effect on the oxidative decomposition of oxalic acid. In the absence of peroxide, ozonation of oxalic acid produces CO_2 very slowly, but when small amounts of H_2O_2 are added, oxalic acid produces CO_2 rapidly.

$$HOOC\text{-}COOH \xrightarrow[\text{slowly}]{O_3} CO_2 + H_2O$$

$$HOOC\text{-}COOH \xrightarrow[\text{rapidly}]{O_3 + H_2O_2} CO_2 + H_2O$$

Reactions With Miscellaneous Compounds

Fremery & Fields[120] studied the reactions of cyclic olefins with ozone in aqueous alkaline emulsions containing hydrogen peroxide. In general, α,ω-dicarboxylic acids were isolated, depending upon the specific cyclic olefin starting material:

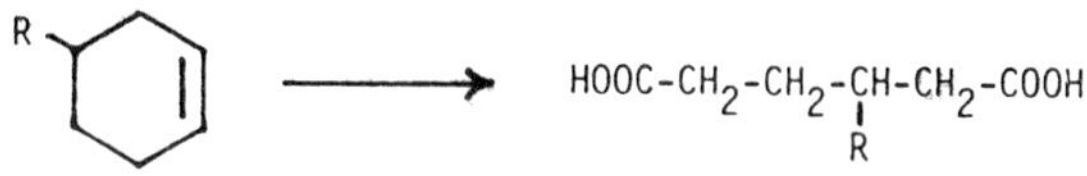

Cyclohexene upon treatment with ozone prepared from oxygen gave a mixture of products containing peroxides and peroxypolymers. These were shown to be side products of the main reaction, for when nitrogen was substituted for oxygen as the ozone carrier gas, cyclohexene gave 20 to 28% adipic acid plus small amounts of δ-hydroxyvaleric acid and its lactone:

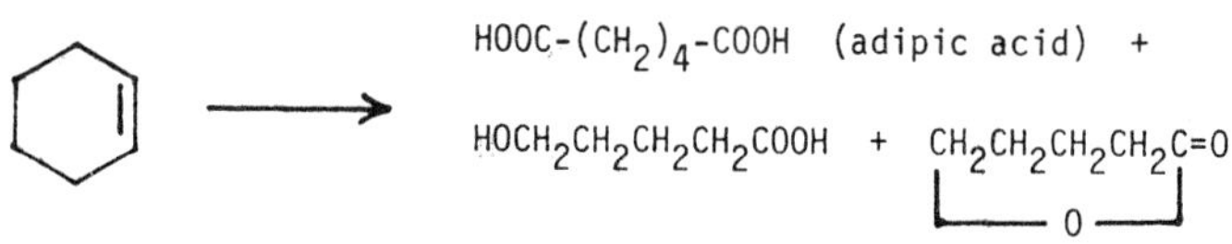

Weber & Waters[121] ozonized aqueous, 0.0005M solutions of dimethyl mercury. After 10 minutes of ozonation, the alkyl mercury compound became undetectable.

Shapiro et al.[122] ozonized aqueous solutions of caffeine (660 mg/l) with 1630 mg of ozone over 90 minutes; 4.2 moles of ozone were consumed/-mole of caffeine. Four major products were isolated (above 5% each) plus 4 minor products. One of the major products was shown to be dimethylparabanic acid by independent synthesis. Caffeine has been shown to be a constituent of sewage treatment plant effluents:

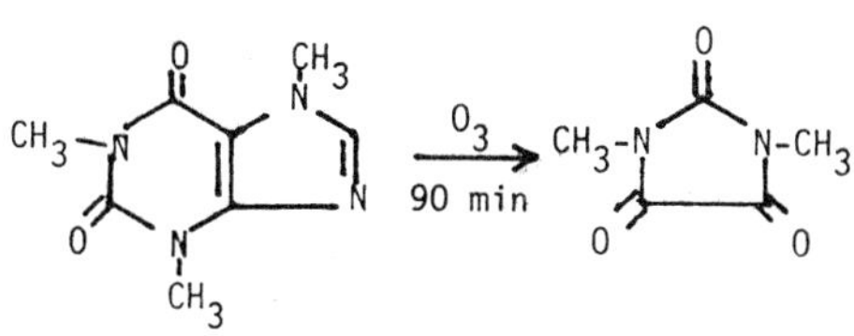

caffeine

dimethyl parabanic acid

Reactions With Pesticides

Robeck, et al.[123] ozonized aqueous solutions of lindane, dieldrin, DDT and parathion and found that dosages of 10 to 38 mg/l of ozone were required to destroy these pesticides to acceptable levels. These dosages were considered to be too high to be practical. These authors also concluded that the more usual drinking water treatment plant ozone dosages of 1 to 2 mg/l probably would oxidize parathion to paraoxon, a compound which is more toxic than is parathion.

Gabovich, et al.[87] ozonized aqueous solutions containing 10 mg/l of malathion. Ozone dosages of 3.5 mg/l (0.5 mg ozone/mg malathion reduced the concentration of malathion to 2 mg/l. Increasing the ozone dosage to 9.8 mg/l (1/1 ozone/malathion) reduced malathion concentration to 1 mg/l. A dosage of 26 mg/l of ozone caused 100% destruction of malathion.

Hoffman & Eichelsdorfer[124] dissolved various pesticides in hexane or acetone, then diluted with water to make aqueous solutions as high in concentration as 2 mg/l of pesticide. These were ozonized over 45 minutes with total ozone dosages up to 240 mg/l. At these dosages, aldrin and heptachlor were "quantitatively" reacted to destruction, but oxidation products were not identified. On the other hand, solutions of dieldrin, heptachlorepoxide, chlordane, lindane, DDT and endosulfan were hardly affected by ozone at all. This raises the question as to whether the ozonation of heptachlor produces heptachlor epoxide. If so, the epoxide will be stable to further ozonation, and itself is a toxic material.

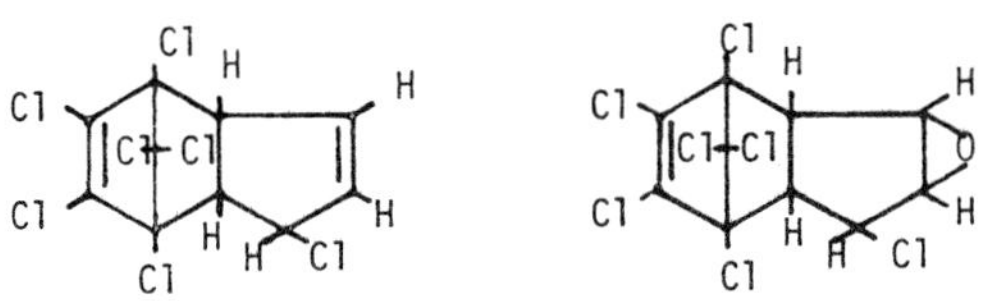

heptachlor heptachlorepoxide

Richard & Brener[125] showed that ozonation of parathion with 3 mg/l ozone dosage forms paraoxon, a more toxic material than parathion itself. The reaction proceeds fastest in acid medium. Continued ozonation of paraoxon (5 mg/l ozone dosage) proceeds slower, with destruction of paraoxon and formation of 2,4-dinitrophenol, picric acid, H_2SO_4 and H_3PO_4:

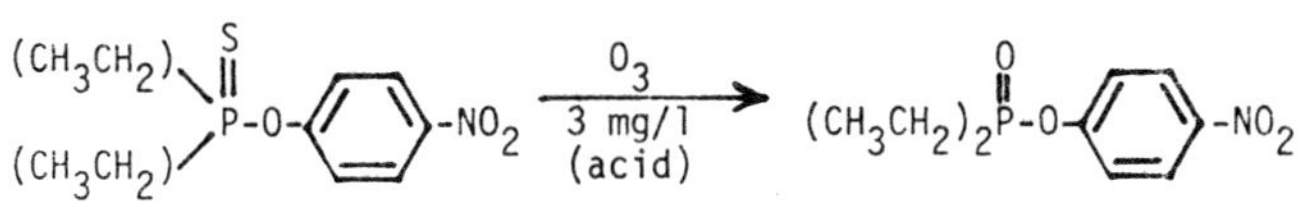

parathion paraoxon

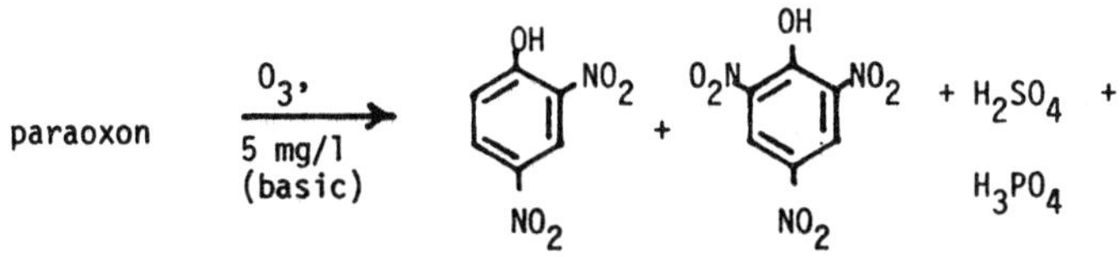

Similarly, Richard & Brener[125] ozonized malathion and isolated malaoxon as the first step intermediate. Continued ozonation destroyed the malathion, producing H_3PO_4 and unidentified, degraded organic compounds.

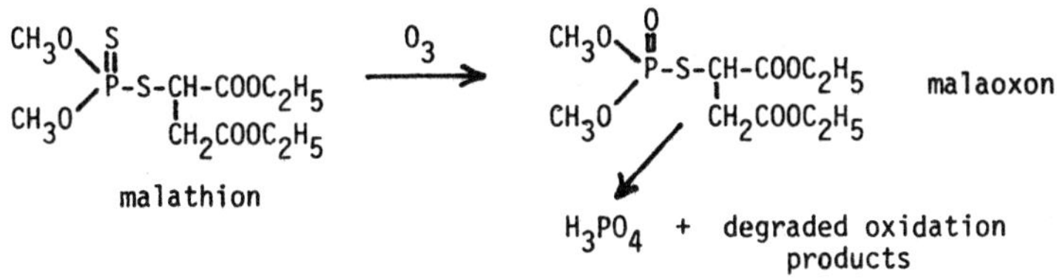

Phosalone upon ozonation did not produce an oxon intermediate. Instead the benzoxalone moiety was cleaved to produce the parent alcohol. This alcohol also underwent self-condensation to produce an ether. Both the alcohol and the ether were isolated and identified by Richard & Brener[125]:

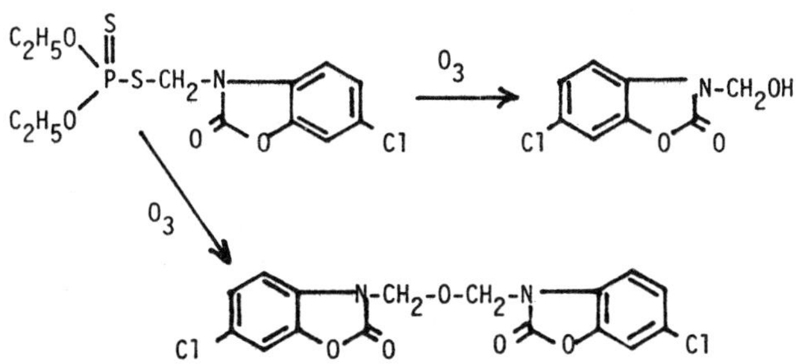

Richard & Brener[125] concluded that under-ozonation of an organic material can produce other organic materials that are toxic, and that it is essential to know the chemical content of waters to be treated with ozone (or any oxidant).

Mallevialle et al.[126] ozonized aqueous solutions of aldrin, and found this compound to be easily degraded by ozone. On the other hand, when aldrin was added to aqueous solutions containing humic acids, 0.45 microg/l of aldrin was detected even after 10 minutes of ozonation. These researchers concluded that ozonation studies on organic compounds conducted in pure solutions can be misleading. It is necessary to know the humics or soils content of water to be ozonized, since these materials can adsorb dissolved organics and thereby "protect" them from the oxidizing action of ozone.

Prengle & Mauk[113] showed that ozonation of DDT in water proceeds very slowly, but the oxidation rate is accelerated by combining UV radiation with ozonation.

Weil et al.[127] ozonized 0.001M solutions of 2,4,5-T with 0.048 mole/hour of ozone and identified oxalic acid, glycolic acid, dichloromaleic acid, chloride ion and CO_2 as oxidation products. No ozonides or polymeric peroxides could be found. The concentration of dichloromaleic acid peaked after 8 to 9 minutes, that of glycolic acid peaked after 12 minutes and that of oxalic acid peaked after 20 minutes of ozonation, after which the concentrations of all three intermediate products decreased with time of ozonation. The concentration of dichloromaleic acid became zero in 25 minutes:

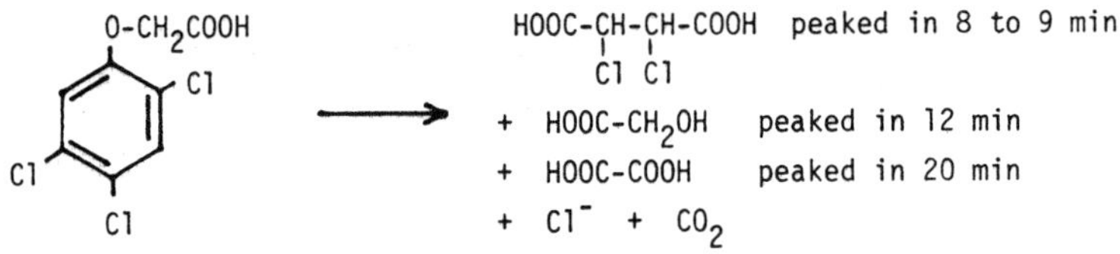

2,4,5-T

Reactions With Humic Materials

Ahmed & Kinney[99] ozonized KOH solutions containing 2 g of humic materials. Water soluble, ozone-resistant acids were isolated, and 65% of the original carbon was isolated as CO_2.

Kinney & Friedman[100] ozonized aqueous alkaline solutions of humic acids and isolated and identified small amounts of acetic acid, terephthalic acid, CO_2 and traces of oxalic acid:

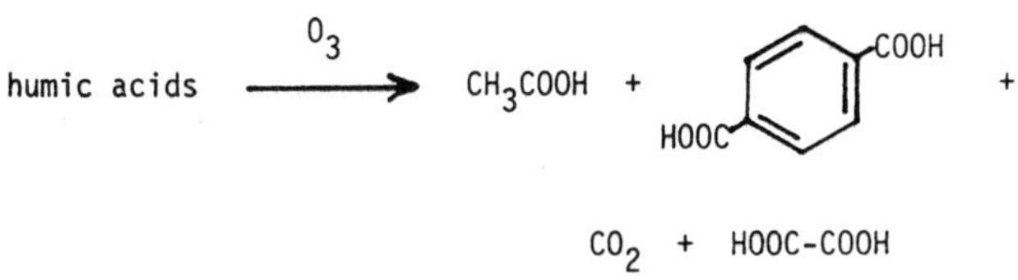

Dobinson & Lawson[128] ozonized solutions of humic acids isolated from coal and identified small amounts of acetic acid and CO_2 as reaction products.

Kinney & Leonard[129] ozonized aqueous solutions of kerogen and isolated water soluble organic acids of molecular weights 200 to 400. The equivalent weight of these acids was 120, and they were shown to be polyfunctional.

Shevchenko & Taran[130] ozonized 1 g/l aqueous solutions of humic acids 80 minutes and identified formic, acetic and oxalic (2.23%) acids.

Buydens[131] ozonized samples of raw water from the River Meuse and found higher suspended solids, higher phenol and higher COD contents after using 0.87 to 1.58 mg/l ozone doses. The author concluded that ozonation decomplexes iron and manganese from organic ligands, thereby liberating more phenolic and COD constituents. In addition, ozone decomposes high molecular weight polymeric humic materials into shorter (phenolic) fragments.

Mallevialle[91] ozonized neutral waters containing 100 to 200 mg/l of humic acids to attain 90% color reduction in 10 minutes. After 20 minutes of ozonation, the solution turned violet, which was ascribed to the decomplexing of manganese followed by oxidation to permanganate. Chromatography of this solution showed the presence of phenolic compounds and formic acid. The author concludes that insufficient ozonation will increase the concentrations of these materials, but for most water supplies, maintaining a residual of 0.4 mg/l ozone over 6 minutes will be a sufficient dosage.

Waters containing 525 mg of humic acids required 100 mg of ozone to destroy 95% of the color and 320 mg of ozone to destroy 95% of the polyhydroxyaromatics[91].

Rook[132] coagulated and filtered River Meuse water, then ozonized it 8 minutes with 2 mg/l doses, then chlorinated. After 8 minutes of ozonation, haloform formation was reduced by 65%. However after the ozonized water had stood for 24 hours, then was chlorinated, the amount of haloforms produced was about the same as without ozonation.

Summary of Ozonation Reactions

With Aromatic Compounds--

- Phenol reacts readily with ozone in aqueous solution to produce the dihydroxybenzenes catechol, hydroquinone and resorcinol. These last two compounds are further oxidized to o- and p-benzoquinone, respectively.

- All of the above oxidized aromatic compounds, upon further oxidation with ozone, undergo ring cleavage to produce the aliphatic unsaturated diacids: muconic, fumaric and maleic, plus the hydroxylated saturated diacids: tartaric and mesotartaric. In addition, glyoxal, glyoxylic acid, glycolic acid and oxalic acid are formed, along with CO_2.

- Oxalic and acetic acids are relatively stable to ozonation in the absence of a catalyst such as UV light or H_2O_2, and thus can be considered to be relatively stable oxidation products of ozonation of phenol in water.

- Ozonation just to the point of destruction of phenol requires 2 to 3 mg of ozone/mg of phenol, but COD is reduced only 50%. To destroy phenol <u>and</u> reduce COD to zero requires 8 to 12 mg of ozone/mg of phenol.

- Ozonation of chloro-substituted phenols ultimately cleaves the aromatic C-Cl linkages, forming chloride ion plus the same types of ring-ruptured aliphatic compounds as does phenol.

- Cresols and xylenols undergo oxidation with ozone at faster rates than does phenol. Before ring cleavage occurs, o-cresol forms salicylic acid.

- Upon ring rupture, ozonized cresols produce the same types of aliphatic products as does phenol, plus mesotartaric, propionic and acetic acids.

- Xylenols produce all these aliphatic products plus hydroxyphthalic acid (before ring-rupture) and diacetyl and glyoxal (after ring-rupture).

- Nitro, amino and sulfonic acid groups on aromatic rings are split off by ozonation, but at much slower rates than is chlorine. Amino groups are converted to ammonia and nitrate ion. Sulfonic acid groups are converted to sulfate.

- Aromatic hydrocarbons such as pyrene, phenanthrene and naphthalene oxidize by ring rupture. Only one ring in phenanthrene opens readily, however. When aliphatic hydrocarbon groups are present on the aromatic rings, these oxidize first, before the ring ruptures.

- Chlorobenzene reacts with ozone slower than does phenol, but gives the same ring ruptured, aliphatic oxidation products as phenol. Intermediate oxidation products include o-, m- and p-chlorophenols plus chlorotartaric acid. Chlorocresols, chlorophenols and thiophenols give similar ozonation products as does phenol.

With Aliphatic Compounds--

- There is no evidence that ozone reacts with saturated aliphatic hydrocarbons under water or wastewater treatment conditions.

- There is also no evidence that ozone oxidizes trihalomethanes. Reduction in concentration of THMs upon ozonation appears to occur by air stripping of aqueous solutions. Ozone combined with ultraviolet radiation <u>does</u> oxidize chloroform to produce chloride ion, but no identified organic oxidation product.

- Unsaturated aliphatic or alicyclic compounds react with ozone, usually at the unsaturated bond, cleaving the molecule into two oxidized fragments (aliphatics) or into diacids or carbonyl-carboxylic acids (alicyclics). The two aliphatic fragments normally are an acid plus an aldehyde or ketone.

- Primary aliphatic alcohols generally are oxidized to aldehydes, then to acids, but at slower rates than phenol oxidation with ozone.

- Secondary aliphatic alcohols produce ketones, then acids plus H_2O_2 upon ozonation.

- Formic acid readily produces CO_2 and water upon ozonation, but oxalic acid is relatively stable to ozonation in the absence of UV radiation or H_2O_2. Acetic acid and propionic acid also are relatively stable to ozonation.

- Oxalic acid oxidizes directly to CO_2 without producing formic acid. Reaction with ozone alone is very slow, but proceeds rapidly in the presence of UV radiation or H_2O_2.

- Maleic acid produces glyoxylic and formic acids initially. Glyoxylic acid then produces oxalic acid, CO_2 and water. These are the sole products of ozonation of maleic acid.

- Propionic acid and 2-propanol produce acetone upon ozonation. Acetone can undergo the haloform reaction and produce chloroform if present during post-chlorination of ozonized water.

- Ozonation of acetate ion and acetic acid ultimately produces glyoxylate ion or glyoxylic acid, respectively, which then form oxalic acid.

- Diethylamine produces acetaldoxime upon ozonation, plus an unidentified nitrogen-containing compound, not a nitrosoamine.

- Prolonged ozonation of a concentrated ethyl alcohol solution in water produces a stable dihydroperoxide which exhibits mutagenic activity.

- Ozonation of Lake of Zürich raw water increased the aldehyde concentrations. Ozonation followed by activated carbon adsorption reduced the aldehyde concentrations below their original concentrations. This indicates that ozonation for oxidation should be followed by a adsorption step.

- UV/ozonation of refractory organic materials increases the rate at which they are oxidized by ozone, but not the nature of the oxidation products.

With Pesticides--

- Ozonation of parathion and malathion produces paraoxon and malaoxon, respectively, as intermediates, which are more toxic than are the starting materials. Continued ozonation degrades the oxons, but requires more ozone than does the initial thion oxidation. Phosalone oxidizes without forming an oxon intermediate. Thus, under-ozonation can produce intermediates which are more toxic than the starting materials.

- Ozonation of heptachlor produces a stable product, not yet identified. Heptachlorepoxide is known to be stable to ozonation. This suggests that ozonation of heptachlor may produce the epoxide, which then would be unaffected by further ozonation.

- Aldrin and 2,4,5-T are readily oxidized by ozone, but dieldrin, chlordane, lindane, DDT and endosulfan are only slightly affected by ozone.

- UV/ozonation destroys DDT, PCBs, malathion and many other pesticides, but requires more extended contact times and ozone doses than normally encountered in drinking water treatment plants.

- Ozonation of aldrin in pure water proceeds rapidly, but at a much slower rate in the presence of humic acids. Similar oxidation rates were observed with solutions of benzopyrene in pure water (rapid) versus water containing colloidal soil particles (slow). This indicates that dissolved organic materials can be adsorbed by humics or soil particles and be "protected" from oxidation, at least partially. This would also indicate that ozonation of water supplies for oxidation should follow a filtration step.

With Humic Materials--

- Humic materials are resistant to ozonation, requiring lengthy times of ozonation to produce small amounts of acetic, oxalic, formic and terephthalic acids, CO_2 and phenolic compounds.

- Ozonation of humic materials in water followed by immediate chlorination (within 8 minutes) reduced trihalomethane formation by 65%. However, when ozonized waters containing humics were allowed to stand 24 hours and then chlorinated, there was no change in the amounts of trihalomethanes formed. This indicates that although ozone changes the chemical nature of trihalomethane precursors, there is continued reaction upon standing, not entirely with residual ozone, to form materials equally capable of producing trihalomethanes upon chlorination.

- Ozonized organic materials generally are more biodegradable than the starting, unoxidized, compounds. If ozonation is used as the terminal treatment step with waters containing dissolved organics, bacterial regrowth in distribution systems is not only likely, but is probable. Thus, Swiss and German water treatment authorities conclude that ozonation always should precede a adsorption step.

- As a general rule (which does not always hold, it should be emphasized), ozonation used specifically for oxidation of organic materials advantageously should follow and/or precede adsorption steps.

****

Dellah[133] conducted extensive pilot plant studies for the new Charles-J. des Baillets drinking water treatment plant being constructed in Montreal, Canada. This plant will use ozone for a major portion of disinfection, and when fully operational will require generation of 15,000 lbs/day of ozone. In assessing the effects of ozonation, followed by chlorination for residual, on the St. Lawrence River raw water, he concluded:

> "With 5 minute contact times and ozone doses sufficient to maintain residual concentrations of 0.4 mg/l, COD reduction was about 50% when the water quality was at its best. When it was at its worst, COD reduction was about 60% after filtration, but little changed after subsequent ozonation. Therefore, ozonation does not eliminate all organics. Ozonation plus chlorination still will produce organochlorine compounds in drinking water."

REACTIONS WITH CHLORINE DIOXIDE

Several excellent review articles on the reactions of chlorine dioxide with organic materials are available. These include Stevens, et al.[134]; Rosenblatt[135]; Gordon[136] and Masschelein[137].

Reactions With Phenol

Masschelein[137] has reviewed the reactions of phenol and chloro-substituted phenolic compounds with chlorine dioxide, and the known oxidation products are shown in Figure 55. Phenol itself reacts either by oxidation (to produce p-quinone), or by chlorination and oxidation, producing 2-chloro-p-benzoquinone and 2,5-dichloro-p-benzoquinone. Chlorophenols unsubstituted in the position para to the aromatic hydroxy group readily form the corresponding p-benzoquinone. 2,4,6-Trichlorophenol produces a 55% yield of 2,6-dichloro-p-benzoquinone, indicating that the 4-chlorosubstituent is first broken away from the aromatic ring by the ClO_2, followed by substitution of oxygen at the 4-carbon. Similarly, 2,4-dichlorophenol gave a 36% yield of 2,6-dichloro-p-benzoquinone, rather than 2,4-dichloro-o-benzoquinone.

Guaicol (o-methoxyphenol) first produced catechol, then o-benzoquinone, then ring-ruptured products, including monomethylmuconic acid as the primary product.

p-Benzoquinone, isolated from chlorine dioxide oxidation of phenol, reacted further with ClO_2 forming the ring-ruptured 4-carbon products, fumaric and maleic acids, and then the 2-carbon oxalic acid.

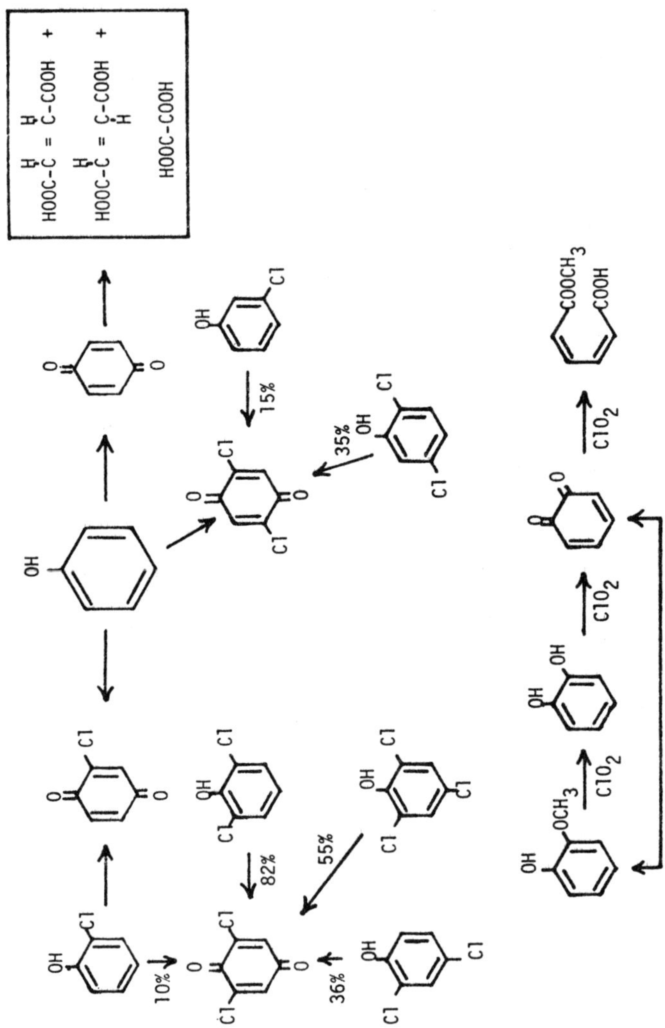

Figure 55. Reactions of chlorine dioxide with phenols[137].

Masschelein[137] has also shown the relationship between reaction of chlorine dioxide with phenolics and the by-product chlorite ion formed. In Figure 56 are plotted the amounts of ClO_2 consumed by phenol, o-chlorophenol and p-chlorophenol versus the amount of ClO_2^- produced. The relationship is linear in all three cases.

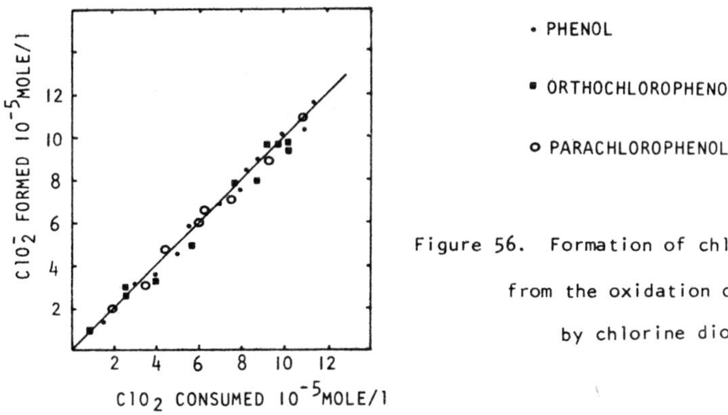

Figure 56. Formation of chlorite ion from the oxidation of phenols by chlorine dioxide[137].

From his literature review of reactions with phenolics, Masschelein[137] concludes that reaction of gaseous ClO_2 (14 to 18%) with phenol at 0°C, in water, produces p-benzoquinone as the major product. Prolonged oxidation with chlorine dioxide then produces ring-chlorinated compounds, then ring rupture, forming aliphatic diacids. Many of the ring-ruptured products do not contain chlorine, and are the same products formed upon ozonation.

p-Nitrophenol treated with ClO_2 produced 8.5% yield of p-benzoquinone[137]:

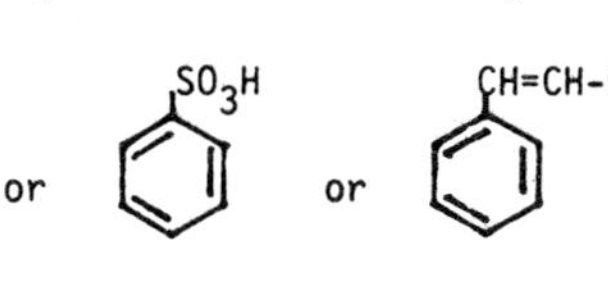

Benzoic, phenylsulfonic and cinnamic acids at pH 7 in water do not react with ClO_2[137]([138] for cinnamic acid):

Cardey[139] reviewed experiments performed by the Ohio River Valley Water Commission in Hamilton, Ohio, on steel mill wastewaters containing phenolic compounds. This work showed that the use of ClO_2 destroyed the phenols, did not form chlorinated phenols, and that the oxidant functioned equally well in acid or basic solution. Reaction times of 15 minutes destroyed 100 to 130 mg/l of phenol, and caused a drop in final pH to 2 to 4, attributed to the formation of organic acids (which were not identified).

Very recent work by Spanggord and co-workers at Stanford Research Institute, under contract to EPA, has been conducted with chlorine dioxide synthesized from sodium chlorite solution and sulfuric acid[140]. This synthesis guarantees the absence of free residual chlorine in solution. Reaction of ClO_2 thus prepared with aqueous phenol has produced mixtures of chlorinated and unchlorinated products. A solution of 0.67 mmole of phenol was treated with chlorine dioxide to allow only 48% reaction. The products isolated and identified are:

2-chlorophenol
4-chlorophenol
2,6-dichlorophenol
2,4-dichlorophenol
2,4,6-trichlorophenol
resorcinol (1,3-dihydroxybenzene)
fumaric acid (<u>trans</u>-2-butenedioic acid)
2-chlorohydroquinone

Reactions With Other Aromatic Compounds

Paluch, et al.[141] found that benzylic acid in water produced small amounts of benzoic acid when treated with ClO_2:

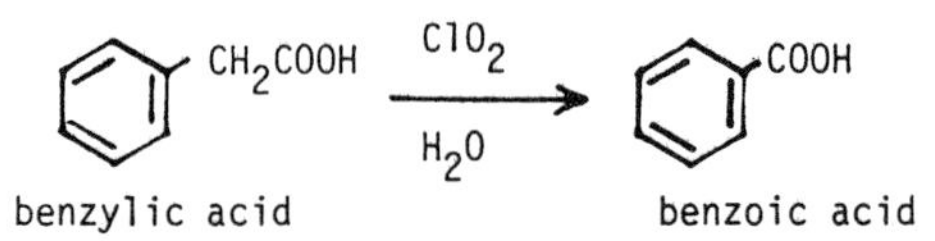

However, chlorite ion in the presence of HCl produced o-, m- and p-chlorobenzoic acids, o- and p-chlorobenzylic acids and o- and p-chlorobenzaldehydes[171]:

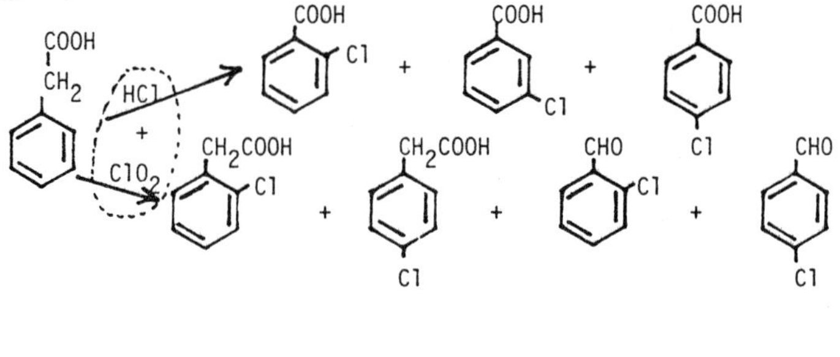

Otto & Paluch[142] reported that an aqueous dispersion of benzaldehyde reacts violently with chlorine dioxide.

Reichert[143] treated 1 to 10 μg/l concentrations of 3,4-benzopyrene in water with 0.11 μg of ClO_2 over 15 to 30 minutes. He observed a logarithmic decrease in benzopyrene concentration to 0.01 μg/l. The lower the original concentration of organic, the longer was the contact time required to reach this concentration.

A subsequent study by Reichert[144] describes the identification of organic compounds produced when 1 to 10 μg/l concentrations of 3,4-benzopyrene were treated with 1 mg of ClO_2 in water. Eight individual compounds were isolated, of which 3 accounted for 90% of the yield isolated. These three compounds are: 3,4-benzopyrene-1, 5-quinone, 3,4-benzopyrene-5,8-quinone and 3,4-benzopyrene-5,10-quinone:

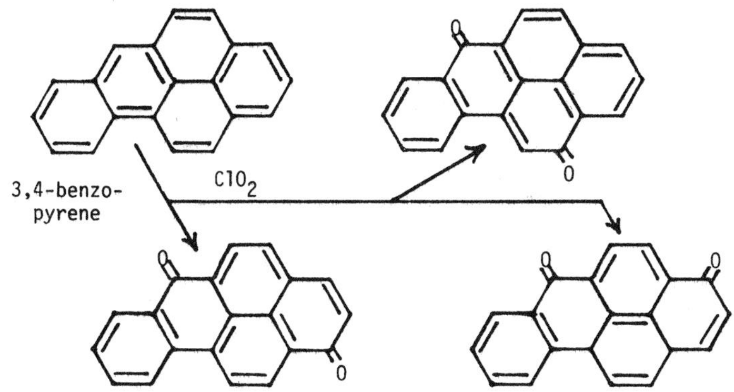

The other 5 compounds isolated by Reichert[144] all were chlorinated, and two of these were identified as 5-chloro-3,4-benzopyrene and 5,8,10-trichloro-3,4-benzopyrene.

Subcutaneous injection of all above pyrene compounds isolated by Reichert[143,144] into rodents gave negative results in all cases.

In 1969 Reichert[145] showed that 10 µg/l of 3,4-benzopyrene was 90% decomposed in 60 minutes when treated with 0.1 mg/l of ClO_2. By way of comparison, 0.5 to 1.5 mg/l ozone dosage caused 99% decomposition in 30 minutes in pure water. In natural water, 99% decomposition of the pyrene was obtained with 0.2 mg/l ozone with 15 minutes contact time.

Spanggord et al.[140] treated aqueous solutions of diphenylamine hydrochloride with ClO_2 (prepared from chlorite and acid so as not to contain free chlorine) and found chlorinated derivatives. The following compounds have been isolated from the reaction mixture:

Diphenylamine
2-chloro-diphenylamine
4-chloro-diphenylamine
2 isomers of a dichlorodiphenylamine
a ring hydroxylated diphenylamine
2 isomers of chloro-hydroxy-diphenylamine
2 isomers of dichloro-hydroxy-diphenylamine

Vanillin reacts with ClO_2 at pH 4 to produce the non-chlorinated β-formylmuconic acid, monomethyl ester[136] which further oxidized to the diacid:

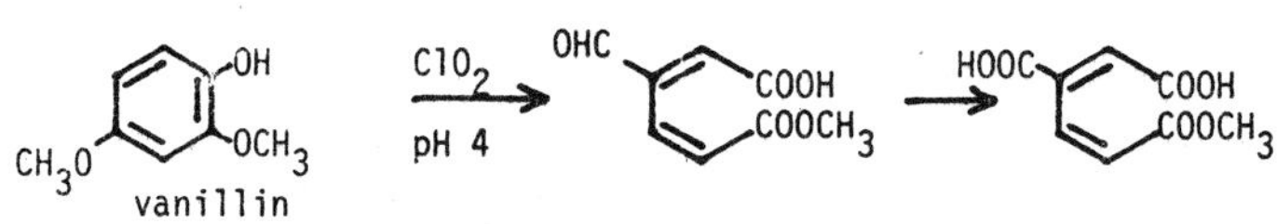

Vanillyl alcohol reacts with ClO_2 at low pH to produce both 2-chloro-5-methoxy-1,4-benzoquinone and a non-chlorinated lactone product of ring cleavage[136]:

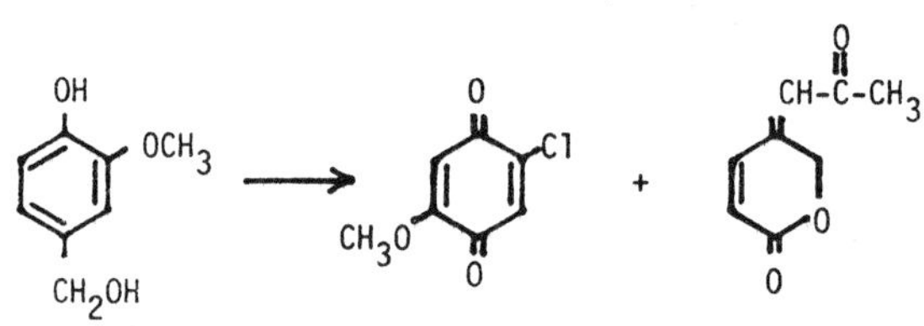

vanillyl
alcohol

Veratryl alcohol reacts with ClO_2 to produce 4,5-dichloroveratrole[147]:

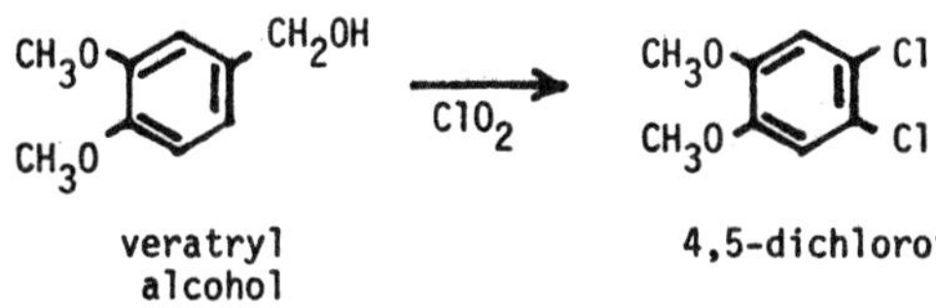

Reactions With Heterocyclics

Kawasaki et al.[147] treated thiamine (vitamin B-1) with ClO_2 in water, at pH below 4.7 under prolonged reaction conditions and isolated 2-methyl-4-amino-5-aminomethylpyrimidine:

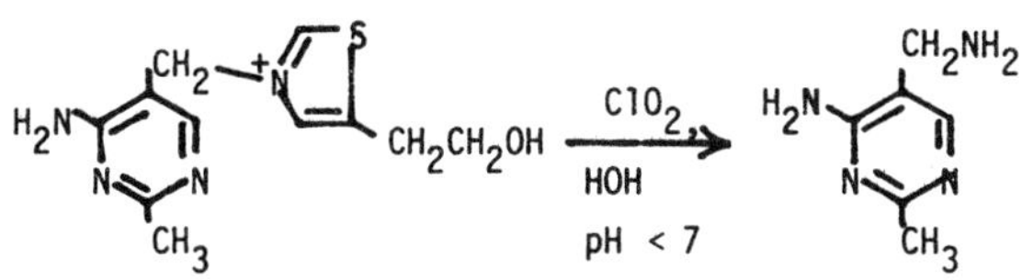

Thianine was found to be stable to ClO_2 or to chlorite under acid conditions[147].

Fujii & Ukita[148] treated tryptophane with chlorine dioxide and isolated a mixture of products including indoxyl, isatine, indigo red plus unidentified, colored, probably polymeric compounds:

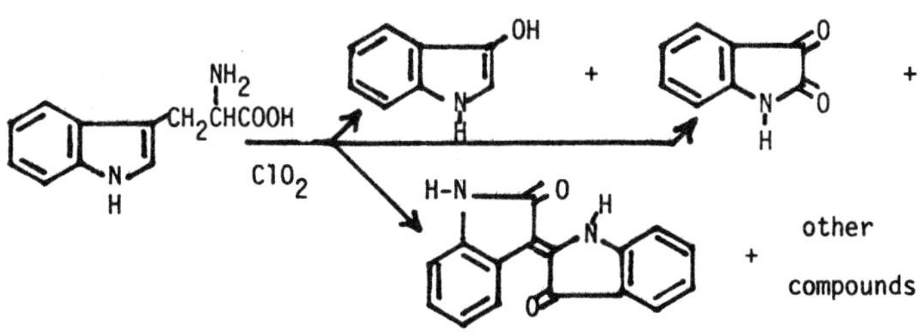

Reactions With Aliphatics

No evidence exists that chlorine dioxide (or chlorine) undergoes reactions with saturated aliphatic hydrocarbons under water treatment conditions[134]. Masschelein[137] states that aliphatic alcohols are often oxidized to aliphatic acids by ClO_2 in aqueous solution:

$$RCH_2OH + ClO_2 \longrightarrow RCOOH + H_2O$$

The stability of aliphatic acids to further oxidation by ClO_2 is supported by the findings of Sarkar[138] that crotonic, maleic and fumaric acids do not react with ClO_2 in aqueous solution. These findings are significant also in that these three acids contain an aliphatic double bond conjugated with a carboxyl carbonyl group. Such a double bond normally is reactive with ozone, hypochlorite and chlorine.

Leopold & Mutton[149] and Lindgren & Svahn[150] treated methyl oleate 60 hours with ClO_2 in aqueous solution under a nitrogen atmosphere and have isolated the compounds shown in Figure 57. It was found that oxidation can occur at the carbon atoms adjacent to the double bond, producing conjugation of the type encountered in crotonic, maleic and fumaric acids, which were found by Sarkar[138] to be stable to further oxidation with ClO_2.

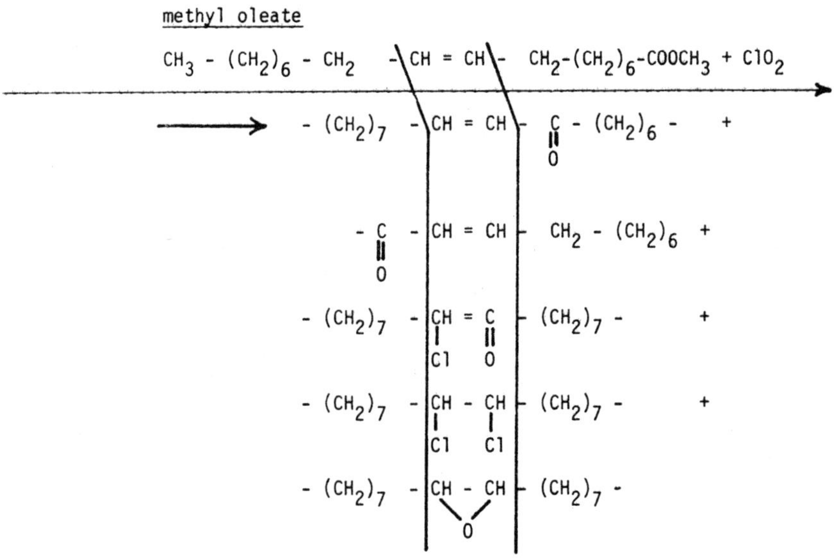

Figure 57. Reactions of methyl oleate with ClO_2[149,150].

Treatment of methyl oleate with ClO_2 also produced a dichloro compound, which appears to have been formed by simple addition of chlorine across the double bond. Other types of compounds include the chloro-carbonyl compounds, which could have formed by addition of HOCl across the double bond to produce the chlorohydrin, followed by oxidation of the alcoholic hydroxy moiety to the ketone by chlorine dioxide.

The possibility of addition of HOCl across the double bond of methyl oleate is supported by isolation of the epoxide, which forms by elimination of HCl from the intermediate chlorohydrin. These products isolated by Mutton[149] and by Lindgren & Svahn[150] confirm the chemistry of chlorine dioxide discussed earlier in this chapter.

Leopold & Mutton[149] consider that ClO_2 cleaves the oleic double bond, producing aldehydic fragments and an equivalent amount of chlorine (not chlorite), which then adds across another double bond.

Somsen[151] found that butane-2,3-diol and diacetyl treated with ClO_2 produced primarily acetic acid:

$$CH_3\text{-}\underset{OH}{\underset{|}{C}H}\text{-}\underset{OH}{\underset{|}{C}H}\text{-}CH_3 \quad \text{or} \quad CH_3\text{-}\underset{O}{\underset{\|}{C}}\text{-}\underset{O}{\underset{\|}{C}}\text{-}CH_3 \; + \; ClO_2 \longrightarrow CH_3COOH$$

Lindgren, Svahn & Widmark[152] treated aqueous suspensions of cyclohexene with gaseous ClO_2 (free of excess chlorine) and isolated and identified 7 products. Three different ratios of cyclohexene/ClO_2 were used, and the product distribution was found to vary. The results are shown in Figure 58. Reaction usually occurred at the double bond, but in most of the compounds isolated the ring was retained intact. Adipaldehyde was isolated, but at pH below 2. Using a 1.5/1 cyclohexene/ClO_2 ratio, glutaric acid was the principal non-cyclic reaction product isolated, along with adipic and succinic acids. The epoxide was not isolated in the final products, evidently having been destroyed by the acids formed during the oxidation.

Flis et al.[153] treated excess glucose with ClO_2 at pH 2 and at 20 to 60°C, and observed oxidation of hydroxymethyl groups to produce aldehydic and carboxylic functions, apparently without ring rupture. They also observed that β,D-glucose reacted faster than α,D-glucose. This lack of rupture of carbodydrate rings by small amounts of ClO_2 may explain why trihalomethanes are not produced from humic materials.

However, Becker, Hamilton & Lucke[154] treated cellotetrose at pH 3 and 56°C over 1 to 7 hours with 10 to 100/1 molar ratios of ClO_2 and isolated gluconic, carbohydrate rings did rupture under the reaction conditions employed (large excesses of ClO_2). (See Figure 59.)

Zienus & Purves[155] treated pectic acid at 70 to 75°C with aqueous 0.14M ClO_2 or $NaClO_2$ at a molar ratio of ClO_2/pectic acid of 10/1, and identified galacturonic acid (most abundant product), muric and d,l-tartaric acids as products. Sodium chlorite produced tartronic acid, which was not found when using ClO_2. Results are summarized in Figure 60. Again, the carbohydrate rings ruptured when treated with large excesses of ClO_2.

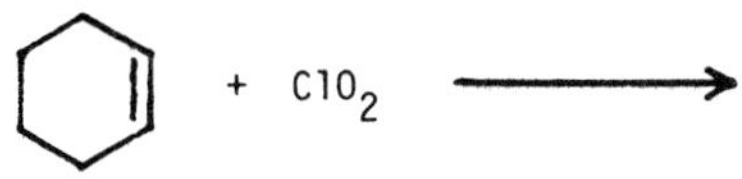

Cyclohexene

	cyclohexene/ClO_2 35/1	cyclohexene/ClO_2 7/1
Cyclohex-1-ene-3-one	16%	13%
2-Chlorocyclohexanone	11	7
1,2-Epoxycyclohexane	3	0
Adipic acid	3	4
trans-2-Chlorocyclohexanol	11	15
3-Chlorocyclohexene	13	14
trans-1,2-Dichlorocyclohexane	5	7

Using 1.5/1 cyclohexene/ClO_2 gave the above compounds plus (in order of importance): adipic acid, glutaric acid, succinic acid, 2-chlorocyclohexane - 1,3-dione

Using excess ClO_2 over cyclohexene gave carboxylic diacids

Figure 58. Reactions of Cyclohexene with ClO_2[152].

Spanggord et al.[140] treated citral with ClO_2 containing no excess free chlorine and identified chlorinated products (apparently chlorohydrins) and oxidation products not containing chlorine:

$$(CH_3)_2C{=}CH(CH_2)_2C(CH_3)CHO \longrightarrow \text{chlorohydrins + non-halogenated oxidation products}$$

(citral)

Reactions With Amino Acids

Shirle[156] treated cystine with excess ClO_2 or ClO_2^- at pH 3.54 and isolated cystine bisulfoxide. Further oxidation produced cysteic acid:

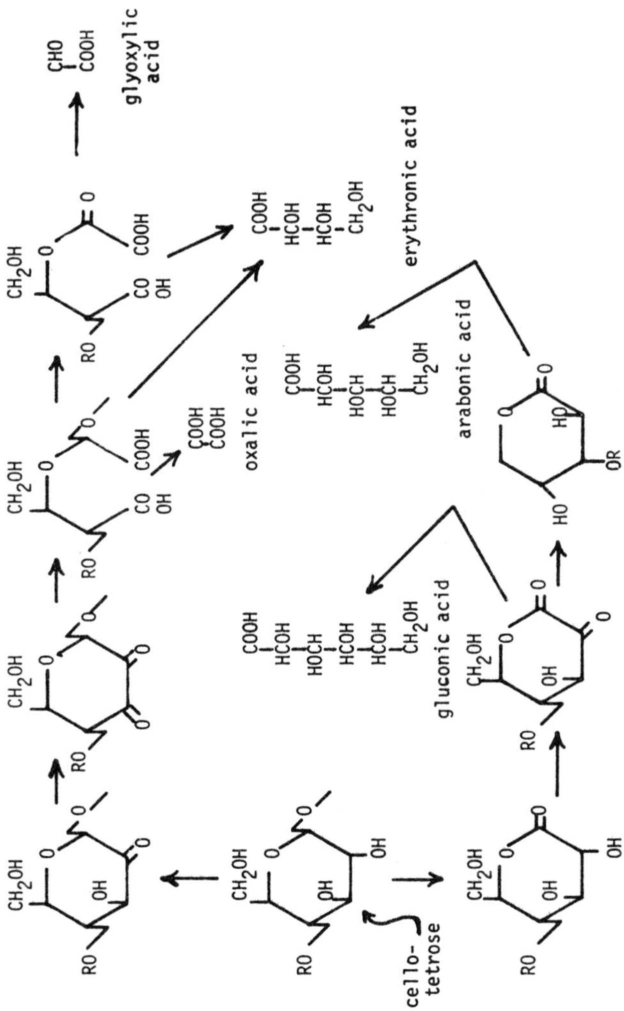

Figure 59. Reactions of cellotetrose with ClO_2[154].

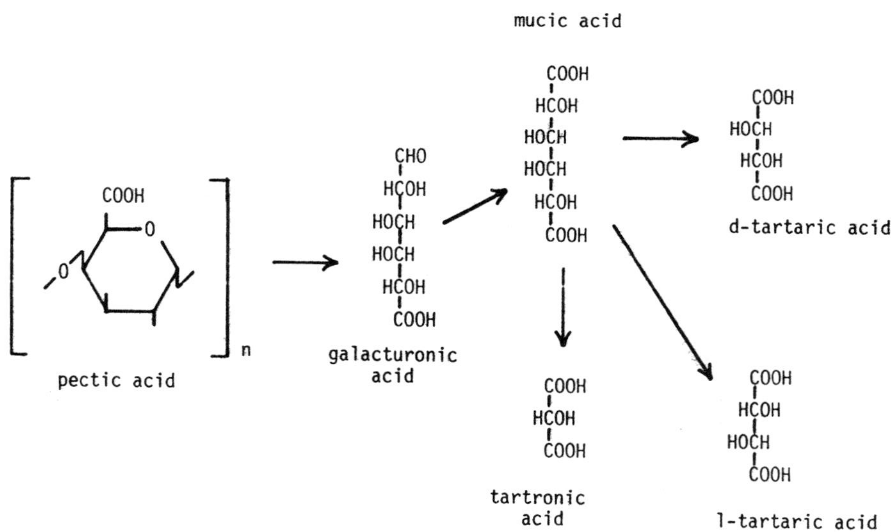

Figure 60. Reactions of pectic acid with ClO_2[155].

$$HOOCCH(NH_2)CH_2\text{-}S\text{-}S\text{-}CH_2CH(NH_2)COOH + ClO_2 \longrightarrow$$

cystine

$$HOOCCH(NH_2)CH_2\text{-}\underset{\underset{O}{\|}}{S}\text{-}\underset{\underset{O}{\|}}{S}\text{-}CH_2CH(NH_2)COOH \longrightarrow$$

cystine bisulfoxide

$$HOOC\text{-}CH(NH_2)\text{-}CH_2SO_3H$$

cysteic acid

Methionine was found to react similarly through the sulfoxide to produce the sulfone[156]:

$$CH_3S(CH_2)_2CH(NH_2)COOH + ClO_2 \longrightarrow CH_3\underset{\underset{O}{\|}}{S}(CH_2)_2CH(NH_2)COOH$$

methionine — methionine sulfoxide

$$\downarrow$$

$$CH_3\text{-}SO_2\text{-}(CH_2)_2CH(NH_2)_2COOH$$

methionine sulfone

Hodgen & Ingols[157] treated tyrosine at pH 4.5 with excess ClO_2 (3 moles/mole of tyrosine) and identified dopaquinone along with isolating other unidentified colored products:

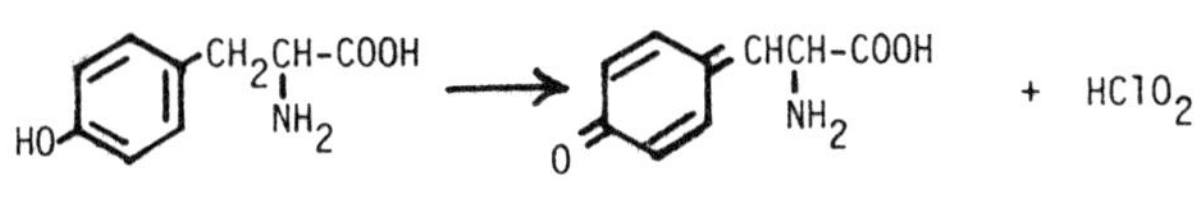

In the presence of free chlorine, Moran et al.[158], found that tyrosine also produces monochlorotyrosine and dichlorotyrosine.

On the other hand, Kennaugh[159] found that the amino acids glycine, leucine, serine, alanine, phenylalanine, valine, hydroxyproline, phenylaminoacetic acid, aspartic acid and glutamic acid are only slightly reactive to chlorine dioxide. Only after standing 3 weeks at ambient temperatures in acetic acid 50% saturated with ClO_2 did these compounds react with chlorine dioxide.

Reactions With Amines

In aqueous solution, primary and secondary amines react very slowly with chlorine dioxide, but tertiary amines react rapidly to produce aldehydes. That primary amines are unreactive with ClO_2 is borne out by the lack of reactivity of amino functions in the aminoacids.

Triethylamine reacts with ClO_2 to produce acetaldehyde and diethylamine, along with chlorite ion, at pH 4.7 to 7.1[160].

$$Et_3N + H_2O + ClO_2 \rightarrow CH_3CHO + Et_2NH + 2H^+ + 2ClO_2^-$$

At lower pH, however, a secondary reaction occurs, by which chlorite reforms chlorine dioxide:

$$H^+ + CH_3CHO + 3ClO_2^- \longrightarrow CH_3COO^- + Cl^- + 2ClO_2 + H_2O$$

Rosenblatt[160] also points out that triethylamine reacts with chlorine dioxide without forming an N-oxide. This is significant since once an N-oxide forms, it is possible that further oxidative degradation might proceed through formation of a nitrosamine.

Rosenblatt[161] reviewed the reactivity of chlorine dioxide with organic substances, and discussed the pertinent chemistry involved with chlorine species. Quinuclidine forms an N-oxide when treated with ClO_2, but triethylenediamine forms piperazine. Ring substituted benzyldimethylamines react faster with ClO_2 the greater is their initial basicity. A summation of the pertinent reactions of amines with ClO_2 is given in Figure 61.

Reactions With Humic Materials

Fuchs & Leopold[162] reported that humic acids react with ClO_2, but they did not identify any of the reaction products. In 1970, Buydens[131] treated River Meuse (Belgium) raw water (containing humic acids) with ClO_2, and found much the same effects as were obtained with pre-ozonation. Phenolic content and COD content increased. However, when ClO_2 was added just ahead of the clarification/flocculation step in the drinking water treatment process, phenols were removed to nearly complete disappearance. Organically complexed iron and manganese also were decomposed by addition of ClO_2 at this step, and the ozone demand also was reduced.

Mallevialle[163] in France found that humic materials treated with pure ClO_2 (containing no free chlorine) do not form trihalomethanes. This finding was confirmed independently in Cincinnati, Ohio by Love et al.[164] with Ohio River waters.

$$NH_3 + ClO_2 \longrightarrow \text{No Reaction}$$

$$RNH_2 + ClO_2 \longrightarrow \text{No Reaction}$$

$$R_2NH + ClO_2 \longrightarrow \text{Slow Reaction}$$

$$(C_2H_5)_3N + ClO_2 \longrightarrow CH_3CHO + (C_2H_5)_2NH + H^+ + 2ClO_2^-$$

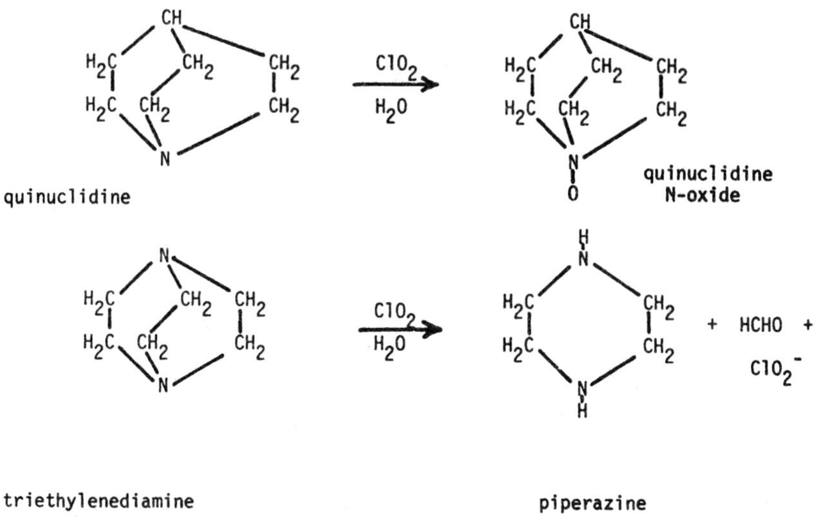

Figure 61. Pertinent Reactions of Amines with ClO_2[82].

Vilagenès et al.[165] studied the use of ClO_2 at the Paris, France, St. Maur drinking water treatment plant where ammonia is removed biologically in slow sand filters. This plant has two treatment lines, and ClO_2 and NaOCl were compared as terminal treatment steps for their ability to produce trihalomethanes. Resorcinol was added as the trihalomethane precursor so that sufficient quantities of trihalomethanes could be produced deliberately.

Trihalomethane production was found to occur only with hypochlorite, and not with ClO_2. Furthermore, the rate of trihalomethane production was dependent only upon the concentration of resorcinol precursor, not upon the concentration of chlorine. When excess chlorine was used with the ClO_2, trihalomethane formation occurred.

These authors also found that chlorodibromomethane was formed in higher concentrations (20 to 35 microg/l) using hypochlorite than was chloroform (1 to 20 microg/l). Since no bromine-containing trihalomethane was produced when chlorine dioxide alone was employed, it was concluded that bromide ion is not oxidized by ClO_2.

Summary of Chlorine Dioxide Reactions

- Treatment of organic compounds with pure ClO_2 containing no excess free chlorine produces oxidation products containing no chlorine in some cases, but products containing chlorine in others.
- When excess free chlorine is present with the ClO_2, chlorinated organics usually are produced, but in lower yield (as compared to the yield when chlorination alone is used, depending upon the concentration of chlorine and its reactivity with the particular organic(s) involved.
- Phenol in water treated with ClO_2 containing no free chlorine produces 2- and 4-chlorophenol, 2,4- and 2,6-dichlorophenol, 2,4,6-trichlorophenol, 2-chlorohydroquinone, resorcinol and p-benzoquinone as the major oxidation products. Prolonged oxidation with ClO_2 then produces ring-chlorinated compounds, then ring rupture, producing aliphatic diacids. Many of the ring-ruptured products do not contain chlorine and are the same as products formed upon ozonation.
- Chlorite ion in the presence of HCl ring-chlorinates and oxidizes benzylic acid to benzoic acid, o-, m- and p-chlorobenzoic acids, o- and p-chlorobenzaldehydes and o- and p-chlorobenzylic acids.
- 3,4-benzopyrene treated with ClO_2 produces the non-chlorinated 1,5-,5,8- and 5,10-benzopyrenequinones (total yield 90%) plus five chlorinated compounds, two of which were identified as 5-chloro- and 5,8,10-trichloro-3,4-benzopyrene. None of these compounds gave positive responses when injected subcutaneously into rodents.
- Using large excesses of ClO_2 over the organic materials appears to favor oxidation reactions (without chlorination), but slight excesses appear to favor chlorination.
- Benzoic acid, phenylsulfonic acid, thianine and cinnamic acid do not react with ClO_2. The first two are oxidized by ozone; the reactivity of the last two with ozone is not known.
- In oxidizing organic materials, ClO_2 can revert back to chlorite ion, ClO_2^-. In the presence of excess chlorine (or other strong oxidant) chlorite can be reoxidized to ClO_2.

- Diphenylamine with ClO_2 containing no excess chlorine produced 2- and 4-chloro-diphenylamine, two dichlorodiphenylamines, a ring-hydroxylated diphenylamine, two chlorohydroxy- and two dichlorohydroxydiphenylamine isomers.

- Substituted nitrogen heterocyclics, such as thiamine and tryptophane, undergo oxidation of the substitutents, retain the heterocyclic ring (at least in the intermediate stages) and products isolated to date are not chlorinated.

- Saturated aliphatic hydrocarbons are not reactive with ClO_2. Alcohols are oxidized to the corresponding acids.

- Crotonic, maleic and fumaric acids (all containing an aliphatic double bond conjugated with a carbonyl oxygen) are unreactive to ClO_2 in aqueous solution. Such unsaturation normally is reactive to ozone, hypochlorite and chlorine.

- Treatment of methyl oleate (containing an isolated aliphatic double bond) produces dichloro, chloroketonic, α-unsaturated ketones, epoxy, and chlorohydrin compounds, all from reactions with or adjacent to the oleic double bond. The methyl ester moiety was retained in all cases.

- Butane-2,3-diol and diacetyl cleave upon treatment with ClO_2 and produce acetic acid.

- Cyclohexene with ClO_2 free of excess chlorine produced 9 ring-containing compounds, chlorinated and non-chlorinated. Functional groups present in the oxidized products included epoxy, chloro, hydroxy and keto. Ring unsaturation was maintained in some compounds, but not in others. Adipaldehyde, glutaric acid and succinic acid were the only ring ruptured compounds isolated, with glutaric acid predominating at pH below 2. The epoxy compound was not isolated in the final products, apparently being destroyed under the aqueous acid conditions.

- Hydroxymethyl groups in glucose produce aldehydic and carboxylic groups, apparently without ring rupture. Stability of the polymeric carbohydrate structure to ClO_2 oxidation may explain why trihalomethanes do not form when humic acids are treated with ClO_2.

- When carbohydrates are treated with large excesses of ClO_2 (10/1 to 100/1 molar ratios) the carbohydrate rings do rupture. However, trihalomethanes have not been reported as products.

- Aminoacids such as glycine, leucine, alanine, phenylalanine and others are stable to oxidation by ClO_2. Sulfur in aminoacids oxidizes to sulfoxides and sulfones and initial -S-S- bonds cleave and oxidize to produce the fragmented organic sulfonic acids.

- Primary aliphatic amines are nearly unreactive with ClO_2 and secondary amines are only slightly more reactive. Tertiary aliphatic amines are reactive, triethylamine forming acetaldehyde and the less reactive diethylamine, along with chlorite ion.

- Triethylamine reacts with ClO_2 without forming an N-oxide. Quinuclidine does form an N-oxide, but triethylenediamine produces piperazine (no N-oxide).

- River Meuse water containing humic materials and treated with ClO_2 showed increases in phenolic and COD contents. This behavior is analogous to that found upon ozonation.

- Under drinking water treatment plant conditions, humic materials and/or resorcinol do not produce trihalomethanes with ClO_2 even when a slight excess of chlorine (1 to 2%) is present.

CONCLUSIONS

A comparative summary of reactions of organic materials with ozone, chlorine dioxide and chlorine is given in Table 28. From this table and the prior discussions, the following conclusions can be drawn:

- Complete oxidation of dissolved organic materials to carbon dioxide and water in aqueous solution is rare, by means of <u>any</u> oxidant.

- In general, if an organic material is resistant to oxidation by ozone (the most powerful oxidant used in water treatment), it will <u>also</u> be resistant to oxidation by other (weaker) oxidants.

- Conversely, ozone will oxidize some materials that chlorine dioxide will not oxidize (such as primary and secondary amines, amino acids, double bonds conjugated with carbonyl groups, etc.).

- Oxidation products formed by ozonation do not contain halogen atoms, unless bromide ion is present. In this case, bromide is oxidized to bromide, which then may react with organic materials present.

- In many cases, oxidation products from chlorine dioxide do not contain halogen atoms. This is not a general rule, however, because chlorinated compounds often are produced, even when chlorine dioxide is synthesized so as to contain no free chlorine.

- Oxidation of phenols with ozone or chlorine dioxide produces oxidized aromatic compounds as intermediates, which undergo ring rupture upon treatment with more oxidant and/or longer reaction times. In many cases, the same, non-chlorinated, ring-ruptured aliphatic products are produced using ozone or chlorine dioxide.

- Oxidation of phenols with chlorine dioxide or chlorine produces chlorinated aromatic intermediates before ring rupture.

TABLE 28. COMPARISON OF OXIDATION OF ORGANIC COMPOUNDS WITH OZONE, CHLORINE DIOXIDE AND CHLORINE

Type of Organic	Oxidation Products Isolated Upon Treatment With		
	ozone	chlorine dioxide	chlorine
Phenol	intermediates: polyhydroxy-aromatics and quinones. Then ring ruptured, non-halogenated difunctional products (alcohol-aldehyde; aldehyde-acid; alcohol-acid). End Products: oxalic acid, CO_2+H_2O	Same as for ozone plus chlorinated phenols and chloroquinones. Upon ring rupture, the same non-chlorinated products as for ozone are obtained, plus some chlorinated aliphatics	chlorophenols plus ring-ruptured products (presumably chlorinated); also non-chlorinated products (aromatic and ring-ruptured)
Cresols and xylenols	Same products as from phenol, plus methyl-substituted compounds (salicyclic acid as aromatic intermediate, propionic and acetic acids after ring rupture). Acetic acid, oxalic acid, CO_2+H_2O end products		probably chlorinated aromatics, then ring rupture.
Chlorinated phenols	Chloride ion plus same products as from phenol. Oxalic acid + CO_2 end products.	Chloroquinones plus ring-ruptured, halogenated and non-halogenated aliphatics	probably more highly chlorinated phenols, then ring rupture.

(continued)

Table 28. Continued

	Ozone	Chlorine Dioxide	Chlorine
Chlorinated benzenes	Chlorophenols, then chloride ion and non-chlorinated, ring-ruptured products, as with chlorophenols, plus some chlorotartaric acid.		probably more highly chlorinated products.
miscellaneous aromatics	nitro, amino and sulfonic acid groups are cleaved, but more slowly than chlorine. Oxalic acid + CO_2 are end products.	benzoic, phenylsulfonic and cinnamic acids are not reactive to ClO_2. Nitro groups are split off during ring oxidation.	
polycyclic aromatics	undergo ring rupture producing polycarboxylic aromatics, which become increasingly resistant to further oxidation.	produces polycyclic non-halogenated quinones plus chlorinated polycyclic aromatics. Eventual ring rupture likely, but at slower rate than with ozone.	Probably same products as with chlorine dioxide.
diphenylhydrazine or diphenylamine	hydroxylamines and ring hydroxylated diphenylamine.	ring-chlorinated, ring-hydroxylated and ring-chlorinated + hydroxylated diphenylamine.	
nitrogen heterocycles	ring rupture to amino acids, then further degradation to aliphatic end products	thianine stable. Pyrimidine and indole rings apparently stable. Substituents oxidize, but do not chlorinate	

(continued)

Table 28. Continued

	Ozone	Chlorine Dioxide	Chlorine
unsaturated aliphatics	cleavage of double bond to aldehydes, ketones and acids. Possible formation of epoxides.	dichloro compounds, chloroketones, chlorohydrins, then epoxides	dichloro compounds, chlorohydrins, then epoxides under alkaline conditions.
primary aliphatic alcohols	yield aldehydes, then acids, then CO_2. Ethanol forms a dihydroperoxide with mutagenic properties under stringent conditions.	yield acids which are stable to further oxidation. Unsaturated acids (crotonic, maleic and fumaric) are stable to ClO_2.	
secondary alcohols	yield ketones, then fragmented acids, then CO_2.	yield ketones, then acetic acid, which is stable.	
primary aliphatic amines		no reaction.	
secondary aliphatic amines	aldoximes + other N-containing compounds, not nitrosamines	very slow reaction	
tertiary aliphatic amines		secondary amine + aldehyde	
chloroform	no reaction except in presence of UV	probably no reaction	no reaction

(continued)

Table 28. Continued

	Ozone	Chlorine Dioxide	Chlorine
humic materials	slowly reactive, producing phenols, ozone - resistant acids (increasing COD) and CO_2.	slowly reactive, producing phenols and increasing COD	trihalomethanes
sugars, carbohydrates		ring substituents oxidize without ring rupture until excess ClO_2 employed.	
trihalomethane precursors	1) O_3 + rapid chlorination 65% lowering of THM 2) O_3 + chlorination after 24 hrs. no effect on THM yield.	no THMs produced from pure ClO_2 (containing no free chlorine)	trihalomethanes
phosalone, aldrin	readily oxidized to destruction.		
2,4,5-T	ring rupture to oxalic acid + CO_2 + chloride ion.		
parathion, malathion	produces oxons, then degradation products.		
dieldrin, chlordane, lindane, DDT, PCBs, PCP, endosulfan	only slightly reactive with ozone, but will oxidize with ozone/UV.		

- Ozonation of chlorinated aromatic compounds ruptures the rings and cleaves carbon-chlorine bonds, forming chloride ion, non-chlorinated aliphatic oxidation products and CO_2.

- Oxidation products formed upon ozonation, and non-chlorinated oxidation products from chlorine dioxide are more biodegradable than are the starting organic materials.

- Oxalic and acetic acids are only slowly reactive with ozone, and are the most stable organic end products of oxidation of organic materials with ozone.

- Combination of ozone with ultraviolet light increases the rate of oxidation of ozone-resistant organic materials, but the same organic oxidation products are obtained as with ozone alone.

- During oxidation, chlorine dioxide is reduced to chlorite ion, which itself is a chlorinating agent. Chlorite ion may also be re-oxidized to chlorine dioxide.

- Epoxide compounds have been isolated from reactions of compounds containing double bonds with chlorine <u>or</u> chlorine dioxide. Heptachlor-epoxide is stable to ozonation, indicating that it <u>may</u> form upon ozonation of heptachlor.

- Oxidation of aldrin or 3,4-benzopyrene with ozone in clean water proceeds rapidly, but proceeds significantly slower when humic materials or soil particles are present. Thus dissolved organics can be adsorbed onto humic or soil materials and be resistant to oxidation.

- Oxidation of parathion and malathion with ozone proceeds through the more toxic oxon intermediates (paraoxon and malaoxon, respectively). These same intermediate <u>may</u> form with other oxidants, but no literature has been found to confirm this.

- Polysaccharides (cyclic sugars, starches, cellulosics, etc.) do not undergo ring rupture with small amounts of chlorine dioxide, but do ring rupture when treated with excess oxidant. This may explain why humic acids treated with ClO_2 free of residual chlorine do not form trihalomethanes.

- Bromide and iodide ions are readily oxidized to the free halogens by chlorine or by ozone. The free halogens then can undergo the haloform reaction to produce trihalomethanes, if the proper organic compounds are present. Formation of bromine-containing trihalomethanes upon treating humic materials with chlorine dioxide (even containing a slight excess of free chlorine) has not been observed.

- Ozonation of humic materials followed by immediate chlorination shows a significant reduction in trihalomethane formation. However, ozonation followed by chlorination 24 hours later shows no reduction of trihalomethane formation.

- Regardless of the oxidant employed, many (possibly all) of the same organic oxidation products will be present in the water or wastewater at the same treatment point. More significantly, in the case of chlorine, these same (non-chlorinated) oxidation products probably have been present all along. A detailed review of the organic oxidation products of chlorine should be made for comparison with the current survey of oxidation products of organic compounds with ozone and chlorine dioxide.

- Trihalomethane formation in drinking waters can be eliminated by changing the disinfecting oxidant, but this will have no effect upon formation of the other, non-chlorinated organics.

- These last two conclusions indicate that in considering an "alternative disinfectant to chlorine" for treating drinking water, "alternative treatment schemes for removing organic materials" before residual disinfectant is added should be considered. Depending on the source of organic pollution and its nature, little may be gained by simply changing the disinfectant.

- There are three basic approaches to minimize the amount of oxidized organic materials remaining in water supplies treated with oxidants:

  -- Add sufficient oxidant (and oxidation catalyst) to convert all organic materials to CO_2 and water. This may be the most costly approach, in terms of oxidant, and may not even be possible, depending upon the specific organic materials present.

  -- Eliminate or significantly reduce the amount of organic materials present before oxidant is added. This would involve better filtration, use of more or improved flocculants, etc.

  -- Oxidation with a non-halogenating reagent (ozone, chlorine dioxide in some cases, permanganate, H_2O_2, etc.) will form more biodegradable organic materials. Following oxidation with a biological filtration step (such as Biological Activated Carbon--see next section) will allow significant reduction of dissolved organics (and ammonia) using smaller amounts of oxidant than in the first two approaches.

- Since halogenated organic materials are both more difficult to oxidize and are less biodegradable, their formation during the early stages of water treatment processes should be avoided, if at all possible. If early-stage emphasis is placed on removal of organics to the maximum degree practicable under specific plant conditions, the following benefits can accrue:

1) chlorine demand will be reduced.

2) amounts of oxidized organics formed later in the process will be reduced.

3) amounts of trihalomethanes formed will be reduced.

4) Detrimental effects on finished water caused by high organic and high chlorine levels will be reduced.

- If the presence of halogenated organic compounds cannot be avoided (for example, if they are present in high concentration in the raw water), pretreatment with the combination of ozone with ultraviolet radiation should be considered, especially as a pretreatment before Biological Activated Carbon columns. Engineering and cost aspects of the ozone/UV approach to destruction of chlorinated organics in drinking water treatment are unknown at this point in time.

SECTION 13

BIOLOGICAL ACTIVATED CARBON

A Review of the Status of Pre-ozonation of Granular Activated Carbon for the Simultaneous Removal of Dissolved Organics & Nitrification of Ammonia

INTRODUCTION

During the November 1976 Workshop on Ozone/Chlorine Dioxide Oxidation Products of Organic Materials, sponsored by the International Ozone Institute and the U.S. Environmental Protection Agency, Kühn, Sontheimer & Kurz[3] discussed the use of pre-ozonation of granular activated carbon columns to more effectively process drinking water supplies. During plant visitations in Europe, the team observed the use of Biological Activated Carbon at the Dohne plant in Mülheim, Germany and at Rouen-la-Chapelle, France, introduced for the primary purpose of removing ammonia biologically, with the secondary purpose of removing organics biologically.

In this section we will describe the concept, and will review the literature which has been published on the process to date as applied to treatment of drinking water supplies. The use of Biological Activated Carbon has been developed independently for sewage treatment at the Cleveland Regional Sewer District in the United States, but this aspect of BAC will not be reviewed here. The reader is referred to Y.A. Hanna at the Cleveland Regional Sewer District, Cleveland, Ohio, and to a chapter by Rice et al.[166] in Carbon Adsorption, for further details on the subject.

In a recent article which discusses the use of granular activated carbon in water treatment, McCreary & Snoeyink[167] state that "beds of granular activated carbon (GAC) are a convenient place for microorganisms to grow because bacteria attach themselves to the irregular external surfaces of the carbon particles and are very difficult to dislodge via backwashing procedures." In the presence of soluble carbonaceous matter, which serves as food for these organisms, and in the absence of oxygen, anaerobic bacteria can develop. There are numerous instances in which sulfidic odors have been reported emanating from granular activated carbon columns used for the removal of dissolved organic materials contained in sewage treatment plant effluents[168,169] and drinking water supplies[170].

On the other hand, with sufficient dissolved oxygen and carbonaceous matter, the bacteria which develop in carbon beds will be aerobic. These do not produce sulfidic odors.

Many of the advantages of biological activated (granular activated) carbon (BAC) were first recognized by German water treatment scientists in the 1960s in drinking water plants along the Rhine River in the Düsseldorf area. Subsequently, BAC processes also have been installed in Swiss and French drinking water treatment plants, and are subjects of active pilot studies in Holland and Belgium. In the United States, the U.S. Environmental Protection Agency's Water Supply Research Laboratory in Cincinnati, Ohio has been testing a pilot BAC column since late in 1976[171].

FUNDAMENTAL PRINCIPLES

Granular activated carbon is made biologically active by the deliberate introduction of sufficient dissolved oxygen (DO) to aqueous streams just before they are passed through GAC columns. As long as the water contains sufficient DO to maintain aerobicity of the bacteria and sufficient dissolved carbon to provide food, the aerobic bacteria will thrive in this environment. Eberhardt[172] has likened bacterial activity in such an ideal environment to a "herd of cows grazing in a luscious meadow".

At our present stage of understanding, there are two mechanisms by which biological activity occurs in the BAC columns. Microorganisms are present both on the surface of the carbon and in the carbon pores. Organics will be adsorbed both at the surface and in the carbon pores, and will be biodegraded in both instances. However, those surface adsorbed organics do not have to be well adsorbed, provided that they are biodegradable. Preozonation can convert larger, less biodegradable organic molecules into smaller, more biodegradable organics, for example, into acetic and oxalic acids.

Sontheimer[173] has summarized the German findings to date which have led to the current theories of operation of BAC.

Although aerobic bacteria are necessary to obtain the benefits from BAC, so also is the adsorptive capacity of the GAC for the dissolved organic materials which will serve as food for these bacteria. This means that the surface area and pore volume of the carbon should be high. Stated another way, it is important that the organic materials present in solution which are not easily biodegradable or easily oxidized, are adsorbable onto the activated carbon column, since the contact times of solutions with the carbon particles in the columns or beds in water and wastewater treatment plants are normally short (15 to 30 minutes -- empty bed). This does not necessarily give the bacteria sufficient time to degrade larger organic carbon molecules, ideally to carbon dioxide and water. Therefore, it is important to be able to retain the dissolved organic molecules in the carbon columns so that the bacteria then will have sufficient time to degrade them, even though the actual contact times involved are rather short.

Many organic materials are readily adsorbed onto GAC, but many others are not. It is well known, for example, that high molecular weight natural humic acids, so prevalent in drinking water supplies, are not readily adsorbed by activated carbon[174].

If solutions of these non-sorbable organic materials are ozonized before passage through the GAC columns, they are converted to more readily biodegradable organic materials[174,175,176]. At the same time, ozonation introduces a large quantity of oxygen into the water which promotes aerobic bacterial growth, and also breaks down the larger molecules, leading to improved adsorption kinetics.

ADVANTAGES OF BIOLOGICAL ACTIVATED CARBON

In European pilot studies and in drinking water treatment plants it has been shown by many workers[172,173,174,177,178,179,180,181,182,183] that preozonation followed by granular activated carbon adsorption results in:

- Increased capacity of the activated carbon to remove dissolved organics,
- Increased operating life of the carbon columns before having to be regenerated (up to 3 years), especially if the GAC can be kept free of halogenated organics,
- Biological conversion of ammonia (to nitrate) in the GAC columns,
- Use of less ozone for removing a given amount of organics than using ozonation alone. (BAC may be more cost-effective over ozonation in removing Dissolved Organic Carbon - DOC),
- Filtrates from BAC columns in drinking water plants can be treated with small quantities (0.1 to 0.5 mg/l) of chlorine or chlorine dioxide, which produces drinking water of acceptable bacterial quality (zero fecal coliforms) and provides a residual disinfectant for distribution systems.

Independent studies on physical/chemical treated sewage at Cleveland Regional Sewer District[184] and in Israel[185] have confirmed these advantages with respect to removing organic materials.

EUROPEAN BACKGROUND

Introduction of granular activated carbon into European drinking water treatment practices occurred after World War II. Its initial application was for dechlorination, then for tastes and odors[186]. Many surface waters containing ammonia undergo breakpoint chlorination at the beginning of the treatment process. This technique effectively removes ammonia, but produces considerable amounts of residual chlorine and chlorinated products in the water[182]. German water treatment objectives are to process surface waters to the same quality as that of natural groundwater (which does not have to be treated in many cases). Therefore, waters treated by breakpoint chlorination have to be dechlorinated before they are treated further or distributed[187,178]. Schalekamp[178] points out that in Switzerland a residual chlorine concentration of only 0.05 mg/l is permitted in drinking water. Therefore, surplus chlorine must be removed when it is present above this level in treated drinking water.

Lower Rhine River water quality is poor, and advantage is taken of biological removal of organic materials by filtration of the river water through the sand banks of the Rhine. Wells are dug into the river banks and water is drawn from these wells as the treatment plant raw water.

It has been found that some 65 to 75% of the organic material in the Rhine (as measured by DOC and/or UV absorbance) is removed by this technique of river bank filtration[188]. These reductions in organic content are attributed to biological degradation that occurs in the sand filters, as well as to some dilution by underground water as it flows through the wells and toward the Rhine.

River bank filtration today is practiced all along the Rhine River; an excellent discussion of river bank filtration is given by Poggenburg[190].

River bank filtration in the Düsseldorf area introduces dissolved iron and manganese compounds into the plant raw water; however, this is removed effectively by ozonation, followed by sand filtration. Passage of the sand filtered water through GAC now removes tastes, odors, ammonia and dissolved organics, to produce high quality drinking water.

Although combinations of ozone and granular activated carbon first were installed in the Düsseldorf area in the early 1960s[186], it was not until nearly ten years later that the biological activity in the carbon columns was recognized as being beneficial. In the intervening years since the introduction of ozone/activated carbon at Düsseldorf, the beneficial effects of biological activity in the carbon columns have been recognized, characterized and optimized. After ozonation, the water is allowed to stand for 20 to 30 minutes to allow the more refractory organic compounds sufficient time to react with residual ozone. This retention time also allows residual ozone to be utilized in performing useful work, rather than simply to be destroyed when passed through the activated carbon column.

On the other hand, the Rhine River also contains considerable amounts of chlorinated organic materials which are not removed during river bank filtration. These halogenated materials also are more resistant to oxidation by ozone than are non-halogenated organics and thus are less likely to be converted into easily biodegradable materials. In addition, halogenated organics are more strongly adsorbed onto GAC[191,192].

Combining the stronger adsorptivity of halogenated organics on granular activated carbon with their lesser reactivity upon ozonation and their lower biodegradability simply means that breakthrough of halogenated organics can occur more rapidly than does breakthrough of non-halogenated organic compounds from GAC columns, even if the GAC columns are biologically active. Thus German water works along the Rhine in the Düsseldorf area monitor their carbon column capacities for Total Organic Chlorine (TOCl)[193,194,195,196] as well as by DOC[197] and/or UV absorption. Carbon columns at three Düsseldorf plants along the Rhein (Flehe, Am Staad, Holthausen) are backwashed every 4 to 6 weeks and regenerated every 5 to 6 months.

When Düsseldorf carbons are regenerated, however, only some 80% of the carbon charge is taken out of the columns. This leaves a portion of biologically active carbon in the column so that the level of bioactivity will not drop significantly when fresh or regenerated carbon is added. With fresh carbon columns, about 15 days of operation usually are required for biological activity to build up to an effective "steady state", particularly for ammonia removal[198].

EUROPEAN DRINKING WATER EXPERIENCES WITH BAC

Switzerland

Granular activated carbon was installed at Zürich initially to protect against oil spillage, later for protection against phenol spills, and for dechlorination[178]. The Zürich Lengg plant raw water comes from the Lake of Zürich, which contains very low concentrations of chlorinated organics and is otherwise a very clean raw water. There is no need for breakpoint chlorination because of very low ammonia content, but a small dose of chlorine (1 mg/l maximum) is added at the intake to prevent growth of mussels. Activated carbon insures dechlorination of this amount of chlorine.

In plant studies[178] (Figure 62) fresh Pittsburgh 400 granular activated carbon was shown to have an organic loading of 4 g of organics/kg of carbon (measured by extraction with dimethylformamide -- DMF). However, after operating in the plant for 2 months, this same carbon had an organic loading of 37 g/kg; after 3 months operation 63.5 g/kg, and after 7 months 68.5 g/kg. At the end of 7 months, both the top and bottom layers of the carbon beds showed equal loadings of organics, thus indicating the need for regeneration. However, regeneration of this carbon was not required because the efficiency of removal of dissolved organic carbon from the aqueous medium by this carbon remained nearly the same as that of the new carbon (Figures 63 and 64). This behavior was attributed by Schalekamp[178] to biological activity within the carbon bed.

At the Zürich Moos plant (also using Lake Zürich water), the slow sand filter was fitted with a 5 cm layer of PKST granular activated carbon. Initially, fresh carbon exhibited an organic loading capacity of 5 g/kg of carbon, but after three years of operation the loading had increased to 29.2 g/kg (Figure 65). Such low total organic loadings again attest to the

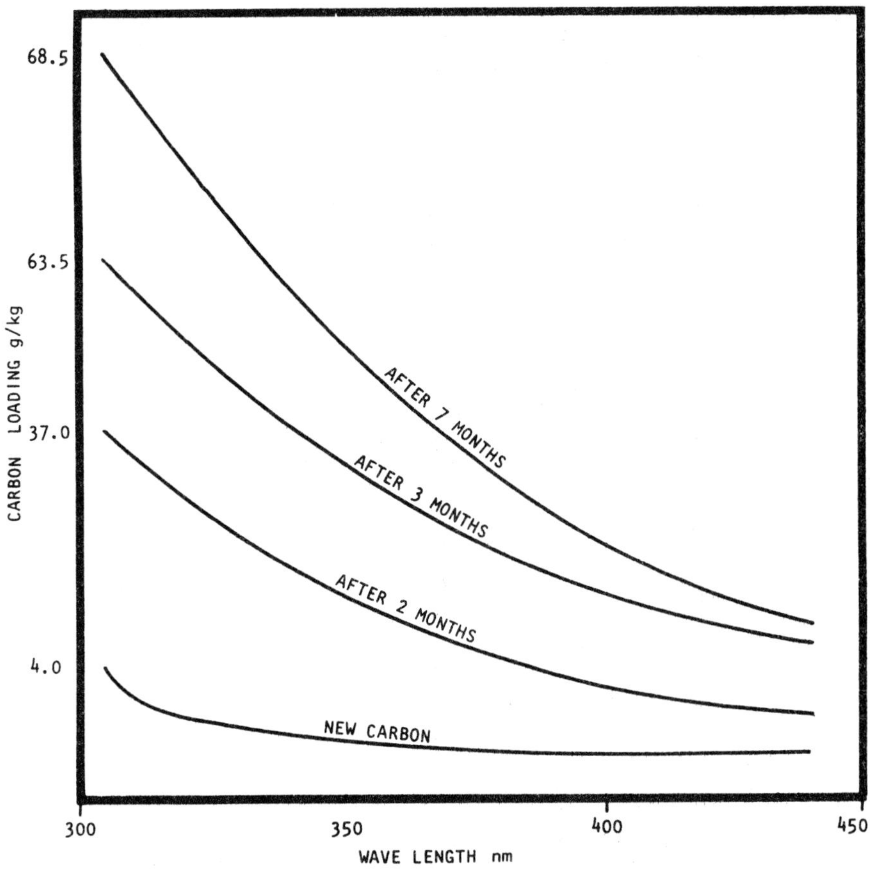

Figure 62. Activated carbon loading. New carbon and carbon after 2,3 and 7 months of operation at the Lengg Plant, Zürich, Switzerland. UV Spectra of Dimethylformamide (DMF) Extracts[178].

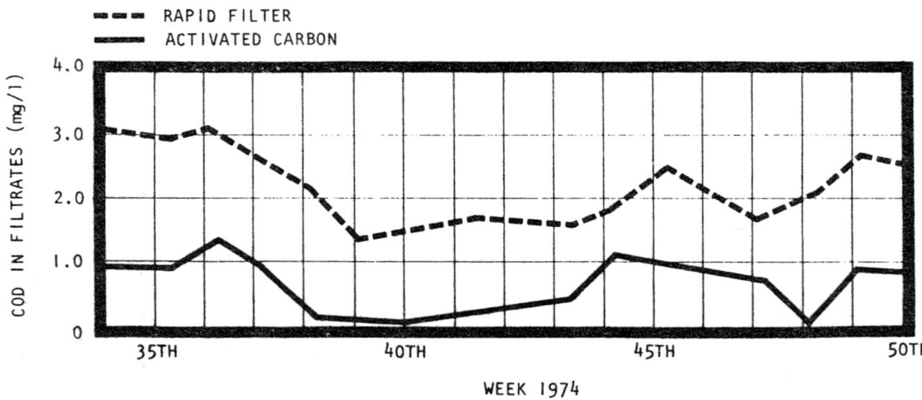

Figure 63. Efficiency of removal of COD from rapid filter and activated carbon at Lengg Plant, Zürich, Switzerland[178].

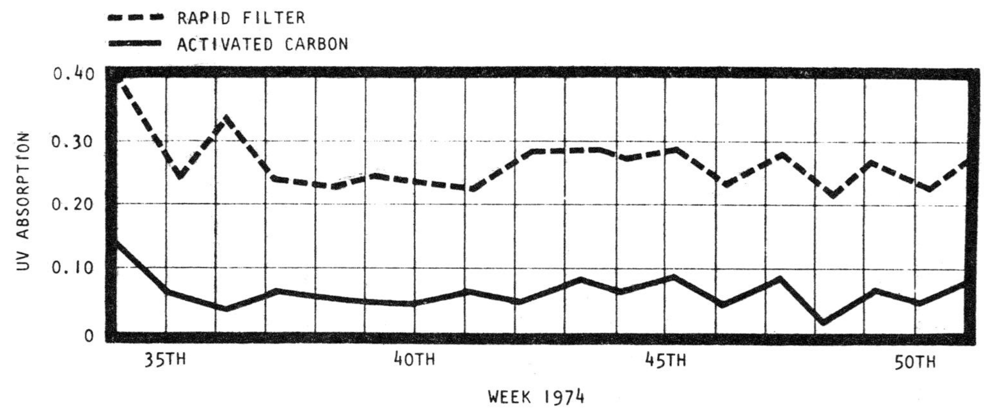

Figure 64. Efficiency of COD removal (measured by UV absorption at 254 nm cell distance = 1 cm) at Lengg plant, Zürich, Switzerland[178].

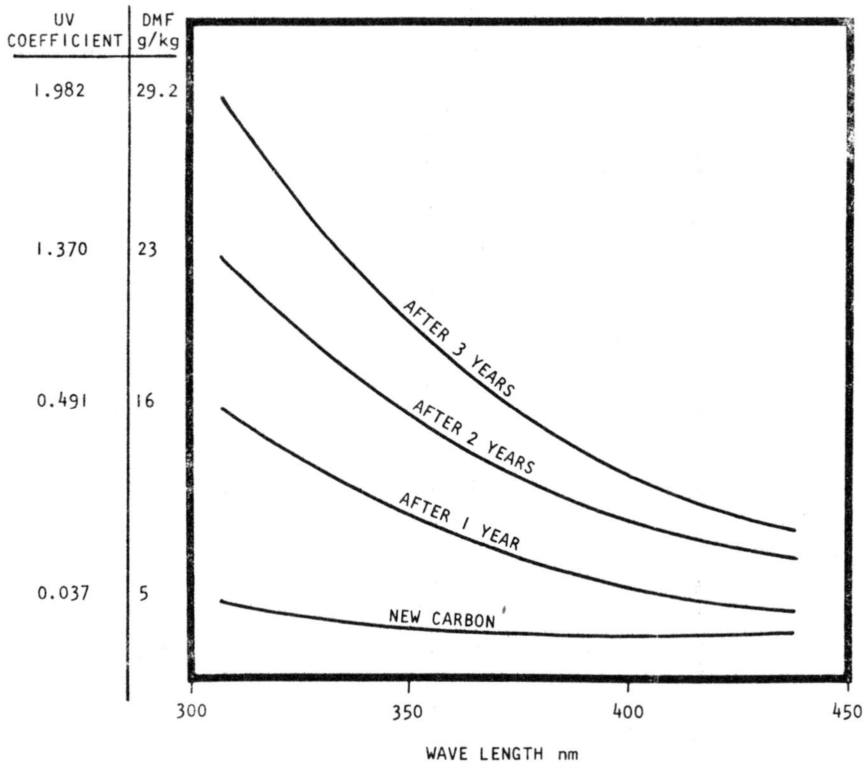

Figure 65. Activated carbon loading from the slow filter at the Moos plant, Zürich, Switzerland. New carbon and after filter operating times of 1, 2 and 3 years. UV spectra of DMF extracts at 254 nm[178].

high quality of the raw water. As with the carbon at Lengg, however, the efficiency of DOC removal from the aqueous solution remained essentially constant over the three year period (about 2.8 mg/l residual COD in the filtrate) (Figure 66). This performance, again attributed to bacterial degradation of the adsorbed organics, was obtained without reactivation of the carbon, although twice weekly backwashes were required.

When one first considers the deliberate use of biological growth in an activated carbon column for treating drinking water, one imagines that the carbon filtrates will contain high quantities of bacteria. That this is not the case with properly designed and operated BAC columns will be shown in a later section discussing a three-year research program at Germany's Bremen water works.

Bacteria need both oxygen and dissolved biodegradable organic carbon for growth. If the activated carbon bed is too large, there will be little, if any, dissolved carbon in the column outlet, thus high bacterial counts within the column will decrease[174,187].

Figure 67 shows the bacterial counts/ml of activated carbon effluents at the Lengg plant after 3 days of incubation at 20°C. Values varied from maximum counts of 42/ml in August to 9 in December.

Operational filters at the Lengg plant are 44 sq meters in area. On the bottoms are 75 plastic nozzles/sq meter. The filters are charged with 50 cm of quartz sand (0.7-1.0 mm grain size) then 1.2 meters of Pittsburgh F 400 granular activated carbon. Operational costs for the carbon beds at the Zürich Lengg plant are $0.016/cu meter (6.05 ¢/1,000 gal) which is 12.5% of the total plant operating costs. Since treated Swiss surface waters must not be inferior to unobjectionable well or groundwater, biological activated carbon filtration is a necessary and also economically feasible component in the treatment process. Pre-ozonation dosages at Lengg are 1 to 1.5 mg/l[8].

Director Schalekamp closes his 1975 presentation with the statement,

> "It is the duty of every water supply to ensure that the quality of the drinking water derived from surface water is in no way inferior to the unobjectionable quality of waters found in wells or groundwater, particularly with respect to the loading of organic substances. Therefore, conditioning of surface water with activated carbon filters can no longer be omitted in Switzerland".

Holland

Van Lier et al.[179] describe experiences with granular activated carbon filters in Dutch pilot studies at Amsterdam. Three carbon pilot trains were studied side by side, using 2 meter column heights, 8 cu m/sq m/hr flow rates through ROW 0.8 Supra (Norit) carbon for 3 months. The water was treated by iron coagulation, rapid sand filtration, chlorination, then:

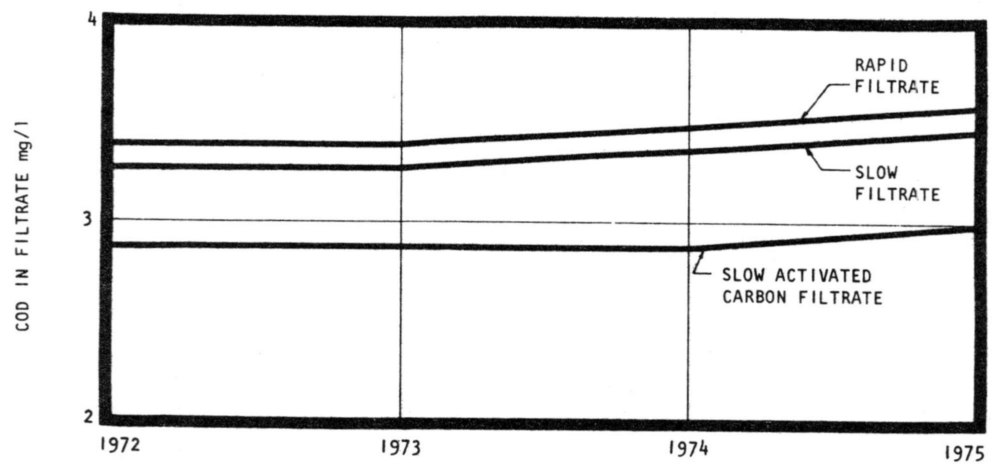

Figure 66. Efficiency of COD removal of BAC over 3 years at Moos Water Works, Zürich, Switzerland[178].

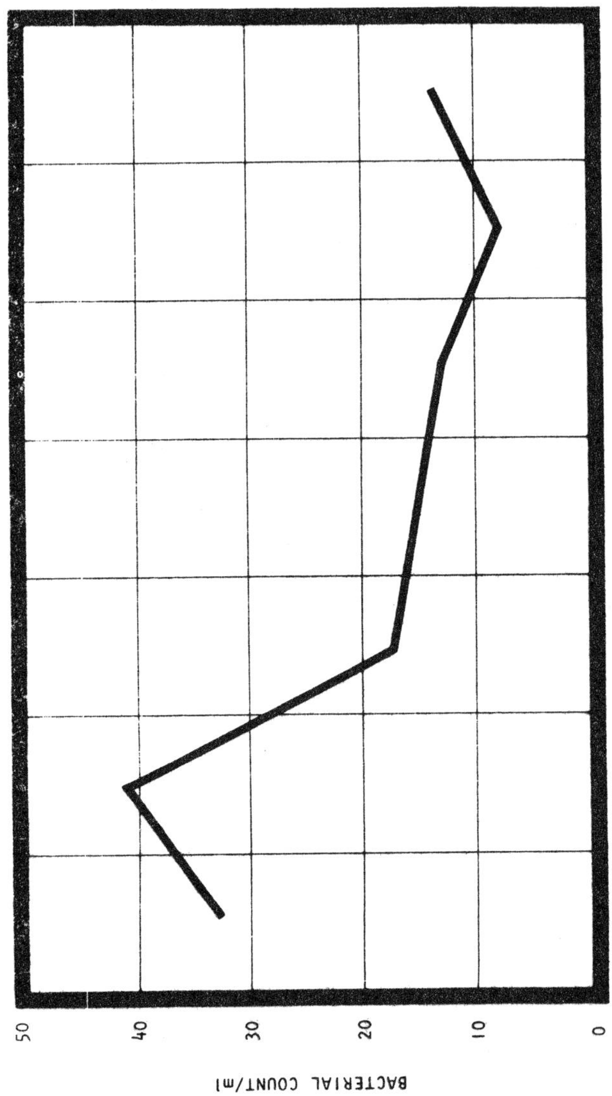

Figure 67. Bacterial count after 3 days incubation at 20°C in 1 ml activated carbon filtrate of the "South" installation, Lengg Water Works, Zürich, Switzerland[178].

Process #1: Ozonation, rapid sand filtration, biological activated carbon, slow sand filtration,

Process #2: Rapid sand filtration, granular activated carbon, slow sand filtration,

Process #3: Granular activated carbon, slow sand filtration.

These researchers concluded (Figure 68) that:

- Ozonation increased bacterial counts considerably after rapid and slow sand filtration,
- Water treated by Process #1 (with ozone) produces water with better color, UV absorption and $KMnO_4$ consumption than waters treated by Processes #2 or #3 without ozone,
- Slow sand filtration in all 3 systems reduces bacterial counts, color and $KMnO_4$ consumption,
- Water qualities by Processes #2 and #3 (without ozonation) are about the same, but the frequency of backwashing of the slow sand filters is significantly different in these processes. Replacement of sand in the rapid filters was not attractive because of the increased backwashing of the following slow sand filters.
- Service time of the granular activated column with ozone was much longer (300 days) than those without ozone (175 days) as measured by UV absorption (Figure 69) of the filtrates.

For the same applied DOC, the amount of DOC removed by the carbon in summer is much greater than in the winter. This is explained on the basis of biological activity in the carbon column.

Oxygen consumption in the winter was found to be 0.006 g/hr/kg of carbon and 0.024 to 0.030 g/hr/kg of carbon in summer.

Longer service times were observed for BAC columns which had optimum empty bed contact times of 20 to 25 minutes.

Costs of operating carbon filters decreased with increasing retention times. Costs of 6 Dutch ¢/cu meter (2.55 U.S. ¢) were estimated for 25 to 30 minute retention times. Reactivation costs were estimated at 3 Dutch ¢/cu meter (1.23 U.S. ¢).

Also in Holland, Van der Kooij[180] investigated the presence and behavior of bacteria on activated carbon filters.

Tests were conducted on three experimental filters:

Filter #1: GAC -- (Norit ROW 0.8 Supra)

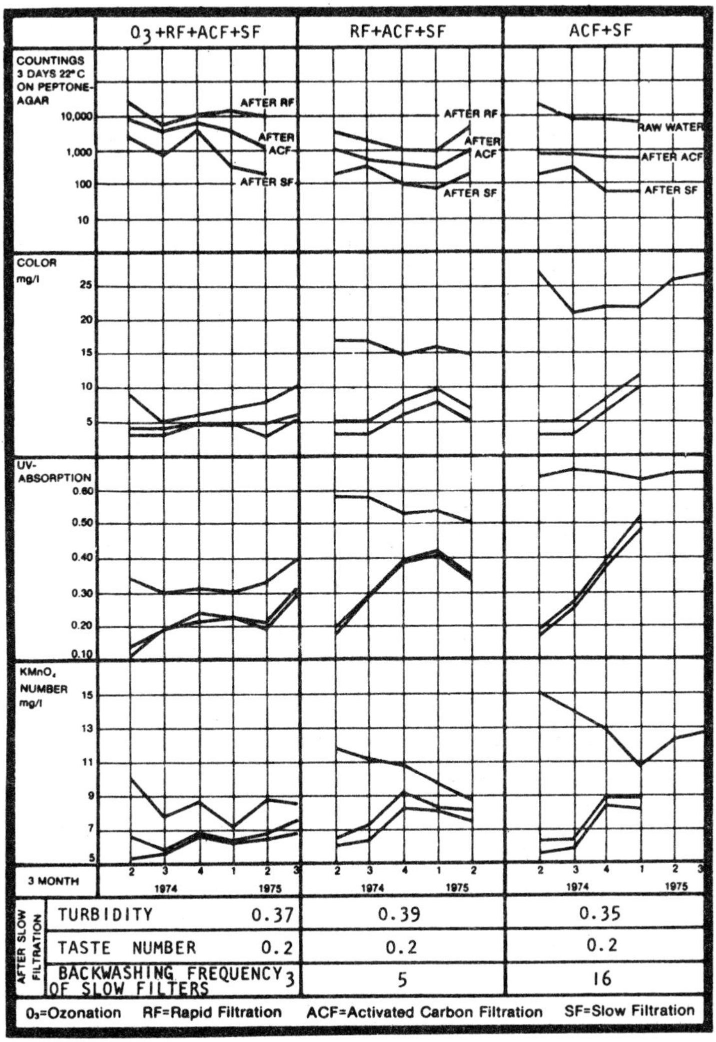

Figure 68. Activated Carbon Studies in Amsterdam[179].

Filter #2: Non-Activated Carbon -- (Norit ROW 0.8)

Filter #3: Sand -- (0.85 to 100 mm in size)

Each filter was fed with tap water at 13 to 17°C and 3.5 m/hour* (3 minutes contact time) flow-through rates over a 10 month period (side-by-side experiments). Bacterial counts were made at regular intervals by the colony count technique on diluted agar (0.35 g/l beef extract, 0.65 g/l peptone, 10 g/l agar) after incubation 10 days at 25°C. Bacterial counts were highest in activated carbon filter #1 (Figure 70). The number of bacteria on the activated carbon column was always 10 times higher than the number found on sand or on non-activated carbon.

From these experiments Van der Kooij[180] calculated the average carbon surface area occupied by a single bacterial cell to be 40 sq microns when the colony count is 10^8/cu cm. Since the surface area of the activated carbon used for these tests was greater than 40 sq cm/cu cm, it was concluded that the density of bacteria on the carbon surface is very low. This was confirmed by electron microscopy. It was also concluded that normal adsorption processes on granular activated carbon are not hindered by the presence of the bacteria on the carbon.

Germany

Wiesbaden--

Klotz et al.[199] studied the microbiology of granular activated carbon filters at the Schierstein plant in Wiesbaden. At this plant, Rhine River water is aerated, settled, chlorinated to the breakpoint, flocculated, filtered through sand, then granular activated carbon, then sent to ground infiltration. There is no pre-ozonation of the activated carbon column.

Bacterial counts were found to decrease between the Rhine River and the entrance to the carbon filters, the largest effect being seen after break-point chlorination. However, fresh populations of microorganisms were formed in the activated carbon filters. Within 20 days the bacterial counts in the carbon filter effluents reached their maximum concentrations of 10^5-10^6/ml in pilot filters (4 to 20 m/hr velocities) but have been found to be as high as 10^6 to 10^7/ml in plant carbon filters. After 3 years of operation, bacterial counts in the carbon columns effluents were 10^3 to 10^4/ml (Figure 71), and were about the same as those in the raw water. Biological activity in the sand filter during the fourth year rose to the level of that in the carbon filter.

Studies of the carbon columns at Wiesbaden over a period of 3 years have shown that seasonal influences are only slight. There is a tendency

---

* The term m/hour is derived from a flow rate of cu m/hr divided by sq m of surface area. Cu m/hr per sq m = m/hr.

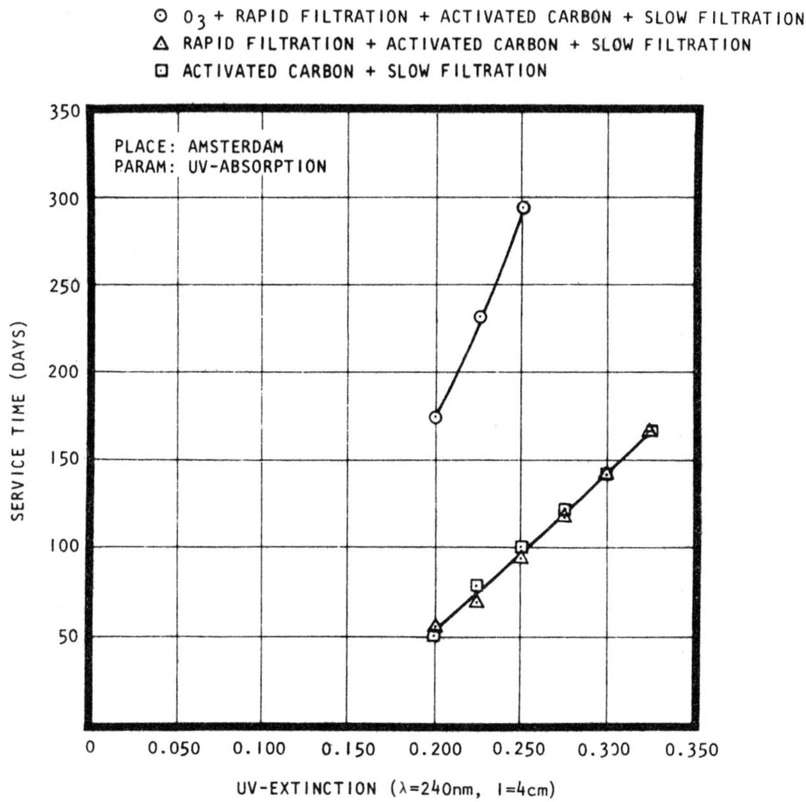

Figure 69. Activated carbon column service time as a function of filtrate quality[179].

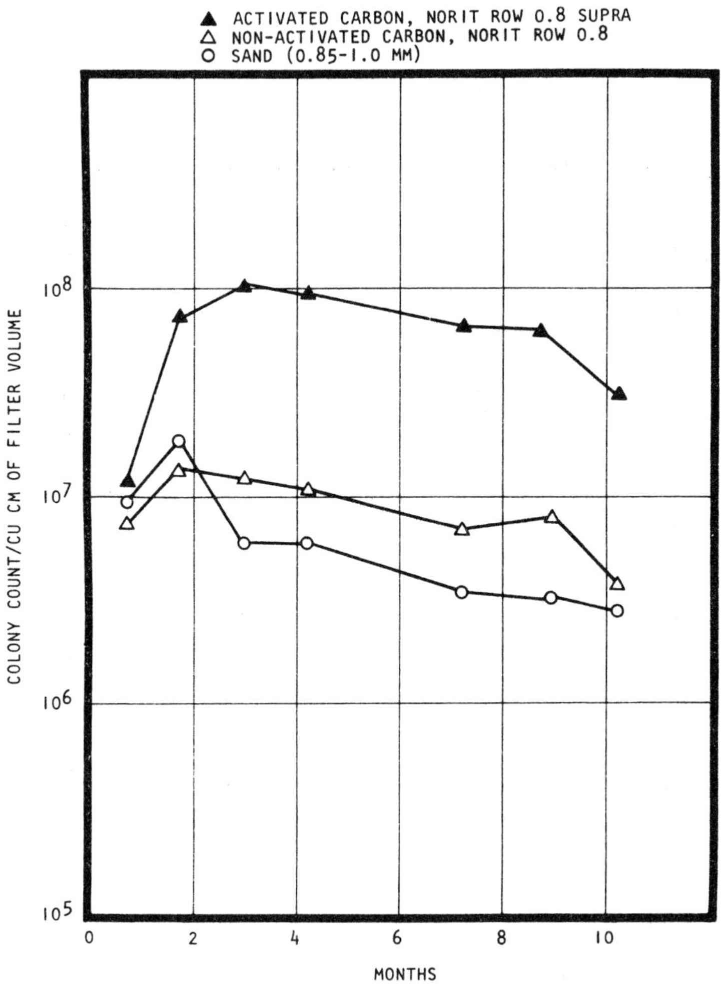

Figure 70. Bacterial counts on ROW 0.8 supra activated carbon, ROW 0.8 (non-activated) and sand (0.85 - 1.00 mm) when fed with tapwater (nominal velocity 3.5 m/hr.) [180].

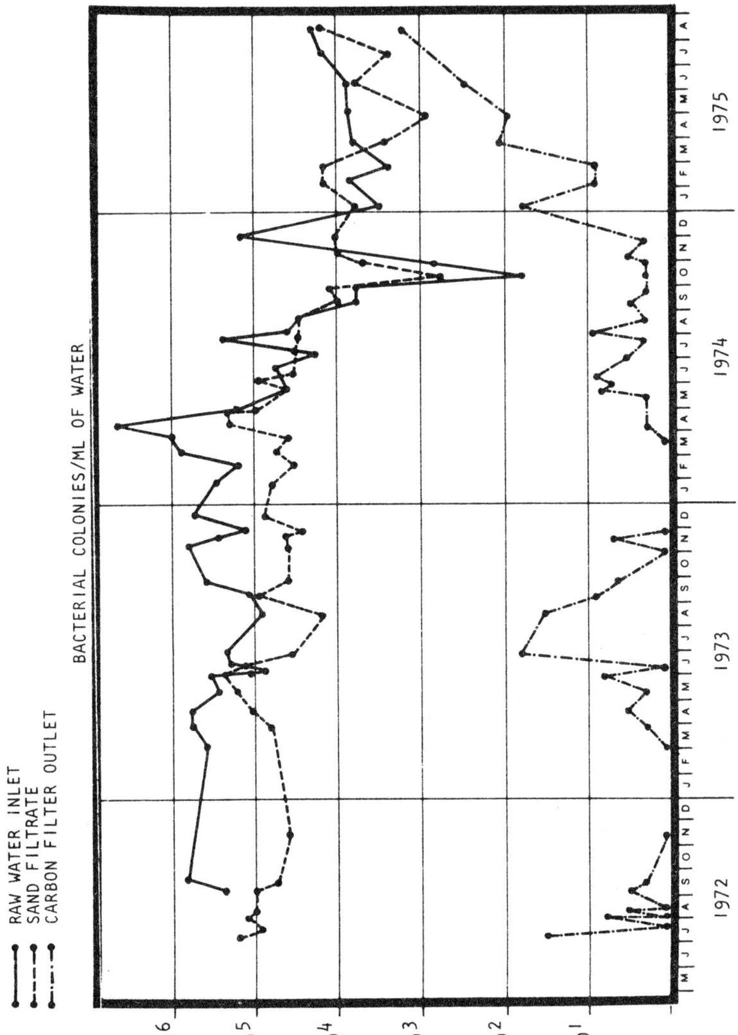

Figure 71. Behavior of microbial populations on activated carbon over three years at Wiesbaden, Germany[199].

for decreased microbial activity in winter, as indicated by oxygen consumption and carbon dioxide production values. However, changes in raw water quality can cause substantial and distinct changes in the colony numbers.

Pilot tests at Wiesbaden were conducted in two granular activated carbon units, each consisting of four series-connected glass tubes, 6 cm inside diameter, with 2 meters of activated carbon layer height. One unit was kept free of bacteria for more than 2 months, by sterile filtration.

In the uppermost carbon layers the chlorine content falls below 0.1 mg/l, where it no longer has any perceptible influence in preventing the growth of bacteria in the columns. During passage through the activated carbon filters, 1.5 mg/l of oxygen is consumed and 4.5 mg/l of carbon dioxide is produced. DOC (measured by UV absorbance) decreases about 55%, but $KMnO_4$ values decrease by about 45%. BOD_2, BOD_5 and BOD_{20} values showed that changes in biodegradability of the solutions before and after passage through the column were slight.

To increase the amount of biodegradable organic substances in the raw water, the inlet was spiked with 50 mg/l of phenol. Biological activity was found to be higher when phenol, a relatively more readily biodegradable material, was present. This means that regrowth of bacteria that can occur in distribution systems (when activated carbon is not used in the treatment processs) can be made to occur in the plant itself by installing granualar activated carbon.

Investigations on the adsorption of microorganisms onto granular activated carbon were made using starved and washed populations (mixed populations) in a nitrogen-free environment. Figure 70 shows an adsorption isotherm for bacterial loading onto granular activated carbon as a function of the concentration of bacteria (adsorptive concentration) in 200 ml of phosphate buffer; pH = 7.2. At high colony counts (larger than 10^{10}/200 ml) the carbon system tends to become saturated. At colony counts of 10^7 to 10^8/200 ml, up to 90% of the bacteria are adsorbed onto the carbon.

After an incubation period of 20 to 30 hours, adsorption and desorption of bacteria were nearing the steady state (Figure 73) of about 10^8 colony numbers/g of activated carbon. In this figure, colony counts refer to living bacteria and cell numbers refer to dead bacteria. The authors[199] state that dead bacteria are slightly better adsorbed than living bacteria.

Electron scan microscopic analysis of granular activated carbons treated differently showed that the bacteria are always present in the form of a single bacterial layer. Thus the carbon surface is only fractionally utilized (about 1%) by the adsorbed bacteria, leaving 99% of the surface area of the carbon free for adsorption of dissolved organic materials and dead bacteria.

Bremen--

Extensive pilot plant studies have been conducted by Eberhardt *et al.*[172]. The original Bremen plant treats Weser River water by flocculation,

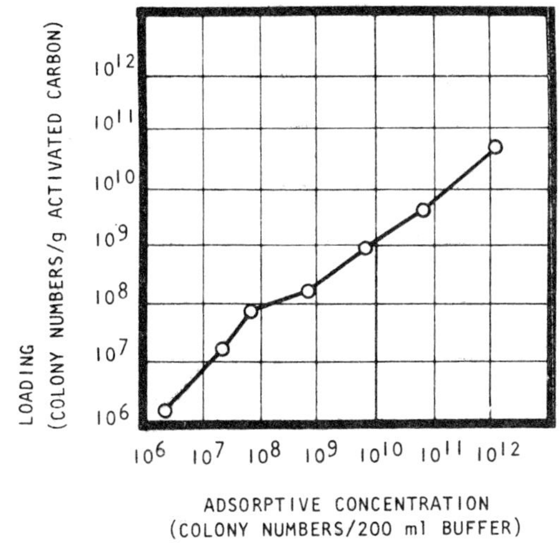

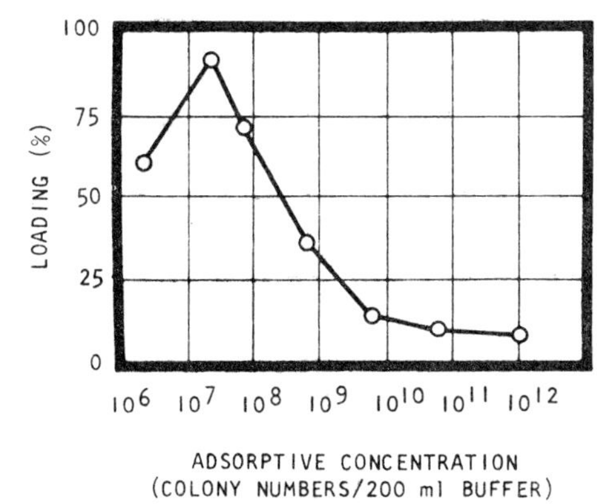

Figure 72. Microbiological loading of activated carbon-dependence on adsorptive concentration[199].

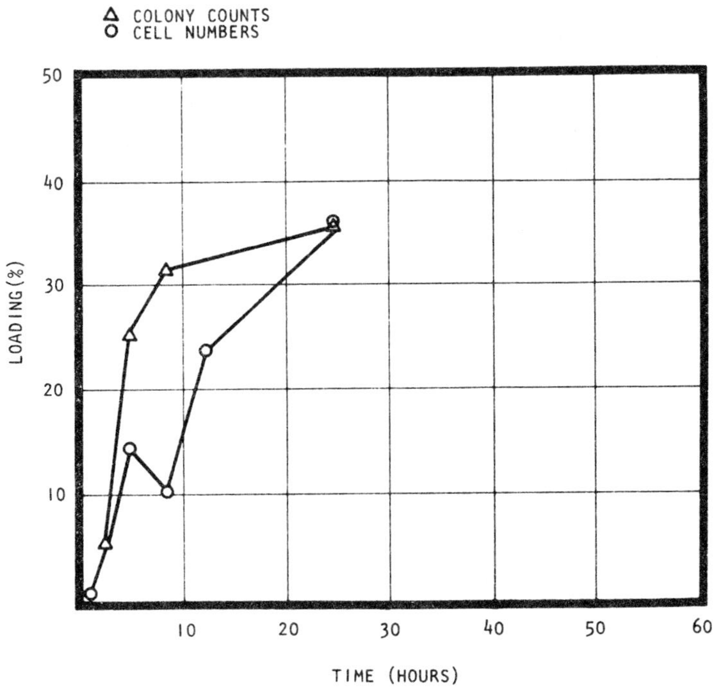

Figure 73. Microbiological loading of activated carbon-dependence on time Wiesbaden studies[199].

rapid filtration, slow filtration and chlorination. A semi-works size pilot facility employing activated carbon with preozonation was constructed in 1969, and research reported in this article was conducted on biological activated carbon over a 3 year period. DOC of the raw water was 5 to 10 mg/l, and the permanganate numbers varied from 10 to 22 mg/l.

The test facility (Figure 74) originally consisted of an ozone contact chamber (0.7 m diameter, 3 m high), two temporary storage tanks (each 2.5 cu m), a slow sand filter (0.7 m diameter, 3 m height), and a granular activated carbon filter (0.8 m diameter, 3 m height). Provision was also made to periodically close off the flow of water through the carbon column. At these times, the water in the carbon columns was recycled back through the carbon beds, and analytical parameters were measured. These included dissolved oxygen, DOC, carbon dioxide, ammonia, nitrate, etc. In this manner, material balances were determined. In later experiments, a rapidly operating slow sand filter was added after the activated carbon filter.

Water from the rapid filter was treated in this pilot facility, at first with the same dosages of ozone as used in water plants along the lower Rhine (2.1 mg/l ozone). Ozonized water was held 1 to 3 hours, then filtered through 1 m of sand, then 1 m of activated carbon at 5 m/hr.

Data in Table 29 (taken over 6 weeks of operation) show that although this low ozone dosage reduced $KMnO_4$ consumption somewhat (15.4 to 13.7 mg/l), no further decrease was observed in the carbon or sand filters (which were placed in parallel for this test). The decisive parameter for comparison, however, is the slow filter performance (last line of data in Table 29). This shows that it is possible to obtain better (lower) values of potassium permanganate consumption without ozonation at the low level used for these tests.

Ammonium ion concentration was lowered significantly in the carbon column, however, from 0.65 to 0.08 mg/l. This reduction in ammonium ion concentration was attributed to nitrification occurring in the biologically active carbon bed. The amount of nitrification which occurred in the carbon under these conditions was significantly higher than that which occurred in the parallel slow sand filter.

Although low colony counts were found after ozonation, after BAC filtration or after parallel sand filtration (48 hours incubation), extraordinarily high colony counts were found on gelatine medium after 72 hours incubation at 22°C. These data indicate that immediate treatment of the BAC column filtrate with chlorine or chlorine dioxide would maintain residual bacterial counts in the effluent at or near zero.

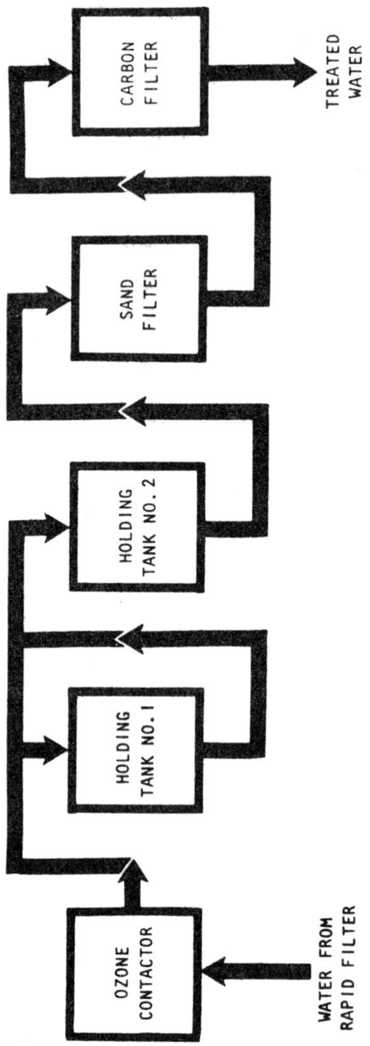

Figure 74. Bremen, Germany water works pilot plant172.

TABLE 29. EXPERIMENTS WITH SMALL DOSES OF OZONE (2/27-4/10/1969) AT BREMEN[172]

	pH	$KMnO_4$ consump mg/l	NH_4^+ mg/l	Turbidity x 10^{-3} abs E	Colony Count/ml after	
					48 hr	72 hr
*Clarifier Outlet	7.85	17.3	1.05	2.49	2,900	----
*Rapid Filter	7.61	15.4	0.65	0.95	1,380	----
After Ozonation	7.59	13.7	----	----	1	1
After Act Carbon	7.40	13.3	0.08	0.92	6	5,200
or						
After Sand Filtrn	7.57	13.2	0.55	0.95	11	53
*Slow Filter (for Comparison)	7.41	11.9	0.06	0.94	45	----

2.1 mg/l ozone dose. 1-3 hr holding tank. Activated carbon and Sand filter in parallel with 5 m/hr filtration time at 1 m height. Raw water temperature 4.1°C.

Geometric averages

*Original treatment process

The next series of experiments was conducted using the treatment train shown in Figure 72 and the ozone dosage was increased considerably. Table 30 shows data obtained when the ozone dosage was increased from 2.1 mg/l to an average 8 to 10 mg/l. A much larger decrease in permanganate consumption was observed after ozonation (14.1 to 10.5 mg/l), but also a further decrease after the carbon column (10.5 to 8.8). The decrease observed after the carbon column was attributed to biological activity within the carbon column, since a similar reduction in permanganate consumption (which would have been due to adsorption only) was not observed in the experiments of Table 29.

As with the first experiment, low colony counts were observed in the activated carbon filtrate after 48 hours incubation, but very high counts were found after 72 hours incubation.

On the basis of good experiences gained in full scale operations at the St. Gallen and Zürich, Switzerland water works, a so-called "fast-driven" slow filter (i.e., a slow filter operated at a flow rate between 0.5 and 1 m/hr) was added downstream of the activated carbon filter. The ozone dosage rate also was reduced. All subsequent data were obtained using the treatment train shown in Figure 72, with the added "fast-driven" slow filter.

TABLE 30. EXPERIMENTS WITH HIGHER OZONE DOSES BEFORE THE ACTIVATED CARBON FILTER (9/1-9/22/1969) AT BREMEN[172]

	pH	$KMnO_4$ consump mg/l	O_2 mg/l	Turbidity x 10^{-3} abs E	Colony Count/ml after		*E. coli* and Coliform B. in 100 ml
					48 hrs	72 hrs	
*Clarifier Outlet	7.77	17.3	7.8	4.22	235	---	---
*Rapid Filter	7.60	14.1	5.7	0.98	104	---	10
After Ozonation	7.50	10.5	12.1	----	0	---	0
After Activated C	7.40	8.8	10.4	0.78	119	2,700	0
*Slow Filter Comparison	7.51	10.3	5.6	0.89	27	---	0

8.5 - 10.0 mg/l O_3 dose. 3 hrs in holding tank. Activated Carbon filtration rate $4 m^3/m/hr$ at 1 m height. Temperature 17.5°C (average).

*Original treatment process

Table 31 shows data obtained with an ozone dose rate of 5.8 mg/l. The desired effects of reduced $KMnO_4$ consumption were achieved with the BAC, but hardly any organic material was removed in the fast-driven slow filter placed downstream. On the other hand, this filter exhibited high efficiency in removing bacterial colony counts, particularly in the values after 72 hour incubation.

It can be concluded from these results that it should be possible to obtain even better results if the residence time in the biologically active carbon filter were to be increased from 15 minutes to 30 minutes, in terms of the empty bed volume. Results from such longer residence time experiments are given in Tables 32 and 33.

TABLE 31. EXPERIMENTS WITH INTERMEDIATE OZONE DOSAGES AND CONSECUTIVE FILTRATION THROUGH ACTIVATED CARBON, THEN RAPIDLY OPERATING SLOW SAND (9/22-10/23/1969) AT BREMEN[172]

	pH	$KMnO_4$ consump mg/l	Colony Count/ml after		*E. coli* and Colif. B in 100 ml
			48 hrs	72 hrs	
*Clarifier Outlet	7.80	17.8	210	---	---
*Rapid Filter	7.59	14.7	65	---	15
After Ozonation	7.52	11.5	>1	---	0
After Act Carbon	7.38	9.2	52	3,100	0
After Rapid/Slow Filter	7.39	9.2	42	400	0
*Slow Filter Comparison	7.52	11.1	25	---	0

5.8 mg/l ozone after holding time. Raw water from rapid filtration step. Activated carbon filtration rate 3.5 m/hr at 1 m height. Filtration rate of rapid/slow filter: 1 m/hr at 1 m height. Termperature: 12.6°C (average).

*Original treatment process

Data of Table 32 were obtained at a filter flow rate of only 2 m/hr in the BAC column (1 m bed height) and a 30 minute theoretical empty bed residence time, corresponding to 12 to 15 minutes effective residence time. An appreciable reduction in $KMnO_4$ consumption was observed, not only after the BAC column, but also after the sand filter which followed. The final values in this respect were considerably better than those obtained after the slow filters of the plant itself.

Data of Table 33 show that very satisfactory levels of water quality can be achieved from the highly polluted Weser River, which sometimes had water temperatures below 2°C. Biological processes occurring in the carbon column are credited with lowering the amounts of contained organics, and particularly in lowering the ammonium ion concentrations. Also noticeable are the very low bacterial colony counts obtained after the BAC column and after the fast-driven slow filter (48 hour incubation). These data also support the conclusion that addition of small amounts of residual disinfectant (chlorine, chlorine dioxide, etc.) after Biological Activated Carbon treatment would keep the so-processed drinking water bacteria-free.

TABLE 32. EXPERIMENTS WITH OZONE, SLOW ACTIVATED CARBON FILTER -- BREMEN[172]

	$KmnO_4$ Consumption mg/l			O_2 mg/l			Colony Counts/ml					
							after 48 hrs			after 72 hrs		
	I	II	S	I	II	S	I	II	S	I	II	S
*Clarifier Outlet	20.1	17.5	18.5	9.6	11.0	10.3	446	1701	871			
*Rapid Filtrate	16.8	15.2	16.0	6.0	9.0	7.5	86	458	198			
After Ozonation	13.7	13.2	13.5	14.2	16.6	15.4	----	----	----			
After Act Carbon	11.3	11.3	11.3	11.8	13.6	12.7	57	134	87	4038	179	2673
After Rapid/Slow Filter	10.7	10.0	10.5	11.0	12.4	11.7	35	12	20	1116	155	416
*Slow Filter for Comparison	12.8	12.6	12.7	7.4	7.8	7.6	39	45	41	---	---	---

Experiment Times: I = 10/23 - 11/27/1969; II = 11/27 - 12/31/1969; S = 10/23 - 12/31/1969

Average Temperature: 9.3°C 3.2°C

4.3 mg/l O_3 dose into the rapid filtrate, then holding time (sample for ozone). Then through 1 m BD activated carbon which had been in service for a long time; 2 m/hr filtration rate. Then rapid/slow filter, 1 m in height, 0.5 and 1 m/hr filtration rate.

*Original treatment process

TABLE 33. EXPERIMENTS WITH OZONE AND ACTIVATED CARBON FILTERS (1/8-2/2/1970) IN BREMEN[172]

	pH	DOC mg/l	COD mg/l	UV-ext (corr.)	$KMnO_4$ mg/l	O_2 mg/l	NH_4^+ mg/l	Turbidity $\times 10^{-3}$ abs. E.	Colony Counts/ml after 48 hrs	Colony Counts/ml after 72 hrs
*Clarifier Outlet	7.75	4.3	12.9	0.278	21.4	11.7	2.68	4.54	6600	---
*Rapid Filtrate	7.63	3.7	11.4	0.257	18.4	10.5	2.37	0.92	2700	---
After O_3	7.59	4.1	10.6	0.163	15.9	16.9	2.16	0.99	0	
After Act C	7.22	2.2	9.1	0.160	13.0	10.1	0.49	0.97	49	4600
After Rapid-Slow Filter	7.18	2.4	7.9	0.151	11.7	8.8	0.21	0.90	9	320
*Slow Filter Comp	7.30	3.2	9.5	0.225	16.5	6.7	1.41	0.89	114	---

4.7 mg/l ozone dose. Activated carbon filtration rate: 2 m/hr at 1 m height. Rapid/slow filtration rate: 0.5 m/hr at 1 m height. Slow filtration as normal in plant. DOC, COD and UV values from only 2 samples. Average Temperature: 1.6°C.

*Original treatment process

Eberhardt, Madsen & Sontheimer[172] concluded that Biological Activated Carbon provides the following performance advantages in producing drinking water from the River Weser:

- About 100 g of dissolved organic carbon was bacterially oxidized/cu m of activated carbon per day. Consumption of oxygen during the summer was 360 g/cu m of activated carbon/day and 240 g/cu m/day during winter.

- After three years of operation without regeneration, the total bacterial count reduction averaged 97% and the E. coli reductions averaged 96%.

- The amount of organic substance removed is dependent upon its concentration in the influent and the residence time in the biological activated carbon filter. Theoretical empty bed residence times of 30 minutes are sufficient for optimum organics removal.

- Higher efficiencies of organics removal are obtained at lower filtration rates.

- Increased efficiencies of organics removal are obtained with smaller grain sizes of GAC.

- Best results are obtained from activated carbons which have high adsorptive capacities (for organics) and high pore volumes.

- BAC columns are 10 to 100 times more biologically active per unit volume of carbon than are slow sand filters, probably because of the high concentrations of organic materials in the carbon pores.

Once the bacterial activity in the granular activated carbon has been fully established, the organics removal process starts with adsorption of organics, biological mineralization of the adsorbed organics and biological regeneration of the activated carbon. The process kinetics at Bremen were found to follow a zero order reaction, but Eberhardt[205] cautions that each water should be checked before applying this as a general conclusion, since different waters and different pretreatments (before the ozone/activated carbon step) may affect the reaction kinetics.

The operative process is adsorption with water containing only organic substances and fresh granular activated carbon. However after 2 to 3 months of operation, biochemical processes have reached a steady state and are in full swing. After biological buildup has occurred, then adsorption and biochemical degradation (mineralization) of the organic substances both occur in parallel, and either adsorption or mineralization processes may predominate. At constant temperature and bed loading, the mineralization rate was found to be constant and independent of the concentration of organics in the feed water.

The following mathematical relationships have been developed by Eberhardt[205].

For the biological conversion of carbohydrates to carbon dioxide and water:

$$C_aH_bO_c + (a + b/4 - c/2) \times O_2 = a \times CO_2 + (b/2) \times H_2O$$

Therefore, the ratio of degraded organic carbon to inorganic carbon produced should equal unity, or:

$$\Delta DOC/\Delta CO_2 = 12a/44a = 3/11 = \text{constant}$$

Therefore, stoichiometrically, 3 g of DOC should produce 11 g of CO_2.

When $\Delta DOC - \Delta(\text{inorganic carbon}) = 0$, then organic materials are being mineralized only. If $\Delta DOC - \Delta(\text{inorganic C}) < 0$, then mineralization processes are predominating.

The following equation has been found to be valid for the reaction kinetics[205]:

$$-d(DOC)/dt = k$$

Therefore,

$$\Delta DOC = DOC_o - DOC = kt$$

where DOC_o = DOC in the influent to the BAC column water and

DOC = DOC in solution after the carbon filter (in g/cu m).

Since: $t = \varepsilon \times l_F/V_F$,

where: ε = ratio of water volume to bed volume

l_F = filter length in meters and

V_F = filter velocity in m/hr.

Therefore: the Specific Biochemical Filter Degradation Effectiveness per cubic meter (in g of DOC/cu m/hr), N_F:

$$N_F = k \times \varepsilon = \Delta DOC(V_F/l_F)$$

N_F increases with increasing temperature, with decrease in the size of activated carbon granules, with improved adsorption capacity, with larger pores in the carbon and when the substances in water are more biodegradable. It would also appear logical that the effectiveness would be dependent upon dissolved oxygen content, at least below a certain critical concentration, although Eberhardt[205] does not discuss this point.

Tests at Bremen showed that the Specific Biochemical Filter Degradation Effectiveness, N_F, is 1 to 4 g DOC/cu m of activated carbon/hour. Knowing N_F and the amount of DOC to be removed from solution per unit time, then the length of the filter medium and the flow rate of water to be treated can be deduced. Since both parameters can vary, local conditions at the site will be determining.

In other tests conducted at Bremen, prechlorination of the water was found to have an adverse effect on the operation and the effectiveness of biological activated carbon columns, even with pre-ozonation[172]. Prechlorination reduced both the growth of bacteria and efficiency of removal of the organics by the carbon. In addition, chlorinated organics are less biodegradable. Detrimental prechlorination effects are not a function of the oxidative powers of chlorine, since pretreatment with $KMnO_4$ did not reduce the biochemical degradation effectiveness.

It now can be appreciated that terminal ozonation of drinking waters which contain organics can promote bacterial growth in distribution systems. When such water is sent to distribution systems without containing a residual disinfectant, the chances for bacterial regrowths are very high. On the other hand, if biological activated carbon filters follow ozonation, then this bacterial regrowth activity is caused to take place within the plant, and is not allowed to occur in the distribution system. Addition of small amounts of residual disinfectant, such as chlorine or chlorine dioxide, now should insure that regrowth does not occur in the distribution system.

NITRIFICATION OF AMMONIA

Under aerobic conditions, ammonia nitrogen is converted biologically to nitrate ion in two discrete steps:

$$NH_4^+ + 1.5O_2 \longrightarrow NO_2^- + 2H^+ + H_2O \qquad \text{(step 1)}$$

$$NO_2^- + 0.5O_2 \longrightarrow NO_3^- \qquad \text{(step 2)}$$

Step 1 is accomplished by means of Nitrosomonas bacteria; step 2 is accomplished by Nitrobacter. During nitrification, some ammonia nitrogen becomes part of the cell tissues of the bacteria.

The nitrification process is known to occur in rapid sand filters if the temperature is above 5°C[205]. It is only necessary to assure that there is sufficient oxygen in the water and sufficient retention time for the bacteria to work (see Tables 29 and 33).

Nitrification of ammonia takes place in activated carbon media at the same rate as in sand filters, which is not surprising since ammonia is not adsorbed by the carbon. There is some indication, however, that activated carbon may be somewhat more efficient at temperatures below 5°C[205], possibly because of the large surface area and increased pore volumes of carbon as compared with sand. These may provide more space or volume for the nitrifying bacteria.

Ammonia nitrification in carbon columns usually requires 3 months to attain steady state conditions, and during this time there may be problems caused by sudden surges of ammonia in surface waters. However, if fresh activated carbon columns are dosed with small amounts of ammonium salts, nitrification is attained more rapidly, sudden surges in ammonia concentrations can be handled better, and the efficiency below 5°C also will improve[205].

Stoichiometrically, total nitrification of 1 gram of ammoniacal nitrogen requires 4.57 g of oxygen. However, Gomella & Versanne[181] found that only 3.2 g of oxygen is required for nitrification at the drinking water treatment plant at Rouen-la-Chapelle, France. Since aerobic bacteria on activated carbon columns and nitrification processes both consume dissolved oxygen, it is therefore necessary to provide oxygen to the water before it enters the carbon columns. This can be done by simple aeration, by addition of oxygen, or by pre-ozonizing the water.

It should be realized also that if all the ammonia nitrogen is converted into nitrate bacteriologically on a 1/1 basis, then the level of nitrate in the finished water will rise as the ammonia concentration decreases.

It is also necessary to prevent the presence of bacterial growth inhibitors, such as toxic heavy metals and halogenated organic micropollutants from entering the BAC columns and interfering with the growth of bacteria and/or the progress of the biological oxidation processes.

Specific examples of removal of ammonia by the use of Biological Activated Carbon in two new European drinking water treatment plants are given in the following sub-section.

CASE HISTORIES

Mülheim, Germany

The Rheinisch-Westfälische Wasserwerksgesellschaft mbH[182] has taken advantage of biological activated carbon to radically change the drinking water treatment process at the 48,000 cu m/day (12.7 mgd) Dohne plant in Mülheim, Germany. Raw water for this plant is the River Ruhr, which until mid-April, 1977 was treated by breakpoint chlorination for ammonia removal, flocculation, sedimentation, then granular activated carbon for dechlorination and ground filtration. Over the years, ammonia concentrations have increased, requiring prechlorination doses of 10 to 50 mg/l. In turn, these high chlorine doses produced large amounts of chlorinated organics (Table 34), which not only were incompletely adsorbed by the carbon columns and passed through the plant into the distribution system, but also caused frequent regeneration of granular activated carbon columns (every 4 to 6 weeks).

TABLE 34. ORGANO-CHLORINE COMPOUNDS FROM BREAKPOINT CHLORINATION TREATMENT, MÜLHEIM, GERMANY (DOHNE) PLANT[182]

	DOCl* ppb	DOClN** ppb	Sum of Haloforms ppb	$CHCl_3$ ppb
Raw water (Ruhr)	17	5	9	<1
After flocc + sedimn			15	6
After filtration	203	30	23	7
After act carbon filters	151	17	21	7
After ground passage and safety chlorination	92	18	23	9

* DOCl = Dissolved Organic Chlorine

** DOClN = Non-Polar Dissolved Organic Chlorine

During a two year pilot study on the use of pre-ozonation of activated carbon for removal of chlorinated organics, it was found that break-point chlorination could be eliminated completely and the BAC operation could be relied upon totally for removal of ammonia. At the same time the DOC was reduced to the desired levels.

This process, involving pre-ozonation of activated carbon, was installed and began operating in mid-April, 1977. After the first five months, the performance of the full scale plant process is as effective as was the pilot process at the same stage of development[173,187].

The newly installed process at Dohne (Table 35) involves pre-oxidation with about 1 mg/l of ozone with addition of poly-aluminum chloride and lime as flocculants. Pre-ozonation oxidizes manganese and aids in flocculating the organics. After flocculation and sedimentation, 2 mg/l of ozone is added to oxidize dissolved organics. After a retention time of 15 to 30 minutes, the ozonized water is preflocculated using 0.2 mg/l aluminum chloride and 0.1 mg/l polyelectrolyte, filtered (rapid sand), then passed through biological activated carbon where the bulk of DOC and ammonia are removed. Filter rates at the start of plant operations were 18 m/hr through 2 m carbon bed depths. However, the carbon bed depths recently have been increased to 4 m to increase the bed retention time to 15 minutes, to

TABLE 35. PROCESS DATA FOR DOHNE WATERWORKS (MÜLHEIM, GERMANY) BEFORE AND AFTER CHANGE OF TREATMENT[182]

Old Process	Treatment step	New Process
10-50 mg/l Cl_2 4 - 6 mg/l Al^{3+} 0.1 kw/cu m Ret time - 0.5 min	Dosing Mixing Preoxidation	1 mg/l O_3 4 - 6 mg/l Al^{3+} 2.5 kw/cu m Ret time - 0.5 min
5-15 mg/l $Ca(OH)_2$ Ret time - 1.5 hr	Flocculation Sedimentation	5-15 mg/l $Ca(OH)_2$ Ret time - 1.5 hr
—	Ozonation	2 mg/l O_3 Ret time - 5 min
v = 10.7 m/hr	Filtration with preflocculation	v = 9 m/hr 0.2 mg/l Al^{3+} 0.1 mg/l Polyelect
v = 22 m/hr h = 2 m	Activated carbon filter	v= 18 m/hr h = 2 m (4 m)
Ret time 12-50 hr	Ground passage	Ret time 12-50 hr
0.4-0.8 mg/l Cl_2	Safety chlorination	0.2-0.3 mg/l Cl_2

further improve process efficiencies and to protect against possible surges in organic pollution or ammonia in the raw water.

After activated carbon filtration the treated water is sent to ground infiltration (12 to 50 hours retention time) after which it is chlorinated (0.2 to 0.3 mg/l) and sent to the Mülheim distribution system.

A comparison of the performance of the new process versus the older one is given in Table 36. The DOC of treated water today is less than half that of water treated by the old process. Even lower DOC values are expected since the carbon column depths have been increased.

Table 37 shows the bacterial content of waters at the various points in the new treatment process. E. coli counts/100 ml are essentially zero after filtration and remain essentially zero after BAC filtration as well.

Pilot plant data are presented in Table 38 which shows the effects of variation of activated carbons on removal of DOC, inorganic carbon, ammonia and dissolved oxygen. In addition, this table also compares the removal of these same parameters with carbon column depths of 2.5 m and 5.0 m for two different activated carbons.

Removal of ammonia and dissolved oxygen are fairly independent of carbon type or column depth. On the other hand, removals of DOC and inorganic carbon are affected by the carbon type.

Most significant, the amount of DOC removed with 5.0 m columns is about 50% higher than with 2.5 m columns, although the amount of inorganic carbon measured increases only slightly.

During pilot studies at the Dohne plant with the BAC process, activated carbon columns were found to have operational lives of at least one year, and in some cases two years, without requiring regeneration. Life of the full scale carbon columns at Dohne is now estimated to be at least 20 months[173,187]. No signs of loss in performance have been noted during the first five months of operation[173,187]

Elimination of breakpoint chlorination at the beginning of the Mülheim process eliminates formation of chlorinated organics which caused the activated carbon columns to have to be regenerated every 2 months under the old process. The 10 to 50 mg/l of chlorine previously required for this step now has been replaced with 3 mg/l of ozone. Additional cost savings associated with this change include the labor which was required with breakpoint chlorination. Formerly, a technician was required to sample water every two hours and to analyze for chlorine and for ammonia. This labor requirement has been eliminated.

Additionally, because the Dohne plant is located in the center of a residential neighborhood, considerable attention had to be paid to the safety aspects of handling such large quantities of chlorine as were required by the former process. This also consumed considerable operating labor[173,187].

TABLE 36. MEAN DOC AND UV-EXTINCTION VALUES FOR THE DIFFERENT TREATMENT STEPS MÜLHEIM, GERMANY (DOHNE) PLANT -- NEW OZONATION PROCESS[182]

	1975			1976			April-July 1977		
	DOC ppm	UV(254 nm) m^{-1}	UV/DOC	DOC ppm	UV(254 nm) m^{-1}	UV/DOC	DOC ppm	UV(254 nm) m^{-1}	UV/DOC
Raw water (Ruhr)	3.9	6.8	1.8	5.0	9.1	1.8	3.6	6.1	1.7
After flocc + sedim	3.2	4.5	1.4	4.0	5.5	1.4	2.9	3.2	1.0
After filtration	3.2	4.4	1.4	3.8	5.6	1.4	2.6	1.8	1.0
After act carbon	3.0	4.0	1.3	3.7	5.3	1.4	2.3	1.6	0.7
After ground passage	1.8	3.1	1.8	2.1	4.0	1.9	0.9	1.4	1.6

TABLE 37. GEOMETRIC MEAN VALUES OF BACTERIAL COUNTS FOR THE MÜLHEIM, GERMANY (DOHNE) PLANT USING OZONE[182]

Sampling place	Bacteria Total count/ml		E-Coli/100ml	
	M_g*	σ_g **	M_g *	σ_g **
Raw water (Ruhr)	14,490	2.0	1,620	1.7
After flocc + sedim	2,340	4.2	6.7	3.2
After filtration	6,010	4.9	<<1	—
After act carbon	3,700	4.0	<<1	—
After ground passage	27	2.3	<<1	—

* M_g = geometric mean

** σ_g = geometric standard deviation

TABLE 38. PERFORMANCE OF BIOLOGICAL ACTIVATED CARBON FILTERS. MEAN VALUES FOR 6-MONTH OPERATION AFTER A 3-MONTH STARTING PERIOD (DOHNE PILOT PLANT, MÜLHEIM, GERMANY[182]

Activated carbon type	Activated carbon bed depth (m)	ΔDOC mg/l	Δinorg C mg/l	ΔNH_4 mg/l	ΔO_2 mg/l
LSS	2.5	0.92	0.83	1.31	6.32
LSS	5.0	1.69	0.96	1.34	6.67
ROW	2.5	1.09	0.97	1.31	6.49
ROW	5.0	1.59	1.05	1.34	6.71
NK12	2.5	0.99	1.36	1.28	6.03
F400	2.5	1.26	1.11	1.32	6.95
BKA	2.5	1.00	0.97	1.28	5.99

Rouen-la-Chapelle, France

At the 50,000 cu m/day (13.2 mgd) plant at la Chapelle St. Etienne de Rouvray in Seine Maritime (west of Paris near the Atlantic Ocean), well waters drawn from near the Seine contain 2 to 3 mg/l of ammonium ion, 0 to 0.2 mg/l of manganese, various micropollutants [detergents, phenols, Substances Extractable With Chloroform (SEC), etc.] and are practically devoid of dissolved oxygen. Since 1968, the ammonia content of the raw water has risen from an average of 0.3 mg/l to an average of 2.6 mg/l. This increase required that the treatment process be improved. Breakpoint chlorination was discarded because it would have required very large contact chambers (close to 7,000 cu m).

After three years of piloting, the following process was developed, was installed and began operating in early 1976[181]:

- Pre-ozonation (0.7 mg/l) for Mn and organics oxidation plus adding dissolved oxygen
- Filtration through quartz sand
- Filtration through BAC
- Ozonation for disinfection (1.4 mg/l)
- Post-chlorination (0.4 to 0.5 mg/l)

Any residual ozone remaining from the pre-ozonation step will be decomposed to oxygen when it enters the carbon column, providing further quantities of DO for the bacteria. This single operation of pre-ozonation assures the following:

- oxygen demands of the materials in water are satisfied,
- water is oxygenated,
- complex, biorefractory molecules are broken down and become biodegradable,
- the content of various micropollutants is lowered,
- manganese is oxidized and precipitated, to be retained on the sand filter so that it does not block adsorption sites on the biological activated carbon.

Initial nitrification is observed in the sand filters. Periodic backwashing of these sand filters to remove oxides of manganese does not upset the action of these bacteria. Similarly, bacterial activity on the activated carbon columns is not displaced during backwashing.

This plant began operating in February, 1976 and showed the performances listed in Table 39 for the first year of operation. The average performance for the first several months of 1977 are stated to be superior to those for the same period of 1976[181].

TABLE 39. ROUEN-LA-CHAPELLE (FRANCE) PLANT OPERATIONAL DATA (1976)[181]

	Raw Water	Pre-ozonized	Filtered (sand & carbon)	Post-ozonized	% Elimination
Turbidity	4	-	-	2	--
Ammonia (mg/l NH_4^+)	1.80	1.80	0.40	0.26	86%
Mn (mg/l)	0.15	0.07	0.04	0.02	87%
detergents mg/l DBS	0.12	0.09	0.06	0.03	75%
phenols (microg/l)	6.5	4.0	1.5	0	100%
SEC (microg/l)	590	470	250	150	75%
Subs. Extble w/cyclohexane	1,335	740	535	410	69%
av. NH_3 content of raw water:	0.3 mg/l in 1968 2.6 mg/l in 1975				

A second French plant, Morsang-sur-Seine (east of Paris), also is piloting a similar biological activated carbon process[200]. Performance data were not yet available as this report was being written.

SUMMARY AND RECOMMENDATIONS FOR BAC OPERATION

Activated carbon columns were installed in West German drinking water treatment plants about 20 years ago, primarily for dechlorination but also for taste and odor removals. Today carbon still is used for taste & odor removal, but this is no longer as important a useage because when the water treatment process is optimized to lower the concentration of total dissolved organic materials or for lowering the concentration of chlorinated organics, tastes & odors are also fully eliminated[183].

For taste & odor elimination only, filtration rates through activated carbon can be as high as 30 m/hour. For removing chlorinated organics, however, 2 to 3 times longer contact times are required, and under such circumstances competitive adsorption processes must be considered[201,202].

Carbon quantities necessary for DOC removal are reduced by at least a factor of 10 upon preozonation of the water. Chlorine has an identical oxidizing effect, but rapidly loads the carbon filter with chlorinated organics, requiring more frequent regeneration of the carbon.

Dosages of ozone normally applied in drinking water treatment plants for effective removal of organics and of ammonia via biological activated carbon columns vary from 1 to 5 mg/l. Dosages of ozone applied to treated municipal wastewater at Cleveland Regional Sewer District prior to biological activated carbon are 5 mg/l[203].

For optimum pollutant removals, activated carbon bed depths should be 4 to 5 meters (13 to 16 feet) in depth, although columns half this height also are effective. Contact times of ozonized water in the beds should be at least 15 minutes (empty bed contact time), preferably 20 to 30 minutes. Carbon quantities should be sufficient to allow for sudden but temporary surges in pollutant concentrations. Dissolved oxygen in the effluent should be at least 2 mg/l and preferably above 3.5 mg/l. This will guarantee sufficient DO to insure optimal bacterial activity within the carbon columns.

Eberhardt[205] suggests using the following formula when designing a biological activated carbon column for removal of dissolved organic carbon:

$$l_F = \Delta DOC(V_F)/N_F$$

where: l_F = length of carbon column

V_F = velocity of water flow (m/hr)

N_F = specific filter efficiency (g/cu m/hr) (conservatively, assume 1 to 5 for BAC)[187]

ΔDOC = desired reduction of DOC concentration by the filter, in g/cu m

If the DOC of the water influent to the BAC column is assumed to be 4 g/cu m, and the DOC of the product is desired to be 1, then ΔDOC is:

$$\Delta DOC = 4 - 1 = 3 \text{ g/cu m (mg/l)}$$

If the rate of flow through the column is selected to be 10 m/hr and N_F = 1, then:

$$l_F = 3 \times 10/1 = 30 \text{ meters}$$

and for $N_F = 4$:

$$l_F = 3 \times 10/4 = 7.5 \text{ meters}$$

Similarly, if the rate of flow through the column is selected to be 3 m/hr, then for $N_F = 1$:

$$l_F = 3 \times 3/1 = 9 \text{ meters},$$

and for $N_F = 4$:

$$l_F = 3 \times 3/4 = 2.25 \text{ meters}.$$

The single carbon column 9 meters in height can be divided into 2 columns, each 4.5 meters in height, with aeration in between (to provide additional dissolved oxygen for added efficiency in the second column.

This equation does not contain a factor relating to the dissolved oxygen content of the influent or the effluent stream, and consideration is not given to parameters which may affect nitrification. It is advisable[187] that dissolved oxygen concentrations of the carbon column effluent be maintained above 2.0 mg/l (preferably above 3.5) and that column detention times of 20 to 30 minutes be provided.

For slow sand filters to provide the same efficiency of COD removal (3 g/cu m), filter volumes 50 to 100 times larger are required[172]. For slow sand filters:

$$N_F = 0.02 \text{ g DOC/cu m of sand/hr}.$$

If insufficient oxygen is being supplied to the bacteria for the dissolved carbon to be converted, more oxygen should be introduced just before the carbon column. This can be accomplished by aeration, by oxygenation, or by increasing the ozone dosage. Specific oxygenation approaches to be used will depend upon the local circumstances, such as organic type and content of the raw water, whether or not a filter is used after ozonation but before the carbon column, etc.

DOC of the influent stream to the carbon column should be monitored as should that of the effluent. In addition, the effluent stream should be monitored for inorganic carbon (carbon dioxide) produced by the biological action. Such monitoring will provide a carbon mass balance which will assure that the operation of the column is proceeding as desired.

With those raw waters which contain dissolved chlorinated organics, or process waters which have been treated by chlorine during the early process stages, the carbon column influent and effluent also should be monitored for Total Organic Chlorine. In addition, the carbon itself should be analyzed periodically for adsorbed organic chlorine[193,194,195]. These analyses will indicate breakthrough of chlorinated organics and the need for regeneration of the carbon.

When regenerating spent BAC, about 20% of the column charge should be left in place, so as to allow the full nitrifying biological activity of the recharged carbon to be re-attained in a period of time shorter than the 90 days required for fresh carbon.

Virgin carbon charges used for biological removal of ammonia should be treated with small doses of ammonium salts until the nitrifying bacteria develop their full activities. Ammonia content before and after the column should be monitored.

Different water pretreatment techniques practiced at different plants require that biological activated carbon pilot studies be performed at each plant so as to determine the optimum column design, carbon type and operating parameters.

Biological Activated Carbon columns, properly designed and operated, can greatly extend the period of carbon column performance. At present, in the absence of chlorinated organics, properly designed and operated BAC columns can be expected to operate effectively for periods of at least 1 to 2 years without requiring regeneration. In the presence of chlorinated organics, however, breakthrough of these halogenated materials will occur sooner, and require the carbon to be regenerated at that time. Plants using breakpoint chlorination before the BAC can expect to have to reactivate BAC every 6 to 8 weeks. With lesser quantities of halogenated organics present, longer useful lives can be expected, but never as long as in the absence of halogenated organics.

BAC columns used for treating drinking water should be backwashed from time to time, when head loss increases (about once per week). The Rouen-la-Chapelle plant in France backwashes the BAC beds once per month[204]. Particles of some carbons have been found to stick together, as a result of bacterial action. In such cases, air scouring before water backwashing has overcome this problem[187]. Other carbons may need such aeration at lesser frequencies during backwashing, to break up clumps which may have formed.

LITERATURE CITED

1. Lamblin, H , "Protection Of An Ozonized Water Network Against Bacteria", at Third International Symposium on Ozone Technology, Paris, France, May 1977. International Ozone Institute, Cleveland, Ohio.

2. White, G.C., Handbook of Chlorination, Van Nostrand Reinhold Co., (1972), p. 596-624.

3. Kühn, W., Sontheimer, H. & Kurz, R., "Use of Ozone and Chlorine in Water Works in the Federal Republic of Germany", in Ozone/Chlorine Dioxide Oxidation Products of Organic Materials, R.G. Rice & J.A. Cotruvo. International Ozone Institute, Cleveland, Ohio (1978), p. 426-441.

4. Coin, L, Hannoun, C. & Gomella, C., "Inactivation of Poliomyelitis Virus by Ozone in the Presence of Water", la Presse Médicale 72(37): 2153-56 (1964).

5. Sontheimer, H. & Maier, D., "Untersuchungen zur Verbesserung der Trinkwasseraufbereitungstechnologie am Niederrhein (1. Bericht)", Wasser/Abwasser 113:187-193 (1972).

6. Poggenburg, W., "Activated Carbon Filters in Water Treatment Plants. Processing Techniques-Engineering Operation", in Translation of Reports on Special Problems of Water Technology, Vol. 9-Adsorption, U.S. Environmental Protection Agency, Report #600/9-76-030, Dec. 1976, p.74-97.

7. Eberhardt, M., Madsen, S. & Sontheimer, H., "Untersuchungen zur Verwendung Biologisch Arbeitender Aktivkohlefilter bei der Trinkwasseraufbereitung", Heft 7, Veröffentlichungen des Bereichs u. des Lehrstuhls für Wasserchemie Leitung: Prof. Dr. H. Sontheimer; Univ. Karlsruhe, Germany (1974); also Wasser/Abwasser 116(6):245-247 (1975).

8. Schalekamp, M., "Experience in Switzerland with Ozone, Particularly in Connection with the Neutralization of Hygienically Undesirable Elements Present in Water", Presented at International Symposium on Ozone & Water, Wasser Berlin, Berlin, Germany, May 1977. Intl. Ozone Inst., Cleveland, Ohio.

9. Nadeau, M. & Pigeon, J.C., "The Evolution of Ozonation in the Province of Quebec", in Proc. 1st Intl. Symp. on Ozone for Water and Wastewater Treatment, R.G. Rice, & M.E. Browning editors, Intl. Ozone Inst., Cleveland, Ohio (1975) p. 167-185.

10. Rideal, E.K., Ozone, Constable and Company, Ltd., London (1920).

11. Vosmaer, A., Ozone, Its Manufacture, Properties, and Uses, D. Van Nostrand Co., New York (1916).

12. Baker, M.N., The Quest for Pure Water, The American Water Works Association, Denver, Colo. (1949).

13. Rosen, H.M. in Ozone in Water and Wastewater Treatment, F.L. Evans, III, editor, Ann Arbor Science Publishers, Inc., Ann Arbor, Michigan (1972).

14. Gagnon, M. in Odor Control and Analytical Aspects, R.G. Rice & M.E. Browning, editors, Intl. Ozone Inst., Cleveland, Ohio (1976), p. 112.

15. Bouchard, J.C. "L'Ozone Depuis Vingt ans dans le Traitment des Eaux Potable", in Proc. Sec. Intl. Symposium on Ozone Technology, R.G. Rice, P. Pichet & M.-A. Vincent, editors, Intl. Ozone Inst., Cleveland, Ohio (1976), p. 705-714.

16. Stahl, D.E., "Ozone Contacting Systems", in Proc. First Intl. Symp. on Ozone for Water and Wastewater Treatment, R.G. Rice & M.E. Browning, editors. Intl. Ozone Inst., Cleveland, Ohio (1975), p. 40-49.

17. Nebel, C., Unangst, P.C. & Gottschling, R.D. "An Evaluation of Various Mixing Devices for Dispersing Ozone in Water", Water & Sewage Works, (1973) p. R-6 to R-12.

18. Sherwood & Pigford, Absorption and Extraction, McGraw Hill, New York, N.Y. (1952).

19. Treybal, Mass Transfer Operations, McGraw Hill, New York, N.Y. (1955).

20. McCammon, W.H., "Ozonation Contacting Systems", presented at 68th Annual Meeting, Am. Inst. Chem. Engrs., Nov. 10-20, 1975, Los Angeles, Calif., 1975.

21. Masschelein, W., Fransolet, G. & Genot, J., "Techniques for Dispersing and Dissolving Ozone in Water, Part I", Water & Sewage Works, Dec. 1975, p. 57.

22. Masschelein, W., Fransolet, G. & Genot, J. "Techniques for Dispersing and Dissolving Ozone in Water, Part II, Jan. 1976, p. 34.

23. Kinman, R.N. "Analysis of Ozone--Fundamental Principles", in R.G. Rice & M.E. Browning, editors, Proc. First Intl. Symp. on Ozone in Water and Wastewater Treatment, Intl. Ozone Inst., Cleveland, Ohio (1976), p. 56.

24. Dellah, A., "Ozonization at Charles-J. des Baillets Water Treatment Plant" in R.G. Rice, P. Pichet and M.A. Vincent, editors, Proc. Sec. Intl. Symposium on Ozone Technology, Intl. Ozone Inst., Cleveland, Ohio (1976) p. 715-725.

25. Richard, Y. & Fiessinger, F., "Emploi Complementaire des Traitments Ozone et Charbon Actif", presented at 3rd Intl. Symp. on Ozone Technol., Paris, France, May, 1977. Intl. Ozone Inst., Cleveland, Ohio.

26. Kühn, W., Univ. of Karlsruhe, Federal Republic of Germany Private Communication (1977).

27. Sontheimer, H., Heilker, E., Jekel, M. Nolte, H. & Vollmer, F.H., "The Mülheim Process - Experience with a New Process Scheme for Treating Polluted Surface Waters", Submitted for publication in J. Am. Water Works Assoc. (1977).

28. Sontheimer, H., "Biological Treatment of Surface Waters in Activated Carbon Filters", Presented at Seminar on Current Status of Wastewater Treatment & Disinfection With Ozone, Cleveland, Ohio, Sept. 15, 1977. Intl. Ozone Inst., Cleveland, Ohio.

29. Nadeau, M. & Pigeon, J.C., "The Evolution of Ozonation in the Province of Quebec", in Proc. 1st Intl. Symp. on Ozone for Water and Wastewater Treatment, R.G. Rice & M.E. Browning, editors, Intl. Ozone Inst., Cleveland, Ohio (1975), p. 167-185.

30. Great Lakes-Upper Mississippi River Board of State Sanitary Engineers, Recommended Standards for Water Works, Health Education Service, Albany, New York, 1968.

31. American Water Works Association, Inc., Water Quality and Treatment-A Handbook of Public Water Suppliers, McGraw-Hill Book Company, New York, N.Y., Third Edition, 1971.

32. Babbitt, H.E., Doland, J.J., & Cleasby, J.H., Water Supply Engineering McGraw-Hill Book Company Inc., New York, N.Y., Sixth Edition, 1962.

33. Mignot, J., "Application of Ozone With or Without Activated Carbon For Drinking Water Treatment" in Proceedings First International Symposium On Ozone for Water and Wastewater Treatment, R.G. Rice & M.E. Browning, editors, Intl. Ozone Inst., Cleveland, Ohio (1975), 134-160. p.

34. Diaper, E.W.J., "Practical Aspects of Water and Wastewater Treatment by Ozone" in Ozone in Water and Wastewater Treatment, Francis L. Evans, III, editor. Ann Arbor Science Publishers, Inc., Ann Arbor, Michigan (1972).

35. Sontheimer, H., "New Experiences With Ozone Treatment in Germany", in Proceedings of First International Symposium On Ozone For Water and Wastewater Treatment, R.G. Rice & M.E. Browning, editors. International Ozone Institute, Cleveland, Ohio, (1975) p. 161-163.

36. Water Treatment Handbook, Degrémont, 1973.

37. Weiner, Arnold L., "Drying Gases and Liquids", Chemical Engineering, Sept. 16, 1974.

38. Technical Bulletin - D113, The C.M. Kemp Manufacturing Company, Glen Burnie, Md., 1973.

39. Zellman, G., The Treatment of Potable Water, Publication No. 6.17.3.05 WO 3/76 1000 en., DEMAG Metallgewinning, Duisberg, Federal Republic of Germany, 1976.

40. Larocque, Ronald, CANOZONE, Montreal, Canada. Personal Communication to C.M. Robson, October, 1977.

41. Bean, E.L., "Ozone Production and Costs" in Ozone Chemistry and Technology, American Chemical Society, Washington, D.C., 1959, P. 430-442.

42. Paré, M., "Production d'Ozone en Moyenne Frequence" in Proceedings Second Ozone Symposium On Ozone Technology, R.G. Rice, P. Pichet & M.-A. Vincent, editors, International Ozone Institute, Cleveland, Ohio (1976), p. 71-81.

43. Horizontal Tube Ozonizers, Technical Bulletin No. 1306C, Soc. Degrémont, Reuil Malmaison, France.

44. Manley, T.C. & Niegowski, S.J., "Ozone" in Encyclopedia of Chemical Technology, Second Edition, Volume 14, John Wiley and Sons Inc., New York, 1967.

45. Johansen, R.P. & Terry, D.W., "Comparison of Air and Oxygen Recycle Ozonation Systems". Presented at Symposium on Advanced Ozone Technology, Toronto, Ontario, Canada, Nov. 17, 1977. Intl. Ozone Inst., Cleveland, Ohio.

46. Rosen, H.M., "Use of Ozone and Oxygen In Advanced Wastewater Treatment", J. Water Poll. Control Fed., 45(12):2521-2536 (1973).

47. Schwartz, M., "Some Practices In Application Of Ozone To Drinking Water" in Forum On Ozone Disinfection, E.G. Fochtman, R.G. Rice & M.E. Browning, editors, Intl. Ozone Inst., Cleveland, Ohio (1977), p. 344-356.

48. Mignot, J., "Practique de la Mise en Contact de l'Ozone Avec le Liquide à Traiter" in Proceedings Second International Symposium On Ozone Technology, R.G. Rice, P. Pichet & M.-A. Vincent, editors, Intl. Ozone Inst., Cleveland, Ohio (1976), p. 15-46.

49. Masschelein, W., Fransolet, G., & Genot, J., "Techniques For Dispersing and Dissolving Ozone In Water; Part 1 and Part 2", Water and Sewage Works, Dec. 1975 and Jan. 1976.

50. Gas Injection Systems, Technical Publication No. 1011, TII Corporation, Ecology Division, Lindenhurst, New York (1975).

51. Torricelli, A., "Drinking Water Purification" in Ozone Chemistry and Technology, American Chemical Society, Washington, D.C., 1959, p. 453-465.

52. Chedal, J., "Modes of Disinfecting Action of Ozone", presented at the International Symposium on Water and Ozone, Wasser Berlin, May 1977. Intl. Ozone Inst., Cleveland, Ohio.

53. "Ozone Processes for Water & Waste Treatments and Air Purification", Technical Bulletin, Canozone, Montreal, Quebec, Canada.

54. "National Dräger Multi-Gas Detector & Detector Tubes", Bulletin ND-100, National Dräger, 48 Inc., Pittsburg, Pa. 1977.

55. "Bendix/Gastec Precision Gas Detector System", Tech. Bull., National Environmental Instruments Inc., Warrick, Rhode Island.

56. Rehme, K.A., "Calibration of Analyzers Designed to Measure Ozone In The Ambient Atmosphere" in Ozone: Analytical Aspects and Odor Control, R.G. Rice & M.E. Browning, editors, Intl. Ozone Inst., Cleveland, Ohio (1976), p.17-41.

57. Legéron, J.P., "Chemical Ozone Demand of a Water Sample by Laboratory Evaluation", presented at Symposium on Advanced Ozone Technology, Nov. 16-18, 1977, Toronto, Canada. Intl. Ozone Inst., Cleveland, Ohio.

58. Paré, M., "Operational Cost of Ozone and Energetic Balance", presented at International Symposium on Ozone and Water, Wasser Berlin, May 1977. Intl. Ozone Inst., Cleveland, Ohio.

59. Coin, L., Hannoun, C. & Gomella, C., "Inactivation of Poliomeyelitis Virus by Ozone in the Presence of Water", la Presse Med. 72(37):2153-56 (1964).

60. Coin, L., Gomella, C., Hannoun, C. & Trimoreau, J.C., "Ozone Inactivation of Poliomyelitis Virus In Water, la Presse Med. 75(38):1883-84 (1967).

61. McCarthy, J.J. & Smith, C.H., "A Review of Ozone and Its Application to Dometic Wastewater Treatment", J. Am. Water Works Assoc. 66(12):718 (1974).

62. Hewes, C.G., & Davison, R.R., "Kinetics of Ozone Decomposition and Reaction with Organics in Water", J. Am. Inst. Chem. Engrs. 71(1):141 (1971).

63. Morris, J.C., "The Role of Ozone in Water Treatment", presented at the 96th Annual Conference, Am. Water Works Assoc., New Orleans, June 20-25, 1976.

64. Le Pauloué, J., Trailigaz, Garges-lès-Gonesse, France. Oral statement (May 1977). French specialists, Drs. Coin, Gomella and Hannoun performed the experimental work.

65. Gomella, C., "Ozone Practices in France", J. Am. Water Works Assoc. 64(1):39 (1972).

66. Maier, D., Bodensee Wasserversorgung, Sipplingen, Federal Republic of Germany, Oral statement (May 1977).

67. Maier, D., Bodensee Wasserversorgung, Sipplingen, Federal Republic of Germany, Oral statement (May 1977). French specialists, Drs. Coin, Gomella & Hannoun performed the experimental work.

68. Jorgensen, J.H., Lee, J.C. & Pahren, H.R., "Rapid Detection of Bacterial Endotoxins in Drinking Water and Renovated Wastewater", Applied and Environ. Microbiol., 32:347 (1976).

69. "Pyrogenic Activity of Carbon-Filtered Waters", EPA Grant No. R-804,420-010. Project Officer, H. Pahren, Cincinnati, Ohio.

70. Gall, R.J., "Chlorine Dioxide: An Overview of Its Preparation, Properties and Uses" in Ozone/Chlorine Dioxide Oxidation Products of Organic Materials, R.G. Rice & J.A. Cotruvo, editors. Intl. Ozone Inst., Cleveland, Ohio (1978), p. 356-382.

71. Granstrom, M.L. & Lee, G.F., "Generation and Use of Chlorine Dioxide in Water Treatment Plants", J. Am. Water Works Assoc. 50:1453-1466 (1958).

72. Hodgen, H. & Taube, H., "The Exchange of Chlorine Dioxide with Chlorite Ion and with Chlorine in Other Oxidation States", J. Am Chem. Soc., 71:2501 (1949).

73. Rosenblatt, D.H., "Chlorine Dioxide: Chemical and Physical Properties", in Ozone/Chlorine Dioxide Oxidation Products of Organic Materials, R.G. Rice & J.A. Cotruvo, editors. Intl. Ozone Inst., Cleveland, Ohio (1978), p. 332-343.

74. Synan, J.F. & Malley, H.A., "Chlorine Dioxide--An Alternative to Chlorine for Disinfection in Water Systems", presented to the Engineering Panel of the Campden Ford Preservation Research Association, Chipping Campden, Flos. Oct. 21, 1975.

75. "Standard Methods for the Examination of Water and Wastewater", 14th Edition. Am. Public Health Assoc., Washington, D.C., 1975.

76. Valenta, J & Gahler, W., "Chlordioxid Anlage", Gas-Wasser-Abwasser, 55(9):566-569 (1975).

77. Rook, J.J., "Formation of Haloforms During Chlorination of Natural Waters", Water Treatment and Examination, 23(2):234-243 (1974).

78. Fair, G.M. "The Behavior of Chlorine as a Water Disinfectant", J. Am. Water Works Assoc. 40:1051 (1948).

79. Sletten, O., "Halogens and Their Role in Disinfection", J. Am. Water Works Assoc. 66:690-692 (1974).

80. Fieser, L.F. & Fieser, M., Organic Chemistry, D.C. Heath & Co., Boston, Mass. (1950).

81. Suffet, I.H., Brenner, L. & Silver, B., "Identification of 1,1,1-Trichloroacetone (1,1,1-Trichloropropanone) Waters: A Known Precursor in Haloform Reaction", Env. Sci. & Technol. 10(13): 1273-1275 (1976).

82. Rook, J.J., "Chlorination Reactions of Fulvic Acids in Natural Waters", Env. Sci. & Technol. 11(5):478-482 (1977).

83. Burttschell, R.H., Rosen, A.A., Middleton, F.M. & Ettinger, M.B., "Chlorine Derivatives of Phenol Causing Taste and Odor", J. Am. Water Works Assoc. 51(2):205-214 (1959).

84. Mills, J.F., "Competitive Oxidation and Halogenation Reactions in the Disinfection of Wastewater", in Ozone/Chlorine Dioxide Oxidation Products Of Organic Materials, R.G. Rice & J.A. Cotruvo, editors, Intl. Ozone Inst., Cleveland, Ohio (1978) p. 400-409.

85. Jolley, R.L., Environ. Letters, 1:321 (1974): Jolley, R.L., J. Water Poll. Control Fed. 47(3):601-618 (1975).

86. Eisenhauer, H.R., "The Ozonization of Phenolic Wastes", J. Water Poll. Control Fed. 40(11):1887-1899 (1968).

87. Gabovich, R.D., Vrochinskii, K.K. & Kurinnyi, I.L., "Decolorization, Deodorization and Decontamination of Drinking Water by Ozonation", Gig y Sanit. 34:336-340 (1969).

88. Bauch, H., Burchard, H. & Arsovic, H.M., "Ozone as an Oxidative Disintegrant for Phenols in Aqueous Solutions", Gesundheits Ingenieur 91(9): 258-262 (1970).

89. Bauch, H. & Burchard, H., "Investigations Concerning the Influence of Ozone on Water With Few Impurities", Wasser Luft & Betrieb 14(7): 270-273 (1970).

90. Smith, G.V., Chen, J.W. & Seyffarth, K., "Catalytic and Sonocatalytic Oxidations of Aqueous Phenol" in Proc. 5th Intl. Congress on Catalysis, Palm Beach, Fla., Aug. 21-25, 1972. Am. Inst. Chem. Engrs., New York, N.Y. (1973), p. 893-903.

91. Mallevialle, J., "Action of Ozone in the Degradation of Simple and Polymeric Phenolic Compounds. Application to Humic Materials Contained in Waters", T.S.M. l'Eau 70(3):107-113 (1975).

92. Spanggord, R.J. & McClurg, V.J., "Ozone Methods and Ozone Chemistry of Selected Organics in Water", in Ozone/Chlorine Dioxide Oxidation Products of Organic Materials, R.G. Rice & J.A. Cotruvo, editors, Intl. Ozone Inst., Cleveland, Ohio (1978), p. 115-125.

93. Gould, J.P. & Weber, W.J., "Oxidation of Phenols by Ozone", J. Water Poll. Control Fed. 48:47 (1976).

94. Throop, W.M., "Alternative Methods of Phenol Wastewater Control", J. Hazardous Materials, 1:319-329 (1977).

95. Hillis, R., "The Treatment of Phenolic Wastes by Ozone", presented at Third Intl. Symp. on Ozone Technol., Paris, France, May, 1977. Intl. Ozone Inst., Cleveland, Ohio.

96. Gilbert, E., "Reactions of Ozone With Organic Compounds in Dilute Aqueous Solution: Identification of Their Oxidation Products", in Ozone/Chlorine Dioxide Oxidation Products of Organic Materials, R.G. Rice & J.A. Cotruvo, editors, Intl. Ozone Inst., Cleveland, Ohio (1978), p. 227-242.

97. Shuval, H. & Peleg, M., "Studies on Refractory Organic Matter from Wastewater by Ozonization", Progress Report of work sponsored by Ges. für Kernforschung, Karlsruhe, Germany and Natl. Council for R&D, Jerusalem, Israel, Dec. 1975.

98. Gilbert, E., "Ozonolysis of Chlorophenols and Maleic Acid in Aqueous Solution", in Proc. Sec. Intl. Symp. on Ozone Technol., R.G. Rice, P. Pichet & M.-A. Vincent, editors, Intl. Ozone Inst., Cleveland, Ohio (1976), p. 253-261.

99. Ahmed, M. & Kinney, C.R., "Ozonization of Humic Acids Prepared from Oxidized Bituminous Coal", J. Am. Chem. Soc. 72:559-561 (1950).

100. Kinney, C.R. & Friedman, L.T., "Ozonization Studies on Coal Constitution", J. Am. Chem. Soc. 74:57-61 (1952).

101. Sturrock, M.G., Cline, E.L., & Robinson, K.R., "The Ozonation of Phenanthrene with Water as Participating Solvent", J. Org. Chem. 28:2340 (1963).

102. Il'nitsky, A.P., Cherkinskii, S.N. & Shabad, L.M., "Effect of Ozonation Upon Aromatic Hydrocarbons, Including Carcinogens", Gig. y Sanit. 33(3):323-327 (1968).

103. Il'nitskii, A.P., "Experimental Investigation of the Elimination of Carcinogenic Hydrocarbons from Water During Its Clarification and Disinfection", Gig. y Sanit. 34(9):317-321 (1969).

104. Gabovich, R.D., Kurinnyi, I.L. & Federenko, Z.R., "The Effect of Ozone and Chlorine on 3,4-Benzopyrene During Water Treatment", Gig. Naselennkh Mest., 88 (1969).

105. Reichert, J., "Examination for the Elimination of Carcinogenic Aromatic Polycyclics in the Treatment of Drinking Water, with Special Considerations of Ozone", Wasser Abwasser 110(18):477-482 (1969).

106. Hoigné, J., "Comparison of the Chemical Effects of Ozone and Irradiation on Organic Impurities in Water" in Proc. Radiation for a Clean Environment, Intl. Atomic Energy Agency, Vienna (1975), p. 297-305.

107. Jürs, R.H., "Die Wirkung des Ozons auf im Wasser Gelöste Stoffe" (The Effect of Ozone on Materials Which are Dissolved in Water"), Fortschr. Wasserchem. Iher Grenzgebiete 41:40-64 (1966).

108. Yocum, F.H., "Oxidation of Styrene With Ozone In Aqueous Solution", in Ozone/Chlorine Dioxide Oxidation Products of Organic Materials, R.G. Rice & J.A. Cotruvo, editors, Intl. Ozone Inst., Cleveland, Ohio (1978), p. 243-263.

109. Chian, E.S.K. & Kuo, P.P.K., "Fundamental Study on the Post-Treatment of RO Permeates From Army Wastewater", Sec. Annual Summary Rept., U.S. Army Medical R&D Command, Washington, D.C., Rept. No. UILU-ENG-76-2019 (1976).

110. Gilbert, E., "Chemical Reactions Upon Ozonation", presented at Intl. Symp. on Ozone and Water, Wasser Berlin, May 1977. Intl. Ozone Inst., Cleveland, Ohio.

111. Prengle, H.W., Jr., Hewes, C.G., III & Mauk, C.E., "Oxidation of Refractory Materials by Ozone with Ultraviolet Radiation", in Proc. Sec. Intl. Symp. on Ozone Technol., R.G. Rice, P. Pichet & M.-A. Vincent, editors, Intl. Ozone Inst., Cleveland, Ohio (1976), p. 224-252.

112. Prengle, H.W., Jr., Mauk, C.E. & Payne, J.E., "Ozone/UV Oxidation of Chlorinated Compounds in Water", in Forum on Ozone Disinfection, E.G. Fochtman, R.G. Rice & M.E. Browning, editors, Intl. Ozone Inst., Cleveland, Ohio (1977), p. 286-295.

113. Prengle, H.W., Jr. & Mauk, C.E., "Ozone/UV Oxidation of Pesticides in Aqueous Solution", in Ozone/Chlorine Dioxide Oxidation Products of Organic Materials, R.G. Rice & J.A. Cotruvo, editors, Intl. Ozone Inst., Cleveland, Ohio (1978), p. 302-320.

114. Dobinson, F., "Ozonization of Malonic Acid in Aqueous Solution", Chemistry and Industry 6:853-854 (1960).

115. Pryde, E.H., Moore, D.J. & Cowan, J.C., "Hydrolytic Reductive and Pyrolytic Decomposition of Selected Ozonolysis Products. Water as an Ozonization Medium", J. Am. Oil Chemists Soc. 45(12):888-894 (1968).

116. Dorfmann, L.M., NSRDS, U.S. Natl. Bur. Stds. - 46 (1973).

117. Kraznov, B.P., Pakul', D.L. & Kirillova, T.V., "Ozonization of Industrial Wastewater", Khim. Prom. 1:28-30 (1974). English Translation in Intl. Chem. Engrg. 14(4):747-750 (1974).

118. Kuo, P.P.K., Chian, E.S.K. & Chang, B.J., "Identification of End Products Resulting From Ozonation of Compounds Commonly Found in Water" in Ozone/Chlorine Dioxide Oxidation Products of Organic Materials, R.G. Rice & J.A. Cotruvo, editors, Intl. Ozone Inst., Cleveland, Ohio (1978), p. 153-166.

119. Schalekamp, M., "Experience in Switzerland with Ozone, Particularly in Connection with the Neutralization of Hygienically Undesirable Elements Present in Water", presented at Intl. Symp. on Ozone & Water, Wasser Berlin, Berlin, Germany, May 1977. Intl. Ozone Inst., Cleveland, Ohio.

120. Fremery, M.I. & Fields, E.K., "Emulsion Ozonization of Cycloolefins in Aqueous Akaline Hydrogen Peroxide", J. Org. Chem. 28:2537-47 (1963).

121. Weber, P. & Waters, W.L., "Ozonation of Aqueous Dimethylmercury", Proc. Montana Acad. Sci. 32:66-69 (1972).

122. Shapiro, R.H., Kolonko, K.J., Greenstein, P.M., Barkley, R.M. & Sievers, R.E., "Ozonization Products From Caffeine in Aqueous Solution", in Ozone/Chlorine Dioxide Oxidation Products of Organic Materials, R.G. Rice & J.A. Cotruvo, editors. Intl. Ozone Inst., Cleveland, Ohio (1978), p. 284-290.

123. Robeck, G.G., Dostal, K.A., Cohen, J.M. & Kreissl, J.F., "Effectiveness of Water Treatment Processes in Pesticide Removal", J. Am. Water Works Assoc. 57:181-200 (1965).

124. Hoffman, J. & Eichelsdörfer, D., "Zur Ozon Einwirkung auf Pestizide der Chlorkohlenwasserstoffgruppe in Wasser", Vom Wasser 38:197-206 (1971).

125. Richard, Y. & Brener, L. "Organic Materials Produced Upon Ozonization of Water", in Ozone/Chlorine Dioxide Oxidation Products of Organic Materials, R.G. Rice & J.A. Cotruvo, editors, Intl. Ozone Inst., Cleveland, Ohio (1978), p. 169-188.

126. Mallevialle, J., Laval, Y., LeFebvre, M. & Rousseau, C., "The Degradation of Humic Substances in Water by Various Oxidation Agents (Ozone, Chlorine, Chlorine Dioxide)", in Ozone/Chlorine Dioxide Oxidation Products of Organic Materials, R.G. Rice & J.A. Cotruvo, editors, Intl. Ozone Inst., Cleveland, Ohio (1978), p. 189-199.

127. Weil, L., Struif, B. & Quentin, K.E., "Reaction Mechanisms Upon Reaction of Organic Substances in Water With Ozone", presented at Intl. Symp. on Ozone and Water, Wasser Berlin, May 1977. Intl. Ozone Inst., Cleveland, Ohio.

128. Dobinson, F. & Lawson, G.J., "Chemical Constitution of Coal. VI. Optimum Conditions for the Preparation of Sub-Humic Acids from Humic Acid by Ozonation", Fuel 38:79-87 (1959).

129. Kinney, C.R. & Leonard, J.T., "Ozonation of Chattanooga Uraniferous Black Shale", J. Chem. Engr. Data 6:474-6 (1961).

130. Shevchenko, M.A. & Taran, P.N., "Investigation of the Ozonolysis Products of Humus Materials", Ukrainskii Khimicheskii Zhurnal 32(5): 532-536 (1966).

131. Buydens, R., "Ozonation and Its Effects on the Method of Purification of River Waters", Trib. du Cebedeau 319-320:286-291 (1970).

132. Rook, J.J., "Haloforms in Drinking Water", J. Am. Water Works Assoc. 68(3):168-172 (1976).

133. Dellah, A., "Study of Ozone Reactions Involved in Water Treatment and the Present Chlorination Controversy" in Proc. Sec. Intl. Symp. on Ozone Technology, R.G. Rice, P. Pichet & M.-A. Vincent, editors, Intl. Ozone Inst. Cleveland, Ohio (1976), p. 161-168.

134. Stevens, A.A., Seeger, D.R. & Slocum, C.J., "Products of Chlorine Dioxide Treatment of Organic Materials in Water", in Ozone/Chlorine Dioxide Oxidation Products of Organic Materials, R.G. Rice & J.A. Cotruvo, editors, Intl. Ozone Inst., Cleveland, Ohio (1978), p. 383-399.

135. Rosenblatt, D.H., "Chlorine Dioxide: Chemical and Physical Properties", in Ozone/Chlorine Dioxide Oxidation Products of Organic Materials, R.G. Rice & J.A. Cotruvo, editors. Intl. Ozone Inst., Cleveland, Ohio (1978), p. 332-343.

136. Gordon, G., Kieffer, R.G. & Rosenblatt, D.H., "The Chemistry of Chlorine Dioxide", in Progress in Inorganic Chemistry, Vol 15, S.J. Lippard, editor, John Wiley & Sons, Inc., New York, N.Y. (1972), p. 201-286.

137. Masschelein, W., "Les Oxydes de Chlore et le Chlorite de Sodium", Monographies Dunod, Dunod Publishers, Paris, France (1969).

138. Sarkar, P.B., J. Indian Chem. Soc. 12:470 (1935).

139. Cardey, F., "Le Traitment des Eaux Residuaires Phenolées par le Chlore", L'Eau 40:75 (1953).

140. Spanggord, R.J., et al., Stanford Research Inst., Menlo Park, Calif., Private communication of results to be published for U.S. EPA, Office of Drinking Water, Washington, D.C., 1978.

141. Paluch, K., Otto, J. & Kozlowski, K., Roczniki Chem. 39:1603 (1965); Chem. Abstr. 64:17467 (1966).

142. Otto, J. & Paluch, K., Roczniki Chem. 39:1171 (1965); Chem. Abstr. 64:1258 (1966).

143. Reichert, J.K., "Kanzerogene Substanzen in Wasser and Boden. XXI. Die Entfernung Polyzyklischer Aromaten bei der Trinkwasseraufbereitung Durch Chlorodioxid: Quantitative Befunde", Arch. Hyg. 152:37 (1968).

144. Reichert, J.K., "Kanzerogene Substanzen in Wasser and Boden, XXIII. Die Entfernung Polyzyklisher Aromaten bei der Trinkwasseraufbereitung Durch ClO_2: Isolierung und Identifizierung der 3,4-benzpyren-folge-produkte". Arch. Hyg. 152:265 (1968).

145. Reichert, J., "Examination for the Elimination of Carcinogenic Aromatic Polycyclics in the Treatment of Drinking Water, With Special Consideration of Ozone", Wasser Abwasser 110(18):477-482 (1969).

146. Dence, C.W. & Sarkanen, K.V., "A Proposed Mechanism for the Acidic Chlorination of Softwood Lignin", Tappi 43:87 (1960).

147. Kawasaki, C. et al., J. Pharm. Soc. Japan 74:907 (1954).

148. Fujii, M. & Ukita, M., Nippon Nogei Kagaku Kaishi 31:101 (1957); Chem. Abstr. 53:20596 (1959).

149. Leopold, V. & Mutton, D.B., "The Effect of Chlorinating and Oxidizing Agents on Derivatives of Oleic Acid", Tappi 42(3):218-226 (1959).

150. Lindgren, B.O. & Svahn, C.M., Acta Chem. Scand. 20:211 (1966).

151. Somsen, R.A., "Oxidation of Some Simple Organic Molecules with Aqueous Chlorine Dioxide Solutions", Tappi 43(2):154-160 (1966).

152. Lindgren, B.O., Svahn, C.M. & Widmark, G., Acta Chem. Scand. 19:7 (1965).

153. Flis, I.E., et al., Tr. Leningr. Techn. Inst.-Tsell. Bumazhu Prom. 16:62 (1955); Chem. Abstr. 64:15964 (1966).

154. Becker, E.S., Hamilton, J.K. & Lucke, W.E., "Cellulose Oligosaccharides as Model Compounds in Chlorine Dioxide Bleaching", Tappi 48(1):60-64 (1965).

155. Zienus, R.H. & Purves, C.B., "Degradation of Pectic Acid by Some Delignifying and Bleaching Agents", Tappi 43:27 (1960).

156. Scherle, C., Bull. Inst. Textile de France 41:21-43 (1953).

157. Hodgen, H.W. & Ingols, R.S., Anal. Chem. 26:1224 (1954).

158. Moran, T., Pace, J. & McDermott, E.E., Nature 171:103 (1953).

159. Kennaugh, J., Nature 180:238 (1957).

160. Rosenblatt, D.H., Hayes, A.J., Jr., Harrison, B.L., Streaty, R.A. & Moore, K.A., "The Reaction of Chlorine Dioxide with Triethylamine in Aqueous Solution", J. Org. Chem. 28:2790-2794 (1963).

161. Rosenblatt, D.H., "Chlorine and Oxychlorine Species Reactivity with Organic Substrates", in Disinfection Water and Wastewater, J.D. Johnson, editor, Ann Arbor Science Publishers, Inc., Ann Arbor, Mich. (1975), p. 249-276.

162. Fuchs, W. & Leopold, H., Brennstoff Chem. 8:101 (1927).

163. Mallevialle, J., "Ozonation des Substances de Type Humique dans les Eaux", in Proc. Sec. Intl. Symp. On Ozone Technol., R.G. Rice, P. Pichet & M.-A.Vincent, editors. Intl. Ozone Inst., Cleveland, Ohio (1976), p. 262-270.

164. Love, O.T., Jr., Carswell, J.K., Miltner, R.J. & Symons, J.H., "Treatment for the Prevention or Removal of Trihalomethanes in Drinking Water", Appendix 3 to "Interim Treatment Guide for the Control of Chloroform and Other Trihalomethanes", U.S. Environmental Protection Agency, Water Supply Research Division, Cincinnati, Ohio, 1976.

165. Vilaginès, R., Monteil, A., Derreumaux, A. & Lambert, M., "A Comparative Study of Halomethane Formation During Drinking Water Treatment by Chlorine or Its Derivatives in a Slow Sand Filtration Treatment Plant and in Wastewater Treatment Plants", presented at 96th Ann. Conf., Am. Water Works Assoc., Anaheim, Calif., May 8, 1977.

166. Rice, R.G., Miller, G.W., Robson, C.M. & Kühn, W., "Biological Activated Carbon", in Carbon Adsorption, P. Cheremisinoff & F. Ellerbusch, editors, Ann Arbor Science Publishers, Inc., Ann Arbor, Michigan (1978).

167. McCreary, J.J. & Snoeyink, V.L., "Granular Activated Carbon in Water Treatment", J. Am. Water Works Assoc. 69(8):437-444 (1977).

168. Guirguis, W.A., Melnyk, P.B. & Harris, J.P., "The Negative Impact of Industrial Waste on Physical-Chemical Treatment", presented at 31st Purdue Indl. Waste Conf., Lafayette, Indiana, May, 1976.

169. Directo, L.S., Chen, C.-L. & Kugelman, I.J., "Pilot Plant Study of Physical-Chemical Treatment", J. Water Poll. Control Fed. 49(10): 2081-2098 (1977).

170. Monsitz, J.T. & Ainesworth, L.D., "Detection and Control of Hydrogen Polysulfide in Water", Public Works 101:113 (1970).

171. Carswell, J.K., U.S. Environmental Protection Agency, Cincinnati, Ohio, Private Communication, 1977.

172. Eberhardt, M., Madsen, S. & Sontheimer, H., "Untersuchungen zur Verwendung Biologisch Arbeitender Aktivkohlefilter bei der Trinkwasseraufbereitung", Heft 7, Veröffentlichungen des Bereichs u. des Lehrstuhls für Wasserchemie Leitung: Prof. Dr. H. Sontheimer; Univ. Karlsruhe, Germany (1974); also Wasser/Abwasser 116(6):245-247 (1975).

173. Sontheimer, H., "Biological Treatment of Surface Waters in Activated Carbon Filters", presented at Seminar on Current Status of Wastewater Treatment & Disinfection With Ozone, Cleveland, Ohio, Sept. 15, 1977. Intl. Ozone Inst., Cleveland, Ohio.

174. Kühn, W., Sontheimer, H. & Kurz, R., "Use of Ozone and Chlorine in Water Works in the Federal Republic of Germany", in Ozone/Chlorine Dioxide Oxidation Products of Organic Materials, R.G. Rice & J.A. Cotruvo, editors, Intl. Ozone Inst., Cleveland, Ohio (1978), p. 426-441.

175. Guirguis, W.A., Jain, J.S., Hanna, Y.A. & Srivastava, P.K., "Ozone Application for Disinfection in the Westerly Advanced Wastewater Treatment Facility", in Forum on Ozone Disinfection, E.G. Fochtman, R.G. Rice & M.E. Browning, editors. Intl. Ozone Inst., Cleveland, Ohio (1977), p. 363-381.

176. Guirguis, W.A., Cooper, T., Harris, J. & Ungar, A., "Improved Performance of Activated Carbon by Preozonation", presented at 49th Ann. Conf. Water Poll. Control Fed., Minneapolis, Minn., Oct. 1976. J. Water Poll. Control Fed. 50(2):308-320 (1978).

177. Scheidtmann, W., "Investigations of the Optimization of Pretreatment When Using Ozone:, in Translation of Reports on Special Problems of Water Technology, Vol. 9 - Adsorption, H. Sontheimer, editor, U.S. Environmental Protection Agency Report No. 600/9-76-030, Dec. 1976, p. 98-111.

178. Schalekamp, M., "Use of Activated Carbon in the Treatment of Lake Water", in Translation of Reports on Special Problems of Water Technology, Vol. 9 - Adsorption, op. cit., p. 128-159.

179. Van Lier, W.C., Graveland, A., Rook, J.J. & Schultink, L.J., "Experiences With Pilot Plant Activated Carbon Filters in Dutch Waterworks", in Translation of Reports on Special Problems of Water Technology, Vol. 9 - Adsorption, op. cit., p. 160-181.

180. Van Der Kooij, D., "Some Investigations Into the Presence and Behaviour of Bacteria in Activated Carbon Filters", in Translation of Reports on Special Problems of Water Technology, Vol. 9 - Adsorption, op. cit., p. 348-354.

181. Gomella, C. & Versanne, D., "Le Role de l'Ozone dans la Nitrification Bacterienne de l'Azote Ammoniacal -- Cas de l'Usine de la Chapelle Banlieue Sud de Rouen (Seine Maritime) France". Presented at 3rd Intl. Symp. on Ozone Technol., Paris, France, May 1977. Intl. Ozone Inst., Cleveland, Ohio.

182. Sontheimer, H., Heilker, E., Jekel, M., Nolte, H. & Vollmer, F.-H., "The Mülheim Process - Experience With a New Process Scheme for Treating Polluted Surface Waters", submitted for publication in J. Am. Water Works Assoc. (1977).

183. Sontheimer, H., "Considerations on the Optimization of Activated Carbon Use in Waterworks", in Translation of Reports on Special Problems of Water Technology, Vol. 9 - Adsorption, op. cit., p. 208-214.

184. Guirguis, W.A., Hanna, Y.A., Prober, R., Meister, T. & Srivastava, P.K., "Reaction of Organics Nonsorbable by Activated Carbon With Ozone", in Ozone/Chlorine Dioxide Oxidation Products of Organic Materials, R.G. Rice & J.A. Cotruvo, editors. Intl. Ozone Inst., Cleveland, Ohio (1978), p. 291-301.

185. Wachs, A., Narkis, N. & Schneider, M., "Organic Matter Removal From Effluents by Lime Treatment; Ozonation and Biologically Extended Activated Carbon Treatment", presented at 3rd Intl. Symp. on Ozone Technology, Paris, France, May 1977. Intl. Ozone Inst., Cleveland, Ohio.

186. Hopf, W., "Versuche mit Aktivkohlen zur Aufbereitung des Düsseldorfer Trinkwassers", Wasser/Abwasser 101(14):330-336 (1960).

187. Sontheimer, H., Univ. Karlsruhe, Germany, Private Communication, 1977.

188. Sontheimer, H. & Maier, D., "Untersuchungen zur Verbesserung der Trinkwasseraufbereitungstechnologie am Niederrhein (1. Bericht)", Wasser/Abwasser 113:187-193 (1972).

189. Robert, A., Degrémont, Reuil Malmaison, France. Private Communication, 1977.

190. Poggenburg, W., "Activated Carbon Filters in Water Treatment Plants. Processing Techniques - Engineering - Operation", in Translation of Reports on Special Problems of Water Technology, Vol. 9 - Adsorption, op. cit., p. 74-97.

191. Kühn, W. & Fuchs, F., "Untersuchungen zur Bedeutung der Organischen Chlorverbindungen und Ihrer Adsorbierbarkeit", Vom Wasser 45:217-232 (1975).

192. Kölle, W., Sontheimer, H. & Steiglitz, L., "Eignungsprüfung von Wasserwerks-Aktivkohlen Anhand Ihrer Adsorptionseigenschaften für Organische Chlorverbindungen", Vom Wasser 44:203-217 (1975).

193. Kühn, W. & Sontheimer, H., "Einige Untersuchungen zur Bestimmung von Organischen Chlorverbindungen auf Aktivkohle", Vom Wasser 41:65-79 (1973).

194. Kühn, W. & Sontheimer, H., "Einflüss Chemischer Umsetzungen auf die Lage der Adsorptionsgleichgewichte an Aktivkohlen", Vom Wasser 40:115-123 (1973).

195. Kühn, W. & Sontheimer, H., "Zur Analytischen Erfassung Organischer Chlorverbindungen mit der Temperaturprogrammierten Pyrohydrolyse", Vom Wasser 43:327-341 (1974).

196. Kühn, W., "Untersuchungen zur Bestimmung von Organischen Chlorverbindungen auf Aktivkohle", Dissertation, Fak für Chemi-Ing. Wesen, Univ. Karlsruhe, Federal Republic of Germany, 1974.

197. Wölfel, P. & Sontheimer, H., "Ein Neues Verfahren zur Bestimmung von Organisch Gebundenem Kohlenstoff in Wasser Durch Photochemische Oxidation", Vom Wasser 43:315-325 (1974).

198. Poggenburg, W., Wasserwerk Düsseldorf, Germany. Private Communication, 1977.

199. Klotz, M., Werner, P. & Schweisfurth, R., "Investigations Concerning the Microbiology of Activated Carbon Filters", in _Translation of Reports on Special Problems of Water Technology, Vol. 9 - Adsorption, op. cit._, p. 312-330.

200. Richard, Y. & Fiessinger, F., "Emploi Complementaire des Traitments Ozone et Charbon Actif", presented at 3rd Intl. Symp. on Ozone Technol., Paris, France, May, 1977. Intl. Ozone Inst., Cleveland, Ohio.

201. Weber, W.J., Jr. & Morris, J.C., "Adsorption in Heterogeneous Aqueous Systems", J. Am. Water Works Assoc. 56(4):447 (1964).

202. Koppe, P., "Untersuchungen über die Konkurrierende Adsorption an Aktivkohle bei der Wasseraufbereitung", Ges.-Ing. 88:312-317 (1967).

203. Hanna, Y.A., Meister, T. & Slough, J., Jr., "Ozone As a Pre-Treatment Step For Physical Chemical Treatment Process", presented at Seminar on Current Status of Wastewater Treatment & Disinfection With Ozone, Cincinnati, Ohio, Sept. 15, 1977. Intl. Ozone Inst., Cleveland, Ohio.

204. Le Pauloüe, J., Trailigaz, Garges-lès-Gonesse, France. Private Communication, 1977.

205. Eberhardt, M., "Experience with the Use of Biologically Effective Activated Carbon", in Translation of Reports on Special Problems of Water Technology, Vol. 9 - Adsorption, op cit., p. 331-354.

BIBLIOGRAPHY

In this section we are listing as much of the published literature dealing with the use of ozone and of chlorine dioxide in treating drinking water as was possible to gather during the study period.

For convenience of the reader, we have listed these literature references in three separate sections entitled:

Each section of this bibliography contains references listed alphabetically by the last name of the senior author.

Technical Articles--Ozone/Drinking Water

Anon., "Ozone Treatment Licks Color Problem", Water & Sewage Works, (April, 1975), p. 52-54.

Anon., "Ozonation of drinking Water", Société Industrielle du Traitement des Liquides et des Gaz. 31:1347 (1958).

Anon., "Ozonation", U.S. Department of Agriculture, Beltsville Md. Farmers Bulletin No. 2248, page 12.

Anon., "Watchgate Treatment Works", British Water Supply 9:7-12 (1972) Manchester, England.

Anon, "Pollution of the Rhine and the Abstraction of Drinking-water," JAWR Memorandum, May (1973). Engler-Bunte Inst., Univ. Karlsruhe, Federal Republic of Germany.

Axt, G., "Über ein indirektes Verfahren zur Wasserozonisierung und dessen apparative und physikalisch-chemische Grundlagen". Vom Wasser 25:93-106 (1958).

Bartuska, J. F., "Ozonation at Whiting, Indiana", J. Am. Water Works Assoc. 33(11):2035-2050 (1941).

Bartuska, J. F., "Ozonation at Whiting (Indiana): 26 years Later" Public Works, (Aug. 1967).

Bays, L.R., Burman, N.P. and Lewis, W.M., "Taste and Odour in Water Supplies in Great Britain: A Survey of the Present Position and Problems for the Future". Water Treatment and Examination. 19(2): 136-160 (1970).

Bean, E.L., "Taste and Odor Control at Philadelphia", J. Am. Water Works Assoc. 49:205-216 (1957).

Benedek, A., "The Effect of Ozone on Activated Carbon Adsorption-A Mechanistic Analysis of Water Treatment Data", presented at Symp. on Advanced Ozone Technol., Toronto, Ontario, Canada, November, 1977. Intl. Ozone Inst., Cleveland, Ohio.

Berger, K.V., "Uber den Betrieb der Ozonanlage Könizberg der Wasserversorgung Bern", Wasser-Abwasser 105(48):1338-1343 (1964).

Bernier, M. "Stérilisation par l'ozone des Eaux d'alimentation de la ville de Nice et des Communes du Littoral". La Technique Sanitaire et Municipale, 45:117 (1950).

Black, A.P. and Christman, R.F., "Chemical Characteristics of Fulvic Acids", J. Amer. Water Works Assoc. 55:897-912 (1963).

Blankenfeld, D., "Keimfreies Wasser durch Ozonanlagen", CZ-Chemie-Technik 1(11):527-528 (1972).

Blazejewski, M., "The Necessity of Using Ozone for Drinking Water in Poland", Gaz, Wada Tech. Sanit., (3)50: (1976).

Bollyky, J., "Ozone Provides Powerful Disinfectant for Water", Water & Sewage Works, 123:66-67 (1976).

Botzenhart, K., "Effects of Ozone on Microorganisms", Presented at Intl. Symp. on Ozone & Water, Wasser Berlin, May 1977. Intl. Ozone Inst., Cleveland, Ohio.

Bouchard, J.C., "Ozone for Water Treatment", The Municipal Utilities Magazine 93(5):58-9, 73-4 (1955).

Bouchard, J.C., "Twenty Years of Ozone in the Treatment of Potable Waters", in Proc. Sec. Intl. Symp. on Ozone Technology, R. G. Rice, P. Pichet & M.-A. Vincent, Editors, Intl. Ozone Inst., Cleveland, Ohio (1976), p. 705-714.

Boucher, P.L., "Micro-Straining and Ozonization of Water and Wastewater", Proc. 22nd Industrial Waste Conf., Purdue Univ., Engrg. Extn. Ser. No. 129:771-787 (1967)

Bryant, A. A., "Saving Money With Pre-pilot Plant Studies. Water and Sewage Works (April 1976), p. 72-75.

Buydens, R. and G. Fransolet, "Etudes et Memoires l'Action de l'ozone sur le chlore, le bioxyde de chlore et le chlorite contenus dans les eaux traitées," Tribune du Cebedeau, 326:4-6, (January 1971).

Buydens, R., "La Reviviscence Microbienne dans les Eaux et, Particulièrement, dans les Eaux Ozonées", Tribune du Cebedeau 25(338):3-9 (1972).

Buydens, R., "L'ozonation et ses répercussions sur le mode d'épuration des eaux de rivières", Tribune du Cebedeau, 319-320:286-291 (1970).

Campbell, R.M., "The Use of Ozone in the Treatment of Loch Turret Water". J. Inst. Water Engrs. 17:333-346 (1963).

Cerkinskij, S.N. and N. Trahtman, "The Present Status of Research on the Disinfection of Drinking Water in the USSR", Bull. WHO 46(2):277-283 (1972).

Chedal, J. "Effect of Ozone on Micropollutants", in Forum on Ozone Disinfection, E. Fochtman, R.G. Rice & M.E. Browning, editors. Intl. Ozone Inst., Cleveland, Ohio (1970) p. 178-185.

Chedal, J. "Modes of Disinfecting Action of Ozone", Presented at Intl. Symp. on Ozone & Water, Wasser Berlin, May 1977. Intl. Ozone Inst., Cleveland, Ohio

Chian, E.S.K., P.P.K. Kuo, "Fundamental Study on the Post Treatement of RO Permeates from Army Wastewaters", Sec. Annual Summary Rept. U.S. Army Medical R & D Command, Washington, D.C., Rept. No. UILU-ENG-76-2019, Oct. 1976.

Cheremisinoff, P.N., F. Valent, D. Wright, R. Fortier and J. Magiliaro, "Potable Water Treatment: Technical and Economic Analysis. Chapt. 3, Water and Sewage Works 123(5):7073 (1976).

Coin, L., C. Hannoun & C. Gomella, "Inactivation of Poliomyelitis Virus by Ozone in the Presence of Water", la Presse Med. 72(37):2153-56 (1964).

Coin, L., C. Gomella, C. Hannoun & J.C. Trimoreau, "Ozone Inactivation of Poliomyelitis Virus In Water", la Presse Med. 75(38):1883-84 (1967).

Cromley, T.J. & J.T. O'Connor, "Effect of Ozonation on the Removal of Iron From a Ground Water:, J. Am. Water Works Assoc. 68:315-319 (1976).

Dahi, E., "Impact of Ultrasonic Disaggregation of Microorganisms in Ozone Water Sterilization Methods". Presented at 3rd Intl. Symp. on Ozone Technol., Paris, France, May, 1977, Intl. Ozone Inst., Cleveland, Ohio.

Damez, F., "Neuilly-sur-Marne/Nosy-le-Grand", in Forum on Ozone Disinfection, E. Fochtman, R.G. Rice & M.E. Browning, editors, Intl. Ozone Inst., Cleveland, Ohio (1977) p. 306-311.

Dellah, A., "Study of Ozone Reactions Involved in Water and the Present Chlorination Controversy", Sec. Intl. Symp. on Ozone Technol., R.G. Rice, P. Pichet, and M.-A. Vincent, editors, Intl. Ozone Inst., Cleveland, Ohio (1975), p. 161-168.

Dellah, A., "Ozonization at Charles-J. Des Baillets Water Treatment Plant", in Proc. Sec. Intl. Symp. on Ozone Technol., R. G. Rice, P. Pichet & M.-A. Vincent, editors, Intl. Ozone Inst., Cleveland, Ohio (1976), p. 715-725.

Diaper, E.W.J., "Microstraining and Ozonation of Water and Wastewater", Water and Wastes Engrg., p. 56-58 (1968).

Dobia, J. and F. Starz, "Ozone Application in Small Potable Water Supply Plants", Presented at Intl. Symp. on Ozone & Water, Wasser Berlin, May 1977. Intl. Ozone Inst., Cleveland, Ohio.

Drobek, W., "Ozonisierung Des Trinkwassers", Arch. Badewesens (16)4:98-101 (1963).

Dyachkov, A.V., "Recent Advances in Water Disinfection". Presented at Sept. 1976 Water Meeting, Amsterdam, Holland. pp. G1-G8.

D'Yakov, V.I., "Organization and Sanitary-Hygienic Evaluation of the Drinking Water Supply of Oil & Gas Regions of the Northern Ob. Area", Gig. y Sanit. 37(10):95-96 (1972).

Evison, L.M., "Disinfection of Water With Ozone: Comparative Study With Entero Viruses, Phage and Bacteria". Presented at 3rd Intl. Symp. on Ozone Technol., Paris, France, May 1977. Intl. Ozone Inst., Cleveland, Ohio.

Falk, H.L. and J.E. Moyer, "Ozone As a Disinfectant of Water", in Ozone/Chlorine Dioxide Oxidation Products of Organic Materials, R.G. Rice, & J.A. Cotruvo, editors. Intl. Ozone Inst., Cleveland, Ohio, (1978). p. 38-57.

Fontalirant, F., M. Paré & J. Guiblais, "Un Exemple De Regulation Automatique de la Production D'Ozone L'Usine de Nantes-La-Roche", Presented at 3rd Intl. Symp. on Ozone Technol., Paris France, May, 1977. Intl. Ozone Inst., Cleveland, Ohio.

Furgason, R. and R. O. Day, "Iron and Manganese Removal with Ozone, Pt. I," Water and Sewage Works, June 1975, 42, 45-47; Part II, ibid., July, 1975, p. 61-63.

Gabovich, R.D., K.K. Vrochinskii and I.L. Kurinnyi, "Decoloration, Dedoration and Decontamination of Drinking Water By Ozone", Gig. y Sanit. 34(6):18-22 (1969).

Gabovich, R. D., G. G. Rudenko, and M. A. Chaikovskaya, "Ozonization of Dnieper Water", Gig. y Sanit. 36(10):16-21 (1971).

Gad, G. & Columbus, C., "Chemische Auswirkung der Wasserozonisierung", Health Engineer 76:268-9 (1955).

Gagnaux, A., "Aspects Techniques du Traitement des Eaux par l'Ozone", p. 435-444, year and publication unknown.

Giebler, G. and Koppe, P., "Methode Zur Bestimmung der Wirkung des Ozons bei der Aufbereitung von Wasser", Wasser-Abwasser 106(8):215-219 (1965).

Glaze, W.H., R. Rawley, F. Huang, and S. Lin, "Ozone and Ozone/UV Destruction of Trihalomethane Precursors and other Refractory Organic Compounds in Water", presented at Symp. on Advanced Ozone Technol., Toronto, Ontario, Canada, November, 1977. Intl. Ozone Inst., Cleveland, Ohio.

Glaze, W.H., R. Rawley and S. Lin, "By-Products of Organic Compounds in the Presence of Ozone and Ultraviolet Light: Preliminary Results", in Ozone/Chlorine Dioxide Oxidation Products of Organic Materials", R.G. Rice, & J.A. Cotruvo, editors, Intl., Ozone Inst., Cleveland, Ohio, (1978). p. 321-331.

Gomella, C., "Ozone Practices in France", J. Am. Water Works Assoc. 64(1):39-45 (1972).

Gomella, C., "Difficultés de l'Affinage des Eaux Dues aux Micropolluants", Chimie and Industrie, Genie Chimique, 104(13):1633-1642 (1971).

Gomella, C., "L'Ozonation Vraie et l'Affinage des Eaux Potables," Sci. Progr. la Nature 3409:161-6 (1969).

Gomella, C., "Pollutions and Micropollutions of Surface Water, Technical and Economic Difficulties In the Production of Drinking Water", l'Eau 55(2):67-73 (1968).

Gomella, C., & D. Versanne, "The Function of Ozone in Bacterial Nitrification of Ammonia Nitrogen - The Plant of la Chapelle at Saint-Etienne-du-Rouvray (Seine Maritime-France). Presented at 3rd Intl. Symp. on Ozone Technol., Paris, France, May 1977. Intl. Ozone Inst., Cleveland, Ohio.

Gomella, C., "Utilization of Ozone When Treating Potable Water in France", Presented at Intl. Symp. on Ozone & Water, Wasser Berlin, May 1977, Intl. Ozone Inst., Cleveland, Ohio.

Grad, B.R., M. Gagnon, R. Charbonneau and C. Doggenweiller, "The Effect of Ozone on Ehrlich Adenocarcinoma Cells in Vitro", in Proc. Sec. Intl. Symp. on Ozone Technol., R.G. Rice, P. Pichet & M.-A. Vincent, editors. Intl. Ozone Inst., Cleveland, Ohio, (1975) p. 465-471.

Gubelmann, H.V., "Bau der Filter- und Ozonanlage Könizberg der Wasserversorgung Bern", 35(12):311-327 (1955).

Gubelmann, H.V., and H. Scheller, "Desinfektion des Wassers durch Ozone", Montasbull Schweiz Verein von Gas - und Wasserfachmännern, 33:99-107 (1953).

Guillerd, J. and R. Leviel, "Quelques Aspects du Traitement des Eaux Potables en France" l'Eau 48:138-142 (1961).

Guillerd, J. and C. Valin, "Traitment Par l'Ozone", l'Eau 48:138-142 (1961).

Hallopeau, J., "Ozonation", General Report No. 4, International Water Supply Congress and Exhibition on Water. Stockholm, Sweden, June 1964.

Hann, V., "Water Quality Improvement With Ozone", Engineering News Record. 139(54):125-127 (1947).

Hann, V., "Disinfection of Drinking Water With Ozone", J. Am. Water Works Assoc. 48(10):1316-20 (1956).

Hann, V., "Ozone Purification of Water". Tappi 35(9):394-397 (1952).

Hann, V., "Ozone Treatment for the Removal of Taste, Odor and Color from Water," J. New England Water Works Assoc., (Sept. 1947),

Harris, W.C., "Ozone Disinfection", J. Am. Water Works Assoc. 64(3):182-83 (1972).

Harris, W. C., "Practical Considerations of Ozone in Water Treatment", Presented at Symposium on New Trends in Water and Sewage Treatment Using Pure Oxygen and Ozone, Denver, Colo., Oct. 12, 1974.

Harris, W., "Ozone Disinfection of the Strasburg, Pennsylvania Water Supply System", in Proc. 1st Intl. Symp. on Ozone for Water & Wastewater Treatment, R.G. Rice, & M.E. Browning, editors, Intl. Ozone Inst., Cleveland, Ohio (1975), p. 186-193.

Heinonen, E., "Use of Ozone in Treatment of Raw Water Affected by Effluents From the Paper Making Industries". Presented at 3rd Intl. Symp. on Ozone Technol., Paris, France, May, 1977. Intl. Ozone Inst., Cleveland, Ohio.

Hill, A.G. and J.B. Howell, "Compression of O_3/O_2 and O_3/Air Gas Mixtures", Presented at Symp. on Advanced Ozone Technol., Toronto, Canada, November 1977. Intl. Ozone Inst., Cleveland, Ohio.

Hoehn, R.C., "Comparative Disinfection Methods", J. Am. Water Works Assoc. 68:302-309 (1976).

Hoigné, J. and H. Bader, "Kinetics and Selectivity of Ozonizing of Organic Matter in Potable Water", Presented at Intl. Symp. on Ozone & Water, Wasser Berlin, May 1977. Intl. Ozone Inst., Cleveland, Ohio.

Hofmann, E., "Betriebserfahrungen mit der Ozonung beim Bodensee Wasserwerk der Stadt St. Gallen", Wasser-Abwasser, 104(44):1261-1271 (1963).

Hofmann, E., "Ozonation of Sea of Constance Water at St. Gallen", Gas Wasserfach 104:1272-1278 (1963).

Hofmann, E., "Von der Ozonanlage des Bodensee-Trinkwasserwerkes St. Gallen", Wasser-Abwasser, 99(18):418-422 (1958).

Holluta, J. "Das Ozone in der Wasserchemie", Wasser-Abwasser, 105(44):1261-1271 (1963).

Holluta, J. and Unger, U., "Die Keimtötung von Bact. Coli Esch. durch Chlordioxyd und Ozon". Vom Wasser 21:129-42 (1954).

Holluta, J., "Das Ozon in der Wasserchemie", Gas/Wasser/Wärme 18:29-35 (1964).

Hopf, W., "Probleme der Wasser Aufbereitung mit Ozon" Kommunalwirtschaft 6:233-38 (1958).

Hopf, W., "Treatment of Water with Ozone and Activated Carbon (Düsseldorf Process) - Part I. Wasser-Abwasser 111(2):83-92 (1970).

Hopf, W., "Zur Wasseraufbereitung mit Ozone und Aktivkohle (Düsseldorfer Verfahren), Teil II. Bemerkungen zur Anwendung von Ozone", Wasser-Abwasser 111(3):156-164 (1970).

Howlett, E. "The Ozone Method of Water Purification". Water and Water Engrg. 50:25-31 (1947).

Ingols, R.S. & T.F. Craft, "Manganese Removal From Potable Water", Rept. No. ERC-1874, to Engrg. Expt. Sta., Ga. Tech. Univ., Atlanta, Ga., Nov. 1974.

Ingols, R.S., and R.H. Fetner, "Some Studies of Ozone for Use in Water Treatment", Water Treatment & Examination 6(1):8-28 (1957).

Jürs, R.H., "Die Wirkung des Ozons auf in Wasser Gelöste Stoffe", Fortschr. Wasserchem. Ihrer Grenzgeb. p. 40-64 (1966).

Keck, K. and Maier, D., "Application of Ozone in the Lake of Constance Water Supply System," in Proc. 1st Intl. Symp. on Ozone for Water & Wastewater Treatment, R. G. Rice & M. E. Browning, editors, Intl. Ozone Inst., Cleveland, Ohio, (1975), p. 164-166.

Keller, J. W., R. Morin, and T.J. Schaffernoth, "Ozone Disinfection Pilot Plant Studies at Laconia, N. H.", J. Am. Water Works Assoc. 66:730-734 (1974).

Keller, J.W. and T.J. Schaffernoth, "The Use of Ozone in Water Treatment Can Mean Lower Construction and Operation Costs", in Proc. 2nd Intl. Symp. on Ozone Technology, R. G. Rice, P. Pichet and M.-A. Vincent, editors, Intl. Ozone Inst., Cleveland, Ohio (1976), p. 650-663.

Killam, E.T., "Microstraining and Ozonation - Pilot-Plant Investigations for Treating a Surface Water Supply at Danvers, Massachusetts". J. New England Water Works Assoc. 73:37-44 (1959).

Korolev, A.A., A.A. Abinder, M.V. Bogdanov, T.A. Zakharova, and N.K. Khitrov, "Hygiene and Toxicological Features of Products of Phenol Degradation During Ozone Treatment of Water", Gig. y Sanit, 28(8):6-10 (1963)

Kjos, D.J., R.R. Furgason and L.L. Edwards, "Ozone Treatment of Potable Water to Remove Iron and Manganese: Preliminary Pilot Plant Results and Economic Evaluation", in Proc. 1st Intl. Symp. on Ozone for Water & Wastewater Treatment, R. G. Rice & M. E. Browning, editors, Intl. Ozone Inst., Cleveland, Ohio (1975), p. 194-203.

Knox, K. A. and K.A. Sankey, "Ozonization Treatment With Ozone at the Watchgate Plant", British Water Supply, about 1973-1974, Details unknown.

Koch, R., "Neue Verfahren in der Wasseraufbereitung". Bohntechnik Brunnenbau Rohrheitung. 11:507-512 (1960).

Kopecky, J., Fortschritte der Wasser Chemie und Ihrer Grenzgebiette. 7:94-100 (1967).

Koppe, P. and G. Giebler, "Untersuchungen Über die Ozonzehrung von Wässern", Wasser-Abwasser 107(8):196-200 (1966).

Kroke, R., "Ozon und Chlor in der Wasser Behandlung". Vom Wasser 30:108-124 (1964).

Kroke, R., "Warum das Ozonverfahren vor 30 Jahren nicht zum Erfolg geführt hat und welche Aussichten heute bestehen", Vom Wasser 20:194-208 (1953).

Kroke, K., "Ozon and Chlor in der Wasserbehandlung", Vom Wasser 31:108-124 (1965).

Kühn, W., H. Sontheimer, and R. Kurz, "Use of Ozone and Chlorine In Water Works in the Federal Republic of Germany", in Ozone/Chlorine Dioxide Oxidation Products of Organic Materials, R.G. Rice, & J.A. Cotruvo, editors. Intl. Ozone Inst., Cleveland, Ohio, (1978), p. 426-441.

Kurzmann, G.E., "Ozone in Drinking Water Preparation", Chemische Industrie, 26(8):502-503 (1974).

Kurzmann, G.E., "Der Einsatz von Ozone zur Aufbereitung von Wasser", in Proc. Sec. Intl. Symp. on Ozone Technol., R.G. Rice, P. Pichet & M.-A. Vincent, editors, Intl. Ozone Inst., Cleveland, Ohio, (1975), p. 514-521.

Kurzmann, G., "Methods of Ozone Generation", Presented at Intl. Symp. on Ozone & Water, Wasser Berlin, May 1977. Intl. Ozone Inst., Cleveland, Ohio.

Lacy, W. & R.G. Rice, "The Status and Future of Ozone for Water Treatment in the United States", U.S. EPA, Washington, D.C., Presented at Intl. Symp. on Ozone & Water, Wasser Berlin, May 1977. Intl. Ozone Inst., Cleveland, Ohio.

Lamblin, H., "Protection d'Un Reseau d'Eau Ozonée", Presented at 3rd Intl. Symp. on Ozone Technol., Paris France, May, 1977. Intl. Ozone Inst., Cleveland, Ohio.

Laszlo, K., "Az Ivoviz Ozonos Kezelese", in Hydraulic Engineering, Ihrig Denes (Editor), No. 4, pp. 441-465 (1963), Muszaki Kőnyvkiado, Budapest, Hungary.

Lebaut, M., "Le Controle de l'Eau d'Alimentation de la Ville de Nice". La Technique Sanitaire et Municipale 45:86 (1950).

Légéron, J.P., "Laboratory Evaluation of Chemical Ozone Demand of a Water Sample", Presented at Symp. on Advanced Ozone Technol., Toronto, Ontario, Canada, November 1977. Intl. Ozone Inst., Cleveland, Ohio.

Leitzke, O., "Instrumental Ozone Analysis in Water- + Gas-Phases", Presented at Intl. Symp. on Ozone & Water, Wasser Berlin, May 1977. Intl. Ozone Inst., Cleveland, Ohio.

Le Page, W.L., "Ozone Treatment at Monroe, Michigan", in Proc. 2nd Intl. Symp. on Ozone Technol., R. G. Rice, P. Pichet and M.-A. Vincent, editors, Intl. Ozone Inst. Cleveland Ohio, (1976), p. 198-210.

Le Viel, R., "La Stérilisation des Eaux par le Chlore et l'Ozone". La Technique Sanitaire et Municipale 45:46-50 (1950).

Lin, S.D,, "Tastes and Odors in Water Supplies; A Review", Water & Sewage Works, Ref. No. 1977, p. R-141-R-163.

Lingel, E.W., "Production and use of Ozone in Water Treatment". Public Works, pp. 155-158, (1964).

Livesey, R.J. & A. Herdman, "The Treatment of Humic and Coloured Moorland Water Containing No Pollution and Low Turbidity". Presented at 3rd Intl. Symp. on Ozone Technol., Paris, France, May, 1977. Intl. Ozone Inst., Cleveland, Ohio.

Mackenzie, E.F.W., "Water Sterilisation - The Choice of Method". The Surveyor, 99(2559):107-110 (1941).

Maier, D., "Wirkung von Ozon auf die Gelosten Organischen Substanzen in Bodenseewasser", Vom Wasser 43:127-60 (1974).

Maier, D., and C.R. Kurz, "Optimizing of Ozone Application When Treating Lake Water", Presented at Intl. Symp. on Ozone & Water, Wasser Berlin, May 1977. Intl. Ozone Inst., Cleveland, Ohio.

Mallevialle, J., M. Roustan and H. Roques, "Determination Experimentale des Coefficients de Transfer de l'Ozone dans l'Eau", in Proc. Sec. Intl. Symp. on Ozone Technol., R.G. Rice, P. Pichet, M.-A. Vincent, editors, Intl. Ozone Inst., Cleveland, Ohio, (1975), p. 47-56.

Mallevialle, J., "Ozonation of Humic Substances in Waters", in Proc. 2nd Intl. Symp. on Ozone Technology, R.G. Rice, P. Pichet & M.-A. Vincent, editors. Intl. Ozone Inst., Cleveland, Ohio (1976), p. 262-270.

Marets, M., "Application d'un Micro processeur dans une Installation d'Ozonation". Presented at 3rd Intl. Symp. on Ozone Technol., Paris, France May 1977. Intl. Ozone Inst., Cleveland, Ohio.

Masschelein, W.J., & R. Gossens, "Continue Meting van Restozon in Water", Journal unknown, p. 100-101 (about 1975).

Masschelein, W.J., "Ozone In Water Treatment", Lecture held at the Seminar of ECOCHEM, 1975.

Masschelein, W.J., G. Fransolet, J. Genot & R. Goossens, "Perspectives de l'Ozonation de l'Eau au Départ d'Air Enrichi en Oxygène", T.S.M.l'Eau 71(8-9):385-398 (1976).

Masschelein, W., "Techniques of Dispersion and of Diffusion of Ozone in Water", Presented at Intl. Symp. on Ozone & Water, Wasser Berlin, May 1977. Intl. Ozone Inst,, Cleveland, Ohio.

Masschelein, W.J., "Mesure de l'Ozone Residual dans l'Eau. Specificité et Automation". Presented at 3rd Intl. Symp. on Ozone Technol., Paris, France, May 1977. Intl. Ozone Inst., Cleveland, Ohio.

Meijers, A.P. "Removal of Organic Water Substances By Ozone", Presented at Intl. Symp. on Ozone & Water, Wasser Berlin, May, 1977. Intl. Ozone Inst., Cleveland, Ohio.

Meijers, A.P., "Quality Aspects of Ozonisation", Water Research. 11:647-652, (1977).

Mignot, J. "Pratique de la Mise en Contact de l'Ozone Avec le Liquide à Traiter", Sec. Intl. Symp. on Ozone Technol., R.G. Rice, P. Pichet, M.-A. Vincent, Editors, Intl. Ozone Inst., Cleveland, Ohio, (1975), p. 15-46.

McLaughlin, M.J., "Ozonation Tests on Philadelphia Water Supply". Water Works Engineering, p. 1131-1158 (1943).

McManman, J.A., "The Use of Ozone In Water Treatment", Presented at the Canadian Institute of Public Health Inspectors, 41st National Conference, p. 1-13, 1975.

Mescery, K., "Wasseraufbereitung Durch Ozonisierung," Gas/Wasser Wärme, 9(8):193-196 (1955).

Middlebrooks, E.J., "Taste and Odor Control", Water & Sewage Works p. R122-R134, 1965.

Mignot, J., "Application of Ozone With or Without Activated Carbon for Drinking Water Treatment", in Proc. 1st Intl. Symp. on Ozone for Water & Wastewater Treatment, R. G. Rice & M. E. Browning, editors, Intl. Ozone Inst., Cleveland, Ohio (1975), p. 134-160.

Morin, R. A., J. W. Keller, T. J. Schaffernoth and R. Paquette, "Ozone Disinfection Pilot Plant Studies at Laconia, New Hampshire," presented at New England Water Works Association Meeting, November 21, 1974.

Morin, R., J. W. Keller, T. J. Schaffernoth and R. Paquette, "Ozone Pilot Plant Studies at Laconia, New Hampshire," Presented at Symposium on New Trends in Water & Sewage Treatment Using Pure Oxygen and Ozone, Denver, Colorado, October 12, 1974.

Nadeau, M. and Pigeon, J.C., "The Evolution of Ozonation in the Province of Quebec", in Proc 1st Intl. Symp. on Ozone for Water & Wastewater Treatment, R. G. Rice and M. E. Browning, editors, Intl. Ozone Inst., Cleveland, Ohio (1975), 167-185.

Naumann, E., "Fachblatt für das Gesamtgebiet der Gastechnik, Gaswirtschaft und des Wasserwesens". Gas und Wasserfach 93:81-88 (1952).

Nobert, J. B., "The Ozonation of Water in the Montreal Region," J. New England Water Works Assoc., 75:53-61 (1961).

O'Donohue, A.T., "The Cleaning of Toronto Waters", Presented at Symp. on Advanced Technol., Toronto, Canada, Nov. 1977. Intl. Ozone Inst. Cleveland Ohio.

O'Donovan, D.C., "Experience in the Operation of Ozonisation Plants", Presented at a meeting in Sligo (Inst. Water Engrs., Ireland Section), Sept. 28, 1973, 17 p.

O'Donovan, D.C., "Treatment with Ozone," J. Am. Water Works Assoc. 57(9): 1167-1192 (1965).

Osheroff, B.J., "Water Treatment: the Background for Public Health Concern," in Ozone/Chlorine Dioxide Oxidation Products of Organic Materials, R.G. Rice & J.A. Cotruvo, editors, Intl. Ozone Inst., Cleveland Ohio, (1978), p. 7-19.

Palin, M.T., "Analyze de l'Ozone Dans l'Eau Seul ou en Presence de Chlore Libre, de Chlore Combine, de Bioxide de Chlore et de Chlorite". Presented at 3rd Intl. Symp. on Ozone Technol., Paris France, May 1977. Intl. Ozone Inst., Cleveland, Ohio.

Paré, M., "Operational Costs of Ozone and Energetic Balance", Presented at Intl. Symp. on Ozone & Water, Wasser Berlin, May 1977. Intl. Ozone Inst., Cleveland, Ohio.

Pasynkiewicz, J., Grossman, A. and Nawara, S., "Application of Ozone in Purification of Drinking Water Containing Phenols". Przemysl Chemiczny 47(4):227-231 (1968). (Czechoslovakian).

Peleg, M., "The Chemistry of Ozone In the Treatment of Water, A Review Paper," Water Research, 10:361-365 (1976).

Powell, M. P., K. G. Drummond, C. Saavedra, F. M. Evans, P. F. Morgan and N. B. Fisher, "Action of Ozone on Tastes and Odors and Coliform Organisms," J. Am. Water Works Assoc. 44:1144-1150 (1952).

Rapinat, M., "French Methods for Evaluating Total Micropollutant Load In Water, and for Determining Its Toxicity", in Ozone/Chlorine Dioxide Oxidation Products of Organic Materials, R.G. Rice, & J.A. Cotruvo, editors. Intl. Ozone Inst., Cleveland Ohio (1978), p. 134-152.

Reichert, J., "Examination for the Elimination of Carcinogenic, Aromatic Polycyclics in the Treatment of Drinking Water, with Special Consideration of Ozone", Water & Sewage (Wasser-Abwasser) 110(18):477-482 (1969).

Reid, L. C. and H. S. Potworowsk, "Ozone Treats Arctic Waters," Water and Pollution Control, 53-55 (March 1974).

Reissaus, K. & W. Rummel, "Wasseraufbereitung mit Ozon bei der Trinkwasser-schönung 2. Mitteilung: Vergleichende Untersuchungen der Schönung-wirkung von Ozon und Chlordioxid", Fortschr. Wasserchem, Grenzgeb. 6:139-159 (1967).

Rice, R.G., G.W. Miller, C.M. Robson & W. Kühn, "Ozone Pre-Treatment of Activated Carbon", Presented at Symp. on Advanced Ozone Technol., Toronto, Ontario, Canada, November 1977. Intl. Ozone Inst., Cleveland, Ohio.

Richard, Y., "Chlorine-Ozone Interferences in Water Treatment", in Proc. 2nd Intl. Symp. on Ozone Technol., R. G. Rice, P. Pichet, and M.-A. Vincent, editors, Intl. Ozone Inst., Cleveland, Ohio (1976), 169-180.

Richard, Y. & F. Fiessinger, "Emploi Complementaire Des Traitments Ozone et Charbon Actif." Presented at 3rd Intl. Symp. on Ozone Technol., Paris, France, May, 1977. Intl. Ozone Inst., Cleveland, Ohio.

Robeck, G.C., K.A. Dostal, J.M. Cohen and J.F. Kreissl, "Effectiveness of Water Treatment Processes in Pesticide Removal", J. Am. Water Works Assoc. 57:181-200 (1965).

Rook, J.J., "Haloforms in Drinking Water", J. Am. Water Works Assoc. 68:168-172 (1976).

Rook, J.J., "Developments in Europe," J. Am. Water Works Assoc. 279-282 (1976).

Rosen, H.M., "Ozonation: A Solution to Potential Carcinogens in Drinking Water", in Proc. 2nd Intl. Symp. on Ozone Technol., R. G. Rice, P. Pichet and M.-A. Vincent, editors, Intl. Ozone Inst., Cleveland, Ohio (1976), 143-160.

Ryan, W. L. and J. W. Grainge, "Sanitary Engineering in Russia," J. Am. Water Works Assoc. 67:285-288 (1975).

Sander, R.V., "Untersuchungen zur Umsetzung von Chlor mit Huminsubstanzen", Wasser-Abwasser 10(5):155-160 (1977).

Sankey, K.A., "The History of the Use of Ozone in the Area Now Controlled by the N.W.W.A". Presented at 3rd Intl. Symp. on Ozone Technol., Paris, France, May, 1977. Intl. Ozone Inst., Cleveland, Ohio.

Schäuble, R., "Purification of Water with Particular Reference to Ozonization", Schweiz., Brau-Runds; 72:1-6 (1961). Abstracted in J. Inst. Brewing (London) 67:209 (1961).

Schäfer, G. "Die Ozonanlage des Neuen Grundwasserpumpwerkes Theinhalde der Stadt Schaffhausen," Wasser-Abwasser 107(30):833-834 (1966).

Schaffernoth, T.J., J.W. Keller, and R.A. Morin, "Viral Disinfection by Ozonation at New Hampshire Pilot Plant", in Proc. 2nd Intl. Symp. on Ozone Technol., R. G. Rice, P. Pichet and M.-A. Vincent, editors, Intl. Ozone Inst., Cleveland Ohio (1976), 472-485.

Schalekamp, M., "Die Wasserversorgung der Stadt St. Gallen-gestern-Heute-morgen," Gas-Wasser-Abwasser, 49(6):191-198 (1969).

Schalekamp, M., "Ozone for Sterilization," Presented at Ninth Congress, Intl. Water Supply Assoc. 11-14 Sept. 1972, New York, N.Y.

Schalekamp, M., "Experiences Suisses Avec l'Ozone Concernant en Particulier les Modifications des Substances Hygieniquement Douteuses". Presented at 3rd Intl. Symp. on Ozone Technol., Paris, France, May 1977. Intl. Ozone Inst. Cleveland, Ohio.

Schalekamp, M., "Experience With Ozone in Switzerland Especially With Respect to Change of Hygienically Doubtful Substances", Presented at Intl. Symp. on Ozone & Water, Wasser Berlin, May 1977. Intl. Ozone Inst., Cleveland, Ohio.

Scheller, H. von, "Betriebserfahrungen mit der Filter-und Ozonanlage Könizberg der Wasserversorgung Bern", Monatsbull., 36(4):73-81 (1956).

Schenk, P., "Die Wasseraufbereitungsanlage des Wasserwerkes Düsseldorf, Am Staad". Wasser-Abwasser 103(30):791-798 (1962).

Schneider, W., "Anforderungen an ein Wasser zur Abfüllung als Quelltrinkwasser und Gedanken zur Abfulltechnik." (Reference Unknown), about 1974.

Schneider, W. and D. Blankenfeld, "Disinfection of Bottled Water", (Chemisohes Laboratorium Fresenius, Demag A-G) Ger. Offen. 2,222,385 (Cl.CO2b), 15 Nov. 1973.

Schinzel, A., "Hygienische Fragen der Wasseraufbereitung mit Besonderer Berücksichtigung der Entkeimung", 14(7):129-136 (1962).

Schwartz, M. and D.J.W. Moncrieff, "Ozone as a Treatment Process for Colour Removal from Drinking Water", in Proc. 2nd Intl. Symp. on Ozone Technol., R. G. Rice, P. Pichet and M.-A. Vincent, editors, Intl. Ozone Inst., Cleveland, Ohio (1976), p. 283-308.

Schwartz, M., "Some Practices in Application of Ozone to Drinking Water", in Forum on Ozone Disinfection, E.G. Fochtman, R.G. Rice & M.E. Browning, editors, Intl. Ozone Inst., Cleveland, Ohio (1977), p. 344-356.

Schwartz, M. & E.A. Lancaster, "An Engineering Approach to Water Treatment Process Selection With Special Emphasis on Halogenated Organics", in Ozone/Chlorine Dioxide Oxidation Products of Organic Materials, R.G. Rice & J.A. Cotruvo, editors, Intl. Ozone Inst., Cleveland, Ohio (1978), p. 264-283.

Sease, W.S., E.W.J. Diaper and W.H. McCammon, "Experiences With Large Ozonators In Europe", in Proc. 1st Intl. Symp. on Ozone for Water & Wastewater Treatment, R.G. Rice, & M.E. Browning, editors, Intl. Ozone Inst. Cleveland, Ohio (1975), p. 186-193.

Sease, W.S., "Ozone Mass Transfer and Contact Systems", in Proc. 2nd Intl. Symp. on Ozone Technol., R.G. Rice, P. Pichet and M.-A. Vincent, editors, Intl. Ozone Inst., Cleveland, Ohio (1976), p. 1-14.

Sease, W.S., "Chiba Plant--Japan", in Forum on Ozone Disinfection", E.G. Fochtman, R.G. Rice & M.E. Browning, editors, Intl. Ozone Inst., Cleveland, Ohio (1976), p. 327-336.

Sigworth, E.A. "Identification and Removal of Herbicides and Pesticides", J. Am. Water Works Assoc. 57:1016-1022 (1965).

Simon, M. & Scheidtmann, H., "Die Neue Ozonlage der Stadtwerke Duisburg", Wasser-Abwasser 109(32):877-882 (1968).

Sommerville, R.C. and G. Rempel, "Ozone for Supplementary Water Treatment," J. Am. Water Works Assoc. 64:377-382 (1972).

Somiya, I. & T. Tokuda, "Model Identification of Ozone and Activated Carbon Treatment Processes Based on a Population Balance Model", Presented at Symp. on Advanced Ozone Technol., Toronto, Canada, Nov. 1977. Intl. Ozone Inst., Cleveland, Ohio.

Sontheimer, H., "New Experiences with Ozone in Germany", in Proc. 1st Intl. Symp. on Ozone for Water & Wastewater Treatment, R. G. Rice & M. E. Browning, Eds., Intl. Ozone Inst., Cleveland, Ohio (1975), 161-163.

Sontheimer, H. "Biological Treatment of Surface Waters in Activated Carbon Filters", presented at Seminar on Current Status of Wastewater Treatment & Disinfection With Ozone, Cincinnati, Ohio, Sept. 15, 1977. Intl. Ozone Inst., Cleveland, Ohio.

Sontheimer, H. "Development Possibilities and Limits of Ozone Application When Treating Potable Water", Univ of Karlsruhe, Presented at Intl. Symp. on Ozone & Water Wasser Berlin, May 1977. Intl. Ozone Inst., Cleveland, Ohio.

Stopka, K., "The Case for Ozonation," Water and Pollution Control, September 1970, 2 pages.

Storey, T.H., "Ozone and Water Purification", J. Maine Water Utilities Assoc. 17:147-162 (1941).

Stumm, W., "Ozone As a Disinfectant for Water and Sewage", J. Boston Soc. Civil Engrs. 45:68-79 (1958).

Stumm, W., "Einige Chemische Gesichtspunkt zur Wasserozonisierung", Schweiz z. Hydrol. 18(2):201-207 (1956).

Suchkov, B.P., "Studies of the Ozonation of Drinking Water Containing Pathogenic Bacteria and Viruses". Gig. y Sanit. 29(6):24-31 (1964).

Sulzer F. "Über das Verhalten Wässeriger Ozonlösungen", Schweiz z. Hydrol. 20(1):16-29 (1958).

Symons, J.M., J.K. Carswell, R.M. Clark, O.T. Love, Jr., R.J. Wiltner & A.A. Stevens, "Interim Treatment Guide for the Control of Chloroform and Other Trihalomethanes," U. S. Environmental Protection Agency, Water Supply Research Laboratory Cincinnati, Ohio, June 1976.

Symons, J. M. and G. G. Robeck, "Treatment Processes for Coping with Variation in Raw-water Quality," J. Am. Water Works Assoc. 67:142-145 (1975).

Symons, J.M., O.T. Love, Jr., and K. Carswell, "Comparison of Practical Alternative Treatment Schemes for Reduction of Trihalomethanes in Drinking Water", in Ozone/Chlorine Dioxide Oxidation Products of Organic Materials, R.G. Rice & J.A. Cotruvo, editors. Intl. Ozone Inst., Cleveland, Ohio, (1978), p. 442-454.

Thompson, G.E., "Ozone Applications in Manitoba, Canada", in Proc. 2nd Intl. Symp. on Ozone Technol., R. G. Rice, P. Pichet, and M.-A. Vincent, editors, Intl. Ozone Inst., Cleveland, Ohio (1976), p. 682-693.

Thompson, H., "Ozone Antidote for Water Pollution", Illinois Water Treatment Company, May-June 1972.

Thorburn, G. W., R. J. Livesey and A. Herdman, "The Ozonisation of Loch Turret Water", in Proc. 2nd Intl. Symp. on Ozone Technol., R. G. Rice, P. Pichet and M.-A. Vincent, editors, Intl. Ozone Inst., Cleveland, Ohio (1976), p. 181-197.

Tomono, K. "The Art of Water Treatment in Japan", J. Am. Water Works Assoc. 69:166-170 (1977).

Torricelli, A., "Drinking Water Purification," in Ozone Chem. & Technol., Advances in Chem. Series, Vol. 21, Am. Chem. Soc., Washington, D.C. (1959), p. 453-465.

Thureson, L.E., "Ozonförsök vid Ealo Vattenverk". Vattenhygien 2:53-56 (1962). In Swedish.

Uhlig, G.W., "Chemical Methods to Determine Ozone in Water Works", Presented at Intl. Symp. on Ozone & Water, Wasser Berlin, May, 1977. Intl. Ozone Inst. Cleveland, Ohio.

Vaillant, C. J., "Ozonation et la Filtration Lente dans le Problème de la Prolifération Microbienne dans le Réseaux de Distribution d'Eau," Gas, Wasser, Abwasser, 50(3):67-70 (1970).

Valenta, J., "Simultaneous use of Chlorine and Ozone in Water Treatment". Gas, Wasser, Abwasser, 55(9):490-494 (1975).

Van Hoof, F., "Some Aspects of Ozone Toxicity", Presented at 3rd Intl. Symp. on Ozone Technol., Paris, France, May, 1977. Intl. Ozone Inst., Cleveland, Ohio.

Vrochinsky, K. K., "Hygienic Problems of Ozonization of Water from the North Donets--Donbas Canal," Gig. y Saint. 28(9):11-17 (1963).

Wallentin, A. and Nyberg, I., "Ozonbehandling av Zygnerns Vatten". Vattenhygien 2:38-52 (1962). In Swedish.

Weil, L., B. Struif and K.E. Quentin, "Mechanisms of Ozone Degradation of Organic Substances", presented at Intl. Symp. on Ozone & Water, Wasser Berlin, May 1977. Intl. Ozone Inst., Cleveland, Ohio.

Weissenhorn, F.J., "Application of Ozone in a Closed Circuit", Düsseldorf, Presented at Intl. Symp. on Ozone & Water, Wasser Berlin, May 1977. Intl. Ozone Inst., Cleveland, Ohio.

Whitson M.T., "The Treatment of Water by Ozone," J. Inst. Civil Engrs., London, 2:83-100 (1943).

Whitson, M.T., "Treatment of Water by Ozone". The Surveyor 103:111-112 (1944).

Würster, E. and Werner, G., "Die Leipheimer Versuche zur Aufbereitung von Donauwasser", Z. für Gas-und Wasserfach 112:193 (1971).

Zeff, J. D., Final Rept No. 1401 of Contract DADA 17-73C-3138, "UV-Ozone Water Oxidation/Sterilization Process," U.S. Army Medical R & D Command, Washington, D.C., Sept. 1974.

Chlorine Dioxide Articles -- Drinking Water

Adamski, H. and Hoffman, W., "Z praktyki zwalczania fenoli w wodzie na stacji wodociagowei Pomorzany w Szczecinie". Gas, Woda I Technika Sanitarna, 34:154 (1960). (In Polish)

Anon., "Water Quality and Treatment--A Handbook of Public Water Supplies," Am. Water Works Assoc., 3rd. ed. (1971), Pages 181, 547-550.

Anon., "Treatment of Water Supplies with Chlorine Dioxide (ClO_2), Olin Water Services Co., Stamford, Conn., March, 1976.

Aston, R. N. and J. F. Synan, "Chlorine Dioxide as a Bactericide in Water Works Operation". J. New England Water Works Assoc. 62:80-94 (1948).

Atkinson, J. W., "The Control of Taste in the River Dee Supply of the West Cheshire Water Board," Water & Water Engrg., pp. 146-149 (April, 1962).

Augenstein, H. W., "Use of Chlorine Dioxide to Disinfect Water Supplies, J. Am. Water Works Assoc. 66:716-717 (1974).

Bays, L. R.., N. P. Burman and W. M. Lewis, "Taste and Odour in Water Supplies in Great Britain: A Survey of the Present Position and Problems for the Future", Water Treatment and Examination 19(2):136-160 (1970).

Bean, E. L., "Taste and Odor Control at Philadelphia", J. Am. Water Works Assoc. 49:205-216 (1957).

Bernardi, M. A., B. A. Israel, V. P. Olivieri and M. L. Granstrom "Efficiency of Chlorine Dioxide as a Bactericide", Appl. Microbïol. 13(5):776-780 (1965).

Black, A. P. and R. F. Christman, "Chemical Characteristics of Fulvic Acids", J. Am. Water Works Assoc. 55:897-913 (1963).

Bouille, S., "La Station de la Ville d'Alencon--Emploi du Bioxyde de Chlore dans le Traitment des Eaux," L'Eau, 44(1): 3-8 (1957).

Buydens, R. and G. Fransolet, "L'action de l'ozone sur le chlore, le bioxyde de chlore et le chlorite contenus dans les eaux traitées," Tribune du Cebedeau, 326:4-6 (1971).

Buydens, R. and G. Fransolet, "Le bioxide de chlore: sa synthèse et son comportement dans les eaux," Tribune du Cebedeau, 316:116-126 (1970).

Cardey, F., "Le Traitement des Eaux Résiduaires Phénolées par le Chlore", L'Eau 40:75 (1953).

Corson, B. I., "Chlorine Dioxide Licked Puzzling Tastes and Odors in Well Supply". Water Works Engrg. 112:372-374 (1959).

Coote, R. "Chlorine Dioxide Treatment at Valparaiso, Ind"., Water & Sewage Works, January, 1950, 13-16.

DeGroot, J. C. "Safe and Palatable Water at Northampton", Public Works, 79:36-37 (1948).

Derby, J. C., "Operating Experience with Chlorine Dioxide". J. New England Water Works Assoc., 69(3):231-235 (1955).

Dowling, L. T., "Chlorine Dioxide In Potable Water Treatment," Water Treatment and Examination, 23(2):190-204 (1974).

Draves, H. J. "Results at Michigan City, Indiana", Public Works, May 1948, p. 36-37.

DuByne, F. T., "Taste and Odor Problems of a Maumee River Supply", J. Am. Water Works Assoc. 55:710-714 (1963).

Gallaher, W. U., Chloro-Phenol Tastes Removed", Public Works, May 1948, p. 38.

Gibbons, M. M., "Experiences with Filtered River Water", Public Works, May 1948, p. 38.

Granstrom, L. and G. Lee, "Generation and Use of Chlorine Dioxide in Water Treatment," J. Am. Water Works Assoc., 50:1453-1466 (1958).

Harlock, R. and R. Dowlin, "Use of Chlorine for Control of Odors Caused by Algae", J. Am. Water Works Assoc. 50:29-32 (1958).

Haynes, L., "Treatment of Industrial Water Pollution at Nitro and Charleston, W. Va.", J. Am. Water Works Assoc. 49:309-312 (1957).

Hermanowicz, O., I. Sikorowska and K. Bernacki, "Stosowanie dwutlenku chloru do usuwania smaku i zapacha chlorofenoli z wody w wodociagu w Szczecinie," Gas, Woda i Technika Sanitarna 33:145-148 (1959) (in Polish).

Holluta, J. and Haberer, K., "Über die Geruchsentwicklung bei der Chlordioxydbehandlung phenolhaltiger Wasser", Wasser-Abwasser 98:552-555 (1952).

Holluta, J. and U. Unger, "Die Keimtötung von Bact. Coli Esch. durch Chlordioxyd und Ozon". Vom Wasser 21:129-42 (1954).

Hopf, E., "Erfahrungen mit chlordioxid zur Trinkwasser-behandlung", Wasser-Abwasser 108(30):852-854 (1967).

Ingols, R. S. and T. F. Craft, "Manganese Removal From Water Supplies", J. Am. Water Works Assoc. 39:561-567 (1947).

Kabler, P. W., N. A. Clarke, G. Berg and S. L. Chang, "Viricidal Efficiency of Disinfectants in Water," Public Health Reports, 76(7):565-70, (July 1961).

Kretch, C. J. H., "Contemporary Chlorine Disinfection Practices in Ontario, Canada," J. Am. Water Works Assoc. 68(4):197-201 (1976).

Lambert, M., and C. Bernard, "Disinfection of Urban Waste Water by Chlorine and Chlorine Dioxide, l'Eau et l' Industrie, Feb., 1977.

Melpas, J. F., "Disinfection of Water Using Chlorine Dioxide," Water Treatment and Examination, 22(3):209-221 (1973).

Malpas, J. F., "Use of Chlorine Dioxide in Water Treatment," Effluent and Water Treatment J., July 1965, p. 370-3.

Masschelein, W., "Progrès de la chimie du bioxyde de chlore et ses applications," Part II. Chemie & Industrie-Genie Chimique, 97(3):346-354 (1967).

Masschelein, W., "Preparation of Pure Chlorine Dioxide", I & EC Prod. Research and Dev. 6:137-142 (1967).

Masschelein, W., Spectrophotometric Determination of Chlorine Dioxide With Acid Chrome Violet K", Anal. Chem. 38:(13)1839-1841 (1966).

Middlebrooks, E.J., "Taste and Odor Control", Water & Sewage Works, 1965, p. R-122-R-134.

Misnakiewicz, W., J. Podkowka, A. Zajac-Chmielewska, "Badania nad dzialaniem dwutlenku chloru na scieki zawierajace fenol," Gaz, Woda i Technika Sanitarna, 32:252-256 (1958), (in Polish).

Mülheim, M. E., "Versuche zur Geschmacksverbesserung von Trinkwasser mit Chlordioxyd". Wasser-Abwasser 101(14):340-342 (1960).

Reichert, J. K., "Kanzerogene Substanzen in Wassen and Boden XXIII. Die Enfernung polyzyklischer aromaten bei der Trinkwasseraufberitung durch ClO_2: Isolierung and identifizierung der 3,4-benzpyren-folge-produkte". Arch Hyg. 152:265 (1968).

Reichert, J. K., Kanzerogene Substanzen in Wasser and Boden XXI. Die Entfernung Polyzyklischer, Aromaten bei der Trinkwasseraufbereitung durch Chlordioxid: Quantative Befunde". Arch Hyg. 152:37 (1968).

Reissaus, K. & W. Rummel, "Wasseraufbereitung mit Ozon bei der Trinkwasserschönung 2. Mitteilung: Vergleichende Untersuchungen der Schönnungwirkung von Ozon und Chlordioxid", Fortschr. Wasserchem, Grenzgeb. 6:139-159 (1967).

Ridenour, G. M. and Armbruster, E. H., "Bactericidal Effect of Chlorine Dioxide" J. Am. Water Works Assoc. 41:537-550 (1949).

Ringer, W. C. & S. J. Campbell, "Use of Chlorine Dioxide for Algae Control at Philadelphia", J. Am. Water Works Assoc. 47:740-746 (1955).

Rook, J. J., "Chlorination Reactions of Fulvic Acids in Natural Waters", Environ. Sci. & Technol. 11(5):478-482 (1977).

Sikorowska, C., "Wptyw rodzaju zanieczyszczeń wody na zapotrzebowanie dwutenoku chloru". Gaz, Woda I Technika Sanitarna, 35:464 (1961). (in Polish).

Simmons, P. D., "Chlorine Dioxide Treatment for Taste and Odor Control". Water Works Engrg. 100:1258-1259 (1947).

Stäheli, Th., "L'utilisation pratique de bioxyde de chlore pour la stérilisation de l'eau potable", Transl. from Monatsbull, Schweiz. Ver. Gas U. Wasserfachm. 42:244-252 (1962).

Synan, J. F., J. D. MacMahon and G. P. Vincent, "Chlorine Dioxide-A New Development in the Treatment of Potable Water", J. New England Water Works Assoc., 58:264-269 (1944).

Synan, J. F., J. D. MacMahon, and G. P. Vincent, "Tastes and Odors Removed by Chlorine Dioxide Treatment", Water Works Engrg. 98:192, 211-12 (1945).

Symons, J. M., "Chlorine Dioxide Use in the United States - January 1964 - From Municipal Water Facilities - Communities of 25,000 Population and Over; U.S. Department of HEW, Public Health Service Publication No. 661," U.S. Environmental Protection Agency, Water Supply Research Division, Cincinnati, Ohio.

Toussaint, M., "Le bioxyde de chlore et le traitment des eaux potables," Centre Belge d'Etude et de Documentation des Eaux, May 1972, pp. 260-266.

Valenta, J., "Simultaneous Use of Chlorine and Ozone in Water Treatment". Gas, Wasser-Abwasser, 55(9):490-4 (1975).

Valenta, J. and W. Gahler, "Chlordioxidanlage", Gas-Wasser-Abwasser 55(9):566-569 (1975).

Wallwork, J. F., M. Bentley and D. C. Symonds, "Identification of Phenols In the River Test and Their Treatment With Chlorine Dioxide", Water Treatment & Examination, Vol. 18 (1969).

Widemann, O., "4 Jahre praktische Erfahrung mit Chlordioxyd," Vom Wasser, 24:50-70 (1957).

Technical Articles -- Drinking Water/Organics

Ahmed, M. and Kinney, C. R., "Ozonization of Humic Acids Prepared from Oxidized Bituminous Coal", J. Am. Chem. Soc. 72:559-561 (1950).

Am. Water Works Assoc., T & P Council Statement: "Organic Contaminants in Drinking Water", Willing Water, December 1974. 3 p.

Arsovic, H. M., and H. Burchard, "Ergebnisse und Neue Erkenntnisse zur Oxidation von o-Chlorophenol mit Ozon, Unter Anwendung des ETIZON-Verfahrens", Gesundheits-Ingenieur 98(9):230-239 (1977).

Balasubramanian, V. and V. Thiazarajan, "Chlorination of Substituted Phenols With Chloramine T. A Kinetic Study", Intl. J. Chem. Kinetics 7:605-623 (1975).

Bauch, H., H. Burchard, & H. M. Arsovic, "Ozone as an Oxidative Disintegrant for Phencls in Aqueous Solutions". Gesundheit Ingenieur 91(9):258-262 (1970).

Bauch, H. and H. Burchard, "Investigations Concerning the Influence of Ozone on Water with Few Impurities," Wasser Luft & Betrieb 14(7):270-273 (1970).

Becker, E. S., J. K. Hamilton & W. E. Lucke, "Cellulose Oligosaccharides as Model Compounds in Chlorine Dioxide Bleaching", TAPPI 48(1):60-64 (1965).

Belevtzev, A. N. and Ju. L. Maximenko, "Studies on Oxidation Processes of Cyanides and Phenols in Waste and Natural Waters by Using Chlorine Dioxide," in Symposium on Physical-Chemical Treatment for Municipal and Industrial Sources, USA-USSR Working Group on the Prevention of Water Pollution from Municipal and Industrial Sources, US/EPA, Cincinnati, Ohio. pp. 105-110, November 12-14, 1975.

Bernatek, E. and A. N. Soteland, "Ozonolysis of Naphthoquinones III. 1,2-Naphthoquinone", Acta Chem. Scand. 16(8):2054-2056 (1962).

Bellar, R. A., J. J. Lichtenberg, and R. C. Kroner, "The Occurence of Organohalides in Chlorinated Drinking Water" J. Am. Water Works Assoc. 66(12):703-6 (1974).

Block, J. C., M. Morlot, and J. M. Foliguet, "Problèmes Liés à l'Evolution du Caractère d'Oxydabilité de Certains Corps Organiques Présents dans l'Eau Traitée par l'Ozone", l'EAU 71(1):29-34 (1976).

Bollyky, L. J. "Reactions of Ozone With Trace Organics In Water and Wastewater", Presented at the Univ. of Michigan Seminar on Viruses and Trace Contaminants in Water and Wastewater, Jan. 1977. 43 pages.

Brooks, J. & G. Shaw, "Ozone in the Chemical Industry. Ozone Reactions with Saturated Organic Aliphatic Compounds", presented at Third Intl. Symp. on Ozone Technol., Paris, France, May 1977. Intl. Ozone Inst., Cleveland, Ohio.

Buescher, L. A., J. H. Dougherty, and R. T. Skrinde, "Chemical Oxidation of Selected Organic Pesticides", J. Water Poll. Control Fed. 36(8):1005-1014 (1964).

Bunn, W. W., B. B. Haas, E. R. Deane, and R. D. Kleapfer, "Formation of Trihalomethanes by Chlorination of Surface Water." U.S. EPA, Region VII; Kansas City, Kansas 66115. August 20, 1975.

Buydens, R., "Ozonation and its Effects on the Method of Purification of River Waters", Trib. du Cebedeau 319-320:286-291 (1970).

Carlson, R. M., "Organic Compounds Produced During Wastewater Chlorination", Seminar presented on the Symp. on the Identification and Transformation of Aquatic Pollutants, Athens, Ga., April, 1974.

Chen, K. Y., and R. W. Okey, "Ozone Effect on Synthetic Rubber Waste Treatment", Indl. Wastes, March/April 1977, pp. 47-49.

Cherkinsky, S. N., and A. A. Korolev, "Comparative Assessment of the Efficacy of Ozonization and Other Means of Treatment of Water Contaminated with Oil Products," Gig. y Sanit. 37(4):14-18 (1972).

Chian, E. S. K., & P. P. K. Kuo, "Fundamental Study on the Post-Treatment of RO Permeates From Army Wastewater". Sec. Annual Summary Rept., U.S. Army Medical R & D Command, Washington, D.C., Rept. No. UILU-ENG-76-2019, Oct. 1976.

Coppock, J. B. M., N. W. R. Daniels and P. W. Russell Eggitt, "Essential Fatty Acid Retention in Flour Treatment", Chem. & Ind., Jan. 2, 1960, p. 17-18.

Dellah, A. "Study of Ozone Reactions Involved in Water Treatment and the Present Chlorination Controversy", in Proc. Sec. Intl. Symp. on Ozone Technol., R. G. Rice, P. Pichet & M.-A. Vincent, editors, Intl. Ozone Inst., Cleveland, Ohio (1976), 161-168.

Dobinson, F., "Ozonisation of Malonic Acid in Aqueous Solution", Chem. & Ind. (6):853-854 (1959).

Dobinson, F. and G. J. Lawson, "Chemical Constitution of Coal VI--Optimum Conditions for the Preparation Of Sub-humic Acids from Humic Acid by Ozonization", Fuel 38:79-87 (1959).

Dowty, B., D. Carlisle and J. L. Laseter: "Halogenated Hydrocarbons in New Orleans Drinking Water and Blood Plasma", Science 187:75-77 (1975).

Dowty, B., D. R. Carlisle, and J. L. Laseter. "New Orleans Drinking Water Sources Tested by Gas Chromatography-Mass Spectrometry." Env. Sci. & Technol. 9(8):762-5 (1975).

Drapeau, A. J., "Du Chloroforme Dans Votre Eau Potable", Eau Du Quebec 8(2):19-23 (1975).

Eisenhower, H. R., "The Ozonization of Phenolic Wastes". J. Water Poll. Control Fed. 40(11):1887-1899 (1968).

Environmental Defense Fund, "The Implications of Cancer-Causing Substances in Mississippi River Water," November 6, 1974, Washington, D.C.

Falk, H. L. & J. E. Moyer, "Ozone As a Disinfectant of Water", in Ozone/Chlorine Oxidation Products of Organic Materials, R.G. Rice & J.A. Cotruvo, editors. Intl. Ozone Inst., Cleveland, Ohio, p. 38-57 (1978).

Fremery, M. I., and E. K. Fields, "Emulsion Ozonization of Cycloolefins in Aqueous Alkaline Hydrogen Peroxide", J. Org. Chem. 28:2537-41 (1963).

Furgason, R. R., H. L. Harding and M. A. Smith, "Ozone Treatment of Waste Effluent". Research Technical Completion Report, U.S. Dept. of Interior, Office of Water Resources Rsch. (Washington, D.C.) Rept. No. UC A-037-IDA, Water Resource Research Inst., Univ. of Idaho, April 1973. NTIS Rept. No. PB 220,008.

Gabovich, R. D., K. K. Vrochinskii, and I. L. Kurinnyi, "Decolorization, Deodorization and Decontamination of Drinking Water By Ozonization," Hygiene and Sanitation (Gig. y Sanit.) 34:336-340 (1969).

Gabovich, R. D., I. L. Kurinnyi, and Z. P. Fedorenko, "Effects of Ozone and Chlorine on 3,4-Benzopyrene During Water Treatment". Gig. Naselennikh Mest. (1969), p. 88.

Gabovich, R. D. and Kurinnyi, I.L., "Ozonization of Water Containing Petroleum Products, Aromatic Hydrocarbons, Nitrogen Compounds and Chloro-Organic Pesticides. Gigiene Naselennykh Mest. (USSR) Izd. "Zdorov'ya", (1967), p. 315.

Gad, G., and C. Columbus, "Chemische Auswirkungen der Wasserozonisierung", Health Engineer 76:268-9 (1955).

Gauntlett, R. B. and R. F. Packham, "The Removal of Organic Compounds in the Production of Potable Water", Chem. & Ind., Sept. 1, 1973, p. 812-817.

Gilbert E., "Ozonolysis of Chlorophenols and Maleic Acid in Aqueous Solution," in Proc. Sec. Intl. Symp. on Ozone Technol., R. G. Rice, P. Pichet and M.-A. Vincent, editors, Intl. Ozone Inst., Cleveland, Ohio (1976), 253-261.

Gilbert, E., "Reactions of Ozone with Organic Compounds in Dilute Aqueous Solution: Identification of Their Oxidation Products", in Ozone/Chlorine Dioxide Oxidation Products of Organic Materials, R.G. Rice & J.A. Cotruvo, editors, Intl. Ozone Inst., Cleveland, Ohio (1978), p. 227-242.

Gilbert, E., "Chemical Reactions Upon Ozonation", Presented at Intl. Symp. on Ozone and Water, Wasser Berlin, May, 1977, Intl. Ozone Inst. Cleveland, Ohio.

Glaze, W. H., R. Rawley and S. Lin, "By-Products of Organic Compounds in the Presence of Ozone and Ultraviolet Light: Preliminary Results", in Ozone/Chlorine Dioxide Oxidation Products of Organic Materials, R.G. Rice & J.A. Cotruvo, editors, Intl. Ozone Inst., Cleveland, Ohio, (1978), p. 321-331.

Glaze, W. H., R. Rawley, F. Huang, and S. Lin "Ozone and Ozone/UV Destruction of Trihalomethane Precursors and Other Refractory Organic Compounds in Water", Presented at Symp. on Advanced Ozone Technology, Toronto, Ontario, Canada, November 1977. Intl. Ozone Inst., Cleveland, Ohio.

Gregersen, J. K., "Evaluation of an Ozonation-Activated Carbon Treatment for a Colored Industrial Waste". Thesis, Iowa State Univ., Ames, Iowa (1971).

Hann, V., "Water Quality Improvement With Ozone", Engr. News Record 139(59):125-127 (1947).

Harrison, R. M., R. Perry and R. A. Wellings, "Effect of Water Chlorination Upon Levels of Some Polynuclear Aromatic Hydrocarbons in Water", Env. Sci. & Technol. 10(12):1151-1160 (1976).

Heertjes, P. M. and A. P. Meijers, "Untersuchung nach organischen Substanzen in Fluss- and Trinkwasser," Wasser-Abwasser 111(2):61-66 (1970).

Helz, G. R., R. Y. Hui and R. M. Block, "Bromoform Production by Oxidative Biocides in Marine Waters", in Ozone/Chlorine Dioxide Oxidation Products of Organic Materials", R.G. Rice & J.A. Cotruvo, editors, Intl. Ozone Inst., Cleveland, Ohio, (1978), p. 68-76 (1978).

Hoehn, R. C., C. W. Randall and F. A. Bell, Jr., "Trihalomethanes and Viruses In a Water Supply", J. Env. Engrg. Div. of Am. Soc. Civil Engrs., Oct. 1977, p. 803-812.

Hoffman, J., and D. Eichelsdörfer, "Zur Ozon Einwirkung auf Pestizide der Chlorkohlenwasserstoffgruppe im Wasser," Vom Wasser 38:197-206 (1971).

Hoigné, J., "Comparison of the Chemical Effects of Ozone and of Irradiation on Organic Impurities in Water", Proc. Radiation for a Clean Environment, Intl. Atomic Energy Agency, Vienna (1975), p. 297-305.

Hoigné, J. and H. Bader, "Beeinflussung der Oxidationswirkung von Ozon und OH-Radikalen durch Carbonat", Vom Wasser 48:283-304 (1977).

Hoigné, J. & H. Bader, "Ozonation of Water: Selectivity and Rate of Oxidation of Solutes", Presented at 3rd Intl. Symp. on Ozone Technol., Paris, France, May 1977. Intl. Ozone Inst., Cleveland, Ohio.

Hoigné, J. and H. Bader, "Rate Constants for Reactions of Ozone With Organic Pollutants and Ammonia in Water", Symp. on Advanced Ozone Technol., Toronto, Ontario, Canada, November, 1977. Intl. Ozone Inst., Cleveland, Ohio.

Hoigné, J. and H. Bader, "Ozonation of Water: Kinetics of Oxidation of Ammonia by Ozone and Hydroxyl Radicals", Environ. Sci. & Technol. 12(1):79-84 (1978).

Hubbs, S. A., "The Oxidation of Haloforms and Haloform Precursors Utilizing Ozone", in Ozone/Chlorine Oxidation Products of Organic Materials, R.G. Rice & J.A. Cotruvo, editors. Intl. Ozone Inst., Cleveland, Ohio. p. 200-209 (1978).

Il'nitskii, A. P., "Effect of Ozonation upon Aromatic Hydrocarbons Including Carcinogens", Hygiene and Sanit. (Gig. y Sanit.) 33(3):323-7 (1968).

Il'nitskii, A. P., "Experimental Investigation of the Elimination of Carcinogenic Hydrocarbons from Water During its Classification and Disinfection", Hygiene and Sanit. (Gig. y Sanit.) 34(9):317-321 (1969).

Ingols, R. S., "Ozonation of Seawater", in "Ozone/Chlorine Dioxide Oxidation Products of Organic Materials", R. G. Rice & J. A. Cotruvo, editors, Intl. Ozone Inst., Cleveland, Ohio, (1978), p. 77-81.

Ishizaki, K., R. A. Dobbs and J. M. Cohen, "Ozonation of Hazardous and Toxic Organic Compounds in Aqueous Solution", in Ozone/Chlorine Dioxide Oxidation Products of Organic Materials, R.G. Rice & J.A. Cotruvo, editors. Intl. Ozone Inst., Cleveland, Ohio, (1978), p. 210-226.

Junk, G. A., and S. E. Stanley, "Organics in Drinking Water. Part I - Listing of Identified Chemicals", Prepared for U. S. Energy Research and Development Administration under Contract W-7405-eng. 82. July 1975.

Junk, G. A., J. J. Richard, M. D. Grieser, D. Witiak, J. L. Witiak, M. D. Arguello, R. Vick, H. J. Svec, F. S. Fritz and F. V. Calder, "Use of Macroreticular Resins in the Analysis of Water for Trace Organic Contaminants," J. Chromatography 99:745-762 (1974).

Jürs, R. H., "Die Wirkung des Ozons auf im Wasser gelöste Stoffe" ("The effect of ozone on materials which are dissolved in water") Fortschr. Wasserchem. Ihrer Grenzgebiete 41:40-64 (1966).

Kennaugh, J., "Action of Diaphanol on Arthropod Cuticles", Nature, 180:238 (1957).

Kinney, C. R. and L. T. Friedman, "Ozonization Studies on Coal Constitution", J. Am. Chem. Soc. 74:57-61 (1952).

Kopfler, F. C., R. G. Milton, R. D. Lingg and W. E. Coleman, "GC/MS Determination of Volatiles for the National Organics Reconnaissance Survey (NORS) on Drinking Water." in Identification and Analysis of Organic Pollutants in Water, 1st ed., L. H. Keith, Ed., Ann Arbor Science Publishers, Inc., Ann Arbor, Michigan, 1975, Chapter 6.

Krasnov, B.P., D.L. Pakul' & T.V. Kirillova, "Ozonization of Industrial Wastewater", Khim. Prom. 1:28-30 (1974). Engl. Transl. in Intl. Chem. Engrg. 14(4):747-750 (1974).

Krause, H., H. Hepp and W. Kluger, "Research on the Destruction of Complexing and Chelating Agents in Radioactive Liquid Wastes by Oxidation", Kernforschungzentrum Karlsruhe (Univ. of Karlsruhe, Germany). AEC Accession No. 32047, Rept. No. KFK-287 (1964).

Kuo, P. P. K., E. S. K. Chian and B. J. Chang, "Identification of End Products Resulting From Ozonation of Compounds Commonly Found in Water", in "Ozone/Chlorine Dioxide Oxidation Products of Organic Materials", R. G. Rice & J. A. Cotruvo, editors, Intl. Ozone Inst., Cleveland, Ohio, (1978), p. 153-166.

Kutepov, E. N., "Experimental Data for Determining Standard Levels of Carbon Tetrachloride in Water Bodies," Gig. y Sanit. 33(1):32-37 (1968).

La Planche, A., G. Martin & Y. Richard, "Etude de la Dégradation des Pesticides par l'Ozone: Cas du Parathion", Centre Belge d'Etude et de Documentation des Eaux, #362 (Jan. 1974).

La Planche, A., G. Martin & Y. Richard, "Contribution a l'Etude de la Dégradation par l'Ozone de Quelques Insecticides du Groupe des Organophosphorés", T.S.M. L'Eau, 271-273, (June 1972)

Lawrence, J., "The Oxidation of Some Haloform Precursors", Presented at 3rd Intl. Symp. on Ozone Technol., Paris, France, May, 1977. Intl. Ozone Inst., Cleveland, Ohio.

Lawrence, J. "Identification of Ozonation Products in Natural Waters", Presented at Symp. on Advanced Ozone Technology, Toronto, Ontario, Canada, November, 1977. Intl. Ozone Inst., Cleveland, Ohio.

Leopold, B. and D. B. Mutton, "The Effect of Chlorinating and Oxidizing Agents on Derivatives of Oleic Acid", TAPPI 42(3):218-225 (1959).

Lindgren, B. O. and C. M. Svahn, "Chlorine Dioxide Oxidation of Cyclohexene", Acta Chem. Scand. 19(1):7-13 (1965).

Lindgren, B. O. and C. M. Svahn, "Reactions of Chlorine Dioxide with Unsaturated Compounds", Acta Chem. Scand. 20(1):211-218 (1966).

Maggiolo, A., "Ozone's Radical and Ionic Mechanisms of Reaction With Organic Compounds in Water", in Ozone/Chlorine Dioxide Oxidation Products of Organic Materials, R. G. Rice & J.A. Cotruvo, editors. Intl. Ozone Inst., Cleveland, Ohio, p. 59-67 (1978).

Maier, D., "Wirkung von Ozon auf die gelösten organischen Substanzen in Bodenseewasser", Vom Wasser 43:127-60 (1974).

Mallevialle, J., "Action of Ozone in the Degradation of Simple and Polymeric Phenolic Compounds-- Application to Humic Materials Contained in Waters", T.S.M. l'Eau 70(3):107-103 (1975).

Mallevialle, J., Y. Laval, M. LeFebvre and C. Rousseau, "The Degradation of Humic Substances in Water by Various Oxidation Agents (Ozone, Chlorine, Chlorine Dioxide)", in Ozone/Chlorine Dioxide Oxidation Products Of Organic Materials, R.G. Rice & J.A. Cotruvo, editors, Intl. Ozone Inst., Cleveland, Ohio, (1978), p. 189-199.

Martin, G. & A. La Planche, "Action of Ozone on Organo-Phosphoric Compounds", Presented at 3rd Intl. Symp. on Ozone Technol., Paris, France, May 1977. Intl. Ozone Inst., Cleveland, Ohio.

Masschelein, W., G. Fransolet & J. Genot., Dispersing and Dissolving Ozone in Water -- Pt. 1, Water & Sewage Works, Dec. 1975, 57-60. Pt. II, Ibid, Jan. 1976, 34-36.

McCammon, W.H., "Ozonation Contact Systems." Proc. 68th Ann. Mtg., Am Inst. Chem. Engrs., New York, N.Y. Nov. 1975.

Mills, J.F., "Competitive Oxidation and Halogenation Reactions in the Disinfection of Wastewater", in "Ozone/Chlorine Dioxide Oxidation Products of Organic Materials", R.G. Rice & J.A. Cotruvo, editors. Intl. Ozone Inst. Cleveland, Ohio (1978). p. 400-409.

Mohamed, A. & A. Osama, "Removal of Phenylamide Pesticides from Drinking Waters - I. Effect of Chemical Coagulation and Oxidants", Water Research, 11:611-616 (1977).

Morris, J.C., "Formation of Halogenated Organics by Chlorination of Water Supplies", US EPA Report No. EPA 600/1-75-002, March, 1975.

Murphy, K.L., R. Zaloum & D. Fulford, "Effect of Chlorination Practice on Soluble Organics", Water Research 9(4):389-396 (1975).

Oehlschlaeger, H.F., "Reactions of Ozone With Organic Compounds", in Ozone/Chlorine Dioxide Oxidation Products Of Organic Materials, R.G. Rice & J.A. Cotruvo, editors. Intl. Ozone Inst., Cleveland, Ohio, (1978), p. 20-37.

Omel'yanets, N.I. & V.N. Kupyrov, "Hygienic Characteristics of Organic Impurities in Tap Water", Gig. y Sanit. 31:420-423 (1968-69).

Paluch, K., Private Communications to R.G.Rice, Oct. 25, 1976.

Paulson, E.G., "Organics in Water Supply", Water and Sewage Works, 110:216 (1963).

Peleg, M., "The Chemistry of Ozone in the Treatment of Water", Water Research 10:361-365 (1976).

Prengle, Jr., H.W., C.E. Mauk and J.E. Payne, "Ozone/UV Oxidation of Chlorinated Compounds in Water", in Forum on Ozone Disinfection, E. G. Fochtman, R. G. Rice and M. E. Browning, editors, Intl. Ozone Inst., Cleveland, Ohio (1976), p. 286-295.

Prengle, H.W., Jr., C.G. Hewes, III & C.E. Mauk, "Oxidation of Refractory Materials by Ozone with Ultraviolet Radiation", in Proc. Sec. Intl. Symp. on Ozone Technol., R.G. Rice, P. Pichet & M.-A Vincent, editors. Intl. Ozone Inst., Cleveland, Ohio (1976), p. 224-252.

Prengle, Jr., H.W. and C.E. Mauk, "Ozone/UV Oxidation of Pesticides in Aqueous Solution", in "Ozone/Chlorine Dioxide Oxidation Products of Organic Materials", R.G. Rice & J.A. Cotruvo, editors. Intl. Ozone Inst., Cleveland, Ohio (1978), p. 302-320.

Pryde, E.H., D.J. Moore and J.C. Cowan, "Hydrolytic, Reductive and Prolytic Decomposition of Selected Ozonolysis Products. Water as an Ozonization Medium", J. Am. Oil Chemists Soc. 45(12):888-894 (1968).

Reichert, J., "Examination for the Elimination of Carcinogenic, Aromatic Polycyclics in the Treatment of Drinking Water, with Special Consideration of Ozone", Wasser-Abwasser 110(8):477-482 (1969)

Richard, Y. and L. Brener, "Organic Materials Produced Upon Ozonization of Water", in "Ozone/Chlorine Dioxide Oxidation Products of Organic Materials, R.G. Rice & J.A. Cotruvo, editors. Intl. Ozone Inst., Cleveland, Ohio (1978), pp 169-188.

Robeck, G.G., K.A. Dostal, J.M. Cohen and J.F. Kreissl, "Effectiveness of Water Treatment Processes in Pesticide Removal", J. Am. Water Works Assoc. 57:181-200 (1965).

Rook, J.J., "Formation of Haloforms During Chlorination of Natural Waters," Water Treatment and Examination 23(2):234-243 (1974).

Rook, J. J., "Haloforms in Drinking Water," J. Am. Water Works Assoc. 68(3):168-172 (1976).

Rook, J.J., "Developments in Europe", J. Am. Water Works Assoc. 68(6):279-282 (1976).

Rosenblatt, D.H., A.J. Hayes, Jr., B.L. Harrison, R.A. Streaty, and K.A. Moore, "The Reaction of Chlorine Dioxide with Triethylamine in Aqueous Solution", J. Org. Chem. 28:2790-2794 (1963).

Rosenblatt, D.H., "Chlorine Dioxide: Chemical and Physical Properties", in "Ozone/Chlorine Dioxide Oxidation Products of Organic Materials", R.G. Rice & J.A. Cotruvo, editors. Intl. Ozone Inst., Cleveland, Ohio (1978), p. 332-343.

Scassellati-Sforzolini, G., A. Savino, S. Monarca and M.N. Lollini, "Decontamination of Water Contaminated With Polycyclic Aromatic Hydrocarbons (PAH). I. Action of Chlorine and Ozone on PAH Dissolved in Doubly Distilled and in Deionized Water", Igiene Moderna 66(3):309-335 (1974).

Scassellati-Sforzolini, G., A. Savino, S. Monarca, "Decontamination of Water Contaminated With Polycyclic Aromatic Hydrocarbons (PAH). II. Action of Chlorine and Ozone on PAH Dissolved in Drinking and River Water", Igiene Moderna 66(6):595-619 (1974).

Schalekamp, M., "Experience in Switzerland with Ozone, Particularly in Connection with the Neutralization of Hygienically Undesirable Elements Present in Water". Presented at Intl. Symp. on Ozone & Water, Wasser Berlin, Germany, May 1977, Intl. Ozone Inst., Cleveland, Ohio.

Shapiro, R.H., K.J. Kolonko, P.M. Greenstein, R.M. Barkley and R.E. Sievers, "Ozonization Products From Caffeine in Aqueous Solution", in "Ozone/Chlorine Dioxide Oxidation Products of Organic Materials", R.G. Rice & J.A. Cotruvo, editors. Intl. Ozone Inst., Cleveland, Ohio (1978), p. 284-290.

Shevchenko, M.A. and P.N. Taran, "Investigation of the Ozonolysis Products of Humus Materials", Ukrainskii Khimicheskii Zhurnal 32(5):532-536 (1966).

Shuval, H. and M. Peleg, "Studies on Refractory Organic Matter from Wastewater by Ozonization", Progress Report of work sponsored by Ges. für Kernforschung, Karlsruhe, Germany and Natl. Council for R&D, Jerusalem, Israel (Dec. 1975).

Sierka, R.A., "The Effects of Sonic and Ultrasonic Waves on the Mass Transfer of Ozone and the Oxidation of Organic Substances in Aqueous Solution", Presented at 3rd Intl. Symp. on Ozone Technology, Paris, France, May, 1977. Intl. Ozone Inst., Cleveland, Ohio.

Sigworth, E.A., "Identification & Removal of Herbicides and Pesticides", J. Am. Water Works Assoc. 57:1016-1022 (1965).

Sletten, O., "Halogens and Their Role in Disinfection", J. Am. Water Works Assoc. 8(13):690-692 (1974).

Smith, G.V., J.W. Chen & K. Seyffarth, "Catalytic and Sonocatalytic Oxidations of Aqueous Phenol", in Proc. 5th Intl. Congress on Catalysis, Palm Beach, Fla., Aug. 1972, p. 893-903 (1973). Am. Inst. Chem. Engrs., New York, N.Y.

Spanggord R.J., and V.J. McClurg, "Ozone Methods and Ozone Chemistry of Selected Organics in Water", in "Ozone/Chlorine Dioxide Oxidation Products of Organic Materials," R.G. Rice & J.A. Cotruvo, editors. Intl. Ozone Inst., Cleveland, Ohio (1978), p. 115-125.

Somiya, I. H. Yamada and T. Goda, "Ozonation of Nitrogenous Compounds in Water", Presented at Symp. on Advanced Ozone Technol., Toronto, Ontario, Canada, November 1977. Intl. Ozone Inst., Cleveland, Ohio.

Somsen, R., "Oxidation of Some Simple Organic Molecules With Aqueous Chlorine Solutions. I. Kinetics", TAPPI 43(2):154-160 (1960).

Stahl, D.S., "Ozone Contacting Systems", in Ozone for Water & Wastewater Treatment, R.G. Rice & M.E. Browning, editors, Intl. Ozone Inst., Cleveland, Ohio, (1975), p. 40-55.

Steelink, C., "Humates and Other Natural Organic Substances in the Aquatic Environment", J. Chem. Educ. 54(10):599-603 (1977).

Stevens, A.A., C.J. Slocum, D.R. Seeger & G.G. Robeck, "Chlorination of Organics in Drinking Water", Appendix 2, of Rept. to U.S. Congress, Water Supply Research Division, Municipal Environmental Research Laboratory, U.S. Environmental Protection Agency, Cincinnati, Ohio 45268, Nov. 1975.

Stevens, A.A., D.R. Seeger & C.J. Slocum, "Products of Chlorine Dioxide Treatment of Organic Materials in Water", in Ozone/Chlorine Dioxide Oxidation Products of Organic Materials, R.G. Rice & J.A. Cotruvo, editors. Int. Ozone Inst., Cleveland, Ohio (1978), p. 383-399.

Stoveken, J. & T. Sproston, "Ozone and Chlorine Degradation of Waste Water Pollutants", U.S. Dept. Interior, Washington, D.C., OWRR Project No. A-017Vt, June 1974.

Sturrock, M.G., E.L. Cline & K.R. Robinson, "The Ozonation of Phenanthrene with Water as Participating Solvent". J. Org. Chem. 28:2340 (1963).

Suchkov, B.P., "Studies of the Ozonation of Drinking Water Containing Pathogenic Bacteria and Viruses", Gig. y Sanit. 29(6):2431 (1964).

Symons, J.M., J.K. Carswell, R.M. Clark, O.T. Love, Jr., R.J. Miltner & A.A. Stevens, "Interim Treatment Guide for the Control of Chloroform and Other Trihalomethanes", U.S. EPA, Water Supply Rsch. Lab., Cincinnati, Ohio, 45268, June 1976.

Tardiff, R.G., "Health Effects of Organics: Risk & Hazard Assessment of Ingested Chloroform", Presented at the 96th Annual Conf. of Am. Water Works Assoc. June 20-25, 1976, pp. 1-20.

Taube, H., "Chain Reactions of Ozone in Aqueous Solution. The Interaction of Ozone and Formic Acid in Aqueous Solutions". J. Am. Chem. Soc. 63:2453-58 (1941).

Tencza, S.J. & R.A. Sierka, "Ozonation of Low Molecular Weight Compounds", presented at Sec. Natl. Conf. on Water Reuse, May 4-8 (1975), Am. Inst. Chem. Engrs., New York, N.Y.

Thielemann, H., "Uber die Einwirkung von Chlordioxid auf einige polycyklische aromatische Kohlenwasserstoffe", Microchimica Acta (1972), 575-577.

Thielemann, H., "Uber die Einwirkung von Chlordioxid auf 1- und 2-naphtholhaltige Testlösungen", Mikrochimica Acta (1972), 669-671.

Thielemann, H., "Uber die papierchromatographische Trennung der Reaktionsprodukte von Polyphenolen mit Chlordioxid", Mikrochimica Acta (1972), 655-657.

Throop, W.M., "Alternative Methods of Phenol Wastewater Control", J. Hazardous Materials 1:319-329 (1977).

U.S. Environmental Protection Agency, "A Report, Assessment of Health Risk from Organics in Drinking Water;" by an ad hoc Study Group to the Hazardous Materials Advisory Committee; Science Advisory Board, April 30, 1975.

U.S. Environmental Protection Agency, "Treatment Measures for Wastes From Industrial Sources", Indl. Tech. Bulletin, 4(11), July 1977.

U.S. Environmental Protection Agency, Report to Congress: "Preliminary Assessment of Suspected Carcinogens in Drinking Water". December 1975.

U.S. Environmental Protection Agency, "Organic Chemical Contaminants, Contaminants, Drinking Water", Federal Register, July 1976.

Vrochinskii, K.K., "Effects of the Color Index of Water on Its Decontamination With Ozone", Vopr. Gigieny Naselin. Mest., Sb., Kiev 5:90-3 (1964)

Weber, J.H., "Organic Matter in Natural Water and Water Treatment." OWRT Rept. No. OWRT-A-035-NH(1), Sept, 1974, U.S. Dept. Interior, Washington, D.C.

Weil, L., B. Struiff, and K.E. Quentin, "Reaktion mechanismen beim Abbau organischer Substanzen im Wasser mit Ozon", Presented at Intl. Conf. on Ozone & Water, Wasser Berlin, May 1977, Intl. Ozone Inst., Cleveland, Ohio.

Yeatts, L.B., Jr. and H. Taube, "The Kinetics of Reaction of Ozone and Chloride Ion in Acid Aqueous Solution", J. Am. Chem. Soc. 71:4100-4105 (1949).

Yocum, F.H., "Oxidation of Styrene With Ozone in Aqueous Solution", in "Ozone/Chlorine Dioxide Oxidation Products of Organic Materials", R.G. Rice & J.A. Cotruvo, editors. Intl. Ozone Inst., Cleveland, Ohio (1978), p. 243-263.

Zienius, R.H. and C. B. Purves, "Degradation of Pectic Acid by Some Delignifying and Bleaching Agents", TAPPI 43(1):27-34 (1960).

APPENDIX A: BLANK QUESTIONNAIRES

OZONATION QUESTIONNAIRE

Section 1: General Plant Information

Note: The purpose of this section is to provide background data for the plant

A. Location:

_______________ (country)

_______________ (city)

Address: _______________

____

Check one: ___publicly owned ___investor owned

B. Person to be contacted for further information:

Name:_______________Title:_______

Address:_______________

Telephone:_______________

C. Plant capacity:

Present design capacity: _______cu m/day
Present average daily flow: _______cu m/day
Present maximum daily flow: _______cu m/day
Present minimum daily flow: _______cu m/day
Original construction cost: _______Year Constructed:_____
Expansion construction cost:_______Year expanded:_____

(continued)

D. Basic process layout:

Provide in the form of a simple block diagram, e.g.:

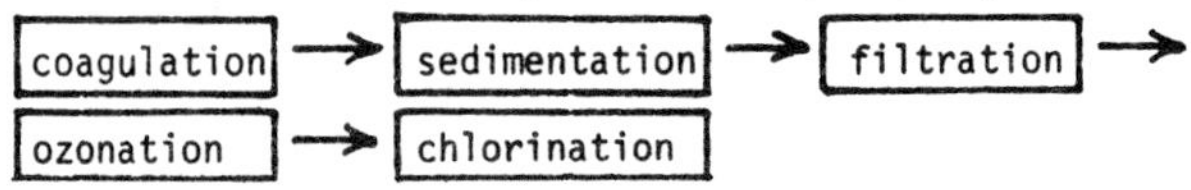

Be sure to show all points in the process where ozone is used.

E. Purpose of Ozonation (check all that apply)

___iron removal
___color removal
___odor removal
___bacterial disinfection
___organic compound removal
___manganese removal
___taste removal
___turbidity removal
___viral inactivation
___other (please specify)________
____

F. History:

Year plant started operation__________________
designed capacity__________________cu m/day
Year ozonation equipment began operation______
ozonation capacity______________________kg/day
Year of latest ozonation plant expansion______
present ozonation capacity________________kg/day

G. Is your plant open for visits by our project team?

___yes ___no

Contact:____________________________________

(address and telephone if different from above)__________________

____

(continued)

Section II: Ozonation System

Note: This section provides more details as to the design and operation of your ozonation system.

A. Feed gas

1. Gas used: ___air ___oxygen ___air/oxygen mixture_____%

2. Gas pretreatment: ___air compressor ___silica gel scrubber ___alumina scrubber ___other ________________
manufacturer__________________________________

B. Ozonation system

1. Design capacity____________________kg/hr
2. Average daily output____________________kg/hr
3. Number of generators____________________
4. Capacity per unit______________________kg/hr
5. Type of unit (plate, tube, air cooled, water cooled, etc.)_____
6. Manufacturer____
7. Cooling water used? ___yes ___no
   source________________________________
   quality:__raw water __chemically treated __solids removed
   __drinking water quality
   maximum input temperature ___________°C.
   minimum input temperature ___________°C.
   average temperature ___________°C.
8. Ozone contacting system:
   number of contacting units:____detention time:____________
   type: ___sparger ___injector ___submerged turbine
   ___packed column ___spray tower ___plate column
   ___other:_______________________________
   number of stages per contactor_________________________
   manufacturer____
   use of supplemental processing with contactor? ___yes ___no
   ___sonics ___ultraviolet ___catalysis __other:_________
   are off-gases processed to remove excess ozone? __yes __no
   describe how:____
   ____

9. Power consumption per kg of ozone produced:____at average daily output or ____________at maximum capacity of unit for
   _gas pretreatment unit _ozone generating unit _ozone contacting system

(continued)

Section III: Process Parameters

Note: This section provides information on the analytical procedures and monitoring used by the plant to control the ozonation facility.

A. Analytical procedures

1. What is the average ozone dosage?__________________
How is it measured?____________________________

2. How is the ozone residual measured?______________
At what point(s) in the process is it measured? (indicate on the block diagram in question I-D)____________________

B. What analytical procedures do you use to determine the effectiveness of the ozone process?____________________________

(What is measured? How are the results reported? Units?)

1. What analyses?__________________________________
2. Where sampled?__________________________________
3. How often?______________________________________
4. For what period of time have you been performing the analyses?____________
5. Comments on ease or difficulty of employing analytical procedure:____________________________________

C. Removal of specific chemical compounds

1. By ozonation only: (Please fill out table)

Chemicals Removed	Verifying Analysis	For What Period of Time Has the Analysis Been Performed

2. By combination of process steps:

Chemicals Removed	Verifying Analysis	For What Period of Time Has the Analysis Been Performed

(continued)

Please show the process steps involved with a block diagram as in question I-D

D. Final products of ozonation

Do you have any information as to the final products of the organic compounds which have been ozonized?__yes __no describe the analytical data: (analyses used, where taken, how often)

____
____
____
____

E. Use of residual disinfectant.

disinfectant used________________________________
where added___________________________________
dosage______________________________________
free residual attained in the plant effluent:___mg/l, at the extremity of the distribution system_________mg/l.

total residual attained in the plant effluent:_____mg/l; at the extremity of the distribution system:________mg/l.

has the residual disinfectant been successful in controlling slime and microbiological growths in the distribution system?
___yes ___no.

how frequently are microscopic examinations for nuisance organisms conducted on distribution samples?_______________

(continued)

F. Ozonation process controls

1. How do you control the ozonation process? (check all that apply)
   _____by water quality parameters (color, COD, etc.)
   _____by ozone residual
   _____by ozone dosage
   _____by ozone contact time
   _____other:____

2. Are controls manual or automatic?__________________________

3. Is a copy of a typical plant operational sheet containing the actual hourly/daily chemical doses fed to the water in all stages of treatment available? If so, please attach. Typical physical, chemical, bacteriological, and microscopic analyses of the raw and finished water would also be appreciated.

G. What effect has ozonation had on quantities of solid waste (sludge) generated by your process and which must be disposed? ____________
____

H. Any observed negative effects of ozone usage? (e.g., corrosion, high cost, workers' objections, development of scum or color after the water has left the treatment plant, etc.)____________________
____

I. If ozone is relied upon exclusively for disinfection, do slime and nuisance biological growths occur in the distribution system?__yes __no.

Section IV: Plant Water Quality

Note: The purpose of this question is to determine quality of plant influent versus plant effluent.

A. Source of raw water:____
Physical, chemical, bacteriological, and microscopic tests on raw water, together with their individual frequency:__________________
____
____

(continued)

B. Physical, chemical, bacteriological, and microscopic tests on product water, together with their individual frequency:_________

Section V: Other Information

A. Does the plant have any unique features? Please describe:

B. Are there restrictions (national, local, public health) on the use of ozone? Are they subject to governmental enforcement?

C. Comments on experience with operation of ozone generation and associated equipment:

1. Maintenance requirements:_________

2. Unusual maintenance problems (e.g., corrosion):_________

3. Any safety problems or accidental exposures to ozone?____

D. Additional comments

(continued)

Section VI: Chlorine Dioxide Use

A. Do you use chlorine dioxide in your facility?____yes ____no

If yes, please complete the following questions.

B. Where in the process is it injected? (Please show this on the block diagram in question I-D)

C. For what purpose is it used?

___iron removal	___manganese removal
___color removal	___taste removal
___odor removal	___turbidity removal
___bacterial disinfection	___provide residual chlorine
___organic compound removal	___other (specify)___________

D. How do you generate your chlorine dioxide?

1. Chemicals used_______________________________
2. Techniques of combination_____________________

____

(attach a copy of your in-plant instructions to operators if you have it)

E. Is there any excess free chlorine present in the chlorine dioxide? ___yes ___no

F. Do you use any other analytical methods to monitor chlorine dioxide addition and results, other than those listed in Section III of this questionnaire? If so, please describe:

____

____

____

____

(continued)

CHLORINE DIOXIDE QUESTIONNAIRE

Section I: General Plant Information

Note: The purpose of this section is to provide background data for the plant

A. Location:

______________________ (country)

______________________ (city)

Address: ______________________

____

Check one: ___publicly owned ___investor owned

B. Person to be contacted for further information:

Name: ______________________ Title: __________

Address: ______________________

Telephone: ______________________

C. Plant capacity:

Present design capacity: __________cu m/day__________mgd
Present average daily flow: __________cu m/day__________mgd
Present maximum daily flow: __________cu m/day__________mgd
Present minimum daily flow: __________cu m/day__________mgd
Original construction cost: __________Year Constructed:______
Expansion construction cost: __________Year expanded:______

(continued)

D. Basic process layout:

Provide in the form of a simple block diagram, e.g.:

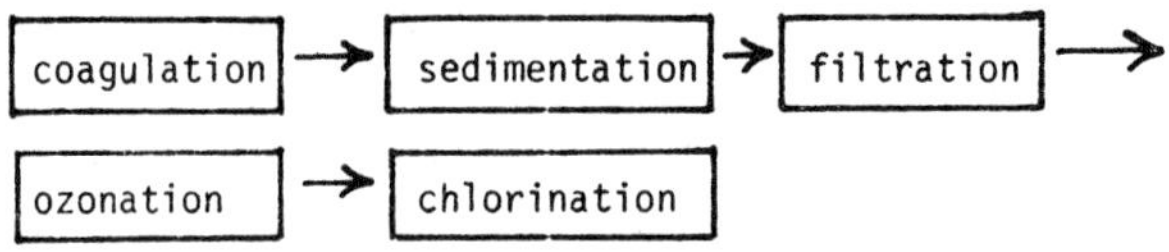

Be sure to show all points in the process where chlorine dioxide is used.

E. Purpose of chlorine dioxide (check all that apply):

___iron removal
___color removal
___odor removal
___bacterial disinfection
___organic compound removal
___manganese removal
___taste removal
___turbidity removal
___chlorine residual
___other (please specify)__________
____
____

1. If chlorine dioxide is fed solely for disinfection purposes:

(a) What is the residual in mg/l leaving the treatment plant?_____mg/l
(b) What is the residual in mg/l prevailing at the extremity of the distribution system?___mg/l
(c) Do slime and nuisance biological growths occur in the distribution system?____yes ____no

2. If chlorine dioxide is fed for taste and odor control:

(a) Is the chlorine dioxide fed intermittently as the taste and odor bodies appear in the raw water, or at a constant dosage as a precautionary measure?_______
(b) What is the range of intermittent dosage in mg/l? ________mg/l
(c) What is the constant dosage in mg/l?________mg/l
(d) Are the taste and odor bodies of a natural or industrial origin?____________________________
(e) Please note their chemical composition, if known or suspected.________________________________
____

(continued)

F. History

Year plant started operation:__________
designed capacity______________cu m/day or________mgd
Year chlorine dioxide equipment began operation:________
chlorine dioxide capacity_______kg/day or_______lbs/day
Year of latest chlorine dioxide plant expansion________
present chlorine dioxide capacity____kg/day or___lbs/day

G. Is your plant open for visits by our project team?

____yes ____no
Contact:____
Address and telephone if different from above:______________
____

Section II: Chlorine Dioxide System

Note: This section provides more details as to the design and operation of your chlorine dioxide system.

A. Please list the chemicals you use to make chlorine dioxide:

______________________ ______________________
______________________ ______________________
______________________ ______________________

B. What is the chemical consumption? (Give units)

maximum:_______________ _____________ _____________
average:_______________ _____________ _____________
minimum:_______________ _____________ _____________

C. Please describe the chlorine dioxide generating system:

____
____
____
____

(Attach a copy of the plant operator's guidelines for chlorine dioxide process operation if one is available)

Who is the manufacturer of the equipment?_______________________
____

(continued)

D. Please describe the chlorine dioxide contacting system:

____
____
____

Who is the manufacturer of the equipment?_______________

number of contacting units:_______________
number of stages per contactor:_______________

Section III: Process Parameters

Note: This section provides information on the analytical procedures and monitoring used by the plant to control the chlorine dioxide facility.

A. Analytical Procedures

1. What is the average chlorine dioxide dosage?__________(units)
   How is it measured?__________________________

2. How is the chlorine dioxide residual measured?____________
   At what point(s) in the process is it measured? (indicate on the block diagram in question I-D)

3. What species are measured? Check one or more if known:
   ClO_2____ ClO_2^-____ HOCl____ Chloramines____ ClO_3^-____ Cl^-____
   Cl_2____ Other______

B. How do you determine if the chlorine dioxide is accomplishing its purpose?

1. What analyses?______________________________________
   ____

2. Where sampled?______________________________________
   ____

3. How often?____
   ____

4. How long have you been performing the analyses?____________

5. Comments on ease or difficulty of employing analytical procedures:_____________________________________
   ____

(continued)

C. Removal of specific chemical compounds

1. By chlorine dioxide only:

Chemicals Removed	Verifying Analysis	How Long Analyzed
________________	________________	________________
________________	________________	________________
________________	________________	________________
________________	________________	________________
________________	________________	________________

2. By combination of process steps:

Chemicals Removed	Verifying Analysis	How Long Analyzed
________________	________________	________________
________________	________________	________________
________________	________________	________________
________________	________________	________________

Please show the process steps involved with a block diagram as in question I-D.

(continued)

D. Final products of chlorine dioxide

Do you have any information as to the final products of the organic compounds treated by chlorine dioxide? ____yes ____no
Describe the analytical data: (analyses used, where taken, how often)

____
____
____

E. Is there any free chlorine formed along with the chlorine dioxide?
____yes ____no
Comments:____
____

F. Chlorine dioxide process controls

1. How do you control the chlorine dioxide process? (Check all that apply)
_____by water quality parameters (color, COD, etc.)
_____by chlorine dioxide residual
_____by chlorine dioxide dosage
_____by chlorine dioxide contact time
_____other____

2. Are the controls manual or automatic?____________________

3. Is a copy of a typical data plant operational sheet containing the actual hourly/daily chemical doses fed to the water in all stages of treatment available? If so, please attach. The physical, chemical, bacteriological, and microscopic analyses of the raw and finished water would also be appreciated.

G. Any observed negative effects of chlorine dioxide? (e.g., corrosion, high cost, workers' objections, etc.)________________
____
____

Section IV: Plant Water Quality

Note: The purpose of this section is to determine quality of plant influent versus plant effluent.

(continued)

A. Source of raw water: ____________________
Physical, chemical, bacteriological, and microscopic tests on raw water, together with their individual frequency: ____________________

B. Physical, chemical, bacteriological, and microscopic tests on product water, together with their individual frequency: ____________________

Section V: Other Information

A. Does the plant have any unique features? Please describe:

B. Are there restrictions (national, local, public health) on the use of chlorine dioxide? Are they subject to governmental enforcement? ____________________

C. Comments on experience with operation of chlorine dioxide generation and associated equipment:

1. Maintenance requirements: ____________________

2. Unusual maintenance problems (e.g., corrosion): ____________________

3. Any safety problems or accidental exposures to chlorine dioxide? ____________________

D. Additional Comments:

APPENDIX B: DESCRIPTIONS OF SELECTED DRINKING WATER TREATMENT PLANTS USING OZONE

BRUSSELS, BELGIUM - TAILFER PLANT

This plant is located 75 km northeast of Brussels, on the river Meuse upstream of Brussels. There are four water treatment trains, each having a design capacity of 65,000 cu m/day. Three are complete and are operating; the fourth was being constructed at the time of plant visitation (May, 1977). Although the original projections were that three lines would be required during the summer and two in winter, it has been found that only two lines are needed in summer to satisfy current water demands and one in winter. The fourth line is being installed as a spare.

The new Tailfer plant was designed, constructed and is operated by the Compagnie Intercommunale Bruxelloise des Eaux (CIBE, a public company) and began operating in 1973. About 40 km upstream of the plant is a nuclear power plant which uses chlorine as the cooling water biocide. Discharges from this power plant are sent to a holding pond where radioactivity (Co-60) and chlorine levels decay. However, the Tailfer plant analyzes for heavy metals and radioactivity in the intake water and on sludges produced during the water treatment process. In addition, a series of four raceways have been installed at Tailfer in which trout are maintained to monitor the intake water for toxic materials. Details of this trout monitoring system (developed by the Netherlands Water Research Institute--KIWA) are described in Section 7.

At the plant there is an extensive analytical laboratory which monitors and analyzes raw, process and treated waters. At each chemical addition step, a portion of the plant water flows continuously into the laboratory where it is sampled and analyzed at regular intervals or monitored constantly. The laboratory has its own analytical control panel where each chemical addition step is controlled. Alarms are set to ring if the addition rates of chemicals exceed set minima or maxima or if turbidities exceed set limits.

Raw water is monitored for turbidity, pH, temperature, conductivity, redox potential and dissolved oxygen. In addition, other analyses are performed from time to time, including radioactivity, ammonia, COD, etc. The Meuse at Tailfer contains 0.2 mg/l of ammonia and 10 to 30 mg/l of COD (with occasional high levels of 70 to 80 mg/l).

Water treatment occurs in three separate blocks at Tailfer (Figure B-1). Block A contains the chemical treatment, Block B decantation and filtration, and Block C the ozonation. As water enters Block A, it is treated, in order, with sulfuric acid (to control alkalinity), chlorine (for predisinfection), chlorine dioxide (to break up organically complexed manganese), alum (to neutralize colloidal charges on suspended materials) and activated silica (as a flocculant aid).

Water containing these chemicals is sent to Block B where powdered activated carbon (7 to 10 mg/l for taste, odor and color causing micro-pollutants) and caustic soda (for neutralization of the sulfuric acid) are added. Chemically treated waters are sent through a Pulsator, which is a unique flocculator-clarifier, manufactured by Degrémont.

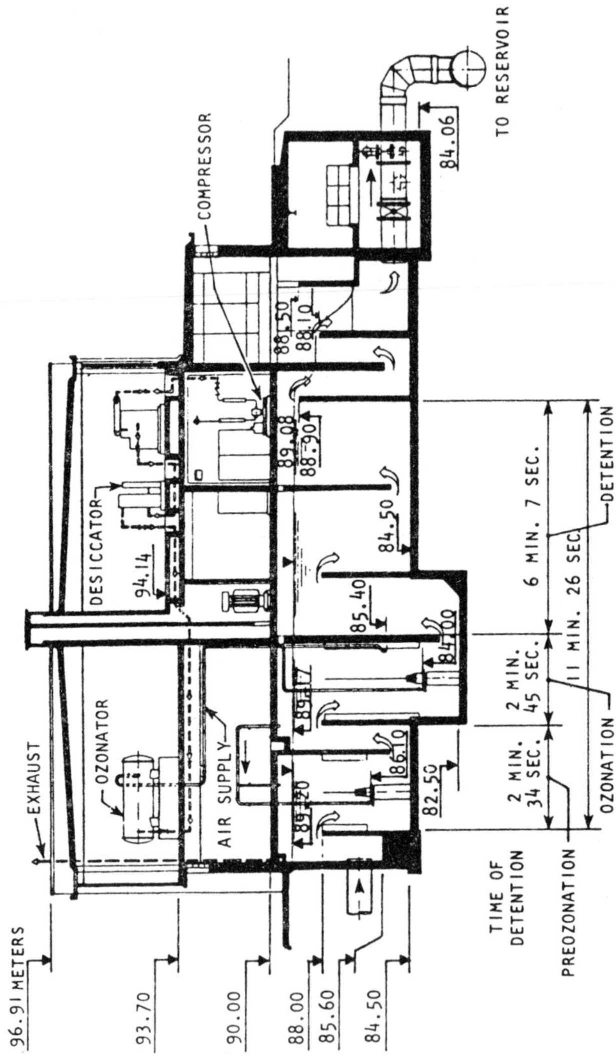

FIGURE B-1. Tailfer water treatment plant, Brussels, Belgium[45].

In the Pulsator, water flows upward through a sludge blanket in a cyclic, pulsating flow. The sludge blanket is maintained in suspension by periodic pulsations. Water passes vertically through this homogeneous blanket which provides efficient contact of chemicals, floc and water. As the size of the sludge blanket increases, excess sludge is drained off. The sludge flows to the center zone of the tank into a sludge concentrator, where it is gravity thickened and discharged at regular intervals by automatic, time controlled valves. The sludge blanket normally is on the order of 2.5 meters thick, although the total water depth may be 5 meters.

Pulsators also were observed at Annet-sur-Marne (France), the Dohne plant in Mülheim (Germany) and at Pierrefonds (Canada).

Following passage through the Pulsator, water then is sent to settling (for 3 hours) and sand filtration. Chemical additions form a 5 m thick sludge layer which expands and overflows into a central sludge tank. Water passes through this blanket over a 30 minute period. Waste sludges are press filtered to 40% dry weight, then 20% lime is added and the sludges are sent to landfill.

Ozonation is conducted in Block C, a separate building. Each water treatment train has two ozone contacting chambers, each chamber containing a submerged turbine. About 90% of the ozone generated is fed to the first chamber and 10% to the second. Off-gases from the second chamber are recycled to the first chamber. Total efficiency of ozone use at Tailfer is 99.5-99.7%.

Ozone dosages vary over the year from 1.5 mg/l in winter to 3.3 mg/l in summer. Contact time in the first turbine chamber is just over 2 minutes and is 2.5 minutes in the second chamber. Ozonized water then passes into a retention chamber where ozone residual is monitored manually at a point representing a total contact time of 6 minutes. Off-gases from this chamber containing small amounts of ozone are thermally destroyed before vent gases are discharged to the atmosphere.

Treated water then is sent to the plant reservoir, then to a pumping station 10 km from the plant where it is treated with 0.2 mg/l of chlorine to produce a chloramine residual of 0.2 mg/l before entering the distribution system to Brussels (the raw water contains 0.2 mg/l of ammonia).

Chlorine dioxide is prepared in a generator designed by Dr. Willy Masschelein, Deputy Director of the CIBE Laboratories. Dr. Masschelein has authored a monograph on Oxides of Chlorine and Sodium Chlorite which was publised in French in 1969. A revised and updated English version will be published in 1978 by Ann Arbor Science Publishers, Inc., Ann Arbor, Michigan, USA, with the title, Chlorine Dioxide.

This chlorine dioxide generator operates on the same general principle as the new Wallace & Tiernan chlorine dioxide generator, described in Section 11. Chlorine gas is added to water under pressure so as to obtain a minumum chlorine concentration of 2 g/l. This solution then is mixed with a solution of sodium chlorite (also under pressure) in a Raschig ring column. Stoichiometric quantities of both reagents are employed so as to obtain 98 to 99% conversion of sodium chlorite to chlorine dioxide. If the 2 g/l concentration of chlorine in water is not attained, then excess chlorine is required to insure conversion of chlorite.

After 1 minute of contacting time in the generator (time of passage through the reactor), the product solution is diluted for feeding to the raw water at the appropriate point in Block A.

Microbiological analyses are conducted every 8 hours on the product water for pseudomonas, total plate counts, actinomycetes and other coliform organisms. The average time for treated water to reach Brussels from Tailfer is 18 hours. During this time the residual 0.2 mg/l of chloramine falls to 0.05 mg/l. The city of Brussels analyzes water at 20 points in the distribution system weekly.

Currently Tailfer has eight Trailigaz water cooled, horizontal tube ozonizers installed, each capable of producing 4 kg/hr of ozone from air at 60 Hertz. More ozone generation capacity will be required for the fourth stage, but CIBE is not planning to install additional ozone generators. Instead, tests have been conducted using oxygen as the feed gas to increase the ozone output of each generator, and this approach is cost-effective at Tailfer. As a result, provisions have been made to install two 25,000 cu m liquid oxygen tanks. One tank was being installed at the time of the plant visit (May, 1977), and as soon as this has been completed, the plant plans to operate at least one water treatment train using ozone generated from oxygen. This treatment will continue for a full year to develop operational data, then a decision will be made to continue constant use of oxygen or to use oxygen only in the summer months when demands for ozone are highest.

Details of the study performed by CIBE of oxygen versus air for generating ozone are described in a recent publication by W. Masschelein, in collaboration with G. Fransolet, J. Genot and R. Goossens entitled, "Perspectives de l'Ozonation de l'Eau au Départ d'Air Enrichi en Oxygène", T.S.M. l'Eau 71(8-9):385-399 (1976). Dr. Masschelein also is studying the feasibility of installing high frequency equipment to allow operation of the Tailfer ozone generators at 600 Hertz, at which more ozone will be generated per unit time than by using the present 60 Hertz frequency.* The objective would be to allow Tailfer to generate ozone at 60 or 600 Hertz and using air or oxygen, without having to invest in additional ozone generators.

Maintenance of the Tailfer ozone generators is performed every six months and an accurate record is kept of the service life of each dielectric tube. Every two years of service, tubes are discarded whether they give evidence of wear or not. In this manner, unprogrammed ozone generator downtime is avoided.

---

* W.J. Masschelein, Private Communication to R.G. Rice & G.W. Miller, January, 1977.

The writers believe that this unusually high rate of tube replacement (15-20% of the tubes are replaced every 6 months) is a result of insufficient ozone generation capacity having been installed originally. This would cause plant personnel to generate ozone at its maximum production rate most of the time, which all ozone generator manufacturers advise will reduce service life of dielectric tubes.

CLAIRFONT WATER TREATMENT PLANT, TOULOUSE, FRANCE

Introduction

The Clairfont water treatment plant provides 110,000 cu m/day of water to the water system that serves the 400,000 people of Toulouse, France (Figure B-2). The plant provides full treatment to 100,000 cu m/day drawn from the Garonne River just downstream from the confluence of the Garonne and the Ariège Rivers, as well as ozonation and post chemical treatment to 10,000 cu m/day of groundwater pumped from the wells in the vicinity of the plant. The second Toulouse water plant, Pech-David, produces an additional 130,000 cu m/day by incorporating semi-rapid filtration in its treatment processes, but not ozonation. The Clairfont plant serves 180,000 Toulouse residents while Pech-David serves the remaining 220,000.

The Garonne River originates in the Pyrenees and has wide variations in flow, becoming torrential at times. A pulp and paper mill is situated some 90 km upstream of the Clairfont plant on the Ariège River. Also on the Ariège River and upstream of the Clairfont plant is an aluminum processing plant, which discharges some heavy metals into the river.

Water from the treatment plant is pumped to the reservoir and distribution system. The distance from Clairfont to the furthermost point in the distribution system is 15 km. The average residence time in the distribution system is 6 to 7 hours, with an average 8 to 12 hours between ozone contacting and the consumer's taps.

Water quality sampling points are located at various points throughout the distribution system. Local public health personnel sample two times each day at various points for bacteriological analyses. The Toulouse Public Health Authority is the municipal laboratory for physical and chemical analyses, while the regional laboratory performs bacteriological analyses.

The Clairfont plant was designed to be constructed in three phases.

- 1970 - 50,000 cu m/day from Garonne River
- 1975 - 50,000 cu m/day from Garonne River
  10,000 cu m/day of groundwater (for ozone disinfection and post chemical treatment)
- Future - 50,000 cu m/day from Garonne River.

The construction cost for the initial 50,000 cu m/day (13.16 mgd) was 13 million French francs ($2.71 million). The second phase cost 27 million

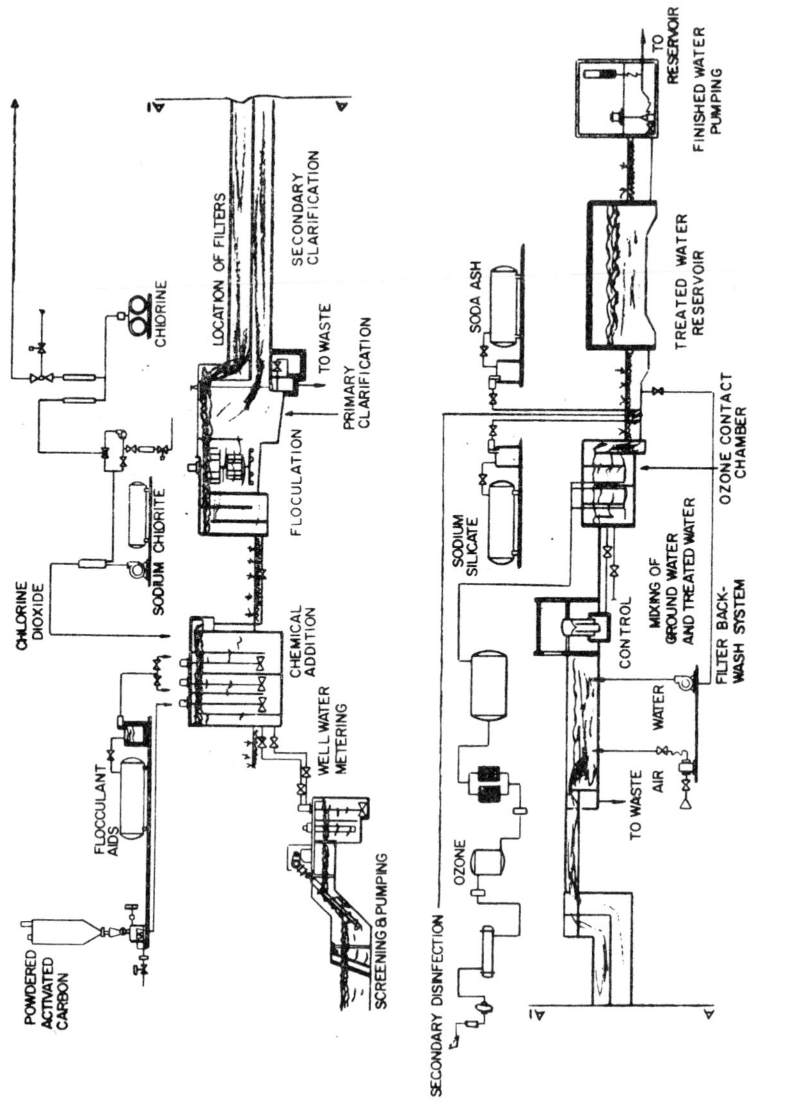

FIGURE B-2 WATER TREATMENT SCHEMATIC OF CLAIRFONT PLANT, TOULOUSE, FRANCE

French francs ($5.625 million*). The increase was partially due to inflation but also because all chemical handling and other basic facilities for the third phase were included in the second phase.

Cost of water treatment at Clairfont, exclusive of the distribution system is 0.216 French francs (Ff) per cubic meter of water (17.1¢/1000 gal*). Of this, 0.0112 Ff (0.9¢) is for ozonation and 0.0098 Ff (0.8¢) is for chlorine dioxide. The selling price of water is 1.35 Ff/cu m ($1.07/1,000 gal*), exclusive of sewage tax and other taxes. The total bill to the householder for water and sewage services, including taxes, is 2.27 Ff/cu m ($1.80/1000 gal*). Part of the applied taxes go to pollution control and financing of rural water distribution networks.

Intake

The plant raw water intake is situated approximately 1.5 km from the plant. While the intake was originally fed by means of three channels in the river, subsequent erosion of the river bottom has necessitated the provision of two 1-cu m/hr Archimedes screw pumps to raise the river water to the intake. The raw water passes through automatic backwashed travelling band screens to be pumped by means of two 2250-cu m/hr pumps through two concrete force mains (700 and 900 mm diameters) to an automatic flow equalization tank situated at the plant.

Raw water is monitored for pH, turbidity, chlorine demand, dissolved oxygen, conductivity, and temperature. Raw water turbidity is monitored ten times per day.

Pretreatment

Water flows from the equalization tank into two 50,000 cu m/day treatment trains and for each train a baffled chemical addition tank (contactor) is provided. A chemical coagulant (either aluminum sulfate or hydrolyzed aluminum chloride) and chlorine dioxide are added at this point. Provision has been made for addition of powdered activated carbon, however, the current water quality has made addition of this chemical unnecessary as yet.

A dosage of 1.5 mg/l chlorine dioxide is prepared from sodium chlorite solution in a CIFEC apparatus. Sodium chlorite solution is transferred from storage to an intermediate tank where it is diluted with water, and thence to the chlorine dioxide preparation unit. The tanks are lined with Ripanol, a German made coating. Chlorine is provided from 1000 liter pressurized containers. Chlorine dioxide is used in pretreatment rather than chlorine for three reasons:

- Taste and odor control
- Color removal

* Assuming 4.8 French francs/U.S. dollar, the prevailing exchange rate in early 1977.

- Considered to be a better predisinfectant

Chlorine dioxide contact time is 50 seconds in the contactor plus 2.5 minutes before clarification.

Clarification

Each 50,000 cu m/day water treatment stream is further divided into two 25,000 cu m/day streams for coagulation and flocculation. Vertical shaft flocculation is used. Each 25,000 cu m/day stream is further divided into three streams which enter the clarification units, each of which contains two compartments. The first compartment discharges to a two level second compartment which also serves as the bottom floor of the filters. Sludge is withdrawn from the first third of the secondary clarifier compartment by means of perforated sludge collection pipes. Sludge is discharged back to the Garonne River downstream of the plant.

Filtration

The clarified effluent proceeds to filtration units, three per 25,000 cu m/day stream or 62 sq m/100,000 cu m of plant flow. Each 75 sq m filter uses a depth of 1.2 m of sand to filter at a rate of 5 cu m/sq m/hr.

An air, then air/water, filter backwash system is used. A short air backwash cycle at a 4500 cu m/hr rate is followed by a ten minute air/water backwash at a 1000 cu m/hr rate followed by a ten minute water rinse at a 1500 cu m/hr rate. Ozonized water is used for backwashing. The interval between filter backwashes varies between one and three days.

The 100,000 cu m/day flow of filtered water is mixed with 10,000 cu m/day of groundwater pumped from adjacent wells.

Ozonation System

The ozonation system includes the following units:

- 4 fixed speed, positive displacement blowers
- 2 water cooled heat exchangers
- 2 refrigerant air coolers
- 2 activated alumina, desiccant drier, two cell units
- 4 horizontal tube dielectric, water cooled ozone generators, each containing 148 tubes
- 4 porous tube ozone diffuser contactors

Inlet air is filtered through paper cartridge filters which are replaced once per year at the same time the ozone generator tubes are cleaned. Discharge pressure from the blowers is 0.8 to 0.9 bar. Total head loss through the ozonation system, including the contactor porous diffusers, is 0.3 bar.

The air temperature entering the water cooled heat exchanger is 70° to 80°C and exiting is 23 to 25°C with a cooling water temperature of 20°C. The air leaving the refrigerant air cooler is at 4 to 6°C. Desiccant unit regeneration cycles are timer controlled and the normal regeneration cycle is 8 to 12 hours, based on normal air flow. If air flow is reduced, time between regeneration cycles is extended to conserve energy.

The air from the desiccant drier and to the ozone generators is monitored by means of a dew point monitor and is controlled at a dew point of minus 60°C. The calibration of the dew point monitor is checked every two months but adjustment has not been necessary for the past two years.

The four Trailigaz ozone generators each are capable of producing 2.25 kg/hr of ozone. These are the only Trailigaz ozone generators in France which have individually fused dielectric tubes. It is not a normal Trailigaz practice to provide individually fused dielectrics. Mr. Michael Croises, Principal Engineer of the Toulouse Water Authority, favors the provision of individual fuses from an operational point of view. He feels that it is better to have a fuse costing approximately 50 French francs ($10) fail than a dielectric tube costing approximately 350 French francs ($70).

Failure of an unfused dielectric in an ozone generator results in the shut-down of the entire ozone generator. If the tube is fused and fails, the ozone generator continues to operate but at reduced capacity (reduced by the capacity of that tube--or 1/148th in this case). Mr. Croises stated that the Clairfont facility replaces many fuses which must withstand 33 kv peak voltage, but replaces very few tubes. Sometimes the plant runs short of fuses. The plant uses approximately 50 fuses per year but replaces only four or five tubes per year, generally during the annual tube cleaning period. Trailigaz claims that tubes fail in service only every two to three years.

Ozonation system maintenance is performed annually by plant operational staff but Trailigaz would provide the same service under contract if called upon. The annual maintenance is performed in June, usually a dry period, but when water consumption is not yet at its peak. The ozonation system is shut down for three days and chlorine is used for disinfection. Although users are advised when chlorination is to be substituted for ozonation (including sensitive industries such as electronic element manufacturers who shut down operations during maintenance of the ozone generators), there are numerous complaints regarding the taste of the chlorinated water.

All piping of the air preparation system is galvanized steel while all piping in contact with ozone is stainless steel. Compressor belts are changed every three months.

Four ozone contactors provide eight minute contact time with a sidewater depth of 4.8 meters. A maximum dosage of 1.8 mg/l ozone is used to provide an almost constant ozone residual of 0.4 to 0.7 mg/l. No residual ozone monitor is used at Clairfont (controlled manually). Each contactor consists of four ozonation compartments. The flow pattern in the ozonation compartments is as follows: countercurrent, cocurrent, countercurrent, cocurrent. Ozone is injected into the contactor through porous diffusers at the bottom of the contactor. Contactor off-gases are destroyed by injection into the exhaust stacks of diesel generators that are used to supply power to the plants.

Post Treatment

Two treatment steps follow ozonation: continuous pH adjustment with soda and intermittent addition of silicate to control corrosion in the distribution system. Silicate addition at a dosage of 1.0 mg/l is practiced for a period of one to two days per month.

ROUEN-LA-CHAPELLE, FRANCE

This plant is situated on the Seine River, about 120 km west (and downstream) of Paris. All water utilities along the Seine below the Paris sewage plant are forbidden to draw water directly from the river for processing into drinking water. Instead, water is drawn from deep wells located near the river banks. This practice is similar, but not equivalent, to river sand bank filtration, prevalent along the lower Rhine River in Germany (see Düsseldorf plant description).

Until 1976, the Chapelle plant at Rouen merely drew groundwater and treated it with chlorine dioxide before sending it to the distribution system. However, increases in pollution, particularly in ammonia content (which has risen to 2 to 3 mg/l today) has forced adoption of more sophisticated treatment. Phenols and detergents also are present in the Rouen well waters.

After pilot plant testing for three years (1968 to 1971), the following treatment scheme was designed, installed, and began operating in January of 1976 (Figure B-3):

- Well water is pre-ozonized for manganese oxidation and for oxidation of micropollutants (dissolved organic materials). This step renders the dissolved organic materials more readily biodegradable and also increases the dissolved oxygen content of the water.

- Preozonized water then is filtered through a 100 cm deep sand bed (to remove the insoluble manganese and some oxidized and flocculated organics) then through a 75 cm deep bed of granular activated carbon which has been rendered biologically active by the preozonation step. Biogrowth also is present in the sand filters, and some of the ammonia is removed here. The balance is removed in the GAC filters, along with some of the organic materials.

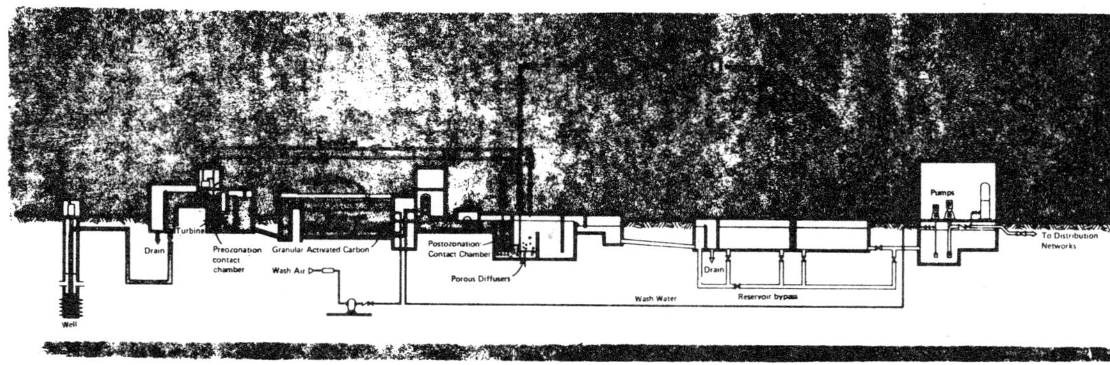

Figure B-3. Rouen-la-Chapelle, Rouen France, plant treatment scheme.

- Post-ozonation follows filtration for disinfection and to destroy any remaining organic micropollution (taste and color bodies).
- Post-chlorination with 0.4 to 0.5 mg/l chlorine prior to distribution.

The pre-ozonation contactors are high speed turbines because porous tubes would clog in the presence of precipitating oxides of manganese and iron. Contact time in the pre-ozonation chamber is short, 3 minutes, and the source of some of the ozone for this step is the off-gases from the disinfection ozonation contactors. This second use of ozone contactor off-gases insures 98% utilization of the generated ozone at Rouen.

The post-ozonation contactors consist of 2 columns, each 4.5 meters deep, containing porous diffusers. Contact time in these diffuser chambers is 12 minutes, after which a residual ozone concentration of 0.4 mg/l is monitored to assure bacterial disinfection and viral inactivation.

Two-thirds of the total amount of ozone generated is sent to post-ozonation, and one-third to pre-ozonation. Total dosage of ozone produced is 2.1 mg/l, so that the pre-ozonation dosage is 0.7 mg/l and the post-ozonation dosage is 1.4 mg/l.

There are six sand filters which operate at 5 m/hour. There are also three ozone generators at the plant, and each generator services the water flow to two sand filters. One sand filter operates while the second is being backwashed. Backwashing is conducted with air and water.

Carbon filters are backwashed once each month. Through January, 1978, after 24 months of service, the activated carbon columns have shown no tendency toward breakthrough and have not yet been thermally regenerated (C. Gomella, SETUDE, Paris. Private Communication). The expected operating time between carbon regeneration is at least two years. Analyses are made quarterly before and after passage through the GAC columns for substances extractable with chloroform and also for substances extractable with cyclohexane to determine the need for regeneration.

The Rouen distribution system is 430 km long, and the most distant water tap is 15 km from the plant. Trihalomethane analyses were conducted once, but none were found.

Four people operate this 50,000 cu m/day plant during the day, and four more people work in the distribution system. Analyses are carried out in a central laboratory in Rouen, and not at the plant. Operation of the plant is paced by the level of water in the plant reservoir. When this level drops, the plant starts operations automatically.

There are three ozone generators at the plant, capable of being operated at 32 kw, but normal operation is at 14 kw. If one of the ozone generators were to shut down, its capacity could be made up by increasing the power to the other two.

In Section 10, Public Health Aspects of Ozone Usage, a table of data is presented which were taken at Rouen two months after startup of the new treatment process. These data show the degrees of removal of a variety of organic materials by the new process at Rouen. In Section 13, Biological Activated Carbon is presented a table of plant operational data for this plant. More than 80% removal of pollutants was obtained by this process during its first year of operation[181].

Rouen-la-Chapelle is the first French plant known to be using Biological Activated Carbon (pre-ozonation of granular activated carbon) without having breakpoint chlorination in the initial stages of treatment. The plant at Morsang-sur-Seine has Biological Activated Carbon after initial breakpoint chlorination, but is piloting BAC without initial chlorination.

DÜSSELDORF, GERMANY

The City of Düsseldorf currently processes water at three treatment plants, Holthausen, Flehe and Am Staad. All three plants are situated along the Rhine River and all three plants draw water from wells located along the river banks. The process of allowing the river water to filter through the sand banks into these wells is called "river sand bank filtration". Passage of river water through these sand banks takes about 20 days. During this time some 60 to 75% of the dissolved organic material is removed. In exchange, however, considerable amounts of manganese and iron salts dissolve into the water. Some portion of the water drawn from these wells (about 25 to 35%) is groundwater flowing in the opposite direction, that is, into the Rhine. River bank filtration wells at Am Staad are over 100 years old.

During the late 1950s, Düsseldorf became the first European city to adopt a water treatment process consisting of ozonation followed by granular activated carbon. In the language of ozone historians, the process has come to be known as "the Düsseldorf process", because the first description of these installations was published under that title. The process was adopted, installed and became operational in the early 1960s.

Ozonation was installed specifically to oxidize the iron and manganese salts and the granular activated carbon was installed to remove dissolved organic compounds which cause taste and odor problems. However, over the intervening years, pollution of the Rhine has increased and the combination of ozone followed by activated carbon has become better understood. When first installed, only 1 g/cu m of ozone dosage was required and only 1 g of carbon/cu m of water treated were required. Today, 3 to 3.6 g/cu m of ozone must be added and 35 g of activated carbon/cu m of water are required to attain the same water quality that was attained in the early 1960s. Because of the large increase in the amount of granular activated carbon required in the Düsseldorf plants today, the city has installed carbon regeneration facilities at the Holthausen plant site. As soon as this facility is operational and has been debugged, Düsseldorf plans to construct carbon regeneration facilities at the two other plants.

During the years following introduction of the ozone/granular activated carbon system in Düsseldorf, treatment plant scientists have noted that the amount of organics removal was more than could be expected simply by summing the known removals of the ozonation and GAC processes. In addition, ammonia is removed by the activated carbon columns, which was unexpected on the basis of adsorption. These observations led to studies of the mechanisms occurring in the activated carbon columns by bacterial action. These studies, conducted at the Bremen Water Works, are described in Section 13, "Biological Activated Carbon". For maximum biological benefit, filtration rates through the carbon columns today are about one-third of the designed rate (10 m/hr today versus 25 to 30 m/hr).

The ozone/activated carbon process has been practiced at the three Düsseldorf plants for 15 to 17 years. The older Am Staad plant was redesigned and rebuilt in the early 1970s, and is the latest operating version of "the Düsseldorf process" of water treatment. It is laid out in a circle with the plant reservoir beneath the plant.

All three Düsseldorf plants treat river sand bank filtered Rhine water by the identical process (Figure B-4). When water enters the plant it is ozonized as the first step. This is accomplished by injecting all of the required ozone into 1% of the treated water. As soon as the ozone has been mixed with 1% of treated water, the ozonized water is mixed with the raw water at the top of a vertical 10 meter long pipe, one-half foot in diameter. This pipe leads to within 1 foot of the bottom of the contact chamber. Details of this type of injector contacting are given in Section 8, Engineering Aspects of Ozonation.

Off-gases from the ozone contactors are sent to wet granular activated carbon beds for destruction. However, each plant has tested catalytic ozone destruction devices and was planning to install these shortly after the site visits were made (May 1977).

The original ozone generators at Düsseldorf were Otto plate type, water cooled generators. These were large (for their ozone generation capacity) and produced only 1 kg/hour per unit. The newer generators are Herrmann tube type, water cooled ozone generators, each capable of generating 1.3 kg/hour. These were installed in 1964.

Many of the alumina glass dielectrics originally installed in 1964 still are operating in the Herrmann generators, which are cleaned every two years. Düsseldorf personnel consider that if one dielectric tube is lost per generator during cleaning, they are disappointed and consider this loss to be high. New tubes cost about $40.

After ozonation, the water flows to a holding tank, where it remains for a period of 20 to 30 minutes. During this time oxidized iron and manganese hydroxides coagulate and begin to precipitate. Residual ozone concentrations also become lower, as the dissolved ozone continues to oxidize dissolved organic material. Finally, any permanganate (which may have been formed during ozonation) also oxidizes dissolved organic compounds, and itself is reduced back to the insoluble manganic state.

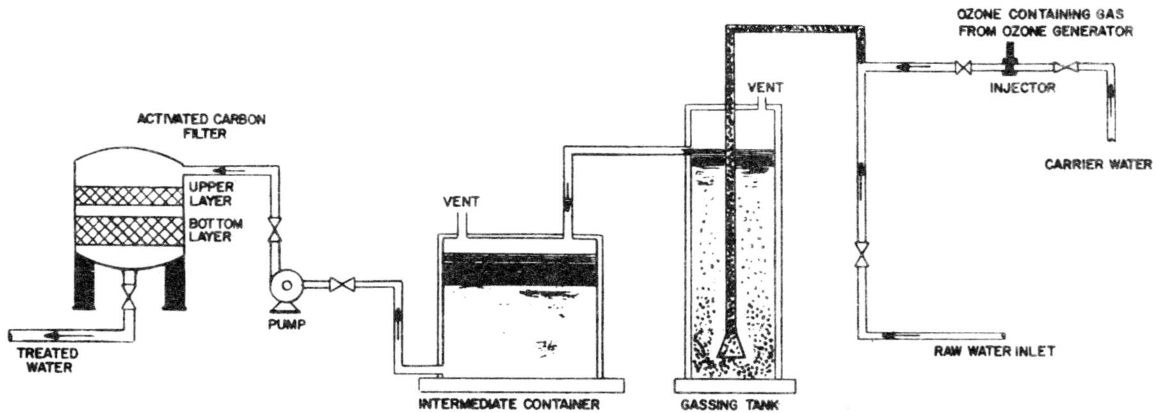

FIGURE B-4: SCHEMATIC OF WATER TREATMENT PLANTS IN DÜSSELDORF, GERMANY (HOLTHAUSEN, FLEHE AND AM STAAD)

After the ozonized water has stood 20 to 30 minutes it is tested manually for ozone by sniffing and for permanganate by observing a slight pink color. If neither ozone nor permanganate can be detected by these methods, then the dosage of ozone is raised.

Water then is filtered through pre-activated carbon (which is a special grade of anthracite coal), to remove iron and manganese precipitates, plus any coagulated organics. This filtered water then is passed through granular activated carbon columns, which are 5 meters high and 2.5 meters in diameter. Filtration rates today are 10 m/hr, versus 25 to 30 m/hr when first installed. The lower filtration rates are optimum for the biological activity contained in the Düsseldorf columns, which remove the 0.2 to 1.2 mg/l of ammonia in the river bank filtered Rhine River water. This raw water also contains 0.3 to 1.2 mg/l manganese and 0.05 to 0.1 mg/l iron.

After carbon contacting, the water is treated with 0.1 mg/l of chlorine dioxide solution (prepared by adding excess chlorine to sodium chlorite solution). This provides a residual in the distribution systems of 0.05 to 0.10 mg/l.

Granular activated carbon columns currently are backwashed every 4 to 6 weeks and are regenerated every 5 to 6 months. When first installed, the Düsseldorf activated carbon columns required reactivation every two years. However the increase in concentration of chlorinated organics in the Rhine River, which are not removed by river sand bank filtration and are not readily degraded biologically in the activated carbon filters, causes the more frequent regeneration. Carbon in the columns is analyzed periodically for total organic chlorine (TOCl) at the top and 6 inches from the bottom, and before this parameter breaks through the column, the carbon is regenerated. If there were not high concentrations of chlorinated organics in the Rhine River water, the need for carbon regeneration at Düsseldorf would not be as frequent as it is. Only 80% of the carbon in a column is removed for regeneration, so as to maintain high bacterial levels in the columns, particularly of the nitrifying bacteria.

Düsseldorf currently has very little reservoir capacity in which to store treated water. Therefore, treated water is produced on demand only. This situation causes wide variations in water flows through the plants during the day. These, in turn, caused changes in air pressures in the older, plate-type ozone generators, which sometimes resulted in cracking of the plate glass dielectrics. Additionally, flow rates of water through the granular activated carbon columns are highly variable, and sometimes cause the bacterial activity within these columns to function non-uniformly.

The City of Düsseldorf has a rule concerning the handling of ozone in water treatment plants. In all rooms through which piping passes which carries ozone, exhaust fans capable of providing 10 air changes per hour must be present. These do not have to operate constantly, only when there is an ozone leak in a pipe. This has not happened very often in the Düsseldorf plants.

Additionally, in each room in which ozone is generated or handled, there must be an ambient atmosphere analyzer to monitor ozone. If the concentration of ozone in plant air reaches the Maximum Allowable Konzentration (MAK) of 0.1 mg/l (by volume), an alarm must sound to warn of faulty operation.

DUISBURG WASSERWERK III WITTLAER PLANT, GERMANY

The Wasserwerk III Wittlaer of the Municipal Public Works of Duisburg, Federal Republic of Germany is the only known municipal water treatment plant which makes ozone from oxygen. It draws raw water from wells situated along the Rhine River through the process of river sand bank filtration. The capacity of the treatment plant was originally 2700 cubic meters per hour (cu m/hr), but in view of decreasing raw water quality, the current peak capacity is now 2000 cu m/hr. The current average flow rate through the treatment plant is 1500 cu m/hr. The Duisburg analytical laboratory takes samples of the process and product drinking water and the quality of the raw water is tabulated as follows:

- Manganese 1.2 to 1.5 mg/l
- Iron 0.3 to 0.6 mg/l
- Dissolved Oxygen 0.5 to 1.0 mg/l
- Ammonia 0.5 to 1.4 mg/l

The Wittlaer plant was constructed in 1964 and began operation in 1965. It is one of three water treatment plants that serve the 370,000 inhabitants of the Duisburg area through a distribution system about 720 km in length. River sand bank filtered water from the Rhine is pumped through the water treatment plant to the distribution system at a pressure of 6 bars from one horizontal and 18 vertical wells. Figure B-5 shows a schematic of the process. The raw water is pumped into the ozone contact chamber where it mixes with an ozone-rich sidestream prepared from 10% of the finished water volume having an ozone concentration of 10 to 25 grams of ozone per cubic meter of water. (The overall ozone dosage is 1.0 to 3.0 mg/l ozone.) The raw water and ozone-rich sidestream mix and flow cocurrently upward through the Raschig rings in the contactor. A contact time of 8 to 10 minutes is provided in the stainless steel chamber with a residual of 0.2 to 0.5 mg/l ozone. An additional five minute contact time is provided in the piping ahead of the pressure filters and pressure granular activated carbon (GAC) columns.

There are sixteen, 3 meter diameter, two-stage pressure vessels in which the functions of a sand filter and a GAC reactor are combined. Figure B-5 illustrates a typical unit. Each unit provides a filter area of seven square meters (sq m). Four new pressure filter/GAC reactors, each having an area of 25 sq m, are under construction. The pressure filter consists of one meter of porous silicate (heat treated pumice) above 30

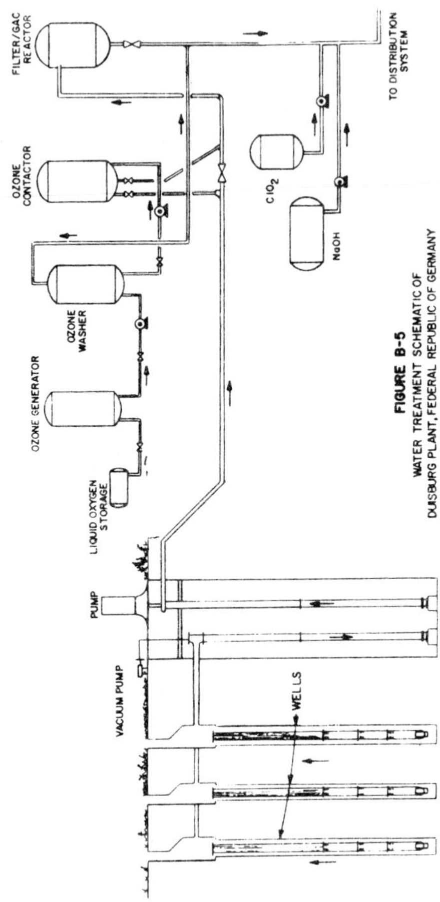

FIGURE B-5
WATER TREATMENT SCHEMATIC OF
DUISBURG PLANT, FEDERAL REPUBLIC OF GERMANY

centimeters of quartz sand. A two meter deep GAC reactor is situated beneath the filters.

Raw water is ozonated so that organics, manganese and iron are oxidized. Flocculated oxidation products are captured in the pressure filters. An aerobic biomass maintained in the GAC bed removes soluble organics and ammonia. The final treatment steps are the addition of chlorine dioxide (ClO_2) to maintain a disinfectant level within the distribution system and the addition of sodium hydroxide (NaOH) for pH control. The ozonation system is unique among the municipal water treatment plants that were inspected in that liquid oxygen (LOX) is employed as the feed-gas for ozone generation. LOX apparently was installed to allow a single pump per treatment train to be used. This is possible because use of LOX requires a closed system in order to recycle oxygen-rich ozone contactor off-gases.

Two suppliers deliver LOX to two separate storage tanks. Figure B-6 illustrates the ozonation system schematic. Each storage tank feeds the ozonation system through an evaporator. The gas emerges from the evaporator at a pressure of 0.5 bar where it meets the recycled contactor off-gas. The mixture of makeup oxygen and recycled gas is fed to the four water-cooled, DEMAG tube-type ozone generators. The ozone-rich feed gas is further pressurized to 1.3 bar pressure by means of a water-sealed compressor and then is injected into the water at the bottom of the ozone washer/contactor at ozone concentrations ranging from 30 to 60 g/cu m. The ozone-rich gas passes upward, countercurrently to the downward flow of a sidestream of finished water in the Raschig ring washer/contactor. The plant has two equal sized process trains.

The contactor off-gases (still containing 10 g/cu m ozone) are passed first through a refrigerant drier and then a silica gel desiccant drier. The dried gas is at an equilibrium concentration and consists of approximately 40% nitrogen and 60% oxygen. The dried and cooled off-gas is recirculated to the point at which high purity oxygen is added as make-up gas. Approximately 5 g of ozone/cu m is lost (by decomposition back to oxygen) in the drier.

The milk-white, ozone-rich sidestream from the washer/contactors is pumped to the ozone contactor at a concentration of 10 to 25 grams of ozone per cubic meter of water. Raw water and the ozone-rich sidestream mix and flow cocurrently upward through Raschig rings in the main contactor providing a contact time of 8 to 10 minutes.

Ozone Generators

Each of the four DEMAG ozone generators contains 206 individually fused dielectric tubes. The ozone generators are installed so that the tubes are in a vertical position, although DEMAG ozone generators are more commonly installed so that the tubes are horizontal. The tubes are 1.5 inches in diameter and approximately 1 meter in length. A steel wool-like electrode is used in place of the metallic tube coating used by other suppliers. The units operate at 200 Hertz (Hz) and 6000 to 8000 volts.

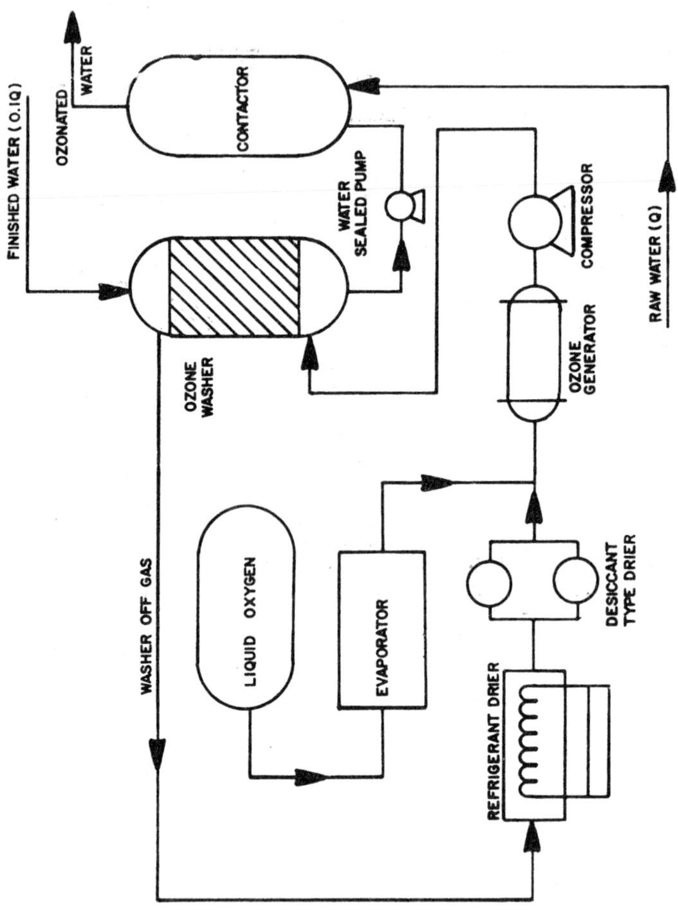

FIGURE B-6 OZONATION SYSTEM SCHEMATIC OF DUISBURG WATER TREATMENT PLANT

Dr. Uhlig, the plant chemist, described the recycle gas treatment system as old and "not the best". The dew point of the feed gas to the ozone generators ranges from minus 35 to minus 45°C. There are problems with the formation of nitric acid as the dew point rises and approaches minus 35°C. Therefore, operating personnel maintain the dew point in the minus 40° to minus 45°C range. Duisburg plant personnel normally wait until 10 or 20 dielectric tubes fail before scheduling maintenance for an ozone generator. The tube failure rate at Duisburg is approximately ten tubes per ozone generator every 6 to 12 months. At the time of the site visit, two of the four ozone generators were out of service for maintenance. One hundred extra dielectric tubes are stored at the plant.

Ozone generator cooling water is circulated in a closed-loop system using phosphate-conditioned finished water. The cooling loop is cooled, in turn, by means of a heat exchanger in the raw water supply. Cooling water temperature rises from 10° to 12°C at the ozone generator inlet to 25° to 30° at the outlet.

Originally, there were corrosion problems with the V2A stainless steel (believed to be the equivalent to U.S. 304 or 304L), particularly at weld points near the ozone generators. No corrosion problems have been experienced since the V2A stainless steel has been replaced by V4A (believed to be the equivalent to U.S. 316 or 316L).

The ozone washer is backwashed regularly, while sludge is removed once a year. The Raschig rings are removed once every two years for cleaning and inspection to determine whether or not the rings should be discarded.

Ozone in the gas phase is monitored by a Bran & Lübbe monitor. The unit was purchased originally to monitor ozone in water, but it proved to be unreliable for that purpose. All ozone residuals in water are now measured manually. The operations staff is convinced that continuous ozone monitoring in the gas phase is practical with any of the numerous instruments available. However, they have no confidence in instruments designed to measure residual ozone in water.

Ambient ozone levels in the ozone generator room are not monitored. However, there are several operating exhaust fans in the room. Operators are instructed not to enter rooms whenever they can smell ozone. They are to notify the plant manager and then use Dräger tubes to check the ozone level. If the ozone level is 0.1 mg/l or lower, the room may be occupied. The operating personnel have never had to use the self-contained breathing apparatuses which are available.

Chlorine dioxide is made by an automatically controlled Wallace & Tiernan Chlorinator system located in a separate building. Two chlorine dioxide reactors are installed. One is in operation, and the other is a standby unit. A third unit will be installed as part of the on-going plant expansion. A strong chlorine odor was noticeable during visitation in the chlorine dioxide reactor room in which the temperature was approximately 25°C.

Chlorine dioxide is produced by mixing an aqueous chlorine solution with a sodium chlorite solution. Each solution is fed from individual "day" tanks which are, in turn, fed by a level-controlled supply from larger storage tanks. Stoichiometrically, 300 grams of chlorine solution per hour should react with 150 grams of sodium chlorite solution per hour to provide the proper chlorine dioxide dose at the lowest water treatment plant flow rate. However, Duisburg plant personnel actually add 4 times the amount of chlorine required by stoichiometry. Chlorine dioxide (ClO_2) dosage is controlled by a continuous chlorine monitor which normally reads 0.20 to 0.24 mg/l. Fifteen minutes after the ClO_2 is added, a chlorine dioxide residual of 0.1 mg/l is measured and the water enters the distribution system (the city is 20 kilometers away). Water pressure in the ClO_2 system is monitored from the control room. If pressure drops below a minimum level, the ClO_2 system automatically shuts down.

After chlorine dioxide is added, sodium hydroxide is added to control pH. Chemical dosage is metered and monitored in the main control room. Two Kalkas pH monitors (based on the principle of Dr. Gunther Axt) are housed in another room.

It is not clear whether the ozone dosage was sufficient to provide oxygen required to maintain aerobic conditions in the GAC columns. However, an additional 10 to 12 mg/l of oxygen apparently is added ahead of the two stage filters, resulting in a dissolved oxygen level of 6 to 8 mg/l in the finished water.

ZÜRICH (LENGG PLANT), SWITZERLAND

Introduction & Overview

The water system for the city of Zürich began in the 15th century with the construction of a cistern system fed by springs in the vicinity of the city. In 1430, the city's first fountain with running water was constructed, fed from the Limmat River.

Between the 15th and 19th centuries improvements to the water system consisted principally of expansion to the well system. In 1868, construction began on a pressurized water system for the entire city, drawing Lake of Zürich water through slow filters in the Limmat River and pumping it to elevated reservoirs. Following a typhus epidemic in 1884, a filtration plant directly drawing water from the Lake of Zürich was constructed. With growing population and pollution of the lake, service was expanded and improved to the present system, shown in Figure B-7, of four main waterworks on a circumferential transmission system.

The Lake of Zürich is a moderate sized catchment of 4 cu km capacity and an average discharge rate of 100 cu m/sec. It is highly significant as a reservoir to communities along its shores, the largest of which is the City of Zürich, drawing 75% of its water supply from the lake. Because of the discharge of domestic and industrial wastewater and runoff, the lake has become mildly eutrophic. A policy of requiring nutrient removal on all wastewater plants has reduced this problem somewhat in recent years. Since 1972, there has also been an intensive program of water quality monitoring in the lake. This, supported by extensive earlier efforts, shows that nitrates and phosphates in the late peaked out in the late 1960s and declined slightly after 1970 from peaks of 0.36 mg/l phosphates and 3.6 mg/l nitrates (annual mean at 30 m depth). Chlorides have risen steadily over the period of record from less than 1 mg/l in 1945, to an average of 3 mg/l in 1973.

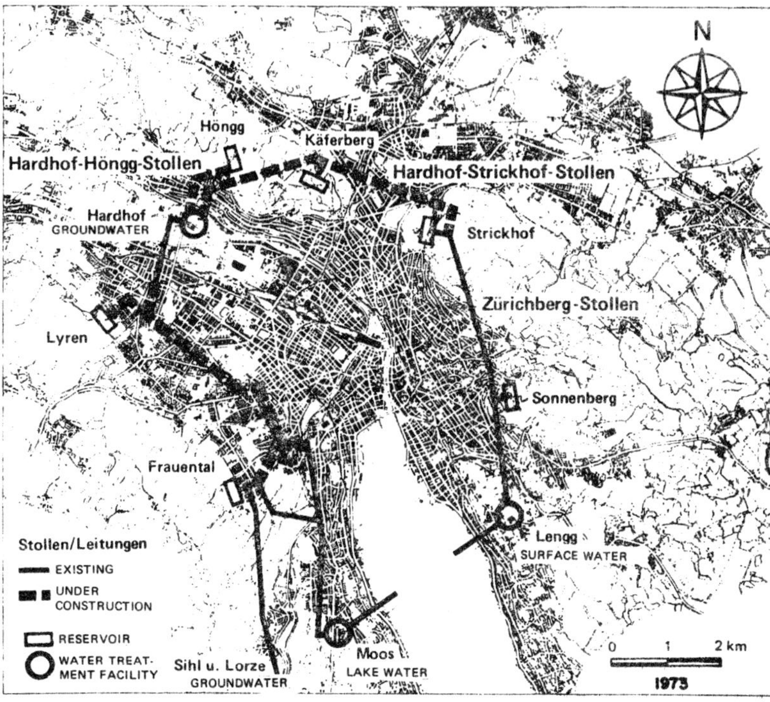

FIGURE B-7 OVERALL MAP OF WATER TREATMENT AND DISTRIBUTION SYSTEM FOR ZÜRICH, SWITZERLAND.

Plant Description

The Lengg plant is one of two plants in the system which draws its supply from the Lake of Zürich. The plant was constructed between 1956 and 1960 with a capacity of 160,000 cu m/day and was expanded from 1972 to 1975 to a capacity of 250,000 cu m/day. The plant was initially constructed to include prechlorination at the intake for destruction of nuisance organisms (molluscs) followed by pumping to the plant site, 60 meters above the lake on the north shore. In the initial plant, rapid and slow sand filtration was followed by final chlorination and distribution.

The expanded Lengg plant process includes improved filtration plus ozonation, granular activated carbon and final chlorination with chlorine dioxide. The plant is constructed with two parallel treatment streams, each with 1200 cu m/hr capacity. Each train has 10 rapid sand filters, 3 ozonators and 3 turbine contactors, 6 granular activated carbon filters, 7 slow sand filters, and one chlorine dioxide production and dosing assembly. The basic elements of the process include the following:

- Intake Prechlorination
- Microflocculation
- Rapid Sand Filtration
- pH Adjustment
- Ozonation
- Granular Activated Carbon
- Slow Sand Filtration
- Chlorine Dioxide Dosing

These are discussed in turn below, with emphasis on the ozonation, granular activated carbon and chlorine dioxide processes.

Intake Prechlorination--

Prechlorination is required to destroy organisms which are native to the lake and which can collect in the pumping and treatment works. Principal among these are small mussels, phytoplankton and zooplankton. Depending on conditions, dosage is normally 1.0 mg/l during the summer and 0.5 mg/l during the winter, added at the intake line. Following this, the raw water is pumped to the water works.

Micro Flocculation--

Chemical micro-flocculation is employed to improve the efficiency of the rapid sand filters. The water quality of the lake of Zürich is quite high, thus coagulation and sedimentation are not required. Rather, 2 mg/l of aluminum sulfate (alum) are added to produce pinpoint flocs which are readily removed on the rapid sand filters. This is supplemented with a polymer added at the rate of 0.2 mg/l. The system utilizes powdered alum which is first mixed in a dissolving tank with water prior to being dosed to the raw water. Powdered activated carbon also can be added before the rapid sand filters, if required.

Rapid Sand Filters--

Twenty rapid sand filters, each 15 m long by 3 m wide by 4 m deep comprise the first filtration step in the Lengg plant. The filters are the dual media type, with 1 meter of quartz sand (0.4 to 0.8 micron particle size) covered by 70 cm of pumice (1 to 3 mm particle size). The normal filtration rate is 20 m/hour and the filters usually require backwashing every 4 days. Clean reservoir water is used for a combined water/air backwash at a rate under 70 m/hour. The rinse water, approximately 2% of total production, itself is filtered and returned to the plant headworks.

Filter washwater from rapid sand and activated carbon filters is filtered in two basins 15 m long, 3 m wide and 6 m deep. A 50 cm layer of quartz gravel, grain size 1 to 2 mm, serves as the filter medium. Filtered backwash water is returned to the plant headworks. The washwater filters are operated at a 7 m/hour rate and backwashed with raw water. Backwash from these filters is sent to the Zürich sewer system.

pH Adjustment--

pH Adjustment is necessary to remove carbon dioxide from the lake water. Lime is used for this purpose and it is added in sufficient quantity to raise the pH of the water to 8.2. Initial lakewater pH ranges between 7.6 and 7.9. As with its alum system, dry lime is mixed with water prior to dosing into the water. The lime is added into the outlet channel of the rapid sand filters.

Ozonation--

Ozonation is used for virus and bacterial control, and oxidation of organic compounds in the filtered water. The Lengg plant has two banks of Kerag ozonators, with three sub-units in each bank. A common cast iron housing surrounds each bank allowing for common cooling of each group of three units. Table B-1 gives known and observed data for the ozonation systems at the Lengg plant. There are two ozone contactors, each having three chambers. Each ozonator feeds a contactor turbine located in series, one to each chamber (Figure B-8). Detention time in the chamber averages 30 minutes at low flow in the daytime and 10 minutes at night. Total volume is 765 cu m divided into three chambers in series of equal volume. There are no internal baffles in the chambers.

Excess ozone from the process is vented at the final chamber, broken down in a thermal destruction system at 250°C and discharged through a tall stack to the outside air.

Maintenance on the ozone process is handled by the manufacturer who services the units once per year, under contract. At that time, 10 to 20 tubes per unit generally require replacing. The cost for this maintenance service ranges from 5000 to 10,000 Swiss francs per year ($2084 to $4100 U.S. dollars -- figured at an exchange rate of 2.4 Sf/$ in early 1977) plus materials, which includes full servicing on the air preparation equipment, the ozone generators, and the ozone contacting equipment. Reported capital costs for the ozonation equipment and contactors were 0.6 cents/cubic meter of capacity. Operating costs as of 1977 are as follows:

Dosage	Cost*
Ozone (mg/l)	¢/cu m of treated water
1.5	40
5 (4 to 6 range)	70

* Figured at an exchange rate of 2.4 Sf/$.

TABLE B-1. LENGG PLANT OZONATION SYSTEM

Manufacturer:	KERAG, CH-8805, Richterswil, Switzerland Telephone 01/784-4451, Telex KES075-498
Type:	Vertical tube ozonators, three sub-units per common water cooled bank, 220 tubes per sub-unit, 660 tubes/-bank
Voltage/Frequency:	275 v input/350 hz (observed) 380 v max/350 hz (maximum)
Voltage on tubes:	6,000 v (observed) 7,200 v (maximum)
Current:	61 amps @ 275 v (observed)
Cooling water:	5 to 6°C product water, which is used once-through, then returned to the lake.
Air Pretreatment:	Air is filtered, dried in two parallel air pretreatment trains in silica gel desiccators (730 kg each), then cooled to 10°C. Dewpoint is not measured. Silica gel is heat regenerated every 2 days.
Ozone Contactor: (See Figure B-8)	Ozonized air is rapid mixed with water in turbines and passed through a 3-section contactor (turbine in each section) for each bank. Minimum detention time is 6 minutes at full design load, 30 minutes at normal low flow (daytime), and 10 minutes at present peak flow (night).
Ozone Dosage:	Normally 1.5 mg/l, but can go to 4.5 mg/l if needed.
Materials of Construction	Ozonator internal parts - glass & stainless steel Ozone cooling vessel - coated cast iron Ozone transmission - PVC line Ozone contactor - stainless steel

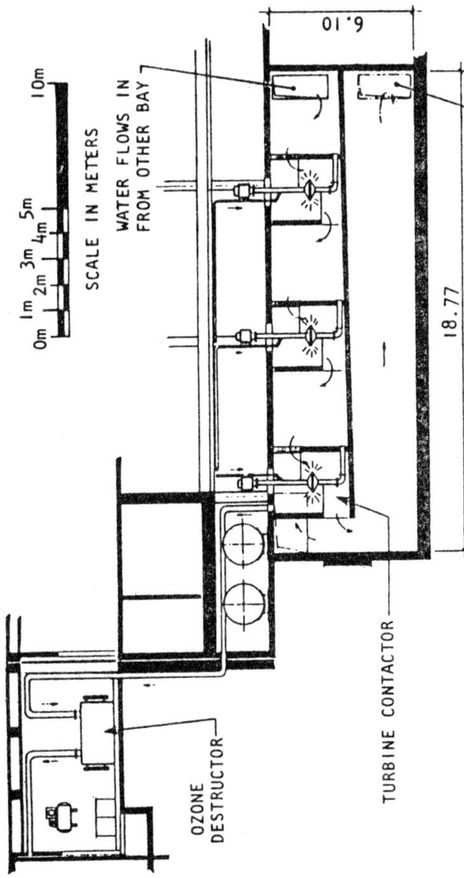

FIGURE B-8. Multi-turbine contactor installation (Lengg Plant, Zürich)[30].

The layout of the ozonator equipment at the Zürich plant is highly compact. Piping and appurtenances in the air drying system are of coated welded steel. Ozone-containing air lines are of PVC plastic pipe. Each bank of ozone generators is housed in a large cast iron vessel through which the cooling water is pumped. Internal parts are of glass or stainless steel. The use of PVC for the ozone-containing air line is unusual. Although the plant personnel did not indicate problems with the PVC, there was some evidence of previous breaks on one line.

Granular Activated Carbon--

GAC removes organic compounds and residual ozone from the potable water stream. The Zürich Lengg plant has 12 activated carbon filters, 6 for each treatment train. The filters measure 5.8 m deep, 9.7 m long and 4.5 m wide, and 130 cm of carbon are placed over 40 cm of quartz sand. The carbon is regenerated annually and the plant has a regeneration furnace for this purpose. The carbon is regenerated at 700°C in a fluidized bed. The contaminating substances are incinerated at 900°C and the effluent gases are water scrubbed. The furnace treats 100 kg of carbon per hour and is fully automatic.

Laboratory analysis indicates that a regenerated carbon will remove TOC (Total Organic Carbon) at the rate of 1 g TOC/4 g of carbon. Exhausted carbon removes TOC at the rate of 1 g TOC/14 g of carbon. TOC breaks through the carbon columns at Lengg before organic chlorine compounds; however, samples are monitored for both at three points in the carbon beds.

Backwashing of the carbon filters is accomplished in three stages. First, the bed is backwashed with air, then with air plus water, and finally with water only. The last phase of backwashing separates the carbon from the sand. For backwashing, air flow is 20 m/hour, air plus water flow is 25 m/hour, and water flow alone is 50 m/hour. The filters must be backwashed approximately every 2 days.

Slow Sand Filtration--

Slow sand filtration is the final filtration step at the Lengg plant. There is a total of 14 slow sand filters, each of which is 48 m long, 24 m wide and 4 m deep. All of the filters are fully enclosed. Rate of filtration is 15 m/hour for these filters. The filters are manually cleaned by laborers who shovel the upper two centimeters of sand into a small wheeled hopper which is connected by hoses to the sand washing machine in another room. The sand is sluiced to the sand cleaning machine by water and returned the same way. The plant presently is testing a floating sand cleaning machine which would accomplish cleaning without shoveling. Addition of a final stage of carbon filtration also is being studied by placing a shallow layer of activated carbon on the filters. The slow sand filters, which are cleaned infrequently and only when their head loss reaches 2 m, serve as the final solids removal step in the Lengg plant.

Chlorine Dioxide Addition--

ClO_2 addition is provided to avoid bacterial regrowth in the product water. It is manufactured on-site by combining sodium chlorite ($NaClO_2$) solution and hydrochloric acid (HCl). The reaction is as follows:

$$NaClO_2 + HCl \longrightarrow ClO_2 + NaCl$$

A schematic of the Lengg ClO_2 system is shown in Figure B-9. Basically, the system dilutes stock solutions of the feed chemicals which are stored in reinforced fiberglass tanks 9 cu m in volume and then combines them to produce an aqueous solution of ClO_2. The stock HCl is diluted to a 7% solution and the stock $NaClO_2$ is diluted to a 9% solution prior to combination, as the reaction is exothermic and potentially explosive in stronger concentrations. The solutions are stored in separate tanks made of reinforced polyester/fiberglass. The chlorine dioxide solution is conveyed to the mixing point through six proportioning pumps operated in conjunction with the six raw/finished water pumps.

Combination of the two chemicals is carried out in a Raschig ring packed column 5 feet tall and 6 inches wide. Pumping rate through the reactor is 40 l/hour, of both solutions. Each solution is pumped through a separate line and into the reactor by the same dual chamber pump. The reactor effluent is stored in the chlorine dioxide storage tank for use. The system is fully automatic and is controlled by the level in the chlorine dioxide tank. Level controls in this tank control the reaction system, and similar level controls in the dilution storage tanks control the dilution process.

Because of the dangerous nature of the process chemicals, safety precautions are numerous. Acid and chlorite solutions are stored in separate concrete rooms below the chlorine dioxide preparation room. There are no drains in the storage rooms to prevent escape of the chemicals in the event of a rupture. Rather, there are leakage alarms in both of the rooms, and another leakage alarm in the room which serves as an entrance to the storage rooms. In the event of a spill, the chemicals would be fully contained and would be pumped out into other tanks for disposal. The control room for this dual chlorine dioxide system is located in a third room. Sprinklers are installed in all chlorine dioxide rooms in case of accidental spills.

The product water from the final filters is dosed to a residual level of 0.05 mg/l of ClO_2 prior to being pumped into the system. Standards call for a minimum of 0.01 mg/l residual chlorine at the most distant customer.

Pilot Plant--

The Lengg plant has a small scale pilot facility which mirrors the larger process, as an aid to study and monitoring of the system. Raw water is run through the pilot plant, which is fully automated, and sampled and analyzed in similar fashion to the full scale process. Three trout aquaria are also present, one containing small, one containing medium, and one containing large fish in a treated drinking water flow. This serves as a living biological monitor for treated water from the plant. At the first sign of general distress in any of the tanks, the system is shut down and analyses performed.

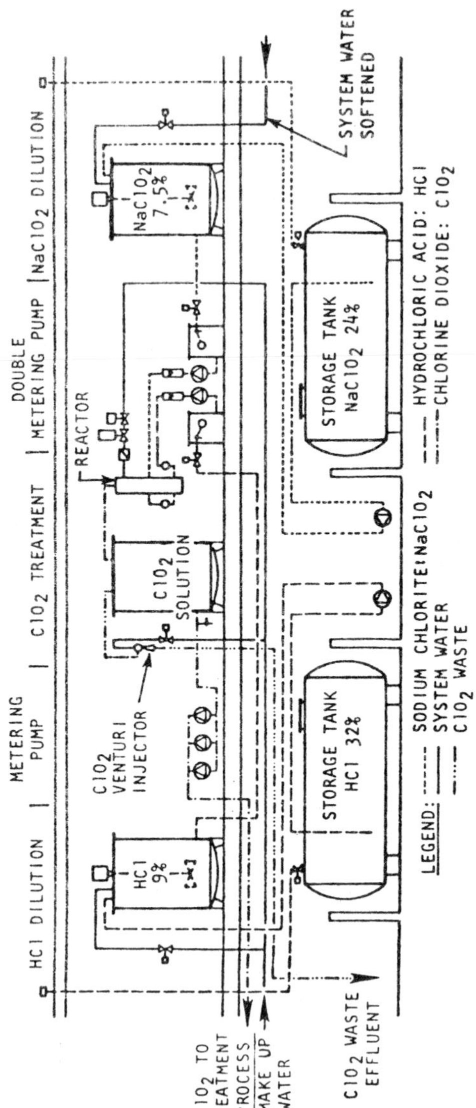

FIGURE B-9. Lengg Waterworks, Zürich, Switzerland. Chlorine Dioxide System.

The entire pilot facility is housed in a separate room near the laboratories and is operated full time as an aid to overall plant operation. It also serves as a study area for proposed process modifications.

SIPPLINGER BERG, SIPPLINGEN, (LAKE OF CONSTANCE) BODENSEE, FEDERAL REPUBLIC OF GERMANY

Introduction & Overview

The Bodensee Wasserversorgung (BWV) water treatment plant was constructed near Sipplingen, West Germany in 1956 to 1958 by a joint consortium of nearly 100 towns and 500 communities in the Baden-Württemberg region of West Germany. This area, located in southern and central regions of West Germany was plagued with repeated water shortages as a result of the geology and climate of the region. Reliance on groundwater was not possible with growing populations because of its unreliability. The nearest large body of fresh water was the Bodensee (Lake of Constance), one of the largest natural reservoirs in Europe and approximately 50 cubic kilometers in volume.

The water from this plant is pumped north through a dual pipeline a distance of over 150 kilometers, terminating considerably north of Stuttgart. The unusual length of this pipeline, coupled with the need to maintain a chlorine residual over its entire length, was one of the major reasons for the addition of ozonation equipment when the plant was expanded in 1966 to 1970. German drinking water standards allow a maximum chlorine dosage of 0.3 mg/l at the outlet of the waterwork, and 0.6 mg/l in special cases. Without ozonation, this amount of residual chlorine would disappear before the product water reaches Stuttgart. With ozonation, however, residual chlorine is present in the Stuttgart mains. Today the plant is the largest single water treatment plant in Germany using ozone, with a capacity of 7.5 cu m/sec (648,000 cu m/day), up from 4.5 cu m/sec at its initial construction.

The water quality of the Bodensee is generally good, particularly in the arm of the lake from which the BWV draws water. There is very little adjacent development and the average 0.08 mg/l phosphorus concentration (as P) and 1 to 2 mg/l TOC (Total Organic Carbon) reflects the relatively good water quality. Nitrates average 3 mg/l, sulfates 35 mg/l, COD 4 mg/l and hardness 8.9 degrees (German measure of hardness). This is a soft water. The age of the lake lies somewhere between the oligotrophic and eutrophic stages. Water is withdrawn at 60 m depth, 420 feet from the shoreline and pumped 312 meters (in elevation) to the hilltop on which the plant is located. The water is subjected to microstraining, ozonation, filtration and chlorination prior to being pumped into the distribution system.

Microstraining

The initial process in the BWV plant is microstraining to remove algae and plankton, and to extend the working capacities and rates of the sand filters. The microstrainers, sized for 40 micron mesh size, are normally stationary, being rotated and backwashed only when head loss exceeds 150 millimeters pressure. Maximum flow rate through all 12 microstrainers is 2200 cu m/hour. Each microstrainer has about 20 cu m filter area and approximately 2/3rds of the screen is exposed to water.

Ozonation

Six Herrmann ozonators (860 tubes/unit) dose the microstrained water to an average level of 1 mg/l. Capacity of the units is 40.5 kg/day total maximum. The ozonators prepare ozone from air dried and cooled in two air preparation units with one on standby. The air is compressed to 0.8 bar, filtered through 3 micron pore size cloth and refrigeration cooled to 16°C before being passed through a silica gel desiccator. Air flow rate is variable, depending on the number of ozonators in service, at a constant rate of 180 cu m/hr per ozonator (thus, ozone concentration varies from 18 g/cu m at the lowest voltage to 32 g/cu m at the highest voltage). After cooling and drying, air enters the ozone generator and its moisture content is monitored continuously by a Panametrics model 3000 dew point meter. The meter is calibrated every six months and plant personnel report satisfactory operation. A maximum dew point of minus 40 degrees is the setpoint on this instrument which will shut down the ozone generators if the dew point rises above this level.

The ozone generators are cooled with 8 cu m/hr (each) raw water which averages 4°C in temperature year-round. After use, the raw water is returned to the inlet well of the plant. Each generator can be operated manually in 10 steps from 10,000 to 18,000 volts (50 Hz). Eighteen watt-hours of power are consumed per gram of ozone produced (18 kwh/kg).

Normal production of the ozone generator is 23 kg/hr, with a maximum of 34 kg/hr. Normal ozone dosage rates are 0.9 mg/l with capacity to achieve 1.2 mg/l dosage if needed. Excess ozone-containing gases are passed through wet activated carbon filters and then vented to the outside air. Off-gases from these carbon units contain approximately 0.8 g of ozone/cubic meter of air. No evidence of ozone odor was detected in the buildings or outside at the time of the plant visit.

Air pressure upon leaving the ozone generators is approximately 0.46 bar, which is approximately 0.08 bar above the pressure at which it enters the contactors.

Maintenance experience on the air cleaning and ozonation train has been good. The silica gel has not required replacement in over seven years of usage and ozone generator tube breakage has been minimal. Some initial difficulty with the drying system was encountered, due to carryover of silica gel dust into the ozonators. This was cured by installing a finer air filter after the desiccator. The ozonators have been cleaned once, after 5 years service, and very little tube breakage had occurred at that time. Ozonized air is moved to the contactors through steel pipes coated with a special corrosion resistant paint at a considerable savings over the cost of the more common stainless steel pipe. Performance of the coating system has been quite satisfactory. Earlier, the facility had tested PVC and polyethylene pipe for this application and found them to be unsatisfactory. Overall, the ozonation facility appeared in like-new condition--there were no overt signs of deterioration.

Ozone Contactors

Ozonation at the BWV plant utilizes packed beds for contacting with the microstrained water. Physically, the 12 contactors are located directly under the 12 microstrainers. The contactors are designed as shown in Figure B-10. Ozone is added at the bottom and meets the countercurrent flow of water falling through the perforated plate onto a packed bed of Raschig rings 2 m in depth. The ozone is fed in at a pressure above ambient of 0.35 bar and total water head loss through the contactor is approximately 3.7 meters. Contact time is about 2 minutes in the contactor. Measured transfer efficiency in the contactors is 98%. The ozonated water passes through two parallel detention reservoirs of 35,000 cu m each, for a minimum detention time of 75 minutes. These chambers serve as final detention basins for the ozonated water and as equalization/storage basins for water from the sand filters.

About 67 cu m of ozonized air is passed through each contactor in one hour. The system is coupled to the operation of the 6 intake pumps such that when only one intake pump is operating, only 2 ozonators are operating. When 2 pumps are operating, 3 ozonators are operating. A maximum of 4 pumps and all 6 ozonators can be operating at the same time, with two pumps on standby.

Safety precautions were numerous throughout the ozonator-contactor system. Every room in which an ozone process was present had ozone detectors coupled to alarms. Exhaust fans also were provided in these areas. In the packed bed contactor, the Raschig ring packing has performed without visible deterioration for 7 years and there is still no apparent need for replacement.

Sand Filtration

Sand filtration at the BWV plant is carried out in a large semicircular building housing 27 rapid sand filters. Each filter has a bed of sand over 0.8 m of pumice, with a thin layer of pebbles at the bottom. Filter surface area totals 3000 sq m and normal flow rate is 9 m/hr. Backwashing is accomplished using product water only at a rate of 37 m/hr. No air backwashing is practiced. There is also provision to place a 20 cm thick layer of carbon on the filters in the event of pollution emergencies.

Injecting the ozone under super-atmospheric pressure has led to shortened filter runs in some cases, due to air bubbles forming in the media. Nevertheless, filter runs are quite long, averaging 3 weeks between backwashings.

Plant personnel stated that ozonation has lengthened filter runs and improved retention of the final chlorine residual. Microflocculation was

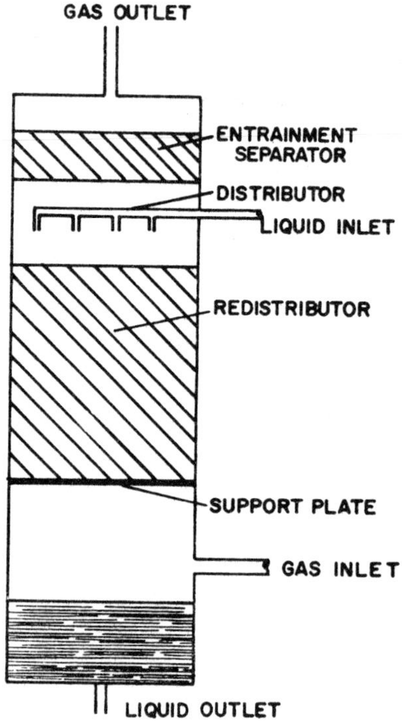

FIGURE B-10
SCHEMATIC OF PACKED BED SIMILAR TO THE ONE USED AT WATER TREATMENT PLANT IN SIPPLINGEN BERG FEDERAL REPUBLIC OF GERMANY

evident in the filter feed water and flocs have been measured at an average diameter of 0.4 mm. The filters can retain particles 0.04 mm and above. Dissolved organic carbon (DOC) measurements indicate that 30 to 40% of the DOC is removed by the sand filters.

Chlorination

The product water is treated with 0.6 mg/l of gaseous chlorine added for residual. Chlorination is accomplished through 6 gas chlorinators fed from multiple cylinders. Near the end of the distribution system in Stuttgart, 150 km to the north, the residual has decreased to 0.05 mg/l. Studies have shown that the chlorine residual will last 500 hours in ozonized water. Prior to the incorporation of ozonation, residuals would last only 150 hours, and problems with regrowth were experienced in the extremities of the distribution system.

After dosing, the water is held in two 20,000 cu m reservoirs prior to pumping into the transmission mains.

Pilot Plant

Similar to the Lengg Plant, the BWV plant has a pilot facility to aid in studies of the basic process and potential modifications. The pilot plant processes raw water at 10 cu m/hr. Considerable research work using biological monitors for detection of toxic substances is being done in the raw water pumping station on the lake shore. Algae which give off oxygen at a specific rate, which changes only with the introduction of toxic substances, and water fleas, whose rate of swimming is affected by toxics, are two of the biological monitors. Still a third technique uses a Grathonemus fish, which is an electric eel from Africa. Change in electrical emissions from these fish is an indicator of the presence of toxic pollutants, and the fish can detect 0.2 mg/l of cyanide. Normal reaction time for these biological monitors is in the two hour range, though early responses can be noted within one hour.

Full laboratory monitoring is provided at this site on the raw water supply. Mercury, lead, copper, arsenic, selenium, chromium, zinc, cadmium, iron, and manganese are monitored to assure compliance with German drinking water standards. This laboratory also monitors nutrients, organic compounds, and biological growth in the raw water. Monitoring of effluents from area wastewater treatment plants is an aid to anticipating changes in lake chemistry. Raw water is also monitored for oil, pH, dissolved oxygen, temperature, redox potential, turbidity, and polyaromatic hydrocarbons (PAH).

ST. DENIS, QUEBEC, CANADA

Background

The St. Denis water treatment treatment plant located in the Province of Quebec, Canada, was constructed in 1972 with a design capacity of 27,300 cu m/day. It is a publicly owned and operated plant which draws its supply from the Richelieu River. Present flows average 9100 cu m/day (2.4 mgd), peak is 15,900 cu m/day (4.2 mgd). Both the raw and finished waters are tested for turbidity, alkalinity and bacteria on frequent schedules.

The plant process involves pre-chlorination, coagulation, sedimentation, filtration, ozonation, and chlorination. Ozonation is used primarily for the control of taste and odors.

Process Description

The water treatment process at St. Denis begins with the injection of 1 mg/l of gaseous chlorine for controlling biological growth in the process. There is also provision for pH adjustment of the finished water as needed, using soda ash. The raw water is then coagulated with alum, settled in clarifiers, filtered, ozonated, and chlorinated.

Two PCI Ozone generators are used at this plant, each of 0.58 kg/hr capacity. They are normally operated at about 65% of capacity. Normal current drawn at this load is 1.5 to 2 amps at 550 volts and 60 Hz. The units are double cooled (oil and water), tube type with 6 tubes per ozonator. The first unit was installed in 1972; the second was put into operation in early 1977. The units are operated 18 hours per day, alternately, on successive days. Air for the units is drawn from within the ozonator room at about 14 cfm at 70°F and 14 psi. The air is refrigerant cooled and then passed through a desiccant dryer. Dew point of the outlet air is not monitored.

The ozonized air is contacted with the filtered water in each of two chambers that are 8 ft wide, 15 ft long and 20 ft deep (18 ft water depth), through 6 porous stainless steel diffusers in each chamber. The chambers are operated in series with differing amounts of ozone added to each chamber. At the time of the plant visit, 9 cfm of ozonized air was flowing to the first basin and 5 cfm was going to the second basin. Total ozone dosage is 2.2 to 2.3 mg/l and detention time is 10 minutes. Contactor off-gases are blown out of the building with a 1 hp fan equipped with a sealed motor. The exhaust gas stack is constructed of stainless steel. In addition, chlorine is added in sufficient amount to equal 0.5 mg/l of total oxidant. Normally this means that 0.3 mg/l of chlorine must be added (the balance being residual ozone).

The plant follows a strict schedule of thorough cleaning of the ozonators after every 2500 hours of operation. In addition, the intermittent operation of the units has lead to considerable other maintenance, mainly fuse and gasket replacement. Tube leakage, however, has not occurred on either unit to date. There is an ozone detector in the ozonator room -- the presence of ozone starts an automatic ventilator fan and another ozone detector with an alarm is placed between the ozone generation room and the contactor.

The raw and finished waters are tested for pH, turbidity, aluminum, calcium, copper, iron, magnesium, manganese, potassium, sodium, zinc, sulfates, chlorides, fluoride, hardness and alkalinity at the plant. Bacteriological samples are taken at 5 points in the distribution system and analyzed by an outside laboratory. Finished water at the plant is not normally analyzed for bacteriological water quality.

SHERBROOKE, QUEBEC, CANADA

The Sherbrooke, Quebec, Canada water treatment plant is a new facility (which had just been turned over to the city about a month before the plant visitation was made) incorporating some equipment from an older plant of the same name. Raw water, pumped from Lake Memphremagog eighteen miles away, passes through microstrainers for algae removal (three new ones in operation and three old ones currently being refurbished). Prechlorination is used when the temperature of the raw water increases or when algae problems appear. The microstrained water (16 mIgd at the time of the visit) flows through the ozone contactors, described below, to a twenty million Imperial gallon reservoir located under an adjacent athletic field. As water is withdrawn from the reservoir, it is chlorinated at a dosage of 1.4 mg/l prior to entering the distribution system. The extremity of the water system is 3.5 to 4 miles from the plant. The average residence time in the distribution system is approximately 24 hours but the system has dead ends in it. Chlorine residuals are monitored at 6 of the 19 pump stations in the system. Water quality samples are taken three times per week for analysis at the Provincial Government Laboratory in Quebec City.

Air Preparation System

The air preparation system was the only one observed which utilizes pressure swing or heatless (desiccant) driers which operate at 100 psi, and one of two to utilize a concept called Ozonazur by the manufacturer, Degrémont, that incorporates an air drier, ozone generator, and controls on a single skid mounted unit (see Figure B-11). The ozonation system was designed and manufactured in Canada by Degremont Infilco Ltd.

Four Gardner-Denver "Electro Screw" compressors, located in a separate room, discharge 100 psi air to four water cooled heat exchangers. High noise levels exist in the compressor room.

Ozone Generators

The cooled air passes to the four modular drier/ozone generator units. Air enters the pressure swing desiccant driers through a filter at 87 psi and leaves at 85 psi. Activated alumina is used as the desiccant material. A dew point of minus 40°C was estimated for the gas under pressure but no dew point monitor or hygrometer was installed. A pressure reducing valve is provided to reduce the pressure of the air entering the ozone generators to 12 psi. At this pressure, the dew point of the gas becomes lower. Ozonized air pressure leaving the ozone generators was 10.5 psi. Inlet air temperature was 31°C and the exit ozonized air temperature 45°C, with

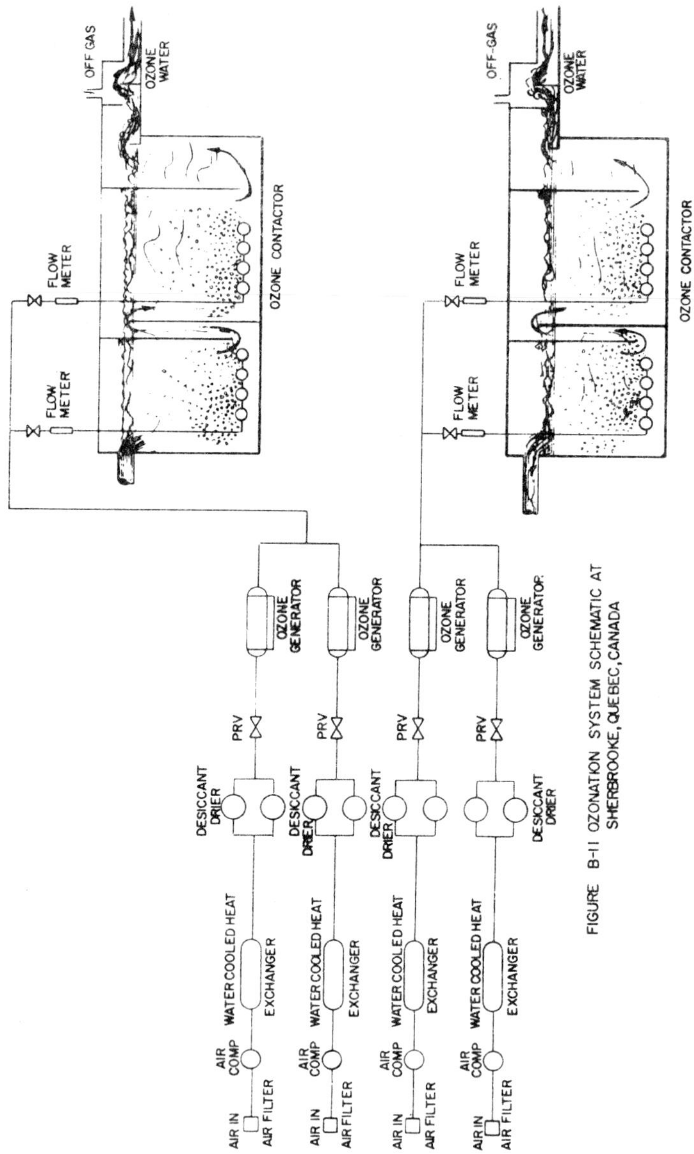

FIGURE B-11 OZONATION SYSTEM SCHEMATIC AT SHERBROOKE, QUEBEC, CANADA

cooling water temperature of 24°C. Ozone generators are constructed entirely of 316L stainless steel except for the bubbler heads which are constructed of type 304L. The heads do not come into contact with cooling water. Each ozone generator contains 126 tubes. Cooling water is finished water which is discharged from the ozone generator to the head of the plant.

Ozone Contacting

Each pair of ozone generators discharges to one of the two ozone contactors through gas flowmeters to porous tubes at a water depth of sixteen feet. Each contactor consists of two compartments operating in the countercurrent mode. A contact time of eight minutes at the design flow rate is provided. Residual ozone levels in solution are monitored at the exit from compartment No. 1 and at the exit from compartment No. 2. The ozone dosage is controlled automatically by the residual ozone monitor reading at the exit of compartment No. 1, while the second monitor is programmed to sound an alarm if the residual ozone level at the exit from compartment No. 2 drops below 0.4 mg/l ozone. The ozone analyzers were Amperazur, amperometric analyzers, Type 5, Series 3, manufactured by Degrémont, France. The signal from the Amperazur is compared to a predetermined set point and the voltage of the ozonator is increased or decreased appropriately to correct the ozonator output. One residual ozone meter was indicating 0.03 mg/l and the other indicated zero. It was not clear whether the units were properly calibrated or not. It is known that the ozone demand of the water is higher than was originally predicted.

Contactor off-gas is discharged to the atmosphere through four rooftop vents. Ozone could be smelled in the vicinity of the plant. There are no ambient ozone monitors in the ozone generator room.

Four, skid mounted, oil filled transformers (one for each ozone generator) are situated in a lower level room that is accessible only from the outside through a door that is normally kept locked. The room temperature was approximately 95 to 100°F.

PIERREFONDS, QUEBEC, CANADA

The Pierrefonds, Quebec, Canada plant draws water from Rivière des Prairies 200 feet off-shore through 1100 feet of 36 inch diameter intake pipe by means of four 7000 U.S. gpm pumps. The design capacity is 42 mIgd but current production is approximately 21 mIgd. The plant began operating in late 1976.

Raw water passes through a magnetic flow meter to a screening room to pass through one of two screens. Prechlorination is used when considered to be required, more frequently in summer than in winter. Screened water passes through a flash mixing chamber to a ribbon type walking beam flocculator. Chemical clarification is provided in two Pulsator units* with

* See description of Tailfer plant (Brussels, Belgium) for discussion of operating characteristics of Pulsator.

activated silica, caustic soda and alum. Provision is made to add chemicals at the intake pipe, in the flash mixing chambers or later in the system. During the visit, chemicals were being added at the mixing chamber.

The clarified effluent proceeds to 6 older single filter units and 4 newer double filter units so that there is the equivalent of 14 single filters. Each filter medium consist of 24 inches of anthracite coal, supported by 6 inches of sand and 6 inches of support gravel. The filter rate ranges from 2 Imperial gpm per sq ft to 4 Imperial gpm per sq ft. The filters are backwashed every 125 hours, when the filtrate turbidity reaches 0.6 JTU, or if the head loss is greater than 0.15 meters. The backwash cycle includes air wash, surface wash, and water backwash. The filtered water passes to the two ozonation contactors for organics oxidation and disinfection. Ozonated water flows to a three million gallon reservoir. Caustic soda (pH adjustment), fluoride and chlorine are added to the water as it is drawn from the reservoir by four finished water pumps, having a maximum combined capacity of 25,000 Igpm.

Approximately 1.0 mg/l of chlorine is added to finished water at Pierrefonds. Water is pumped to the Dolar reservoir that is eight miles from the plant. Chlorine residual at the Dolar reservoir is maintained at approximately 0.4 mg/l, which maintains the chlorine residual level in the distribution system at approximately 0.2 mg/l. There are approximately 23 miles of 24 inch and 36 inch diameter distribution pipes in the system. Average residence time of water in the system is approximately 3 days.

Ozonation System

The Pierrefonds ozonation system is illustrated in Figure B-12. Blowers, water cooled heat exchangers and refrigerant coolers are located in a lower floor room, apparently to isolate the blowers and refrigerant coolers from the remainder of the plant. The rotary valve blowers take suction from the interior of the room through individual inlet filter-silencers. The blowers also are equipped with discharge silencers and were selected for a minimum rotational speed to minimize noise generation. Three blowers are fixed speed (750 RPM) and one variable speed (up to 1400 RPM). Air temperature leaving the air blowers is at 100°C. Four water cooled heat exchangers are provided to reduce the air temperature from the blowers to 24°C. Finished water from the high pressure pumps is used for cooling. Two refrigerant (glycol) driers are used to reduce the air temperature to the desiccant driers to 8°C.

The cooled air passes upward to the ozone generator room which also houses the desiccant driers. Two dual-cell, silica gel desiccant driers discharge gas at approximately 28°C and at an equivalent dew point of minus 80°C (2 to 3 mg/l of moisture measured by hygrometer). The dried air is fed to three Trailigaz, water cooled, horizontal tube ozone generators. Each generator contains 240 tubes. Cooling water is supplied from the discharge side of the high pressure pumps and returned upstream of the gravity filters. Cooling water is approximately 1°C in winter and as high as 24°C in summer.

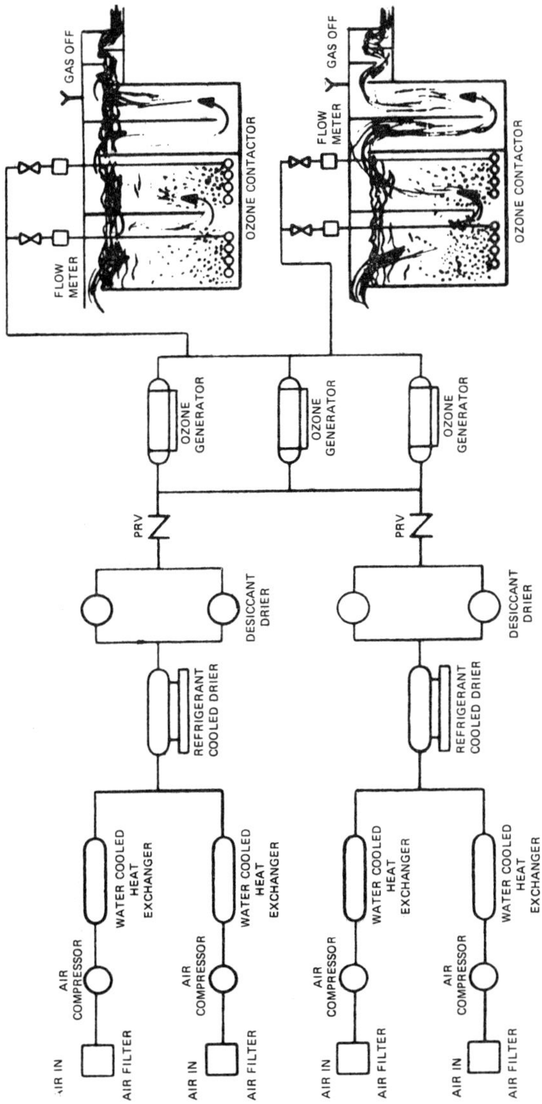

FIGURE B-12
OZONATION SYSTEM SCHEMATIC
PIERREFONDS WATER TREATMENT PLANT QUEBEC, CANADA

Two ozone contactors are provided with a water depth of approximately 18 feet. Each contactor has two ozonation compartments, the first with countercurrent ozonation and the second with cocurrent ozonation. A third compartment is provided for degassing. 1.6 mg/l of ozone is added to the first chamber and 0.3 mg/l is added to the second chamber. A total ozone dosage of 1.5 to 2.0 mg/l ozone is required to maintain a residual of 0.4 mg/l. The ozone residual in the treated water is measured by equipment supplied by Chloro Residual Co. of Milano, Italy (Chloresiduometro -Chlorosis - Brevatato - Sis, Milano). The monitor indicated a residual of 0.47 mg/l during the plant visit. The unit is recalibrated every two weeks by using ozone-free water to establish a zero point.

The ozone contactor off-gases are discharged to a rooftop stack without complaints from the neighbors in the rural surroundings. No off-gas monitoring is used. Dasibi (Model 1003AH) ambient air ozone monitors were installed in both the blower room and the generator room. The monitors in the generator room were not in service. They had just returned from repair as they had become plugged with paint when the room had been spray painted. The alarm was set for 0.08 mg/l ozone. The Dasibi monitor samples ambient air every 20 seconds and provides an instantaneous digital reading. The Dasibi monitor in the blower room indicated C.07 to 0.08 mg/l during the plant visit.

The ozonation system has been in operation only since 1976 but the plant manager plans once per year maintenance of the system. Three to four dielectric tubes were broken during start-up but none in subsequent operation so far. The plant manager has ordered 10 tubes as spare parts. Dielectric tubes cost approximately $120 each, or $200 for each unit complete with high tension connection. A feature of this plant was the provision of three mobile carts for tube storage. Each wheeled cart has 64 vertical spaces to provide tube storage during generator servicing.

QUEBEC CITY

Introduction and Overview

The Quebec City water treatment plant is located adjacent to the St. Charles River, approximately 8 miles north of the city. The plant was designed to treat 220,000 cu m/day (58 mgd); it can be readily expanded to 264,000 cu m/day (72 mgd). It is semi-automated and is operated by 14 people. Raw water is pumped from the river and prechlorinated to 1 mg/l with 181 kg/day of chlorine, followed by chemical addition. After flash mixing, the water is flocculated, clarified and filtered. Ozonation follows filtration and the treated water is stored in 4 reservoirs totalling 31,820 cu m of storage capacity.

The first Quebec water works was built in 1840 and pumped essentially untreated water a distance of 18 miles to the city. Chlorination was instituted in 1930, but no other treatment was used until 1969 when the present plant came on line.

The plant tests daily for COD, turbidity, color, pH, chlorine residual, alkalinity and hardness. Bacteriological tests are not performed at the plant, but rather are checked at 14 points (including the plant) and analyzed at a laboratory in Quebec City.

Pretreatment, Coagulation, Flocculation and Sedimentation

Processing of the raw water at the Quebec City plant begins with addition of 0.5 to 1.5 mg/l (depending on season; higher in summer) of chlorine, used specifically to kill the larvae of river flies. This is followed by the addition of lime and alum coagulants. Flash mixing of the chemicals is carried out in 6 flash mixing basins of 1 minute detention time, based on design flow. Polyelectrolyte is added after flash mixing and the water then is flocculated in 7 ribbon flocculator equipped basins. Sedimentation is carried out in 4 circular basins with 4 hour retention time, based on design flow, prior to filtration. Water treatment sludges are returned to the river. Chlorine in the amount of 0.1 mg/l is also added prior to filtration.

Filtration

Sixteen multimedia rapid sand filters of 65 sq m surface area are provided for filtration. Bed design is 61 cm of anthracite (0.9 to 1.2 mm) over 15 cm of sand (0.75 to 0.55 mm) with 30 cm of graded rock and gravel on the bottom. Design flow rate in treatment is 9 m/hr (9 cu m/sq m/hr) and backwash rate is 36 m/hr with clear water only. No air washing is used at this plant. Backwash times average 7 minutes and approximately 273 cu m are consumed with each washing. Backwash water is returned to the river. Better raw water conditions in winter often allow filtration rates exceeding 18 to 20 m/hr during winter months.

Ozonation

Ozonation at Quebec City is provided by 12 Otto plate ozone generators, in banks of 3 each. The units are water cooled (37 cu m/hr total air flow), and air is filtered and dried by two parallel air treatment trains, one operating and one on standby. Air is filtered, compressed, dried in silica gel columns and then fed to the ozonators. Four sets of silica gel beds are provided, two for each treatment train. The desiccators are regenerated each 8 hours with hot air at 220°C. Output humidity is monitored after the desiccators with a lithium bromide relative humidity meter. The unit will sound an alarm and shut the ozonator down when its set point is exceeded.

Negative pressure is maintained in the ozonators to reduce leakage, which has been a problem. At the time of the plant visit, the ambient ozone level in the ozone room read 0.78 mg/l and was strongly noticeable. This level dropped to 0.15 mg/l later in the visit.

Ozone is generated in a voltage range of 9,000 to 15,000 volts, adjusted in a stepwise manner. Normally, the units are set at 11,500 volts. Total installed power for the 12 ozonators is 760 kw and the plant normally draws 300 to 350 kw. Ozone residual in the contact chamber is 0.4 mg/l after 4 minutes which appears as 0.2 mg/l at the sampling point in the laboratory. The ozone is contacted with the water supply for 6 to 8 minutes. The contacting is done in an injector system which contacts the ozone with a small portion of the water flow, then reinjects it into a down-flowing tube. The system is similar to those at Düsseldorf and Konstanz (Germany) plants. Ozone is drawn into the water flow through a venturi under negative pressure, hence maintaining a slightly negative pressure on the ozonators. Off-gases from the contact basin are vented untreated to the atmosphere.

In general, the ozonation equipment has performed satisfactorily. Maintenance problems arise from the non-continuous operations of the ozonator as they are only operated 18 hours per day. It has been found necessary to flush an ozonator for at least 1 hour with dried gas after maintenance prior to restarting at the lowest voltage setting.

Final Chlorination

Following processing, the treated water is dosed with 0.5 to 1.0 mg/l of chlorine (more in summer) and stored in two on-site 3-day retention reservoirs. As needed, additional chlorine is added to the reservoirs to maintain a residual of 0.6 mg/l. The overall operational objective is to maintain a minimum residual of 0.15 mg/l at a distance 7 miles from the plant. Lime and polyelectrolytes are also added after ozonation for pH adjustment and corrosion control in the distribution system.

APPENDIX C: FIGAWA TECHNICAL PAPERS ON OZONE TECHNOLOGY

In this Appendix we are including translations of three documents developed in Germany by collaboration of the FIGAWA and the DVGW.

FIGAWA is the Technischen Vereinigung der Firmen im Gas- und Wasserfach, founded in 1926 and headquartered in Köln, Federal Republic of Germany (Marienburger Strasse 15; 5000 Köln 51). Its members are equipment supplier companies who market to gas and water works.

The DVGW (Deutscher Verein des Gas- und Wasserfaches e.V.) is a technical-scientific organization, founded in 1859 and located in Eschborn, Federal Republic of Germany (DVGW-Hauptgeschäftsführung; Postfach 5240, Frankfurter Allee 27; 6236 Eschborn 1). Members of DVGW are scientists, engineers and organizations in the fields of gas, water or wastewater treatment.

The three technical documents included in this appendix are:

(1) Ozone in Water Treatment (May, 1975), which discusses the use of ozone in drinking water treatment, swimming pools and wastewater.

(2) Ozone and Corrosion - Safeguard Systems (July, 1975), which discusses the methods of protecting treatment plant materials and components.

(3) Ozone Production Plants for Water Treatment (December, 1976), which contains minimum requirements for the structure, operation and safety equipment for ozonation systems used in water treatment plants.

We are including these documents because they represent the experiences and considered opinions of organizations who have been involved with the generation and application of ozone in water treatment for many years.

FIGAWA
Technical Report #1

Ozone in Water Treatment (May, 1975)

Recently papers were compiled in the work groups of FIGAWA which deal with special and actual technical problems in the different research areas and in which material collected by members of these groups concerning these questions are contained. As is the practice of FIGAWA, these papers are given to the appropriate committee of experts of DVGW for further revision. There they serve as a basis for discussion for the compilation of DVGW work and information papers. In that these documents deal with actual technical questions and that FIGAWA is constantly asked for such information, the results of this work should be published as "technical reports" in the group's journal. This would occur only as these papers become available.

The FIGAWA "Ozone" work group is now offering technical report #1 "Ozone in Water Treatment". This information was collected over a period of two years and contains basic information concerning the use of ozone in different areas of water treatment such as drinking water, swimming pool water, and treatment of wastewater. It is the basis of discussion in the newly formed DVGW "Ozone" work group. In connection with this research, other types of papers are being developed which will deal with the question of corrosion in regard to the use of ozone. Suggestions and improvements are greatly appreciated by the management of FIGAWA.

The following firms participated in the compilation of the FIGAWA "Ozone" work group.

1. Akdolit-Werk GmbH, 5605 Hochdahl
2. Bamag-Verfahrenstechnik GmbH, 6308 Butzbach
3. Benckiser Wassertechnik, 6905 Schriesheim
4. Chemie und Filter GmbH, 6900 Heidelberg-Wieblingen
5. Clouth-Gummiwerke AG, 5 Köln 60
6. DEMAG-Metallgewinnung, 41 Duisburg
7. Th. Goldschmidt AG, 68 Mannheim 81
8. Gebr. Herrmann, 5 Köln 30
9. Lechler Chemie GmbH, 7 Stuttgart 40
10. Messer-Griesheim, 4 Stuttgart 1
11. Secova GmbH, 504 Bruhl
12. Dr. Starck & Co., 52 Siegburg
13. Philipp Muller, 7 Stuttgart 1
14. WABAG, 865 Kulmbach

OZONE IN WATER TREATMENT

1. Ozone

1.1 Definition

The term "ozone" is understood to be the triatomic modification of oxygen and is designated by the formula O_3 with a molecular weight of 47.998. Three oxygen atoms form a triangle in the ozone molecule with an angle of 127° in which the invervals between the atoms amount to 1.26-1.26-2.26 10^{-11} m.

Under normal conditions (p = 1.012 bars and T = 0°C) ozone is a distinctly blue-colored gas with the following constants:

Boiling Point	(Kp) $\hat{=}$	-110.5°C
Melting Point	(Fp) $\hat{=}$	-251.4°C
Density	(L) $\hat{=}$	2.144 g/l

With regards to the equation:

$$O_3 + 2H^+ + 2e \longrightarrow O_2 + H_2O$$

it yields a normal potential of Eo = 2.07 v. Ozone is, therefore, one of the strongest means of oxidation. Ozone decays very rapidly at room temperature, but it decays much slower at this temperature in gas mixtures which are produced by normal ozone producing equipment. The decay rate is increased by traces of particular elements such as silver, platinum and manganese oxides. The decay of ozone in liquid solutions is accelerated at higher pH values.

1.2 Formation and Production

The formation of ozone is obtained by numerous processes: such as electrolysis of sulfuric acid and by the decomposition of oxygen-rich mixtures such as permanganates or dichromates. Of the others, ozone is formed by energy-rich radiation of air. Therefore the earth is surrounded by an ozone layer at an altitude of approximately 15 to 30 km which is formed there by the unimpaired ultraviolet rays of the sun. The maximum levels of ozone contained in air are found at an altitude of approximately 22 km and amount to approximately 10^{-6} volume parts.

For the production of greater ozone quantities one must make use of the silent electric discharge as the only efficient and acceptable method. Werner Siemens' principle of the so called "ozone canal" (tube, pipe, shaft) developed over one hundred years ago still has value for us today.

The electrical discharge devices were, in the course of development, simplified and their performances improved so that today one has available tubular and plate discharge elements. There is a principle common to all such units by which two metal electrodes (non-corrosive metal) are separated from one another by a dielectric medium as well as an air gap or space. High tension alternating current of an appropriate frequency is conducted to the electrodes whereby the dielectric serves the purpose of a series resister, thereby hindering the direct charge flow from electrode to electrode leading to a short circuit. Within the air gap there is, with the transmission of an oxygen-containing gas, silent electric discharge which results in ozone formation.

Ozone is formed according to these equations:

$$0.5O_2 \rightleftharpoons O \qquad \Delta H = +59.1 \text{ Kcal}$$

$$O + O_2 \rightleftharpoons O_3 \qquad \Delta H = -24.6 \text{ Kcal}$$

$$1.5O_2 \rightleftharpoons O_3 \qquad \Delta H = +34.5 \text{ Kcal}$$

The process is endothermic, which depends upon the use of a relatively large quantity of energy. Therefore, a large part of the conducted energy is released in the form of heat. The latter is removed by a suitable means of cooling because it influences in a negative manner the performance efficiency of an ozone generator.

The equilibrium is influenced by the moisture content of the oxygen-containing gas to be injected into the ozone generator so that desiccation is necessary. Thereby a dewpoint of under -45°C for the desiccated gas is necessary in order to prevent the formation of nitrogen oxides and ultimately nitrogen oxide acids which damage the construction elements of the ozone generator.

The energy consumption of modern ozone generators for the production of 1 g of ozone lies between the following values:

a. with the introduction of pure oxygen 7 to 15 watt hours

b. with the introduction of desiccated atmospheric air 15 to 30 watt hours

The actual energy required depends on different construction parameters as well as desired ozone concentration.

1.3 Reactions

In Section 1.1 the oxidation capacity of ozone was indicated, characterized by its normal potential of 2.07 V. It follows that all reactions with ozone are oxidative. Reactions are possible with all materials which are accessible to corrosion due to determined functional groups or by a low valence state (capable of combining with oxygen).

The effect of ozone takes place mostly with reactions with inorganic materials through the utilization of an oxygen atom. The reaction with organic materials in water is dependent upon the structure of the organic substances as well as the pH value. The employment of ozone in water treatment techniques has acquired a wider range. In this area its introduction as a means of disinfection has developed to a large and varied importance.

It is employed today for:

- separation and precipitation of heavy metal ions such as iron and manganese
- decomposition of odor and taste elements
- elimination of colors
- eradication of toxic materials
- flocculation of colloids

The action of ozone on organic substances is designated as "ozonating"; analogous to the terms "chlorination" and "chlorinating".

1.4 Physiological effect

Ozone is detectable by its characteristic odor even in the smallest concentrations (greatly diluted). The odor threshold lies at 0.02 mg/l = 0.04 mg ozone/cu m air.

Physiologically ozone affects one as an irritant and it has a general anesthetizing effect. Particular weak points for acute effects are the mucous membranes of the eyes, nose and lungs. Long periods of exposure in an atmosphere with an ozone concentration above 0.2 mg ozone/cu m will cause coughing; higher than 0.2 mg ozone/cu m will cause chest and head pains, and at even greater concentrations, circulatory disturbances and salivation.

The Accident Prevention Ordinances (Professional Society of Gas and Water Works, Düsseldorf) allows a maximum concentration at a work location (MAK value) of 0.1 mg/l = 0.2 mg ozone/cu m air. More detailed information may be obtained from the named organization.

2. Effect On Elements Found In Water

2.1 Microorganisms

The introduction of ozone for the destruction of microorganisms (disinfection) has been recognized for years. Through work carried out in recent times the bactericidal and, above all, the virucidal effects were reconfirmed. Ozone is discussed accordingly with all the other disinfection means found in water chemistry and the following table will make points of comparison clear:

DISINFECTION EFFICIENCY OF OZONE VS. CHLORINE

Germ Count	Means of Disinfection	Add mg/l	Lethal Effect/sec
60,000	chlorine	0.1	15,000
coliforms/ml	ozone	0.1	5
350 spores	chlorine	1.4	9,000
of B. subtilis/ml	ozone	0.05	30
PM-Virus strains	chlorine	0.25-1.0	180*
MV and Le Virus suspension 1:1000	ozone	0.05-0.46	2*

*From Bringman [Z. Hyg. Infektionskronkh. 139:130 (1954); 139:333-337 (1954)]

With regard to French work, it is further recognized that a residual of 0.4 mg/l ozone over a period of four minutes produces a sufficient assurance for the inactivating of polio virus. This result also applies to river water from various sources as well as to variable levels of contamination. A residual of ozone is needed for a guaranteed disinfection of the water in which the bactericidal and viricidal effects extend themselves over a wide range of pH values.

2.2 Heavy Metals

Heavy metals such as iron and manganese are transformed by ozone into their highest valence. The oxidation of Fe^{2+} and Mn^{2+} ions occurs in accordance with the following equations:

$$2Fe^{2+} + O_3 + H_2O \longrightarrow 2Fe^{3+} + O_2 + (OH)^-$$

$$2Mn^{2+} + 5O_3 + H_2O \longrightarrow 2MnO_4^- + 5O_2 + 6H^+$$

The employment of ozone for removal of iron and manganese is applicable especially in the case of waters found difficult to treat; for example in the presence of protective colloids (humus elements). In accordance with the first equation iron is in the form, after ozonizing, of ferric oxide hydrate, e.g., a form capable of filtration. Its precipitation occurs in a suitable filter. The permanganate ion formed in the second equation can be reduced, insofar as it is not converted in the water treatment process, in activated carbon and disposed of. These reactions are continuously dependent upon pH value which is especially important for the removal of heavy metals from weakly acidic water.

2.3 Organic substances

The substances of an organic nature found in water are identified in only a few cases. Therefore one is assured that substances which impair the odor and the taste of water (to yield turbid colored water) are predominantly organic in nature. Ozone is effective here too for the breaking down of these substances. The basic investigations of Harries, Rieche, Staudinger and Criegee are gathered for the reaction mechanism. The existence of "ozonides" is accepted by many authors who support the treatment effect of oxidizing and disinfecting. These assumptions are contested because the reaction mechanism of ozonation requires, for the formation of ozonides, water free solution.

For the reaction of, for example, phenol (carbolic acid) with ozone in an aqueous solution, the intermediate products of pyrocatechol (catechol, pyrocatechin), o-quinone, and muconic acid are produced as well as the end products maleic acid, fumaric acid and oxalic acid.

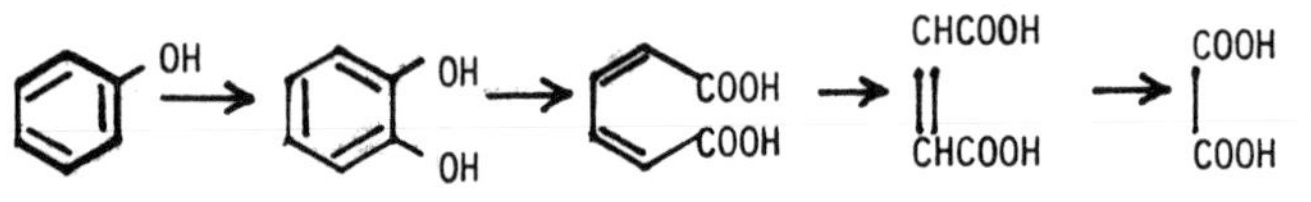

phenol | pyro-catechin | muconic acid | maleic acid | oxalic acid

Generalizing, one can say that the reaction of ozone with organic materials found in water leads to the destruction of the original molecule, whereby, in many cases, the reaction product results in a lower molecular weight. This deals also with carcinogens and polycyclics which are transformed into non-carcinogenic forms.

By the oxidation process with ozone certain organic substances may become toxic or increase, at least their virulence, as is the case with pesticides such as Aldrin and Parathion which are to be removed through further treatment.

3. Areas of Application (techniques)

3.1 Regarding the provisions of DIN 2000, drinking water must be free of all disease agents and is not permitted to contain elements dangerous to health. Along with elements allowed in the Drinking Water Treatment Ordinance (TAV) in toxicologically harmless concentrations, ozone represents, as is described in section 2.1, an effective means of disinfection. Further, ozone serves as a means of disinfection for water difficult to treat as described in sections 1.3, 2.2, 2.3.

The ozone requirement is dependent upon the type and quantity of the material contained in water. If the occasion arises, the necessary treatment steps are to be added to the ozone treatment.

3.2 Mineral water

Ozone has proven itself for the treatment of mineral water. In essence, the provisions of Section 3.1 are to be fulfilled. Fundamentally the table water ordinance (TWV) is to be followed.

The effects of ozone are dealt with in sections 1.3, 2.1, 2.2 and 2.3. Particular attention applies to the content of bromides and/or iodides which are oxidized by ozone. With a combination of ozone and ultraviolet it is to be observed that ozone can be subjected to decay by photochemical reactions.

3.3 Swimming Pool Water

The experiences of ozonizing drinking water have also been applied to swimming pool water since the middle of the 1960s. The following is the goal of the sixties:

- breaking up of produced organic material
- breaking up of the components of urine
- destruction of disease agents

At the same time the typical odor of the indoor swimming pool is eliminated by ozone input, as oxygen deficit is replenished and a visual improvement of the water is noted.

The redox potential is presented as a criterion for the effect of ozonizing in the area of swimming pool water. A smaller surplus of ozone produces a redox potential of $E = +800$ mV [against saturated calomel (mercurous chloride)] electrodes with a pH of 7. With this potential, disease agents are eliminated in the shortest amount of time (inactivated) (see section 2.1).

An average input of 0.5 to 1.5 g ozone into each cubic meter of water is sufficient for the treatment (of the water).

Because organic contaminations are continually eliminated and/or flocculated by ozone, the addition of flocculating agents as well as the quantity of chlorine needed for the assurance of good water can be reduced. A reduction of the latter (chlorine) results in a decreased formation of chlorine substitution products which are responsible for the irritation of the mucous membrane and for the typical odors of the indoor swimming pool. Following the guiding principle for "Water Treatment for Swimming Pool Water, June, 1972", ozone, due to toxicological reasons, is not permitted to be present in pool water. With the addition of suitable reducing agents, a content of residual ozone is avoided. For the treatment of thermal and/-or salt water a content of bromides and iodides is, with the addition of an ozone stage (of treatment), to be taken into account (see Section 3.2).

3.4 Wastewater

The oxidation strength of ozone allows for a widespread input when treating wastewater. Because of the relatively large amounts of ozone necessary for this process it is seldom employed.

The advantages of ozonizing for the decoloration, deodorizing, and reduction of concentrations of organic substances as well as disinfection are also to be applied, as they were in previous paragraphs, to wastewater.

A particular equation is the decontamination of cyanide:

$$2KCN + H_2O + 5O_3 \longrightarrow 2KHCO_3 + N_2 + 5O_2$$

Further, the oxidation of nitrite and sulfite is known to follow the equations:

$$NO_2^- + O_3 \longrightarrow NO_3^- + O_2$$

$$SO_3^{-2} + O_3 \longrightarrow SO_4^{-2} + O_2$$

The possible elimination of phenols already was discussed in Section 2.3.

In conclusion, one may say that ozone may be employed in wastewater techniques. Here too, investigations are necessary which:

a) have as their goal the effective reduction of ozone requirement

b) eliminates a possible toxicity of wastewater treated with ozone (see Section 2.3).

FIGAWA
Technical Report #2

Ozone and Corrosion - Protective Systems (July, 1975)

Once the technical report #1 of FIGAWA was presented as publication 5 May 1975, the FIGAWA "Ozone" work group today submits, in reference to the forward of bbr 5/75, Technical Report #2, "Ozone and Corrosion - Protective Systems". The information was collected after more than two years of work and contains the most up-to-date information on this subject. It is the basis of discussion by the committee of experts of the DVGW "Internal Corrosion-Water" group. Suggestions and improvements are greatly appreciated by the management of FIGAWA.

The following participated in the compilation of this report by the FIGAWA "Ozone" work group:

1. Akdolit-Werk GmbH, 5605 Hochdahl
2. Bamag-Verfahrenstechnik GmbH, 6308 Butzbach
3. Benckiser Wassertechnik, 6905 Schriesheim
4. Chemie und Filter GmbH, 6900 Heidelberg-Wieblingen
5. Clouth-Gummiwerke AG, 5 Köln 60
6. DEMAG-Metallgewinnung, 41 Duisburg
7. Th. Goldschmidt AG, 68 Mannheim 81
8. Gebr. Herrmann, 5 Köln 30
9. Lechler Chemie GmbH, 7 Stuttgart 40
10. Messer-Griesheim, 4 Stuttgart 1
11. Philipp Müller, 7 Stuttgart 1
12. Secova GmbH, 504 Brühl
13. Dr. Starck & Co., 52 Siegburg
14. WABAG, 865 Kulmbach

OZONE AND CORROSION - PROTECTIVE SYSTEMS

1. Definition

The behavior of diluted liquid ozone solutions makes the safeguarding of metallic plant materials necessary; particularly the unalloyed carbon steels used in many types of equipment.

Ozone, in smaller concentrations in a liquid phase, acts as a depolarizer much like oxygen in a water solution, i.e., in one way it is able to form protective oxide films quickly. Corrosion affects unalloyed steel to a particular degree. This steel is used primarily for the water area of the container. For this reason the safeguarding of these structured elements by the use of coating systems is necessary.

Metallic substances are not the only substances which are affected by corrosion. Non-metallic substances are also affected; of the latter, those with double bonds are most endangered. The presence of water causes an accelerated corrosion. The Dechema material table "Ozone" (ref. 21 of this document) offers a survey of this point.

The mode of operation of a corrosion-resistant coating system is divided into four sections:

--It must adhere well to the undercoating

--It must demonstrate a great resistance capability against corroding materials - ozonizing water - and against mechanical effects

--It must be absolutely safe in a physiological sense. In drinking water it is not permitted to emit either taste, odor, or colour or materials dangerous to health or hygiene (in suspicious amounts). Algoid formation and bacteria growth are not to take place in this water.

--Along with a high diffusion resistance it must possess pore "freedom" in order to also prevent local corrosion.

2. Area of Application and Requirements

The protective systems employed due to the input of ozone during water treatment must be absolutely safe in a physiological sense corresponding to recommendation XXXI of the commission dealing with synthetics (of the BGA), i.e., they must possess the quality of food. Its safeness is to be proven. Its resistance capability must extend to aqueous ozone concentrations in the area of 0.5 of 3.0 mg/l of ozone.

The amount of suitable coating materials is dependent upon:

--materials contained in water and their concentration, for example carbonic acids, salts

--pH value

--temperature

The safeguard systems must possess a sufficient mechanical stability in the pressure area up to 16 bars and resist abrasion, for example, the filter material in the backwashing of a filter.

Additional requirements for the protective systems are:

--ordering possibility of the materials in question; in particular metallic, but also those of concrete

--adequate adhesion to the undercoating

--sufficient diffusion resistance to the medium used in the operation

--ability to test for coating defects

--ability to repair mechanical defects

3. Protective Systems

The following described systems serve as a safeguard against the corrosion of metallic and non-metallic materials. Consideration (of these systems) depends on the present state of technology.

3.1 Thin Coating Systems

Included here are such systems which are applied in one or more coats and have a dry coating thickness up to 160 μm.

3.2 Thick Coating Systems

One understands these systems which are applied in one or more coats and has a dry coating thickness over 160 μm.

3.3 Foil Containing Systems

Foil coatings (linings) are prefabricated and attached to the material to be safeguarded.

With respect to their resistance against ozonized water, various coating materials (found in 3.2 and 3.3) have proven themselves in practice. These are systems that have been proved:

--chlorinated rubber

--mixed polymers of polyvinylidene chlorides/polyvinychloride

--unsaturated polyester

4. Performance

The protecting value of a coating system depends not only on its quality but to a great degree on the following factors:

--pre-treatment and condition of the surface material to be protected

--conditions of the environment during the coating procedure

--post-handling of the protective casing

4.1 Pre-treatment

Metallic surfaces are to be sand-blasted to metallic sheen with a quality ASa3, BSa3, or GSa3 (following the Swedish standard SIS 055900). Surfaces with a quality DSa3 are excluded for use as plant parts.

Washed quartz sand with a grain size of 0.5 to 1.2 mm, sharp-edged steel grain with a grain size between 0.5 and 0.75 mm (No. 34 - No. 24) or corundum with a grain size of 0.5 to 1.2 mm are recommended as sand blasting materials.

The sand blasting material must be dry and is not permitted to contain an "aggressive" component.

After the sand blasting the dust is to be removed from the surfaces.

4.2 Surface Condition

The rough depth of the metallic surface to be protected should amount to between 30 and 50 μm. In such a condition the surface is not permitted to exhibit any evidence of rust and no "aggressive" qualities are to be displayed.

4.3 Environmental Conditions

During the preparation of the protective systems the temperature (metal and air) should be between 10° and 30°C. The dew point of the air must be at least 5° lower than the surface temperature of the material to be protected and the apparatus used in the process (of baking).

The environment must be free of "aggressive" gases or dust (for example: salt dust).

4.4 Preparation and Post-Handling

The preparation and post-handling of the protective system depend on special qualities. The recommendations of the manufacturer are to be followed.

The one applying the coating must be continuously careful not to contaminate the cleansed undercoating (hand sweat may be removed with pure Xytol whereby the solution must be allowed to vaporize from the surface).

5. Testing

Concerning non-destructive tests the following are considered:

--measuring coating thickness	magnetic, electro-magnetic, eddy current, capacitive, β-rays
--pore examination	high voltage, electrolytic

These examinations can be combined.

6. Transport and Storage

These depend upon the special qualities of the coating system. The regulations of the manufacturer are to be followed.

Thermoplastics should be transported only at temperatures above +5°C due to the danger of tearing. During storage the temperature is not permitted to go below 0°C.

STANDARDS, REGULATIONS AND RECOMMENDATIONS

1.	DIN 18 364 (VOB-Part C)	Efforts dealing with the protection of steel surfaces
2.	DIN 50 902	Handling of metal surfaces for protecting against corrosion
3.	DIN 53 151	Testing of coating materials-barbed wire test
4.	DIN 53 210	Demonstration of rust percentage of coating
5.	DIN 55 928	Protective coating of steel structures
6.	SIS 055900	Rust percentage of steel surface and materials used in the preparation of steel surfaces for rust preventative coating
7.	VDI 2532	Form and performance of structures to be protected
8.	VDI 2533	Surface protection with organic materials
9.	VDI 2535	Surface protection with organic materials in liquid form
10.	VDI 2537	Surface protection with strips of natural and synthetic (India) rubber

STANDARDS, REGULATIONS AND RECOMMENDATIONS (continued)

11.	AGI Work Report K 20	Protection of steel constructions (Arbeitsgemeinschaft Industriebau eV)
12.	Report concerning use of steel Nr. 269	Surface protection of steel by coating
13.	DB DV 807 (rust)	Technical regulations for rust protection of steel structures of the German Federal Railroad
14.	Report concerning use of steel Nr. 197	Flaming steel structures
15.	Report concerning use of steel Nr. 329	Surface protection by flame and coating
16.	British Standard Code of Practice CP 2008	Protection of iron and steel structures from Corrosion
17.	SSPC-Steel Structure	U.S. Regulations for the handling of undercoating and recommendations of Painting Council coating systems

18. Report - official report of the BMV, volume 6/1965: Regulations for the performance and characteristics of rust preventive coatings for steel structures

19. Wastewater - technical cleansing (ATV -Essen): Regulations for the protective coating of iron components in canalization and cleansing plants

20. Comité Européen des Associations des fabrikants de peintures et d'encres d'imprimerie

21. Dechema - materials table "Ozone"

FIGAWA
Technical Report #3

Ozone Generators for Water Treatment December 15, 1976

Since November, 1975 the FIGAWA "Ozone" work group has dealt with the drafting of a standard for ozone generators. This outline should contain the minimum requirements for structure, operation and safety equipment of such generators.

The representatives of well-known manufacturers of ozone generators, as well as those who employ ozone, worked together with the FIGAWA "Ozone" work group, in particular the following:

Mr. Blankenfeld
Mr. Bredtmann
Mr. Bürgel
Mr. Ensenauer
Mr. Goslau
Mr. Hoelscher
Mr. Koepf
Mr. Kurzmann, Chairman
Dr. Leitzke
Dr. Roennefahrt
Mr. Suppan
Mr. Voss
Mr. Böhme, group leader

Through the work dealing with analysis of ozone in the gas phase by the DVGW "Ozone" work group (Chairman, Dr. Maier) drafting of the standard can be correspondingly expanded so that the now existing uncertainties in this area can be removed.

This outline which is now presented for public consideration was discussed with the representatives of those who employ "ozone" who do not belong to the aforementioned work groups.

We also extend our thanks to Dr. Geiner and Dr. Wolf.

FIGAWA "Ozone" work group

OZONE GENERATORS FOR WATER TREATMENT

1. Scope of Investigation

This standard applies TO ozone GENERATORS which employ the principle of the silent electric discharge and which are employed for the treatment of water (drinking water, industrial water, pool water, wastewater).

2. Goal

The ozone gas mixtures produced in the ozone production plants are employed for the oxidation of inorganic and organic materials as well as for the elimination of microorganisms.

3. Definition

Ozone is created by the effect of a silent electric discharge on oxygen or oxygen-containing gases. The electric discharge results in a "gas space" between two electrodes which are separated from each other by a dielectric medium. One electrode is connected to high voltage while the opposing electrode is connected to a grounded return line to the source of the high voltage or to earth potential.

The gas employed in ozone production must be mechanically purified and "dried" to a dew point under 228°K. At this dew point an economic operation free from disturbance is produced.

4. Terminology

Ozone production element: plant unit in which oxygen or oxygen-containing gases are exposed to a silent electric discharge.

Ozone generator: plant unit in which the ozone production elements are formed.

Ozone generator: complete operation which is necessary for the production of ozone.

5. Characteristics of the Ozone Generator

Ozone generators are distinguished by the following characteristics:

5.1 Generators of a compact or separated construction style:

i.e., spatially grouped or separate placement of structure groups (see Section 7).

5.2 Form of the ozone generating elements:

Tubular or plate-type form of ozone generating elements.

5.3 Working pressure

with negative pressure or positive pressure ozone generators.

5.4 Working frequency

Ozone generators operating with line frequency (50/60 Hz) or higher frequencies.

6. Technical Data of the Ozone Generator

The ozone generator is to be characterized by the statement of the following technical data which are to be presented in the operational instructions:

--manufacturer, type and year of construction of the ozone generator.

--gas used (composition), i.e., gas added.

--ozone production in grams per hour (O_3 g/hr) at a concentration of O_3 g/cu m related to a pressure of 1013 mbar and a temperature of 0°C (see Section 12).

--volume flow in cu m/hr under operational conditions.

--working pressure of the ozone generator and the highest allowable negative pressure or excess pressure in bars.

--working pressure of the air preparation system and highest allowable pressure in bars.

--connection value of the ozone generator in volts (V), amperes (A), and hertz (Hz).

--working frequency in the ozone generator element in hertz.

--maximum voltage in the ozone generator element in kV (kilovolts).

--energy requirement for the ozone generator in kW.

--connection value of ozone generator in kVA.

--cooling agent.

--volume flow of the cooling agent in cu m/hr.

--entrance temperature of the cooling agent in degrees Kelvin (°K).

--maximum and minimum pressure of the cooling agent in bars.

--dimensions of the plant in meters (length, width, height).

--space requirement for the generator in cu m.

--operational weight of the generator in kg.

7. Structure groups, construction, and materials of the ozone generator

The following components belong to an ozone generator in the meaning of this standard and independent of the "form" of the ozone production elements and the spatial placement of structure groups:

--equipment for the supply of gas added during production of ozone

--ozone generators

--electrical equipment for the production of high voltage

--equipment for the operation and supervision (of the plant).

Besides this, the necessary safety equipment stated in Section 8 must be present.

7.1 Equipment for the Supply of Necessary Gas

7.1.1 Construction

The equipment needed is established by the type and condition of the gas to be used in the ozone production. Usually this structure group encompasses the use of air, a double absorption system which is filled with a desiccant and through which the gas to be used is directed, whenever the occasion arises, during the insertion of a cooling aggregate.

The regeneration of the absorbent occurs according to the type of drying system with dry and/or warm gas. For an operation with dry oxygen (dew point 228°K) without circulation control, a drying unit is not necessary.

According to the construction of the ozone generator, this plant unit is operated with negative or positive pressure, whereby the gas production equipment is stipulated.

7.1.2 Materials

These are governed by the type and condition of gas to be used in the ozone production.

During the employment of oxygen as the gas to be used for ozone production, the safety regulations for oxygen (VBG 62) are to be observed.

7.2 Ozone Generators

7.2.1 Construction

The ozone generator elements are to be ordered in a vertical or horizontal manner within a closed container (ozone generator). Cooling equipment is necessary for the removal of heat produced during the production of ozone whereby the removal of this heat can be effected by gases or liquids.

7.2.2 Materials for the Parts Affected by Ozone

Stainless steel (for example material #1,4571), aluminum (Al 99.8), suitable synthetic materials (PTFE-Teflon), ceramics and glass.

7.3 Electrical Equipment for the Production of High Voltage

The electrical equipment must correspond to the VDE regulations.

High voltage transformers and, when the occasion arises, throttle valves and the corresponding regulatory apparati, serve for the production of high voltage. For frequencies higher than the line frequency an acceptable transducer is needed in the primary circuit.

7.4 Equipment for Operation and Supervision

The control equipment for the operation of an ozone generator is to be easily accessible as well as arranged. The control panel for control and supervision must contain everything necessary for the safe operation of the switch, regulatory, guidance, measurement and control parts of the generator.

The primary voltage contained in the high voltage transformer must be adjustable for the control of the ozone production. Voltage and current intensity are indicated visibly on the appropriate measuring instruments. This also applies for generators with frequencies higher than the line frequency. If the ozone production is regulated by frequency, this too is to be indicated.

A flow and pressure indicator must be built into the gas control system so that the present operating conditions are readily evident.

8. Safety Equipment

The safety equipment must shut off or prevent the initiation of operation of any ozone generator whose particular limit values are either exceeded or, on the other hand, not attained.

To this end, the following safety equipment is necessary:

The protection or safety of:

--gas control (quantity and/or pressure)

--cooling (pressure, quality and/or temperature)

--high voltage source (over-current and/or temperature)

Further the following items should be present:

--ozone gas warning indicator

--operational hours counter

--dew point indicator

--high voltage indicator

--ozone concentration indicator

The safety equipment are to correspond with the most up-to-date VDE and DIN regulations. For the connection of the ozone generator with circuitry, one must follow the regulations for the operation of a high voltage production plant of the conditional electricity maintenance undertaking (EVU). Electrical indicators and/or machines which are easily accessible and whose locations do not occasion the need for a greater degree of safety are to at least carry out a system of safety in regard to IP 23 (see DIN 40 050, part one, safety types: contact, foreign body, and water protection for electrical operations; definition and application August, 1970). If these indicators and/or machines are not easily accessible, a safety type such as iP 00 (see part one of DIN 40 050) will suffice whenever the VDE regulations for this area are carried out.

9. Location

Different types of electrical machines and/or indicators are located in the ozone generator which necessitate certain minimum requirements as to location (of these machines, etc.). The arrangement necessitates the type of safety system for the installed electric machines and/or indicators (see Section 8). For the protection of a measured life span of the electrical equipment a temperature above 303°K and a relative humidity over 60% is not permissible and it is suggested that the manufacturer of the ozone generator add additional treatment. The location should be continuously dust free (maximum dust content in the air: 2.5 mg/cu m).

Unstable gases and contaminants which oxidize, such as chlorine, sulfuric acid, and carbon monoxide, etc., are not permitted to be present in the environment.

For ventilation, the environment's air is to be changed at least ten times per hour whereby this ventilation must be allowed to operate outside of the danger zone (MAK value for O_3 = 0.1 ppm). (See the publication concerning the employment of ozone as a means of water sterilization; published as UVV of the Professional Society of Gas and Water Works, Düsseldorf, 1970). In order to avoid damage to a water-cooled ozone generator, a frost-free location is necessary.

The location's area should be designed so that maintenance and repair work can be carried out without any hindrance.

10. Cooling Water

The water should not surpass a temperature of 293°K when an ozone concentration of over 20 g/cu m is to be produced (with a pressure of 1013 bar and a temperature of 273°K).

The following standard values apply (methods for the determination of the standard value followed by the DEV -- Deutschen Einheitsverfahren):

pH value	over 6.5
content of displaced materials	under 0.1 ml/l
content of iron (Fe)	under 5 mmole/cu m
content of manganese (Mn)	under 2 mmole/cu m

The employment, by the manufacturer of the generator, of stainless steel applies for the additional chlorine content.

With the employment of aluminum as a material to be in contact with the water, its solubility with a pH value greater than 7.5 is to be indicated.

Recycling or reuse of the cooling water is possible.

11. Operating Information

For the operation of an ozone generator, the technical data as well as information concerning installation, servicing and maintenance by the manufacturer are necessary.

After the start of operation of an ozone generator, evidence of its performance must be gathered by way of testing the ozone product in the manner stated in Sections 12 and 13.

This performance evidence applies as part of the sample testing. This is to be carried out by competent personnel. A contract agreement for maintenance work is recommended. The necessary data concerning operation is to be entered into an operations book on a daily basis.

12. Determination of Ozone Concentration

The following described methods for the determination of the ozone concentration and, therefore, the ozone production per unit time, relate to the input gas.

12.1 Titrimetric determination following the potassium iodide method (see article 10, October 1976).

12.1.1 Theory

The reaction of ozone with potassium iodide takes place only in a neutral buffered aqueous solution (pH = 7) in a stoichiometric manner following the equation:

$$O_3 + 2I^- + H_2O \longrightarrow I_2 + O_2 + 2(OH)^-$$

The liberated iodine is titrated by the acidified sodium thiosulfate:

$$I_2 + 2(S_2O_3)^{-2} \longrightarrow 2I^- + (S_4O_6)^{-2}$$

12.1.2 Indicators and equipment

--means of transmission	high quality steel, glass or PTFE (Teflon)
--wash bottles	500 ml content after Muenke
--three way spigot	with plugs out of PTFE (Teflon)
--membrane pump	ozone tight [necessary only in plants employing low pressure (negative pressure)]
--gas meter	wet design with needle counter as well as thermometer fittings for gas and liquid, consumption performance per drum revolution: 1 liter
--barometer	
--stop watch	
--general accessories	glass beads, used tubing out of PE or PVC for short elastic connections

--retort stand WE 500 (DIN 12 835)

--graduated cylinder 250 (DIN 12 680)

--burette S25 x 0.05 A (DIN 12 700)

--pipettes VPA 1 and VPA 5 (DIN 12 690)
(recommended: magnetic stirrer with stirring rod made of PTFE--Teflon).

12.1.3 Chemicals

Potassium iodide solution:	20.0 g Potassium iodide (KI), 7.3 g disodium hydrogen phosphate (Na_2HPO_4-x$2H_2O$) and 3.5 g potassium dihydrogen phosphate (KH_2PO_4) are dissolved in 1000 ml of double distilled water.
Sodium thiosulfate solution:	0.1N standardized titrimetrically
Sulfuric acid, diluted:	25 ml sulfuric acid (H_2SO_4), d = 1.84 are carefully added to double distilled water and mixed well
Zinc iodide-starch solution:	4 g starch are triturated with a little distilled water and added to a simmering solution of 20 g zinc chloride in 100 ml of water. The solution is boiled until it becomes clear by the restoration of the evaporated water, diluted, displaced with 2 g zinc iodide, filled up to 1000 ml and filtered. The solution is stored in a brown bottle; it is not permitted to turn blue after dilution with 50 times the volume of water during acidification with diluted sulfuric acid.

12.1.4 Construction

The structure of the regulating area (of the plant) is represented in Illustration 1 (not included in publication). The gas feeder [A] should be as short as possible and connected directly to the ozone generator.

All parts of the measuring apparatus which come into contact with the gas mixture containing ozone are not permitted to exhibit any type of ozone consumption. Necessary elastic tube connections are to be as short as possible. Every other wash bottle (placed one after the other) is supplied with 200 ml of potassium iodide solution each and employed corresponding to the experimental set-up in the gathering of samples with regard to Figure 1. Both of the three-way spigots are to be set up next so that the gas used in the experiment is directed through the by-pass area. In this position the flow speed of the measuring gas is adjusted by the throttling valve so that the volume flow is fed through at approximately 1 liter/ min.

During the switching over of the three-way spigot in the measuring area the location of the indicator needle of the gas meter should be watched. A renewed measuring time during the measuring is not necessary because the tolerance in the inserted flow speed of ± 20% can be accepted without hesitation. The total input of gas volume should amount to more than 1 liter, but not more than 5 liters, and is determined exactly. After the completion of the measuring the gas flow is switched again to the by-pass area and the measuring is repeated. During the gathering of samples the barometric levels and measuring gas temperature should be read off.

The freed iodine is titrated with 0.1 N sodium thiosulfate solution after acidification with 5 ml of diluted sulfuric acid. At the conclusion of the titration 1 ml zinc iodide-starch solution is added to the bright yellow colored solution and titrated until it is colorless. Both wash bottles are to be considered, particularly with high concentrations of ozone.

12.1.5 Evaluation

1 ml 0.1 N sodium thiosulfate corresponds to 2.4 mg ozone (O_3). Calculation is possible following the formula:

$$G = \frac{2.4af(1 + 0.00367t)(1013)}{bp}$$

G = content of ozone (O_3) in g/cu m under normal conditions (273°K, 1013 mbar)
a = consumption of 0.1 N sodium thiosulfate solution in ml
b = gas volume which was fed through (in liters)
t = temperature of measured gas in °C
p = air pressure at measuring location in mbar
f = normality factor of the sodium thiosulfate solution

The results are rounded off to the next whole number.

12.1.6 Errors

The sum of errors is to be estimated with the ± 5% (relative error) of the measured values.

12.2 Information Concerning Further Methods of Determination Dependent Upon Calibration

12.2.1 Photometric Determination in the UV Region

Ozone exhibits a wide absorption band between 200 and 300 nm (known as the Hartley Band). At 253.7 nm the absorption coefficient of ozone passes through a maximum. The principle of the method is based on the measurement of the optical absorption of a monochromatic ray with a wave length of 253.7 nm within a gas absorption bulb with quartz glass windows.

12.2.2 Caloric Determination by the Measurement of Enthalpy Decay

During the decay of ozone, the release of heat quantities of 144.41 KJoule per mole of ozone represents a measurement of the ozone concentration of an ozone-containing gas. The principle of the method is based on the exact measurement of a temperature difference between gas containing ozone and gas free of ozone which is brought about by the released heat quantities (through catalytic ozone decay) depending on the ozone concentration of the gas which was fed through.

13. Performance of the Ozone Generator

The performance of the ozone production plant is worked out by the following formula:

$$A = G \times Q_L$$

where:

A = performance of the ozone production plant in g O_3/hr

G = g O_3/cu m (after section 12.1.5)

Q_L = volume flow in cu m/hr (after section 12.1.5)

The result is influenced above all by the variation of:

--volume flow

--net primary voltage and frequency

--cooling water temperature

--multiplication factor

APPENDIX D: OZONE QUESTIONNAIRE SUMMARY TABLES

This appendix provides summary tables drawn from the ozone questionnaire responses from Europe and Canada. The data reported herein are as provided by the individual plants. "NA" signifies "Not Available", and that the respondent did not provide data for the specific question.

The tables are arrayed on two pages and read from left to right. The treatment process is described in order of the design, beginning with the first process step. Data are provided for ozone dosage, power consumption, and chlorine or chlorine dioxide use in final disinfection. The reader is cautioned, however, that the bases on which some of these data have been reported are not always known; hence, data which vary widely from the norm should be discounted. In particular, power consumption data were not always adequately defined by the responding plants, and very low or very high data can be misleading.

TABLE D-1. EUROPEAN OZONE QUESTIONNAIRE SUMMARY

Plant, Location, Water Source & Year Ozone Installed	Design Capacity (cu m/day)	Ozone Generation Capacity (kg/day)	Treatment Process	Purpose(s) of Ozonation	Type of Contacting
Tullich Treatment Works, Oban, Scotland Loch River (1977)	10,000	72	microstraining, ozonation, filtration ozonation, chlorination, pH adjustment	Color removal; additional benefits are Fe & Mn removal, viral inactivation, & bacterial disinfection	3 sparger contactors, 2 stages, 11 min total contact time
Loch Turret Treatment Works, Crieff, Scotland,	81,800	168	microstraining, prechlorination, ozonation, post-chlorination	Color removal, bacterial disinfection	7 units, single stage, contacting by injectors, 5
Sunnyhurst Treatment Plant, Blackburn, Great Britain (1969)	5,000	14.4	lime treatment, slow sand filtration, ozonation, chlorination	Color removal	1 unit, single stage, sparger
Watchgate Treatment Plant, Kendal, Great Britain, Haweswater Lake (1971)	450,000	800	microstraining, prechlorination, rapid sand/anthracite filtration, ozonation, lime addition, chlorination, sulphur dioxide for dechlorination	Color removal	9 units, single stage, injector, 30 minute contact time
Barmby Treatment Works, near Howden, North Humberside, England, Derwent River (to begin in 1978)	55,000	288	rapid gravity filtration, ozonation, slow sand filtration, chlorination	Color removal	3 units, single stage, submerged turbine
Durleigh Filter Station, Durleigh Surface Reservoir 1974	30,000	192	microstraining, chlorination, ozonation (3 stages), coagulation, chlorination, pressure filtration	Iron & manganese, color, taste & odor removal	3 units (number of stages unknown), diffusers, 1-5 min contact time

(continued)

TABLE D-1 (continued)

Plant Name	Ozone Dosage	Energy Demand, Kwh/kg	Contactor off-gas Treatment	Type of Generators	Conjunctive Use of Cl_2 or ClO_2 in Pretreatment	Conjunctive Use of Cl_2 or ClO_2 in Post-Treatment
Tullich	plant has not begun operation	23 estd.	reinjected at plant inlet	Trailigaz tube type, water cooled	-	Cl_2 used, concentration unknown
Loch Turret	2.5	37 total 2 air prep 22.6 generation 6.6 contacting 5.3 O_3 destruction	thermally destroyed at 300°C	CEO Trailigaz, Hollow Plate, water cooled	-	0.7 mg/l Cl_2 applied; 0.1 mg/l free residual in plant effluent
Sunnyhurst	1.5 to 2.0	not available (NA)	none	Trailigaz tube type, water cooled	-	0.2 to 0.3 mg/l Cl_2 added at plant outlet
Watchgate	up to 3.0, avg. 1.25	25.5	activated carbon, vent gas absorber	Trailigaz, tube type, water cooled	-	Cl_2 added, up to 0.4 mg/l
Barmby	not yet in operation	27 estd.	catalytic destruction	Gebruder Herrmann tube, water cooled	-	Cl_2 to be used when plant begins operation
Durleigh	4	36 total 12 for air preparation 18 for generation, 6 for contacting	activated carbon, destruction units to be added in 1977	Trailigaz tube type, water cooled	4 mg/l	residual in plant effluent is 1.0 mg/l

(continued)

TABLE D-1 (continued)

Plant, Location, Water Source & Year Ozone Installed	Design Capacity (cu m/day)	Ozone Generation Capacity (kg/day)	Treatment Process	Purpose(s) of Ozonation	Type of Contacting
Hondsweg Onddorp, Netherlands Haringvliet Water (surface), 1976	15,000	44	ground infiltration (60 days retention time), sand filtration, ozonation, dual media filtration (coal, sand), hypochlorite disinfection	color, taste & odor, organics removal	2 submerged turbine contactors; 10-20 min contact time; stages of contactor unknown
Noordendijk Dordrecht, The Netherlands Rhine River, 1968	28,800	86.4	breakpoint chlorination, coagulation, upflow filtration, downflow filtration, ozonation, chlorination	taste & odor removal, color removal	6 units, single stage, dispersion by rotating cavitating pumps; 10 min contact time
Engelse Werk Zwalle, Netherlands, groundwater plus bankfiltrate, 1972	30,000	9	river sand bank filtration, dry filtration, deacidification tower, filtration, ozonation	taste removal	1 unit, single stage, dispersion by injector; 5 sec contact time
Provinciale Amsterdam, Netherlands Bethunepolder, December 1976	82,200	600	coagulation, settling, rapid filtration, ozonation, coagulation, rapid filtration, slow filtration, chlorination	taste, color, organic compound removal, viral inactivation, bacterial disinfection	1 unit, 3 stages dispersion by sparger; 4 min contact time
Pumpstation Buren, Buren (The Isle of Ameland) Netherlands	1,080	9	rapid filtration, slow sand filtration, ozonation, holding basin, rapid sand filtration	iron removal, color removal	1 unit, single stage, dispersion by submerged turbine; contact time of 6-18 secs; detention time in holding basin of 40-115 min

(continued)

TABLE D-1 (continued)

Plant Name	Ozone Dosage	Energy Demand, Kwh/kg	Contactor off-gas Treatment	Type of Generators	Conjunctive Use of Cl_2 or ClO_2 in Pretreatment	Conjunctive Use of Cl_2 or ClO_2 in Post-Treatment
Hondsweg Onddorp	3 (av)	17-20	thermal destruction	Kerag tube type, water cooled	--	sodium hypochlorite added at concentration of 0.2 mg/l
Noordendijk Dordrecht	2	35	thermal destruction	Kerag tube type, water cooled	breakpoint chlorination	chlorine added at 1.0 mg/l concentration; 0.4 mg/l free residual in effluent; 0.05 mg/l at extremity of distribution system
Engelse Werk Zwalle	0.23 (0.2 kg/hr)	50	destruction by activated carbon treatment	Argentox tube type, water cooled	none	none
Amsterdam	2.5	25	thermal destruction	Trailigaz tube type, water cooled	--	Cl_2 applied as terminal step at concentration of 0.4 mg/l. Free residual of 0.1 mg/l obtained in plant effluent
Pumpstation Buren	5	NA	none	Kerag tube type, water cooled	--	none

(continued)

TABLE D-1 (continued)

Plant, Location, Water Source & Year Ozone Installed	Design Capacity (cu m/day)	Ozone Generation Capacity (kg/day)	Treatment Process	Purpose(s) of Ozonation	Type of Contacting
Cantineweg, Katwijk aan de Rign, Netherlands, 1972	70,000	120	aeration, rapid sand filtration, ozonation, slow sand filtration	taste, odor, color & organic compound removal	1 unit, dispersion by submerged turbine; 2 min contact time
Schaardijk, 150 Rotterdam, Netherlands, Meuse (3 months storage) 1977	210,000	576	coagulation, ozonation, secondary coagulation, filtration, activated carbon, chlorination	taste, odor & color removal, bacterial disinfection, turbidity removal, viral inactivation, filtration aid	6 unit, single stage, submerged turbine; 4 min contact time
Stadtbetriebe Linz G.m.b.h. Linz a.d., Donau, Austria, 1972	26,000	27.36	ozonation, degasing	bacterial disinfection, viral inactivation	4 units, single stage, dispersion by injector; 10 min contact time
Pumpwerk Seibersdorf, Sudstadtzentrum, Austria, horizontal filtered well, 1972	4,320	5.04	ozone is added by injector in a partial stream for sterilization. This is only treatment.	bacterial disinfection	1 unit, dispersion by injector
Stadtwerke, St. Pölten, St. Pölten (Wellfield 3, Well equipment 16, 17) Austria, Groundwater (well) 1972	8,640	10.8	ozonation, hypochlorite	bacterial disinfection	1 unit, 10 min contact time, schwimmbegaser
Stadtwerke St. Pölten, St. Pölten (Wellfield 3, Well equipment 14) Austria, Groundwater (well) 1965	8,640	8.64	ozonation	bacterial disinfection	1 unit, 10 min contact time, schwimmbegaser

(continued)

TABLE D-1 (continued)

Plant Name	Ozone Dosage	Energy Demand, Kwh/kg	Contactor off-gas Treatment	Type of Generators	Conjunctive Use of Cl_2 or ClO_2 in Pretreatment	Conjunctive Use of Cl_2 or ClO_2 in Post-Treatment
Cantineweg	2.4	24.5 total, 3.0 air preparation, 21 O_3 generation, 0.6 contacting	thermal destruction at 300°C	Trailigaz tube type, water cooled	---	None
Schaardijk	3.0	26	combination of thermal destruction plus catalyst	Kerag tube type, water cooled	---	1 mg/l Cl_2 in summer; total residual of 0.3-0.6 mg/l in plant effluent
Linz	0.06	45 total, 14 air preparation 21 generation, 10 contacting	partial reinjection	Sauter plate type, water cooled	---	Not necessary
Seibersdorf	Not given	40 total, 20 air preparation, 20 generation	activated carbon filter	Trailigaz plate type, water cooled	---	None
St. Pölten	0.37-0.56	16	surface area vacuum pump	Kerag tube type, water cooled	---	0.8 mg/l of hypochlorite applied
St. Pölten	0.17-0.21	16	surface area vacuum pump	Kerag tube type, water cooled, 1 unit	---	None

(continued)

TABLE D-1 (continued)

Plant, Location, Water Source & Year Ozone Installed	Design Capacity (cu m/day)	Ozone Generation Capacity (kg/day)	Treatment Process	Purpose(s) of Ozonation	Type of Contacting
Salzburger, Stadtwerke-Wasserwerk, Salzburg, Austria, deep wells with much intermediate dilution of high quality groundwater	58,000	3.6	ozonation, holding tank, distribution	color removal, bacterial disinfection organics removal, viral inactivation	9 units, dispersion by injector; 4 min contact time; 3 injectors per chamber
Seewasserwerk Riet, St. Gallen, Switzerland, Lake of Constance 1968	60,000	288	neutralization, flocculation, 2-stage rapid filtration, slow sand filtration, ozonation, activated carbon filtration, chlorine dioxide	color removal, taste & odor removal, bacterial disinfection, organics removal, viral inactivation	8 submerged turbines; 27-50 min contact time (includes time in piping after contact chamber & until reaching GAC filter)
City of Bern Waterworks, Bern, Switzerland, Wellwater, 1955	24,000	21.4	filtration, ozonation, reservoir, residual ozone measurement, chlorine dioxide addition	odor removal, bacterial disinfection	1 unit, dispersion by injection; 5-8 min contact time
Waterworks Arbon, Arbon, Switzerland, Lake of Constance, 1964	30,000	30	prechlorination, filtration, ozonation activated carbon, chlorination	taste & odor removal, bacterial disinfection, viral inactivation	1 unit, dispersion by injector, contact time unknown
Waterworks Rohrschach, Rohrschach, Switzerland, raw water source unknown 1969	15,000	35	prechlorination, filtration, ozonation, chlorination	bacterial, disinfection, viral inactivation, taste removal	3 units, single stage, injector; 10 min contact time

(continued)

TABLE D-1 (continued)

Plant Name	Ozone Dosage	Energy Demand, Kwh/kg	Contactor off-gas Treatment	Type of Generators	Conjunctive Use of Cl_2 or ClO_2 in Pretreatment	Conjunctive Use of Cl_2 or ClO_2 in Post-Treatment
Salzburger	1.2	55	None	Argentox tube type, water cooled, 3 generators	---	chlorine used only if the ozone equipment is not in service; 0.5 mg/l
Riet, St. Gallen	0.3 - 0.58	26.7	thermal destruction	Kerag tube type, water cooled, 4 generators of 3 kg/hr capacity	---	ClO_2 added as terminal step in concentration of 0.10 to 0.15 mg/l
Bern Waterworks	0.3 - 0.5	81.1 total, 24.1 for air preparation, 25.1 generation, 30.9 contacting	None	Welsbach & Degrémont/ van der Made; tube type, water cooled, 6 generators	---	ClO_2 added at reservoir outlet in 0.04 mg/l concentration
Waterworks Arbon	ca. 0.8	17 (ozone generation only)	None	CEO Trailigaz plates, water cooled, 5 generators	prechlorination	Cl_2 added after activated carbon, 0.1 mg/l
Waterworks Rohrschach	0.3	NA	None	Sauter plate type, water cooled, 3 generators of 0.6 kg/hr capacity	prechlorination	Cl_2 added at reservoir outlet, 0.2 mg/l

(continued)

TABLE D-1 (continued)

Plant, Location, Water Source & Year Ozone Installed	Design Capacity (cu m/day)	Ozone Generation Capacity (kg/day)	Treatment Process	Purpose(s) of Ozonation	Type of Contacting
Waterworks Biel, Biel, Switzerland, Bieler Lake (38 m deep)	36,000	130 (based on 20 hr/day operation). Total capacity is 155.5	prechlorination, alum coagulation, dual media (pumice, sand) filtration, ozonation, activated carbon/sand filtraiton, pH adjustment with NaOH, chlorine dioxide addition	taste & odor removal, color removal, organics removal, viral inactivation	36 units, 2 stages, dispersion by sparger; 10 min contact time
Waterworks Thal, Kauton, St. Gallen, Switzerland, Lake of Constance, 1967	12,000	27	filtration, chlorination, ozonation, chlorine dioxide addition	taste & odor removal, bacterial disinfection, viral inactivation	3 units, dispersion by turbine
Waterworks Altstatten, Altstatten, Switzerland, wellwater	2,500	3.79	ozonation only	bacterial disinfection, organics removal	1 unit, single stage, mixing by turbine
Villingen, Baden-Württemberg, Federal Republic of Germany (F.R.G.), wellwater, 1974	13,000	26	ozonation, flocculation, filtration, chlorine dioxide addition	bacterial disinfection	injector
Süsel, Schleswig Holstein, FRG, groundwater, 1969	30,000	20	ozonation, double layer filtration, (activated carbon/sand)	iron removal organics removal	Wabag contactor, 1 min contact time, type of unit unknown
Solingen, Nordrhein-Westfälen, Bank filtrate from Ruhr River, 1977	2,000	48	aeration, softening, double layer filter, ozonation, filtration plus flocculation, activated carbon, chlorine dioxide	organics removal	washer w/Raschig rings

(continued)

TABLE D-1 (continued)

Plant Name	Ozone Dosage	Energy Demand, Kwh/kg	Contactor off-gas Treatment	Type of Generators	Conjunctive Use of Cl_2 or ClO_2 in Pretreatment	Conjunctive Use of Cl_2 or ClO_2 in Post-Treatment
Biel	1.6	23 total, 2.3 air preparation, 17 generation, 3.7 contacting	dilution with air	Sauter plate type, water cooled, 6 generators of 1.08 kg/hr capacity	prechlorination	ClO_2 added, 0.18 mg/ℓ
Thal	NA	20	none	Kerag tube type, water cooled, 3 generators of 0.375 kg/hr capacity	chlorination, prior to ozonation	ClO_2 post-treatment
Altstatten	NA	NA	withdrawn by suction pump	Kerag tube type, water cooled, 1 generator of 0.158 kg/hr capacity	---	Cl_2 equipment available
Villingen	0.3	NA	destruction by activated carbon	unknown	---	ClO_2 added, 0.3 mg/ℓ
Süsel	2.0	NA	none	Herrmann tube type, water cooled; 2 units	none	---
Solingen	0.4-0.8	50	NA	tube type, water cooled, manufacturer unknown, 2 units	---	ClO_2 added, 0.2 mg/ℓ

(continued)

TABLE D-1 (continued)

Plant, Location, Water Source & Year Ozone Installed	Design Capacity (cu m/day)	Ozone Generation Capacity (kg/day)	Treatment Process	Purpose(s) of Ozonation	Type of Contacting
Siegburg, Nordrhein-Westfälen, wells, 1968	70,000	48	aeration, filtration, ozonation, activated carbon, chlorination	organics removal, color, taste & odor removal	injector, 1 column
Osterode, Niedersachsen, FRG, wellwater, 1972	119	0.24	ozonation only	bacterial disinfection	injector
Düren, Nordheim, Westfälen, FRG, Ruhr River water, 1970	36,000	18	microstraining (5μ), ozonation, granular activated carbon, softening	Fe, Mn removal disinfection, organics removal	2 Frings mixers; 4 min detention
Duisburg, Nordheim, Westfälen, FRG, Rhine River 1966-72	72,000	192	sand bank filtration, ozonation, granular activated carbon, chlorine dioxide, softening	disinfection, taste & odor removal, oxygenation, color removal, viral inactivation, Mn removal, organics removal	2 packed columns; 6 min detention, Wabag
Diez/Lahn, Rheinland, Pfalz, FRG, groundwater, 1966	2,000	1.0	ozonation, hypochlorination	disinfection	2 injectors; 30 min detention
Wieslautern/Dahn, Rheinland, Pfalz, FRG, groundwater 1975	1,500	1.5	ozonation	disinfection	1 injector
Bad Neustadt, Bavaria, FRG, groundwater, 1965	9,000	4.8	ozonation	disinfection	1 injector; 5 min detention

(continued)

TABLE D-1 (continued)

Plant Name	Ozone Dosage	Energy Demand, Kwh/kg	Contactor off-gas Treatment	Type of Generators	Conjunctive Use of Cl_2 or ClO_2 in Pretreatment	Conjunctive Use of Cl_2 or ClO_2 in Post-Treatment
Siegburg	4-5	20	treated, method unknown	Herrmann tube type, water cooled	---	Cl_2 added, 0.3 mg/l
Osterode	57	28	none	Kluber tube type, water cooled; 1 generator	---	none
Düren	0.3	30	activated carbon	Herrmann tube type, water cooled; 1 generator	none	none
Duisburg	2.0	13	recycling	Demag tube type, water cooled; 4 generators	---	chlorine dioxide 0.5 mg/l
Diez/Lahn	0.15	100	none	Benckiser tube type, water cooled; 2 generators	---	hypochlorination 0.25 mg/l
Wieslautern/Dahn	0.3	20	yes	Schade tube type, water cooled; 3 generators	none	none
Bad Neustadt	0.25	100	none	Kluber tube type, water cooled; 10 generators	none	none

(continued)

TABLE D-1 (continued)

Plant, Location, Water Source & Year Ozone Installed	Design Capacity (cu m/day)	Ozone Generation Capacity (kg/day)	Treatment Process	Purpose(s) of Ozonation	Type of Contacting
Bad Honnef, Nordheim, Westfälen, FRG, Rhine River water, 1975	13,000	24	river sand bank filtration, ozonation, filtration, chlorine dioxide	taste & odor control, disinfection, organics removal, viral inactivation	1 Ozotech injector; 10 min detention
Annweiler, Rheinland, Pfalz, FRG, groundwater, 1977	480	0.5	ozonation	disinfection	injector
Albstadt, Baden-Württemberg, FRG, groundwater, 1976	18,000	40	preozonation, CO_2 degassing, double layer filtration, granular activated carbon, chlorine dioxide	disinfection, viral inactivation, taste & odor, turbidity & organics removal	2 Sauter gas turbines; 8 min detention
Neustadt/Sisch, Bavaria, FRG, wellwater, 1959	2,600	2.6	ozonation only	bacterial disinfection	turbulence chamber w/rotor; 20 min contact time
Neuffen, Baden-Württemberg, FRG, wellwater, 1970	1,600	1.56	flocculation, sand filtration, ozonation	bacterial disinfection, organics, turbidity, viral inactivation	Argentox injector; 14 min contact time
Mülheim/Ruhr, Nordrhein-Westfälen, FRG, Ruhr River 1977	50,000	192	preozonation, flocculation, sedimentation, ozonation, filtration, carbon adsorption, biological filtration, ground passage, chlorination	color & turbidity removal, Fe & Mn removal, taste & odor control, flocculation, disinfection & viral inactivation	2 diffuser systems; 1 turbine system

(continued)

TABLE D-1 (continued)

Plant Name	Ozone Dosage	Energy Demand, Kwh/kg	Contactor off-gas Treatment	Type of Generators	Conjunctive Use of Cl_2 or ClO_2 in Pretreatment	Post-Treatment
Bad Honnef	3.0	--	catalytic	Herrmann tube type, water cooled; 2 generators	---	chlorine dioxide 0.1 mg/l
Annweiler	0.3	20	yes	Schade tube type, water cooled	none	none
Albstadt	0.8	--	catalytic	Sauter plate type, water cooled; 1 generator	---	chlorine dioxide 0.3 mg/l
Neustadt	2.0	78	none	Kerag tube type, water cooled; 1 generator	none	none
Neuffen	0.4	107	none	Argentox tube type, water cooled; 1 generator	none	none
Mülheim/Ruhr	4.0	--	catalytic	Trailigaz tube type, water cooled; 2 generators	---	0.3 mg/l Cl_2

(continued)

TABLE D-1 (continued)

Plant, Location, Water Source & Year Ozone Installed	Design Capacity (cu m/day)	Ozone Generation Capacity (kg/day)	Treatment Process	Purpose(s) of Ozonation	Type of Contacting
Münchengladback, Nordrhein-Westfälen, FRG, groundwater, 1965	20,000	14.4	preaeration, gravel filtration, ozonation, hypochlorination	disinfection, taste control	Argentox injector; 3-5 min detention
Lindau, Bavaria, FRG, Lake of Constance, 1972	20,000	44	microstraining, ozonation, filtration, final chlorination	disinfection, organics removal, turbidity	2 VAR injectors; 10 min detention
Limberg, Hessen, FRG, groundwater, 1969	7,700	2.4	ozonation	disinfection	1 injector
Grunstedt, Rheinland Pfalz, FRG, wellwater, 1970	240	0.4	ozonation	disinfection	1 injector
Düsseldorf, Nordheim, Westfälen, FRG, Rhine River, sandbank, 1969-77	403,000	874	sand bank filtration, ozonation, prefiltration, granular activated carbon, softening, chlorine dioxide	Fe, Mn removal, taste & odor control, organics removal, disinfection	1 injector; 5 min detention plus 20 min holding tank
Aachen, Nordheim, Westfälen, FRG, groundwater, 1968	7,200	3.6	ozonation, filtration, chlorine dioxide	taste, Fe & Mn, color, turbidity, disinfection	Wabag packed column; 3.7 min detention
Bodensee Wasserversorgung (Sipplinger Berg), Baden-Württemberg, FRG, Lake of Constance water, 1969	650,000	780	microstraining, ozonation, filtration, chlorine dioxide	taste & odor, disinfection & viral inactivation, micro-flocculation, organics removal	Wabag packed column; 12 min detention

(continued)

TABLE D-1 (continued)

Plant Name	Ozone Dosage	Energy Demand, Kwh/kg	Contactor off-gas Treatment	Type of Generators	Conjunctive Use of Cl_2 or ClO_2 in Pretreatment	Conjunctive Use of Cl_2 or ClO_2 in Post-Treatment
Münchengladback	0.9	30	none	Argentox tube type, water cooled; 12 generators	---	0.1 mg/l NaOCl
Lindau	0.4	--	none	CEO plate type, water cooled; 2 units	---	0.2 mg/l Cl_2
Limberg	2.7	28	yes	Benckiser tube type, water cooled; 2 units	none	none
Grunstedt	0.3	30	yes	Schade tube type, water cooled; 1 unit	none	none
Düsseldorf	3.0	30	activated carbon & catalytic	Herrmann tube type, water cooled; 28 generators	---	chlorine dioxide 0.2 mg/l
Aachen	0.35	30	none	Herrmann tube type, water cooled; 1 generator	---	chlorine dioxide 0.3 mg/l
Sipplinger Berg	1.0	29.5	activated carbon	Herrmann tube type, water cooled; 6 generators	---	chlorine 0.6 mg/l

(continued)

TABLE D-1 (continued)

Plant, Location, Water Source & Year Ozone Installed	Design Capacity (cu m/day)	Ozone Generation Capacity (kg/day)	Treatment Process	Purpose(s) of Ozonation	Type of Contacting
Wuppertal, Nordheim, Westfälen, FRG, Rhine River, 1967	150,000	360	sand bank filtration, aeration, ozonation, gravel filtration, granular activated carbon, chlorine dioxide	Fe, Mn removal, taste & organics removal, viral inactivation	spray tower within filter
Königswinter, Nordheim, Westfälen, FRG, Rhine River, 1962	4,800	5.6	sand bank filtration, granular activated carbon, ozonation, granular activated carbon, chlorination	Fe, Mn removal, taste removal	injector, 40 min detention
Altheimingen, Rheinland Pfalz, FRG, groundwater, 1964	400	0.9	ozonation	disinfection	injector
Oberhessen, Hessen, FRG, groundwater, 1970	50,000	32.4	ozonation, chlorination	disinfection, viral inactivation	VAR injectors, 7 min detention
Floremberg, Hessen, FRG, groundwater, 1970	2,500	1.2	ozonation	disinfection	Argentox injector
Friedrichshafen, Baden-Würrtemberg, FRG, Lake of Constance water, 1970	30,000	52.4	ozonation, flocculation, granular activated carbon over sand filter, chlorination	taste removal, disinfection, organics & turbidity removal	VAR injectors, 4-8 min detention
Meersburg, Baden-Württemberg, FRG, Lake of Constance water, 1969	4,000	9.0	microstraining, ozonation, sand filtration, chlorination	color, taste & odor removal, disinfection	1 injector, 1 rotor, VAR

(continued)

TABLE D-1 (continued)

Plant Name	Ozone Dosage	Energy Demand, Kwh/kg	Contactor off-gas Treatment	Type of Generators	Conjunctive Use of Cl_2 or ClO_2 in Pretreatment	Post-Treatment
Wuppertal	0.8-1.0	35	none	Herrmann tube type, water cooled	---	chlorine dioxide
Königswinter	0.2	--	none	Demag tube type, water cooled	---	chlorine dioxide
Altheimingen	0.3	50	none	Schade tube type, water cooled; 2 generators	none	none
Oberhessen	0.6	35	none	VAR plate type, water cooled; 2 generators	---	chlorine 0.08 mg/l
Floremberg	0.35	--	none	Argentox, water cooled	---	---
Friedrichshafen	0.8	22.5	activated carbon	Trailigaz plate type, water cooled; 3 generators	---	chlorine dioxide 0.3 mg/l
Meersburg	1.6	--	activated carbon	VAR Sauter plate, water cooled; 1 generator	---	chlorine 0.4 mg/l

(continued)

TABLE D-1 (continued)

Plant, Location, Water Source & Year Ozone Installed	Design Capacity (cu m/day)	Ozone Generation Capacity (kg/day)	Treatment Process	Purpose(s) of Ozonation	Type of Contacting
Bad Mergentheim, Bad-Württemberg, FRG, 1976	6,000	6	flocculation, filtration, ozonation, filtration, chlorination	disinfection	2 injectors, Schneider, 10 min detention
Mathay, Commune de Mathay, France, Doubs River	75,000	270	prechlorination, coagulation, sedimentation, filtration, ozonation, chlorination	color, taste & odor, bacterial disinfection, turbidity, viral inactivation	porous plate diffuser, 2 compartments per chamber
Alencon, France, surface water, 1965	20,000 Avg Daily Flow	80	prechlorination, coagulation, sedimentation, filtration, ozonation, chlorine dioxide	taste & odor, color removal	2 units, 3 stages, porous plate diffuser
Usine Francois Duroy, Louvigny-Caen, France, Orne River, 1976	30,000	108	sodium hypochlorite, coagulation, sedimentation, powd. act. carbon, filtration, ozonation, chlorination	taste & odor, color, organics removal, bacterial disinfection, turbidity reduction, viral inactivation	2 units, porous plate diffusers, 10 min contact
Le Mans, France, river water 1970	30,000	240	prechlorination, coagulation, sedimentation, filtration, ozonation, chlorine dioxide	iron oxidation, taste & odor control, organics removal, bacterial disinfection, viral inactivation	3 units, 15 min contact time, porous tubes

(continued)

TABLE D-1 (continued)

Plant Name	Ozone Dosage	Energy Demand, Kwh/kg	Contactor off-gas Treatment	Type of Generators	Conjunctive Use of Cl_2 or ClO_2 in Pretreatment	Conjunctive Use of Cl_2 or ClO_2 in Post-Treatment
Bad Mergentheim	0.6	--	yes	Argentox tube type, water cooled; 1 generator	---	chlorine dioxide 0.1 mg/l
Mathay	1.5	21.9	none	Degrémont tube type, water cooled; 3 generators of 3.75 kg/hr	prechlorination	Cl_2 added, 0.3 to 0.5 mg/l. 0.2 mg/l free residual at plant exit
Alencon	3	NA	therman destruction at 220°C	Degrémont-Welsbach	prechlorination	ClO_2 added, 0.2 mg/l
Louvigny-Caen	1.3	18-20	none	Degrémont tube type, water cooled; 2 generators of 2.25 kg/hr capacity	sodium hypochlorite addition	Cl_2 added, 0.15 mg/l
Le Mans	3 to 4	24	dilution & ventilation	Trailigaz tube type, water cooled; 4 generators--2 of 1.5 kg/hr & 2 of 4.5 kg/hr capacity	prechlorination	chlorine dioxide; 1.0-1.5 mg/l

(continued)

TABLE D-1 (continued)

Plant, Location, Water Source & Year Ozone Installed	Design Capacity (cu m/day)	Ozone Generation Capacity (kg/day)	Treatment Process	Purpose(s) of Ozonation	Type of Contacting
Rosporden, France, river plus spring water, 1974	3,600	7.2	prechlorination, coagulation with alum, sedimentation, filtration, ozonation, lime addition, chlorination	bacterial disinfection, organics, taste, turbidity reduction	1 unit, porous tubes, 5.5 min contact time
Langres, France, water from dam, 1958	1,200 avg daily production	4.56	prechlorination, alum coagulation sedimentation, ozonation	bacterial disinfection, viral inactivation	1 unit, dispersion by injector, 2 stage contactor
Saint-Charles Waterworks, Nancy, France, Moselle River, 1935	140,000	187.2	prechlorination, coagulation, sedimentation, filtration, lime addition, chlorination, ozonation	bacterial disinfection	dispersion by injector
Du Rupt du Mad Waterworks, Metz, France, 1971	90,000	199.68	prechlorination, coagulation, sedimentation, filtration, ozonation, post-chlorination	taste & odor removal, bacterial disinfection, Mn removal	1 unit, porous tubes, 20 min contact time
Planques Waterworks, Montauban, river water plus spring water, 1973	19,200	4.8	prechlorination, coagulation, sedimentation, filtration, ozonation	bacterial disinfection	2 units, porous tubes, 7 min maximum contact time
Toulon, Perigueux, springs of Cluzeau and Abime, 1971	30,000	36	coagulation, sedimentation, filtration, ozonation	taste & odor, organics, turbidity reduction, bacterial disinfection	porous tubes, 10 min contact time

(continued)

TABLE D-1 (continued)

Plant Name	Ozone Dosage	Energy Demand, Kwh/kg	Contactor off-gas Treatment	Type of Generators	Conjunctive Use of Cl_2 or ClO_2 in Pretreatment	Post-Treatment
Rosporden	NA	NA	none	Trailigaz tube type, water cooled	prechlorination	Cl_2 added at 0.4 mg/l maximum
Langres	1.2	23.85	none	CEO Trailigaz plate type, water cooled, 2 generators	prechlorination	none
Saint-Charles (Nancy)	NA	NA	none	CEO Trailigaz plate type, water cooled, 11 generators	prechlorination	Cl_2 added prior to ozonation
Du Rupt du Mad	NA	NA	none	Degrémont tube type, water cooled, 1 generator of 0.2 kg/hr capacity	prechlorination	none
Planques	NA	NA	thermal destruction	Degrémont tube type, water cooled, 4 generators of 2.08 kg/hr capacity	prechlorination	0.1 mg/l free Cl_2 residual at plant exit
Toulon	0.30 to 0.50	47 total; 11 air preparation 26 generation 12 thermal destruction	thermal destruction	Degrémont tube type, water cooled, 1 generator	prechlorination	0.15 to 0.20 mg/l Cl_2 added

(continued)

TABLE D-1 (continued)

Plant, Location, Water Source & Year Ozone Installed	Design Capacity (cu m/day)	Ozone Generation Capacity (kg/day)	Treatment Process	Purpose(s) of Ozonation	Type of Contacting
Anglet, France, le Nive River, 1967	25,000 avg daily production	50.4	prechlorination coagulation, sedimentation, filtration, ozonation	bacterial disinfection	2 spray tower units, 5 minute contact time
Les Vans, France le Chassezac River, 1973	4,320	8.2	prechlorination, alum and lime coagulation, sedimentation, filtration, ozonation, sodium carbonate	bacterial disinfection	dispersion by a submerged tube
Pont Aven, France (Moelan-sur-Mer), l'Aven River, 1976	2,000	3.6	chlorine dioxide, coagulation, sedimentation, filtration, ozonation, chlorine dioxide	bacterial disinfection	dispersion by spray tower
Pont Aven, France (Route de Bannalec), 1974	4,800	24	chlorine dioxide, coagulation, sedimentation, filtration, ozonation, chlorine dioxide	bacterial disinfection, organics removal	porous tubes
Barrage des Rivières, le Longeron, raw water from dam, 1952	8,000	18	chlorine/activated carbon, lime and alum coagulation, sedimentation, sand filtration, ozonation, chlorination, lime addition	bacterial disinfection, viral inactivation	porous tubes, 1 chamber, single stage, 6 min contact time
Annet-sur-Marne, France, Marne River, 1973	25,000	162	prechlorination, coagulation, sedimentation, ozonation, chlorine dioxide	taste & odor removal, bacterial disinfection, viral inactivation	3 units, porous tubes, 7 min contact time

(continued)

TABLE D-1 (continued)

Plant Name	Ozone Dosage	Energy Demand, Kwh/kg	Contactor off-gas Treatment	Type of Generators	Conjunctive Use of Cl_2 or ClO_2 in Pretreatment	Conjunctive Use of Cl_2 or ClO_2 in Post-Treatment
Anglet	10	NA	none	CEO Trailigaz plate type, water cooled	prechlorination	none
Les Vans	NA	9.4	none	Welsbach-Degrémont tube type, water cooled, 1 generator	prechlorination	none
Pont Aven (Moelan-sur-Mer)	0.3	NA	none	Degrémont tube type, water cooled, 1 generator	ClO_2 addition	ClO_2 addition
Pont Aven (Route de Bannalec)	NA	NA	none	Degrémont tube type, water cooled, 1 generator	ClO_2 addition	ClO_2 addition
Barrage des Rivières	1 to 2.5	105	none	Degrémont tube type, water cooled, 1 generator	prechlorination	1.0 to 1.5 mg/l Cl_2 added
Annet-sur-Marne	2	NA	recycling	Trailigaz tube type, water cooled, 3 generators of 2.25 kg/hr capacity	prechlorination	0.5 mg/l ClO_2 added

(continued)

TABLE D-1 (continued)

Plant, Location, Water Source & Year Ozone Installed	Design Capacity (cu m/day)	Ozone Generation Capacity (kg/day)	Treatment Process	Purpose(s) of Ozonation	Type of Contacting
Chattellerault, France, Vienne River, 1925	23,000	72	coagulation, sedimentation, filtration, ozonation	taste & odor, color, bacterial disinfection, organics removal, viral inactivation	2 chambers, 4 injectors, 6 min contact time, 3 stages per chamber
Aubusson, France, dam water, 1974	2,880	7	coagulation, sedimentation, filtration, ozonation	taste & odor, color, bacterial disinfection, organics removal, viral inactivation	2 chambers, single stage, injector, 8 min contact time
Bourbon Lancy, France water from alluvial water table of the Loire River, 1974	4,800	9.9	ozonation plus filtration (for demanganization), neutralization, chlorination	iron & manganese oxidation, bacterial disinfection	1 chamber, dispersion by injector, 8 min contact time
Gueugnon (Saone et Loire), France, Arroux River, 1976	8,000	15	prechlorination, coagulation, sedimentation, filtration, ozonation, chlorination	taste removal, bacterial disinfection	1 chamber, dispersion by injector, 8 min contact time
Ste. Cecile, France, 1970	2,400	3.96	prechlorination, coagulation, sedimentation, filtration, ozonation, chlorination (intermittent)	taste & color removal, bacterial disinfection	2 chambers, porous diffusers
Lescure D'Albigeois, France, groundwater from alluvial water table of Tain River	3,350	NA	ozonation only	bacterial disinfection, viral inactivation (sterilization)	1 chamber, porous diffusers

(continued)

TABLE D-1 (continued)

Plant Name	Ozone Dosage	Energy Demand, Kwh/kg	Contactor off-gas Treatment	Type of Generators	Conjunctive Use of Cl_2 or ClO_2 in Pretreatment	Conjunctive Use of Cl_2 or ClO_2 in Post-Treatment
Chattellerault	3	18	none	CEO Trailigaz plate type, water cooled, 3 generators of 1.2 kg/hr capacity	none	none
Aubusson	2	14	none	CEO Trailigaz plate type, water cooled, 2 generators of 0.18 kg/hr capacity	none	none
Bourbon Lancy	twice the concentration of manganese in the raw water	NA	none	Degrémont tube type, water cooled	none	0.5 mg/l Cl_2 added
Gueugnon	0.8	NA	none	Degrémont tube type, water cooled	prechlorination	Cl_2 added, approximately 0.1 mg/l free residual at plant exit
Ste. Cecile	0.2	NA	none	Degrémont tube type water cooled	prechlorination	Cl_2 added intermittently
Lescure D'Albigeois	NA	NA	none	Degrémont tube type, water cooled	none	none

(continued)

TABLE D-1 (continued)

Plant, Location, Water Source & Year Ozone Installed	Design Capacity (cu m/day)	Ozone Generation Capacity (kg/day)	Treatment Process	Purpose(s) of Ozonation	Type of Contacting
Ste.-Maries-de-la-Mer, France, river water	3,600	15	prechlorination, coagulation, sedimentation, filtration, ozonation	taste & odor, color, bacterial disinfection, viral inactivation	2 chambers, porous diffusers, 4 min contact time
Saint Gilles du Gard, France, river water, 1970	4,800	5.8	prechlorination, coagulation, sedimentation, filtration, ozonation	taste & odor, color, bacterial disinfection, viral inactivation	2 chambers, porous tube, 5 min contact time
La Roche waterworks, Nantes, France, Loire River, 1975	240,000	720	3 to 17 mg/l chlorine, alum and sodium silicate coagulation, sedimentation with pulsators, filtration, ozonation, chlorination	taste & odor, color, bacterial disinfection, organics removal, viral inactivation	3 chambers, 2 stages, porous tube diffusers, 8 min contact time
Villeneuve-la-Garonne, France, well water, 1970	40,000 max daily production	96	nitrification/deferrization, filtration, ozonation	bacterial disinfection	1 chamber, 2 stage, injector, 15 min contact time
Flins, France, well water, 1968	120,000	264	coagulation, powd. act. carbon, sedimentation, nitrification/deferrization, ozonation, chlorination	bacterial disinfection	1 chamber, 2 stage, porous tubes, 15 min contact time
Le Pecq, France, well water, 1967	30,000	36	nitrification/deferrization, filtration, ozonation	bacterial disinfection	1 chamber, 2 stage, porous tubes, 15 min contact time

(continued)

TABLE D-1 (continued)

Plant Name	Ozone Dosage	Energy Demand, Kwh/kg	Contactor off-gas Treatment	Type of Generators	Conjunctive Use of Cl_2 or ClO_2 in Pretreatment	Conjunctive Use of Cl_2 or ClO_2 in Post-Treatment
Ste.-Maries-de-la-Mer	0.6	31.6	none	Trailigaz tube type, water cooled	prechlorination	none
Saint Gilles du Gard	0.5	32	recycled	Trailigaz tube type, water cooled, 1 generator of 0.240 kg/hr capacity	prechlorination	none
La Roche	1.5	50	thermal destruction	Degrémont tube type, water cooled, 6 generators of 5 kg/hr capacity	prechlorination 3 to 17 mg/l	0.5 to 2 mg/l Cl_2 added
Villeneuve-la-Garonne	1	30 total 12 air preparation, 18 generation	NA	Degrémont tube type, water cooled, 2 generators of 2 kg/hr capacity	NA	none
Flins	1	30 total 12 air preparation, 18 generation	thermal destruction	Degrémont tube type, water cooled, 1 generator of 11 kg/hr capacity	none	0.2 mg/l Cl_2 added
Le Pecq	1	30	NA	CEO plate type, water cooled, 1 generator of 1.5 kg/hr capacity	NA	none

(continued)

TABLE D-1 (continued)

Plant, Location, Water Source & Year Ozone Installed	Design Capacity (cu m/day)	Ozone Generation Capacity (kg/day)	Treatment Process	Purpose(s) of Ozonation	Type of Contacting
Croissy, France, well water, 1975	30,000	72	nitrification/deferrization, filtration, ozonation	bacterial disinfection	4 chambers, 2 stage, porous tubes, 15 min contact time
Huez-en-Oisans, France, Lake Blanc	2,500	7	mineralization (carbon dioxide gas addition), filtration, ozonation, sodium carbonate neutralization	bacterial disinfection	1 chamber, 2 stage, porous tubes, 8 min contact time
Clairfont Plant, Toulouse, France, Garonne River, 1970	110,000	216	chlorine dioxide, coagulation, double sedimentation, sand filtration, ozonation, lime/sodium silicate addition	taste, odor, organics, bacterial disinfection, viral inactivation	4 chambers, 4 stage, porous tubes, 8 min contact time
Neris Les Bains, Marcoing, France, surface water (dam), 1974	780	8.2	prechlorination, coagulation, settling, filtration, ozonation, final chlorination	bacterial disinfection	diffuser
Aubagne, France, Marseille Canal (Purance River), 1967	25,000	20.4	prechlorination, coagulation, settling, filtration, ozonation, final chlorination	bacterial disinfection, viral inactivation	4 diffusers, 20 min detention
Marseille, France, Provence Canal (Verdon River), 1976	90,000	200	prechlorination, coagulation, filtration, ozonation, final chlorination	taste & odor, organics, bacterial disinfection, viral inactivation	4 diffusers, 20 min detention

(continued)

TABLE D-1 (continued)

Plant Name	Ozone Dosage	Energy Demand, Kwh/kg	Contactor off-gas Treatment	Type of Generators	Conjunctive Use of Cl_2 or ClO_2 in Pretreatment	Post-Treatment
Croissy	1	30	thermal destruction	Degrémont tube type, water cooled, 1 generator of 1.5 kg/hr capacity	NA	none
Huez-en-Oisans	0.7	20	none	Degrémont tube type, water cooled, 1 generator of 0.342 kg/hr capacity	none	none
Clairfont	1.8	43	cycled into diesel generators	Trailigaz tube type, water cooled, 4 generators of 2.25 kg/hr capacity	ClO_2 added	none
Neris Les Bains	1-2	NA	none	Degrémont tube type, water cooled	yes	chlorine, 0.5 mg/l
Aubagne	0.9	NA	none	2 tube type, water cooled	chlorine	chlorine, 0.4 mg/l
Marseille	1.0	NA	none	Degrémont tube type, water cooled, 2 units	chlorine	chlorine, 0.2 mg/l

(continued)

TABLE D-1 (continued)

Plant, Location, Water Source & Year Ozone Installed	Design Capacity (cu m/day)	Ozone Generation Capacity (kg/day)	Treatment Process	Purpose(s) of Ozonation	Type of Contacting
Annecy-le-Vieux, France, Lake of Annecy, 1975	7,500	15	NA	taste & odor, organics, bacterial disinfection, viral inactivation	3 diffusers
Ferel, France, river water, 1971	30,000	60	prechlorination, coagulation, settling, filtration, ozonation, chlorine dioxide	taste & odor, organics, viral inactivation, bacterial disinfection	2 diffusers, 15 min detention
St. Jean-sur-Mayenne, France, Mayenne River, 1976	2,400	2.7	prechlorination, coagulation, settling, filtration, ozonation, final chlorination	taste & odor, organics, viral inactivation, bacterial disinfection	1 diffuser
Ferel, France, river water, 1971	30,000	120	prechlorination, coagulation, settling, filtration, ozonation, chlorine dioxide	taste & odor, organics, bacterial disinfection, viral inactivation	2 diffusers, 15 min detention
Couesmes-en-Froulay, France, river water, 1973	600	1.2	prechlorination, coagulation, settling, filtration, ozonation, neutralization, final chlorination	Fe removal, color, taste & odor, bacterial disinfection, viral inactivation	1 diffuser

(continued)

TABLE D-1 (continued)

Plant Name	Ozone Dosage	Energy Demand, Kwh/kg	Contactor off-gas Treatment	Type of Generators	Conjunctive Use of Cl_2 or ClO_2 in Pretreatment	Conjunctive Use of Cl_2 or ClO_2 in Post-Treatment
Annecy-le-Vieux	NA	29	thermal destruction	Degrémont tube type, water cooled, 2 units	NA	0.1 mg/l Cl_2
Ferel	2.0	25	none	Welsbach, tube type, water cooled, 4 units	Cl_2	2 mg/l ClO_2
St. Jean-sur-Mayenne	2.0	NA	none	Degrémont tube type, water cooled, 12 units	Cl_2	0.5 mg/l Cl_2
Ferel	2.0	25	none	Welsbach, tube type, water cooled, 4 units	Cl_2	2 mg/l ClO_2
Couesmes-en-Froulay	3.0	NA	none	Degrémont tube type, water cooled, 3 units	Cl_2	0.5 mg/l Cl_2

(continued)

TABLE D-1 (continued)

Plant, Location, Water Source & Year Ozone Installed	Design Capacity (cu m/day)	Ozone Generation Capacity (kg/day)	Treatment Process	Purpose(s) of Ozonation	Type of Contacting
Port Brillet, France, lake water, 1975	2,400	6.0	prechlorination, coagulation, settling, filtration, ozonation, neutralization, final chlorination	taste removal, bacterial disinfection, viral inactivation	2 diffusers
Ancteville, France, river water, 1970	1,000	2.0	prechlorination, coagulation, settling, filtration, ozonaticn, final chlorination	bacterial disinfection, viral inactivation	1 plate column, 20 sec
Le Poiroux, France	16,000	32	prechlorination (chlorine dioxide), $KMnO_4$ addition, coagulation, settling, filtration, ozonation, neutralization, chlorination	taste, odor, color, bacterial disinfection, viral inactivation, turbidity, organics	diffuser
Coutances, France, river water	2,000	7	chlorine dioxide, coagulation, settling, filtration, ozonation, chlorine dioxide	taste removal, viral inactivation, bacterial disinfection	1 plate column, 20 sec
Pinel Hauterive (Lot et Garrone) Lot River, 1968	7,200	12	prechlorination, flocculation, settling, filtration, ozonation	organics removal, viral inactivation, bacterial disinfection	2 injectors, 4 min detention
Crissey, France, groundwater, 1974	7,200	34	ozonation, coagulation, filtration	iron & manganese	diffuser

(continued)

TABLE D-1 (continued)

Plant Name	Ozone Dosage	Energy Demand, Kwh/kg	Contactor off-gas Treatment	Type of Generators	Conjunctive Use of Cl_2 or ClO_2 in Pretreatment	Conjunctive Use of Cl_2 or ClO_2 in Post-Treatment
Port Brillet	0.15	20-30	none	Degrémont tube type, water cooled, 7 units	Cl_2	1.4 mg/l Cl_2
Ancteville	1.6	20-30	none	Degrémont tube type, water cooled, 5 units	Cl_2	0.2 mg/l Cl_2
Le Poiroux	NA	NA	NA	Degrémont tube type, water cooled, 2 units	ClO_2	Cl_2
Coutances	3.4	20-30	none	Degrémont, 19 units, water cooled	ClO_2	0.2 mg/l ClO_2
Pinel Hauterive	1.0	NA	none	Otto plate type, water cooled, 2 units	Cl_2	none
Crissey	0.3-0.5	48	thermal destruction	Degrémont tube type, water cooled, 2 units	none	none

(continued)

TABLE D-1 (continued)

Plant, Location, Water Source & Year Ozone Installed	Design Capacity (cu m/day)	Ozone Generation Capacity (kg/day)	Treatment Process	Purpose(s) of Ozonation	Type of Contacting
Bonnetable, France, 1948	2,000	3	prechlorination, coagulation, settling, filtration, ozonation, final chlorination	bacterial disinfection, viral inactivation	2 injectors, 5 min detention
Gueret, France, 1967	2,500	12.0	chlorine dioxide, coagulation, settling, filtration, ozonation	color, taste, odor, organics, bacterial disinfection	diffusers
Tintry Pont Duroi, France, river water, 1973	7,500	14.0	chlorine dioxide, coagulation, settling, preozonation, filtration, ozonation	bacterial disinfection, viral inactivation	4 diffusers, 6 min detention
Branden, France, lake water, 1967	10,000	12.0	chlorine dioxide addition, coagulation, settling, preozonation, filtration, final ozonation	Fe, Mn removal, bacterial disinfection, viral inactivation	2 diffusers, 12 min
Blanzy, France, lake water, 1974	10,000	30.0	chlorine dioxide, coagulation, settling, preozonation, filtration, ozonation	Fe, Mn removal, color removal, bacterial disinfection	diffusers, 12 min detention
Salgues, (Aveyron), France, river water, 1977	NA	NA	chlorine dioxide, coagulation, settling, filtration, ozonation, chlorine dioxide & polyphosphate addition	color, taste, odor, turbidity, organics, bacterial disinfection	NA

(continued)

TABLE D-1 (continued)

Plant Name	Ozone Dosage	Energy Demand, Kwh/kg	Contactor off-gas Treatment	Type of Generators	Conjunctive Use of Cl_2 or ClO_2 in Pretreatment	Post-Treatment
Bonnetable	0.15	21.3	none	CEO Trailigaz plate type, water cooled, 10 units	Cl_2	1 mg/l Cl_2
Gueret	0.5	NA	none	Trailigaz tube type, water cooled, 1 unit	ClO_2	none
Tintry Pont Duroi	0.2	20	none	Trailigaz tube type, water cooled, 7 units	chlorine dioxide	chlorine dioxide (as needed)
Branden	NA	20	none	Trailigaz tube type, water cooled, 2 units	chlorine dioxide	none
Blanzy	1.8	17	none	Trailigaz tube type, water cooled, 1 unit	chlorine dioxide	chlorine dioxide (as needed)
Salgues (Aveyron)	NA	NA	NA	Trailigaz tube type, water cooled	chlorine dioxide	chlorine dioxide

(continued)

TABLE D-1 (continued)

Plant, Location, Water Source & Year Ozone Installed	Design Capacity (cu m/day)	Ozone Generation Capacity (kg/day)	Treatment Process	Purpose(s) of Ozonation	Type of Contacting
Morsang-sur-Seine, France, Seine River, 1970	150,000	316.8	prechlorination, coagulation, settling, filtration, ozonation, granular activated carbon filtration, final chlorination	color, taste, odor, bacterial disinfection	4 diffusers, 10 min detention
Oll'Ioules, France, Verdon River, 1976	37,000	58	prechlorination, coagulation, settling, filtration, ozonation, final chlorination	turbidity, odor, organics, bacterial disinfection, viral inactivation	diffuser
Chalon/Saone, France, 1972	21,600	32	preozonation, filtration, final ozonation, neutralization	Fe, Mn, taste, odor, organics, bacterial disinfection	diffuser, injector
Villeneuves/Lot, France, Lot River, 1971	10,000	21	chlorine dioxide, rapid mixing, coagulation, settling, filtration, ozonation, chlorine dioxide	color, odor, taste, organics, viral inactivation, bacterial disinfection	1 diffuser, 11 min detention
Penne d'Argenais, France Lot River, 1972	1,200	3	prechlorination, pulsator, filtration, ozonation, chlorination	color, odor, taste removal, bacterial disinfection, viral inactivation	1 diffuser, 15 min detention

(continued)

TABLE D-1 (continued)

Plant Name	Ozone Dosage	Energy Demand, Kwh/kg	Contactor off-gas Treatment	Type of Generators	Conjunctive Use of Cl_2 or ClO_2 in Pretreatment	Post-Treatment
Morsang-sur-Seine	2.85	37.5	thermal destruction	Welsbach-Degrémont tube type, water cooled, 4 units	chlorine	0.2 mg/l chlorine
Oll'loules	NA	NA	recycling	tube type, water cooled	chlorine	chlorine
Chalon/Saone	NA	NA	NA	tube type, water cooled	NA	NA
Villeneuves/Lot	2.0	NA	none	Trailigaz tube type, water cooled, 2 units	chlorine dioxide	chlorine dioxide 0.2 mg/l
Penne d'Argenais	2.0	NA	none	Trailigaz tube type, water cooled, 1 unit	chlorine	chlorine dioxide 0.2 mg/l

(continued)

TABLE D-1 (continued)

Plant, Location, Water Source & Year Ozone Installed	Design Capacity (cu m/day)	Ozone Generation Capacity (kg/day)	Treatment Process	Purpose(s) of Ozonation	Type of Contacting
Muret, France (Haute Garonne), Garonne River, 1970	10,000	20	prechlorination, coagulation, settling, filtration, ozonation	color, odor, taste removal, bacterial disinfection, viral inactivation	1 diffuser, 14 min detention
Buzet-sur-Tarn, France, Tarn River, 1968	4,800	9.6	prechlorination, coagulation, settling, filtration, ozonation	taste & odor, bacterial disinfection, viral inactivation	1 plate column, 15 min detention
Blagnac, France, Garonne River, 1969	4,800	24	settling, filtration, ozonation	taste, odor, color removal, viral inactivation, bacterial disinfection	1 diffuser
Vichy, France	22,000	96	prechlorination, chlorine dioxide, coagulation, settling, filtration, ozonation, chlorine dioxide	NA	2 diffusers, 8 min detention
Mougins, France, (Alpes Maritimes), river water, 1959	100,000	129.6	coagulation, settling, filtration, ozonation	organics removal, bacterial disinfection, turbidity removal	2 injectors, 4 min detention
Moulin de Ferguilyen, Bodilis, France	3,000	4.8	prechlorination, coagulation, settling, filtration, ozonation, final chlorination	color, odor, taste, turbidity, organics removal, viral inactivation, bacterial disinfection	2 diffusers, 8 min detention

(continued)

TABLE D-1 (continued)

Plant Name	Ozone Dosage	Energy Demand, Kwh/kg	Contactor off-gas Treatment	Type of Generators	Conjunctive Use of Cl_2 or ClO_2 in Pretreatment	Post-Treatment
Muret	2.0	28	none	Trailigaz tube type, water cooled, 2 units	chlorine	none
Buzet-sur-Tarn	0.4	NA	none	tube type, water cooled, 1 unit	chlorine	none
Blagnac	3.0	NA	none	Trailigaz tube type, water cooled	none	none
Vichy	0.4	NA	NA	Welsbach, Degrémont tube type, water cooled, 2 units	prechlorination/ chlorine dioxide	chlorine dioxide
Mougins	0.3-0.4	26	none	CEO, plate type, water cooled, 12 units	none	none
Moulin de Ferguilyen	1.2-1.5	NA	NA	Welsbach-Degrémont, tube type, water cooled, 1 unit	chlorine	0.6-0.8 mg/l chlorine

TABLE D-2. CANADIAN OZONE QUESTIONNAIRE SUMMARY

Plant, Location, Water Source & Year Ozone Installed	Design Capacity (cu m/day)	Ozone Generation Capacity (kg/day)	Treatment Process	Purpose(s) of Ozonation	Type of Contacting
l'Assomption, Rivière l'Assomption, 1966	9,988	NA	screening, chlorination, coagulation, sedimentation, filtration, ozonation, post-chlorination, pH adjustment	odors, tastes, phenols, organics, bacterial disinfection, viral inactivation	injector (no contact time given)
Buckingham, Rivière du Lièvre, 1976	18,900	28.6	prechlorination, lime, alum + polyelectrolyte, coagulation, sedimentation, filtration, caustic soda, reservoir, ozonation, chlorination	tastes & odors	porous diffuser (5-10 min)
Drummondville, Rivière St.-Francois, 1966	54,000	40.8	prechlorination, coagulation, sedimentation, filtration, ozonation, post-chlorination	tastes & odors throughout winter	injector (3-5 min)
l'Epiphanie, Rivière l'Achigan, 1962	4,540	NA	screening, chlorination, coagulation, sedimentation, filtration, chlorination, ozonation, adjust pH	tastes, odors, phenols, organics, bacterial disinfection, viral inactivation	injector
Ile Perrot, Lac St.-Louis, 1963	6,800	9	chlorination, alum, SiO_2, coagulation, sedimentation, filtration, ozonation, lime, fluoride	color, tastes & odors, bacterial disinfection	submerged turbine (20 min)
Lac Etchemin, Northern part of Lake, 1966	2,383	7.2	microstraining, ozonation, chlorination	color, tastes, odors, "sterilization"	injector

(continued)

TABLE D-2 (continued)

Plant Name	Ozone Dosage	Energy Demand, Kwh/kg	Contactor off-gas Treatment	Type of Generators	Conjunctive Use of Cl_2 or ClO_2 in Pretreatment	Conjunctive Use of Cl_2 or ClO_2 in Post-Treatment
l'Assomption	NA	NA	none	Trailigaz, plate type, water cooled, 1 unit	0.5 mg/l	0.4 mg/l
Buckingham	1-2 mg/l	24 total	none	Degrémont horizontal tube, water cooled, 1 unit	NA	1 mg/l dosage; 0.2 mg/l at plant exit; 0.02 mg/l at extremity of distribution system
Drummondville	0.6 mg/l	22 total; 2.2 air preparation; 15.4 O_3; 4.4 contacting	none	Trailigaz, plate type, water cooled, 3 units	1.5-3.2 mg/l	0.2-0.3 mg/l dosage; 0.2-0.3 at plant exit; 0.1 mg/l at extremity of distribution system
l'Epiphanie	NA	NA	none	Trailigaz, plate type, water cooled, 1 unit	NA (8 lbs/day total)	3 mg/l at plant exit, 0 at extremity of distribution system
Ile Perrot	0.8-1.0 mg/l	45 total; 9 air preparation; 21 O_3; 15 contacting	none	Welsbach, horizontal tube, water cooled, 1 unit	1.3 mg/l at coagulation basin	trace of Cl_2 leaving the plant
Lac Etchemin	NA	28 total	none	Trailigaz, plate type, water cooled, 1 unit	none	5 lbs/day; 0.1 mg/l at plant exit, 0 at extremity of distribution system

(continued)

TABLE D-2 (continued)

Plant, Location, Water Source & Year Ozone Installed	Design Capacity (cu m/day)	Ozone Generation Capacity (kg/day)	Treatment Process	Purpose(s) of Ozonation	Type of Contacting
Laval (Pont Viau), Rivière des Prairies, 1957	95,850	72	SiO_2 + alum, coagulation, sedimentation, filtration, reservoir, ozonation (2 stages), lime, fluoride, chlorination	tastes, odors, bacterial disinfection, viral inactivation	injector (5-10 min)
Laval (Ste-Rose), Rivière des Mille-Iles, 1961, 1968	25,560	48	SiO_2 + alum, coagulation, sedimentation, filtration, reservoir, ozonation, lime, fluoride, chlorination	tastes, odors, bacterial disinfection, viral inactivation	injector (2-5 min)
Oka, Lac des Deux-Montagnes, 1958	2,192	0.3	prechlorination, lime, alum, coagulation, sedimentation, filtration, reservoir, ozonation, lime. No chlorine residual	tastes, viral inactivation, bacterial disinfection	injector (5 min)
Pierrefonds, Rivière des Prairies, 1976	95,500	250	soda, chlorine, $(NH_4)_2SO_4$, Na silicate, alum, rapid mixing, flocculation, decantation (pulsator), filtration, ozonation, chlorination	tastes, odors, bacterial disinfection	porous diffusers (4 min)
Quebec City, Rivière St.-Charles, 1969	218,000	360	prechlorination, coagulation, flocculation, sedimentation, filtration, ozonation, post-chlorination, pH adjustment	tastes, organics, viral inactivation, bacterial disinfection	injector (pulverization column) (4 min)
Repentigny, Rivière l'Assomption, 1973	22,730	45.5	screening, prechlorination, coagulation, sedimentation, filtration, ozonation, adjust pH, post-chlorination. If ozone generators don't work, then PAC, not Cl_2	tastes, organics, viral inactivation, bacterial disinfection	porous diffusers

(continued)

TABLE D-2 (continued)

Plant Name	Ozone Dosage	Energy Demand, Kwh/kg	Contactor off-gas Treatment	Type of Generators	Conjunctive Use of Cl_2 or ClO_2 in Pretreatment	Conjunctive Use of Cl_2 or ClO_2 in Post-Treatment
Laval (Pont Viau)	1.5-2 mg/l	21.4 total (6 kw/mlg)	none	Trailigaz, plate type, water cooled, 12 units	none	3 mg/l dosage; 0.3 mg/l at plant exit; trace at extremity of distribution system
Laval (Ste-Rose)	1.5-2 mg/l	13.3 total (5 kw/mlg)	none	Trailigaz, plate type, water cooled, 3 units	none	0.85 mg/l dosage; 0.25 mg/l at plant exit; trace at extremity of distribution system
Oka	2 mg/l	23.3 total	none	Trailigaz, plate type, water cooled, 1 unit	NA	none
Pierrefonds	2 mg/l	18 total; 6 air pretreatment; 12 O_3	none	Trailigaz, horizontal tube type, water cooled, 2 units	0.8 mg/l	0.5 mg/l dosage; 0.5 mg/l at plant exit
Quebec City	1.3 mg/l	30 total	none	Trailigaz, plate type, water cooled, 12 units	1.2-2.4 mg/l	0.35 mg/l at plant exit; 0.1 mg/l at extremity of distribution system
Repentigny	NA	NA	none	Trailigaz, horizontal tube, water cooled, 1 unit	yes, but NA	0.83 mg/l total (pre + post) dosage. 0 mg/l at plant exit

(continued)

TABLE D-2 (continued)

Plant, Location, Water Source & Year Ozone Installed	Design Capacity (cu m/day)	Ozone Generation Capacity (kg/day)	Treatment Process	Purpose(s) of Ozonation	Type of Contacting
Rivière du Loup, Rivière du Loup, 1977	15,120	36	prechlorination, rapid mixing, coagulation, sedimentation, filtration, ozonation, chlorination	tastes & odors	porous diffusers
Roberval, Lac St.-Jean, 1971	7,715	43.18	ozonation, coagulation, filtration, ozonation, chlorination. During period when raw water is colored, but low turbidity, ozonation, filtration, ozonation, chlorination	color	porous diffusers
St. Denis, Rivière Richelieu, 1972	27,300	28	prechlorination, coagulation, sedimentation, filtration, ozonation, post-chlorination	tastes, odors, bacterial disinfection	porous diffusers
St. Eustache, Rivière des Mille-Iles; Lac des Deux-Montagnes, 1957	1,060	NA	coagulation, sedimentation, filtration, ozonation, chlorination	tastes & odors	injector (20 min)
Sherbrooke, Lac Memphremagog, 1965, 1977	98,262	163	occasional prechlorination, microstraining, ozonation, chlorination	tastes, odors, bacterial disinfection	porous diffusers
Terrebonne, Rivière des mille-Iles, 1963	15,120	28.5	prechlorination, alum, polyelectrolyte, coagulation, sedimentation, filtration, fluoride, reservoir, ozonation, chlorination, lime	tastes, odors, viral inactivation, bacterial disinfection	injector (5-10 min)

TABLE D-2 (continued)

Plant Name	Ozone Dosage	Energy Demand, Kwh/kg	Contactor off-gas Treatment	Type of Generators	Conjunctive Use of Cl_2 or ClO_2 in Pretreatment	Post-Treatment
Rivière du Loup	1 mg/l	25.3 total	catalytic	PCI-Ozone, horizontal tube, water & oil cooled, 1 unit	2.5 mg/l	5 lbs/day dosage; 0.3 mg/l at plant exit. NA at extremity of distribution system
Roberval	4.5 mg/l	NA	none	Welsbach, horizontal tube, water cooled, 1 unit	none	3.1 mg/l dosage; 0.4 mg/l at plant exit; 0.05 mg/l at extremity of distribution system
St. Denis	0.6 mg/l and less	NA	none	PCI-Ozone, horizontal tube, water & oil cooled, 2 units	1.0 mg/l	0.2-0.3 mg/l dosage; 0.1 mg/l at plant exit; trace at extremity of distribution system
St. Eustache	NA	NA (144 kg/day, total)	none	Trailigaz plate type, water cooled, 3 units	none	2 mg/l dosage; 0.7 mg/l at plant exit; trace at extremity of distribution system
Sherbrooke	2 mg/l	29.1 total 7.9 air treatment, 21.2 O_3	none	Degrémont, horizontal tube, water cooled, 4 units	NA	about 88 kg/day total dosage; 0.2 mg/l at plant exit; none at extremity of distribution system
Terrebonne	1.15 mg/l	23.3 total	none	Trailigaz, plate type, 2 units	NA	2 mg/l dosage at entrance to reservoir; 0.4 mg/l at plant exit; 0.2 mg/l at extremity of distribution system

APPENDIX E: DESCRIPTIONS OF SELECTED U.S.A. DRINKING WATER TREATMENT PLANTS USING CHLORINE DIOXIDE

COLUMBUS, OHIO

The 35 million gallon per day (mgd) (133,000 cu m/day) plant at Columbus uses ClO_2 seasonally for the control of taste and odor problems which are generally caused by phenols and algae (70% natural and 30% industrial). The industrial sources are identified as highway runoff and washdown wastes from a nearby railroad switching yard. The plant obtains its raw water from the Scioto River through a series of two surface reservoirs. The ClO_2 operation began in 1975 and has since solved the taste and odor (T/O) problem.

The Columbus plant (Dublin Road) has the following process steps:

- chemical addition - lime/alum
  - $KMnO_4$ (optional)
- flocculation
- sedimentation
- prechlorination - chlorine
- softening - Na_2CO_3
- recarbonation
- chemical addition - fluoride
  - sequestration
- filtration
- chemical addition - chlorine dioxide
- clearwell - distribution system

The ClO_2 is generated by mixing aqueous Cl_2 and 37% aqueous $NaClO_2$ in the ratio of 1:1 by weight. The addition of ClO_2 is seasonal. Liquid chlorine is delivered to the plant in one ton (900 kg) cylinders. The $NaClO_2$ is shipped to the site in a 4,000 gallon (15.2 cu m) tank truck and pumped into two 1,300 gallon (4.94 cu m) fiberglass storage tanks. These containers are located inside a large room which adjoins the chlorine storage room. The $NaClO_2$ room is heated in the cold months because of the relatively high freezing point of 37% concentrated $NaClO_2$ (49°F). Olin Corporation manufactures the $NaClO_2$ but an area distributor delivers the chemical to the plant.

The ClO_2 generation system consists of one Fischer and Porter ClO_2 reactor vessel, two BIF proportional metering pumps for $NaClO_2$ feed (one as standby), and one Fischer and Porter chlorinator. A chlorine sensor is located inside the clearwell to measure the chlorine residual which results from the addition of ClO_2 to the finished plant water. A signal is transmitted to the chlorinator for purposes of increasing or decreasing the chlorine feed. A set point establishes the desired chlorine residual in the finished water. Because the chlorine feed may vary whereas the $NaClO_2$ feed is constant, the 1:1 ratio may not always be achieved. It was stated however, that the chlorine feed does not vary appreciably and there are rarely "spikes" in the feed rates. Piping for the ClO_2 system is schedule 80 PVC except for the chlorine gas piping which is black steel.

The chlorine room contains the plant chlorinators, one ton (900 kg) chlorine cylinders and the ClO_2 reactor vessel. A Fischer and Porter chlorine leak detection system automatically activates the exhaust fan system if a leak occurs inside the room. The air is sampled for chlorine every 60 seconds. The two exhaust fans appear to be adequately sized for the 40 ft x 80 ft (12 m x 24 m) room. The entrance doors have glass panels for visual inspection of the room interior. Entrance to the chlorine room is through an interior wall only -- there is no exterior entrance to the chlorine room. The one ton (900 kg) chlorine cylinders are delivered to the chlorine storage room through a common overhead door which opens to the adjoining $NaClO_2$ room. The overhead crane carries the ton cylinders from the loading platform outside the $NaClO_2$ room to the chlorine room.

The $NaClO_2$ room has no forced exhaust system, although the exterior windows provide ventilation during warmer months. The area surrounding the $NaClO_2$ storage tanks is bordered by a 12 inch high (0.3 m) concrete step. If the tanks or transfer pipes develop a rupture, the liquid $NaClO_2$ is momentarily contained. The liquid chemical is emptied through the floor drain and directed into the plant coagulation basin. The $NaClO_2$ feed pumps have a similar retaining wall and drainage system.

The production of ClO_2 is monitored visually by the color that is generated inside the ClO_2 reactor vessel. A site glass mounted at the discharge port of the opaque reactor allows the operator to inspect the ClO_2 color. Plant personnel also monitor the taste and odor of the finished water every two hours as a means of evaluating the ClO_2 performance. When the ClO_2 unit is operating, the average concentration of $NaClO_2$ in the plant water is 0.4 to 0.5 mg/l.

Plant supervisors expressed a concern for toxic effects of chlorite ions in the finished water. Although ClO_2 is used sparingly because of its high cost, the supervisor is considering replacing chlorine in the prechlorination process with ClO_2.

The chemical cost for chlorine is \$0.086/pound (\$0.191/kg). The cost for $NaClO_2$ was not immediately available. Costs for the $NaClO_2$ tanks and chemical feed pumps have been promised. Chemical costs for the production of finished water were not available at the time of this writing.

The plant expects to install temperature probes in its largest water surface reservoir in order to determine seasonal turnovers. This, in turn, can assist the plant in predicting a possible rise in algae or manganese concentration.

NEWARK, OHIO

The 10 mgd (38,000 cu m/day) plant at Newark uses ClO_2 to control the presence of phenols and magnesium in its raw water source -- the Licking River. Taste and odor problems, which are attributed to phenols, were at their peak in the early 1960s, but the origin of the phenols was never determined. The plant subsequently began adding ClO_2 in 1964 as a "pre-chlorination" step, and the process and has worked well since.

The Newark plant has the following process steps:

- raw water intake - chlorine dioxide
- chemical addition - lime/alum
  - powdered activated carbon
- flash mix
- flocculation
- sedimentation
- recarbonation/excess lime for iron removal
- filtration
- chemical addition - chlorine
  - fluoride
- clearwell - distribution system

Water quality data for both raw and finished water have been obtained.

ClO_2 is generated by mixing aqueous Cl_2 and aqueous $NaClO_2$ in the ratio of 4.18:1 by weight [100 pounds (45 kg) Cl_2/day and 24 pounds (10.8 kg) $NaClO_2$/day]. This is a constant feed dosage. Post-chlorination requires about 40 pounds/day of chlorine. Technical grade anhydrous $NaClO_2$ is delivered in 100 pound (45 kg) drums. The $NaClO_2$ is manufactured by Olin Corporation and distributed by a local retailer.

The ClO_2 generation system is Wallace and Tiernan equipment: one reactor vessel for ClO_2 production, one diaphragm pump for $NaClO_2$ feed and one chlorinator. There is also a W&T chlorinator for the post- chlorination step. The piping is schedule 80 PVC except for the chlorine gas line which is black steel. The ClO_2 generating equipment is housed within the plant building, but in a room with an outside entrance. The room where the one ton (900 kg) chlorine cylinders are stored adjoins the ClO_2 room. The entrance to this room is also from the exterior wall of the building.

Both rooms have forced exhaust systems which are installed at floor level. The capacity of the units appears adequate. The switches for both the lights and fans are located on the exterior wall of the entrance doorways. There are floor drains in the ClO_2 room which empty into french* drains beneath the building. An ammonium hydroxide bottle has been placed inside the chlorine room for emergency neutralization. Both entrance doors have glass panels for inspection of the room interior. There are ABC fire extinguishers in both rooms and two self-contained breathing apparatuses in the main control room about 75 feet (22.5 m) away.

Production of ClO_2 is monitored visually by the color that appears inside the reactor vessel. The chemical feed rates of Cl_2 and $NaClO_2$ also are checked daily. The effectiveness of ClO_2 is measured by the absence of T/O problems in the finished water. The concentration of magnesium in the raw water can jump to above 25 mg/l while the finished water may contain 10 to 12 mg/l. The estimated concentration of $NaClO_2$ in plant water is around 0.3 to 0.4 mg/l. The ClO_2 dosage remains constant year-round.

Oxidants in the finished water are analysed by the DPD method. The total chlorine residual leaving the plant is 0.1 to 0.2 mg/l. The extremities of the distribution system have a total chlorine residual of 0.1 to 0.05 mg/l.

The chemical cost of $NaClO_2$ is $83/100 pounds ($1.84/kg). Chlorine costs 9¢/pound (20¢/kg) when delivered in ton cylinders. Chemical costs for the finished water are 5.4¢/1,000 gallons (1.42¢/cu m) (lime, alum, carbon, sodium chlorite, chlorine, natural gas, etc).

Plant management is pleased with performance of ClO_2. As reported, the chemical is quick to respond to phenols or manganese with no T/O carryover to the filters or finished water. When the turbidity rapidly rises in the raw water (an example offered was from 40 to 4,400 turbidity units in less than two hours), the lime and alum feed are increased in greater proportions in order to avoid an increase in the dosage of expensive chlorine dioxide.

BETHESDA, OHIO

The 79,000 gallons per day (gpd) plant at Bethesda uses ClO_2 for the control of taste and odor problems and for biological disinfection. The plant water also is prechlorinated (with chlorine) upstream of the ClO_2 addition. The T/O problems are generally attributed to manganese (autumn and winter) and algae (summer). The raw water source is a surface reservoir located on the plant premises. The addition of ClO_2 has been in operation since 1972-1973.

The Bethesda plant has the following process steps:

---

* A french drain is a gravel pit that drains into the ground.

- prechlorination - chlorine
- chemical addition - lime/alum
  - powdered activated carbon (optional)
- flash mix
- flocculation
- sedimentation
- chemical addition - sodium tripolyphosphate blended with $KMnO_4$ (1% by weight of chemical feed)
- chemical addition - chlorine dioxide
- filtration
- clearwell - distribution system

Manganese (Mn) in the raw water averages 0.7 mg/l while iron (Fe) averages about 0.6 mg/l. The finished water has only trace amounts of Mn but 0.4 mg/l Fe. The reservoir is seasonally treated with copper sulfate ($CuSO_4$) for control of algae growth. There are no immediate records for $CuSO_4$ loading rates for the past five years.

ClO_2 is generated by mixing aqueous $NaClO_2$ and aqueous Cl_2 in the ratio of 5.6:1 by weight [20 pounds (9 kg) $NaClO_2$/day and 3.6 pounds Cl_2/day (1.62 kg)]. Total Cl_2 consumption, including prechlorination, is approximately five pounds per day. The concentration of $NaClO_2$ in the plant water is 17.4 mg/l. Plant personnel feel that this is a desirable dosage to prevent T/O complaints. The $NaClO_2$ is manufactured by Olin Corporation and delivered to the plant by an area distributor. The anhydrous $NaClO_2$ is technical grade and is supplied in 100 pound (45 kg) drums. The chlorine gas is supplied to the plant in 150 pound (67.5 kg) cylinders.

The ClO_2 generation system consists of one Wallace and Tiernan reactor vessel for ClO_2 production, one Gorman-Rupp bellow pump and one Wallace and Tiernan chlorinator. The piping for the $NaClO_2$ and ClO_2 solution is schedule 80 PVC. The chlorine gas is piped to the chlorinator with black steel. The tubing from the chlorinator to the ClO_2 reactor vessel is a flexible plastic pipe of unknown specifications. Several of the joints are secured with hose clamps. There have been no reports of chemical leakage. The $Cl_2/NaClO_2$ chemical feed equipment and ClO_2 reactor unit are housed in one building separate from the plant.

The building has no forced exhaust system. There is also no alarm system or a chlorine gas leak detection system (a bottle of ammonium hydroxide solution for detecting chlorine gas was not present). There are, however, windows in this 10' x 20' (3 m x 6 m) building that offer some ventilation. The entrance door has a glass panel for easy inspection of the room interior. A light switch is located on the inside wall beside the doorway. There are no floor drains inside the chemical equipment room.

The production of ClO_2 is monitored visually by the color that appears inside the reactor vessel and also by checking the $NaClO_2/Cl_2$ feed rates daily. ClO_2 is generated seasonally when the need arises. The effectiveness of ClO_2 is evaluated by the absence of T/O problems in the finished water. The DPD method is used to check chlorine residual for the state, but the OTA method is used around the plant for "spot checking" chlorine residual in the process water. The finished water has a combined chlorine residual of 2.5 mg/l (1.8 mg/l free chlorine). At the extremity of the distribution system, there is a combined chlorine residual of 0.82 mg/l (0.63 mg/l free chlorine).

The cost of $NaClO_2$ is $105/100 pounds ($2.33/kg). Chlorine costs are not available. The overall chemical cost for plant water is about 43¢/1000 gallons (11.3¢/cu m) when powdered activated carbon is used.

HAMILTON, OHIO

The 15 mgd (57,000 cu m/day) plant at Hamilton uses ClO_2 as the only oxidant for disinfection. This lime softening plant uses groundwater (18 wells - 200 ft deep) as its raw water supply. Chlorine was used when the plant started operation in 1956, but because of customer complaints with the chlorine residual in the finished water, the plant switched to ClO_2 in 1972. The complaints reportedly have ceased. The groundwater has only trace amounts of iron and manganese.

The Hamilton plant has the following process steps:

- aeration
- chemical addition - lime/alum
- flash mix
- sedimentation
- recarbonation
- filtration
- chemical addition - sodium silicofluoride
- chlorine dioxide
- clearwell - distribution system

The ClO_2 is generated by mixing 37% aqueous $NaClO_2$ and aqueous Cl_2 in the ratio of 1:1 by weight [20 pounds (9 kg)/day $NaClO_2$ (dry weight) and 20 pounds (9 kg)/day Cl_2]. Chlorine gas is delivered to the site in 150 pound (67.5 kg) cylinders. Aqueous $NaClO_2$ is manufactured by Olin Corporation and shipped to the site by an area distributor. Each drum contains 200 pounds (9 kg) of 37% aqueous $NaClO_2$.

The ClO_2 generation system consists of one plant fabricated reactor vessel for ClO_2 production, one BIF peristaltic pump for liquid $NaClO_2$ and two Fischer and Porter chlorinators with one serving as a standby. Two 150 pound (67.5 kg) liquid chlorine cylinders are positioned next to the chlorinators. One is in service while the other serves as a backup. The weight of the chlorine cylinder contents is measured by a scale. Switchover from one tank to the other is manual. Fischer and Porter specifies schedule 80 PVC tubing for the line between the chlorinator and the reactor vessel. Heavy Tygon* tubing is used for transporting the liquid $NaClO_2$ from the drum to a small plastic day tank and to the reactor vessel. The Tygon tubing reportedly loses its rigidity after a month's use and is subsequently replaced. The semi-transparent day tank allows visual inspection of the level of liquid $NaClO_2$ and thereby enables the operator to maintain an acceptable suction head on the peristaltic pump.

The ClO_2 reactor vessel is made from what appears to be schedule 80 polyvinyl chloride (PVC) pipe material -- 18 inches (45.7 cm) high and approximately 6 inches (15.2 cm) in diameter. The vessel is filled with 1 inch (2.54 cm) diameter rings of PVC pipe. The chamber is opaque except for the sight glass which is mounted "in-line" on the discharge piping. A white card is placed behind the sight glass for proper inspection of the ClO_2 color.

The ClO_2 generation equipment is housed in a room with no direct access to the outside. A corridor connects the room to the outside where 150 pound (67.5 kg) chlorine cylinders are stored. The cylinders are protected from direct sunlight. There is an exhaust fan system in the ClO_2 room. There is no chlorine leak detector in the room. Both entrance doors to the ClO_2 room have glass panels for visual inspection of the room interior.

Production of ClO_2 is monitored visually by the color of the ClO_2 that appears inside the reactor vessel. Chemical feed rates for $NaClO_2$ and Cl_2 are also checked to ensure the proper mix of chemicals for ClO_2 production. Both color and feed rates are checked hourly.

The effectiveness of ClO_2 is determined by bacteriological tests and the absence of T/O problems in the finished water. Plant water is analyzed three times daily and once daily in the distribution system. The ClO_2 in the finished water is measured spectrophotometrically to levels of less than 0.2 mg/l as ClO_2. It was noted that the H-acid method requires a higher level of skill than other tests such as OTA or DPD. The concentration in the plant water is 0.2 mg/l as ClO_2. The ClO_2 feed concentration generally is constant year-round. The ClO_2 residual leaving the plant is 0.15 mg/l and the residual at the extremities of the distribution system is 0.10 mg/l ClO_2.

---

* Tygon is a registered trademark of the Dupont Company.

The cost for ClO_2 addition is about 3.6¢/capita/year (1977). Chlorine and $NaClO_2$ together cost about $6,540/year (1977). Chemical costs for finished water average 19¢/1,000 gallons (5¢/cu m). The fraction that is attributed to ClO_2 reportedly is negligible. The operating and maintenance (O/M) costs for ClO_2 generation are estimated to be less than $50/year (1977). The plant-fabricated ClO_2 reactor, piping, hardware and installation are estimated by the plant supervisor to cost around $400 in the 1977 market. Installation was done by plant personnel. The BIF peristaltic pump for $NaClO_2$ addition reportedly costs less than $200 (1977).

It was also reported that the addition of ClO_2 abated the problems caused by iron bacteria in the distribution system. Consumers were frequently complaining about the staining effects from tap water. When the plant switched from Cl_2 to ClO_2 as a disinfectant, brown slimes apparently loosened from the mains and were subsequently flushed from the system. Shortly thereafter, the complaints stopped. The plant attributed the source of the problem to the biological formation of crenothrix and leptothrix that were present in the extremities of the distribution system before introduction of ClO_2.

TOLEDO, OHIO

The 78 mgd (296,400 cu m/day) plant at Toledo uses ClO_2 for the control of taste and odor problems in its raw water supply. It also uses ClO_2 along with the separate addition of chlorine for bacterial disinfection. The T/O problems are caused from phenols and algae in Lake Erie. Phenols are both natural and industrial in origin. The use of ClO_2 has been in operation since 1956.

The Toledo plant has the following process steps:

- raw water intake -powdered activated carbon
- chemical addition - lime/alum
- flash mix
- flocculation
- chemical addition -lime/soda softening
- sedimentation
- recarbonation
- chemical addition - polyphosphates
- filtration

- chemical addition - fluoride
  - chlorine
  - chlorine dioxide
- clearwell - distribution system.

The ClO_2 is generated by mixing aqueous Cl_2 and aqueous $NaClO_2$ in the approximate ratio of 6.5:1 by weight [935 pounds (421 kg) Cl_2/day and 147 pounds (66 kg) $NaClO_2$/day]. Liquid chlorine is supplied to the plant in one-ton (900 kg) cylinders. The anhydrous $NaClO_2$ is supplied as technical grade in 100 pound (45 kg) drums. Toledo formerly purchased the $NaClO_2$ from France through a United States distributor (named Pettibone) located in Chicago. The chemical was marketed under the name of "Cloritane" and was packaged in 50 and 80 kg containers (110.25 and 176.4 pounds, respectively). The $NaClO_2$ used at the plant presently is manufactured by Olin Corporation and distributed by a local retailer. The plant is scheduled to switch to aqueous $NaClO_2$ because of the potential flammability of $NaClO_2$ in the flake form. Delivery of the aqueous $NaClO_2$ will be made by tank truck.

The ClO_2 generation system consists of four plant fabricated ClO_2 reactor vessels (two serve as standby), four Wallace and Tiernan chlorinators (two serve as standby) and two BIF proportional feed pumps. The standby chlorinators are rotated periodically with the operating chlorinators in order to equalize wear on the equipment. Piping is schedule 80 PVC except for black steel pipe which is used for the transport of chlorine gas from the ton cylinders to the chlorinators. Most of the PVC joints [sizes 2" (7.08 cm) and smaller] are sealed with Teflon tape. There was no report of chemical or water leakage from the piping system. The $NaClO_2$ and ClO_2 chemical equipment are housed in one room whereas the chlorine cylinders are stored in an adjoining room.

Both the ClO_2 and Cl_2 rooms have forced air exhaust systems mounted at floor level. The switches for both the exhaust system and lights are located next to the entrance door on the outside of each room. Only the chlorine room has an exit to the outside of the building. This is an overhead door which provides access to the ton cylinder storage area from the delivery yard outside. The other doors have glass panels for visual inspection of the room interior. There are no chlorine leak detectors in either room, although one is being considered for the chlorine room. There are two MSA gas masks and four self-contained breathing apparatuses located throughout the plant building. The two MSA masks are located in the hall corridor outside the ClO_2 room. Any chemicals that spill onto the floor are washed into the floor drains which are in each room. The drains feed into the sanitary sewer system.

Production of ClO_2 is monitored visually by the color of ClO_2 that appears in each reactor vessel. The $NaClO_2/Cl_2$ chemical feed rates also are checked periodically. Finished water is analyzed hourly by the threshold odor test. The plant uses powdered activated carbon at the raw water lift station to reduce the carry-over of T/O problems into the plant. Chlorine residual is analyzed by the DPD method and in the future will be measured by the amperometric method. The total chlorine residual (chlorine + ClO_2) in the finished water is 1 mg/l (0.4 mg/l free chlorine) and a 0.1 mg/l free chlorine residual in the distribution system. The average concentration of $NaClO_2$ in the plant water is 0.23 mg/l. The ClO_2 is manually proportioned to plant flow and is added year-round.

The chemical cost for solid $NaClO_2$ is $72/100 pounds ($1.60/kg). The cost for the "chloritane" brand was $65.75/100 pounds ($1.46/kg). The French product apparently has dropped from the area market for reasons not immediately known. Chlorine costs $0.09/pound. The O/M costs for ClO_2 production are nominal and are not considered in the plant budget.

ATLANTA, GEORGIA (CHATTAHOOCHEE PLANT)

The 110 mgd (418,000 cu m/day) Chattahoochee plant at Atlanta uses ClO_2 for seasonal problems due to iron and manganese. The plant obtains its raw water from the Chattahoochee River which flows from a surface reservoir 50 miles (80 km) upstream. ClO_2 has performed satisfactorily since its first operation in 1960.

The Atlanta plant has the following process steps:

- chemical addition raw water intake - chlorine
  - lime
  - powdered activated carbon (optional)
- chemical addition at plant - chlorine
  - chlorine dioxide (optional)
  - lime
  - powdered activated carbon
  - alum
- flash mix
- flocculation
- sedimentation
- filtration
- chemical addition - chlorine
  - lime
  - fluoride
- clearwell - distribution system

The ClO_2 is generated by mixing aqueous Cl_2 and aqueous $NaClO_2$. Chlorine feed rates for ClO_2 production were not readily available because ClO_2 was not being added at the time of the visit. The anhydrous $NaClO_2$ is a technical grade manufactured by Olin Corporation. A local distributor delivers the 100 pound (45 kg) drums to the plant. Liquid chlorine is delivered in one-ton (900 kg) cylinders.

The ClO_2 generation system consists of Wallace and Tiernan equipment: two reactor vessels for ClO_2 production, two plunger pumps and two chlorinators. One ClO_2 reactor serves as a backup unit to the ClO_2 reactor that is operating. The piping from the $NaClO_2$ mixing tanks to the chemical feed pumps is Tygon tubing. The remaining piping of the ClO_2 system is schedule 80 PVC. The only exception is the piping from the one-ton (900 kg) chlorine cylinders to the chlorinators, which is black steel. There is ample floor space around the mixing tanks, pumps and ClO_2 reactors. The equipment area is serviced by floor drains which empty into the municipal sanitary sewer system. The $NaClO_2$ mixing tanks are one floor below the ClO_2 reactors, all of which are located in the filtration building. Piping is exposed along the walls and ceilings for easy inspection and maintenance.

The chlorine storage room is also part of the filtration building. Approximately 16 one-ton (900 kg) chlorine cylinders are housed in this room whereas 24 one-ton (900 kg) cylinders are stored in the chlorine building at the raw water intake. Both facilities have forced exhaust systems with motor operated dampers. Exhaust systems are mounted at floor level and the switches for these units (and for the lights) are mounted on the wall outside the entrance doorway. The doors have glass panels for visual inspection of the room interior. Each chlorine storage room has a Wallace and Tiernan leak detector. Air samples are monitored for chlorine gas continuously.

The production of ClO_2 is monitored by the color of ClO_2 that appears inside the transparent reactor vessel. Chemical feed rates are closely checked during each shift. The efficiency of ClO_2 performance is evaluated hourly by measuring the concentration of iron and manganese in the plant effluent. The plant adds ClO_2 only when chlorination does not satisfactorily reduce the levels of iron and manganese in the finished water. For a typical summer day, the plant uses approximately 800 to 1,000 pounds (360 to 450 kg) of chlorine at the raw water intake, another 250 pounds (112.5 kg) at the plant for prechlorination and another 210 pounds (94.5 kg) of chlorine for post-chlorination. Residual chlorine is measured at six different locations throughout the treatment works. A 0.5 mg/l free chlorine residual is maintained at the raw water lift station.

The plant uses the amperometric technique for analyzing chlorine residual. The DPD method is used to measure chlorine residual in the water distribution system. The finished water has a 1.5 mg/l free chlorine residual, whereas the extremities of the distribution system have a 0.5 mg/l free chlorine residual.

The cost of chlorine is 8¢/pound (17.8¢/kg). Anhydrous technical grade $NaClO_2$ costs $70/100 pounds ($1.55/kg).

The Chattahoochee plant has a well developed watershed monitoring program which was implemented in 1963. There are 8 sampling stations upstream of the plant which have in-line sensors for determining the following river quality parameters: pH, electrical conductance, turbidity, dissolved oxygen, temperature and water stage height. There are 50 other stations along the river where grab samples are taken and analyzed for additional water quality data. This monitoring program alerts the plant to any adverse quality river water that is flowing toward the plant intake. The addition of ClO_2 is used primarily to supplement the other chemicals when the river water quality becomes very poor.

ATLANTA, GEORGIA (HEMPHILL PLANT)

The 70 mgd (266,000 cu m/day) Hemphill Plant at Atlanta uses ClO_2 to control the presence of manganese in its raw water supply. Water from the Chattahoochee River is diverted into two reservoirs which in turn supply the Hemphill treatment works. ClO_2 has been employed since the 1960s.

The Atlanta plant has the following process steps:

- chemical addition - lime/alum
  - chlorine
  - chlorine dioxide
- flash mix
- chemical addition - powdered activated carbon
- flocculation
- sedimentation
- filtration
- chemical addition - lime
  - chlorine
  - fluoride
  - sodium tripolyphosphate
- clearwell - distribution system.

ClO_2 is generated by mixing aqueous Cl_2 and aqueous (29%) $NaClO_2$ in the ratio of 0.75:1 by weight [400 pounds (180 kg) of chlorine per day and 524 pounds (236 kg) of $NaClO_2$ per day]. Liquid chlorine is delivered in one-ton (900 kg) cylinders; aqueous $NaClO_2$ is delivered to the plant by 4,000 gallon (15.2 cu m) tank trucks and pumped to an 8,000 gallon (30.7 cu m) storage tank inside the chemical feed building. The 29% $NaClO_2$ is manufactured and delivered by Olin Corporation.

The ClO_2 generation system consists of two Wallace and Tiernan ClO_2 reactor vessels, one Wallace and Tiernan chlorinator and two Wallace and Tiernan diaphragm pumps. Piping is schedule 80 PVC except for the feed line to the chlorinator; this pipe material is black steel. The ClO_2 reactors are wall-mounted one floor above the $NaClO_2$ chemical feed pumps. The ton cylinders of chlorine are stored in a separate room.

Inside the chlorine storage room are two exhaust fans mounted at floor level. Switches for both the exhaust system and the ceiling lights are located just inside the doorway. The room has two chlorine leak detectors positioned on opposite sides of the room. The air is monitored for chlorine gas every 30 to 60 seconds. The chlorine gas sensors are mounted on the side of a channel drain which runs below the metal floor grating. Two self-contained breathing apparatuses are stored outside the chlorine room. The floor drains inside the chlorine room as well as the floor drains around the $NaClO_2$ tanks empty into the municipal sanitary sewer system.

The production of ClO_2 is monitored visually by the color that appears inside the ClO_2 reactor vessel. Chemical feed rates also are checked during the day. The efficiency of ClO_2 addition is monitored by the absence of manganese in the finished water. When ClO_2 is being generated, the estimated concentration of $NaClO_2$ in the plant water is 0.9 mg/l.

Chlorine residual in the plant water is analyzed by the amperometric method. Field checks along the distribution system are performed more frequently with the OTA than by the DPD method. The finished water leaving the plant has a free chlorine residual of 1.3 mg/l. The ends of the distribution system have free chlorine residuals of 0.5 mg/l.

The chemical cost for 29% aqueous $NaClO_2$ is \$74.59/100 pounds (\$1.66 per kg). Chlorine costs \$0.076/pound when delivered in one-ton cylinders. For the month of August 1977, the plant was generating ClO_2 to control a high concentration of manganese in the raw water supply. Chemical costs for that month were as follows: $NaClO_2$ - 47,869 pounds (21,541/kg) @ \$20,780; Cl_2 - 60,391 pounds (27,176 kg) @ \$4,590. The latter figures are for the entire use of chlorine at the plant. A breakdown of Cl_2 for ClO_2 production for the month of August was not available.

The plant operator is satisfied with the performance of ClO_2. The plant had tried $KMnO_4$ but the high chemical costs and chemical feed problems persuaded management to discontinue its use. The ClO_2 reportedly removes the manganese early in the treatment process and the manganese is almost totally absent at the sedimentation tanks. Chlorine was found to be not effective for removing manganese, thus the plant began adding ClO_2.

CARROLLTON, GEORGIA

The 4.5 mgd (17,034 cu m/day) plant at Carrollton uses ClO_2 for the removal of iron and manganese. The raw water supply is the Little Tullapoose River. Since 1953, ClO_2 has been used at the plant, although potassium permanganate was used on a trial basis for replacing ClO_2. The $KMnO_4$ did not perform satisfactorily.

The Carrollton plant has the following process steps:

- chemical addition - chlorine dioxide
  - lime
  - alum
  - powdered activated carbon

- flash mix
- coagulation
- sedimentation
- filtration
- chemical addition - chlorine
  - lime
  - fluoride
- clearwell - distribution system.

ClO_2 is generated by mixing aqueous Cl_2 and aqueous $NaClO_2$ in the ratio of 4.7:1 by weight [280 pounds (126 kg) of Cl_2/day and 60 pounds (27 kg) of $NaClO_2$/day]. This is a constant feed year-round. Liquid chlorine is delivered to the site in one-ton (900 kg) cylinders. Technical grade $NaClO_2$ is manufactured by Olin Corporation and delivered to the site by an area distributor. Anhydrous $NaClO_2$ is shipped in 100 pound (45 kg) drums.

The ClO_2 generation system uses Wallace and Tiernan equipment. There are two ClO_2 reactor vessels, two diaphragm pumps and two chlorinators. There is also a standby chlorinator. The piping from the ton cylinders to the chlorinators is black steel. The remaining chemical feed piping is schedule 80 PVC. Tygon tubing is used for the suction line from the diaphragm pumps to the two $NaClO_2$ day tanks. There are reportedly no problems with the pumps or piping. The chemical feed system for ClO_2 generation is located in the chlorine storage room.

There is a motor operated damper unit inside the chlorine room for forced ventilation. The exhaust fan is mounted at floor level. The switch to activate the exhaust system and the interior lights are located outside the doorway. A glass panelled door provides for visual inspection of the room interior. There is no chlorine gas leak detector inside the room. A MSA mask is stored outside the chlorine room along with a chlorine ton cylinder emergency repair kit. A self-contained breathing apparatus is located in the main control room.

Production of ClO_2 is monitored visually by the color of ClO_2 that appears inside the reactor vessel. Chemical feed rates of Cl_2 and $NaClO_2$ also are checked. The efficiency of ClO_2 is evaluated by the absence of Fe and Mn in the finished water. Typical concentrations in the raw water are 1 to 1.5 mg/l for iron and 0.5 mg/l for manganese. The estimated concentration of $NaClO_2$ (as dosed) in the plant water is 1.6 mg/l.

The concentration of chlorine in the finished water is measured hourly using a "color wheel". Finished water has a free chlorine residual of 1.5 mg/l while the extremities of the distribution system have 0.5 mg/l free residuals.

The operator is concerned about the high cost of chlorine dioxide. $NaClO_2$ costs $103/100 pounds (46¢/kg) delivered whereas chlorine costs 15¢/pound when delivered in ton cylinders.

The operator is pleased with the secondary benefits from using ClO_2 for disinfection and T/O control, but despite the success of ClO_2, he is still considering using $KMnO_4$ in place of ClO_2 for a second trial period. Since the plant was redesigned in 1971, he believes that $KMnO_4$ may perform better than it did before.

FAYETTEVILLE, GEORGIA

The 0.15 mgd (570 cu m/day) plant at Fayetteville uses ClO_2 as the sole chemical oxidant in the treatment process. Although excess chlorine is added for ClO_2 production, the ClO_2 is used for disinfection and also to reduce manganese, color and T/O problems. The source of raw water supply is Ginger Lake Creek. The plant has been adding ClO_2 since the early 1960s.

The Fayetteville plant has the following process steps:

- chemical addition - chlorine dioxide
  - lime
  - alum
- coagulation
- sedimentation
- filtration
- chemical addition - chlorine dioxide (optional)
- clearwell - distribution system

ClO_2 is generated by mixing aqueous Cl_2 and aqueous $NaClO_2$ in the ratio of 2:1 by weight [10 pounds (4.5 kg) Cl_2/day and 5 pounds (2.25 kg) $NaClO_2$/day]. This is a constant dosage year-round. Liquid chlorine is supplied in 150 pound (67.5 kg) cylinders. Technical grade $NaClO_2$ is manufactured by Olin Corporation and delivered in 100 pound (75 kg) drums by an area distributor.

The ClO_2 generation system consists of Wallace and Tiernan equipment. There is one ClO_2 reactor vessel, one chlorinator and one diaphragm plunger pump. The piping from the chlorine tank to the ejector is black steel. Heavy gauge rubber tubing is used between the ejector and the ClO_2 reactor vessel as well as from the reactor vessel to the discharge point in the chemical addition chamber. The $NaClO_2$ pump suction line is Tygon tubing and the discharge line is of a heavier gauge rubber of undetermined specifications. The 150 pound (67.5 kg) chlorine tanks, $NaClO_2$ drums and mixing tank are located in the small, crowded ClO_2 room.

The ClO_2 room has two windows but no air exhaust system. The room has a solid wood door which opens only to the filtration gallery. The switch for the interior lights is located inside the ClO_2 room. There is no chlorine leak detector inside. There are several holes that have been chiseled into the concrete floor which allow wash down water to drain from the ClO_2 room to the service pump room below. No floor drains were noted in the pump room.

The piping for the ClO_2 generation system appears disorganized and disjointed. The operator reported that some of the piping has been patched and replaced with material around the plant. The present piping arrangement, however, reportedly presents no operational problems.

The mixing tank for $NaClO_2$ is a plastic 30 gallon (114 liter) drum. Mixing is provided by opening a water spigot and using the plant water pressure to mix the chemical. There is evidence of "caking" around the periphery of the tank. The hose from the spigot is submerged in the $NaClO_2$ solution. No back flow prevent valve was noted on the plant water line.

There is no flow meter for the finished water discharge pumps. The operator therefore approximates the flow through the plant (presumably) using pumping curves provided by the manufacturer). Based on a reported flow rate of 0.15 mgd (520 cu m/day) and a range of $NaClO_2$ addition from 5 to 22 pounds (2.25 to 9.2 kg) per day, the plant water has an estimated concentration of $NaClO_2$ from 4 mg/l to 17.6 mg/l. The plant operates 16 hours a day (at a constant flow rate), and shuts down the remaining 8 hours.

The production of ClO_2 is monitored visually by the color of ClO_2 that appears in the ClO_2 reactor vessel. The chemical feed rates of the $Cl_2/NaClO_2$ systems are also checked periodically. Manganese is monitored in the finished water every 2 hours. The concentration of manganese in the raw water may range from 1.5 to 3.0 mg/l; iron concentration is negligible. The finished water has a manganese concentration of 0.1 mg/l or less. Both manganese and chlorine are analyzed colorimetrically.

The free chlorine residual in the chemical addition tank is maintained at 1.2 to 2.0 mg/l, while the finished water has a concentration of 1.0 to 1.4 mg/l. The OTA method is used to analyze for chlorine residual. (It was noticed that there is an area of concentrated bubbling inside the chemical addition tank. Presumably this is due to poor hydraulics in the tank and an excessive volume of chlorine gas entering the tank. The smell of chlorine was sharply noticeable at the chemical addition tank.)

MARIETTA, GEORGIA (WYKOFF PLANT)

The 22 mgd (83,600 cu m) plant at Marietta uses ClO_2 primarily for removal of iron and manganese from its raw water. Lake Allatoona, which is fed by the Etowah River, is the raw water supply for the plant. ClO_2 is used also for disinfection and abatement of T/O problems. The plant has been using ClO_2 since 1958.

The Marietta plant has the following process steps:

- raw water lift station - chlorine
- chemical addition - chlorine dioxide
  - lime
  - alum
- flash mix
- coagulation
- sedimentation
- filtration
- chemical addition chlorine
  - lime
  - fluoride
- clearwell - distribution system.

The ClO_2 is generated by mixing aqueous Cl_2 and aqueous $NaClO_2$ in the approximate ratio of 1.24:1 by weight [80 pounds (36 kg) of chlorine per day and 64 pounds (28.8 kg) of $NaClO_2$ per day]. The $Cl_2/NaClO_2$ ratio can increase to as high as 7.8:1 depending upon the quality of water that is pumped to the filtration plant. ClO_2 is used to pretreat the water before chemical coagulation in a constant feed system. Liquid chlorine is delivered to the site in one-ton cylinders. Technical grade $NaClO_2$ is manufactured by Olin Corporation and delivered by an area distributor in 100 pound drums.

The ClO_2 generation system consists of equipment manufactured by Wallace and Tiernan. There is one ClO_2 generator, one plunger pump and one chlorinator. $NaClO_2$ is mixed in a large holding tank in a room above the ClO_2 generation room. The $NaClO_2$ day tank in the ClO_2 room is gravity filled each day from the larger $NaClO_2$ tank. Anhydrous $NaClO_2$ is stored on the floor above along with the larger $NaClO_2$ mixing tank. Piping from the chlorine tanks to the ejector is black steel; piping from the ejector to the ClO_2 reaction vessel is a heavy gauge rubber hose. The $NaClO_2$ chemical system piping from the large mixing tank to the day tank one floor below is schedule 80 PVC. The suction line from the $NaClO_2$ plunger pump is Tygon tubing; the discharge piping from the pump to the ClO_2 reactor vessel is a heavy gauge rubber hose - supplied by the manufacturer. The chlorinators and the one ton (960 kg) chlorine cylinders are located in separate rooms next to the ClO_2 generation room.

The chlorine storage room has forced ventilation by way of a motor generated damper unit. The exhaust fan is mounted at floor level and appears to provide for adequate circulation. The switches for the exhaust system as well as for the ceiling lights are located outside the entrance doorway to the chlorine room. The door, which opens to the outside grounds, has a glass panel for visual inspection of the room interior. The room where the chlorinators are located is provided with the same general equipment. The chlorinator room adjoins the chlorine storage room and has an exit to the outside yard. There are no chlorine leak detectors. There are, however, gas masks positioned outside of these rooms in the hallway.

The ClO_2 generation room houses the ClO_2 reaction vessel, the $NaClO_2$ feed pumps and day tank. The room has a floor drain that connects to the local municipal sewer system. There are spigots and hoses for washdown. The day tank consists of a plastic 40 to 50 gallon (152 to 190 liter) tank supported inside a metal 55 gallon (209 l) drum. The tank has a plastic lid that fits securely on top to prevent spills. The plastic tanks are replaced every 4 to 5 years for aesthetic purposes. The metal drum has a one-inch open slit along the side to allow the operator to see the liquid level of $NaClO_2$ inside the day tank. The volume is measured in gallons as determined by the markings painted along the open slit.

Production of ClO_2 is monitored visually by the color of ClO_2 that appears inside the reactor vessel. Chemical feed rates of Cl_2 and $NaClO_2$ also are checked. The efficiency of ClO_2 addition is evaluated hourly by confirming the absence of iron and manganese in the finished water. The typical influent concentrations of these metals during the late summer is 0.15 mg/l. The approximate concentration of $NaClO_2$ in the plant water is 0.35 mg/l.

The concentration of free chlorine in the finished water is 1 mg/l. The free chlorine residual before the flash mix ranges from 0.1 to 0.75 mg/l. Chlorine residual is measured frequently by the OTA method, but when recording information for State files, the DPD method is employed. The plant supervisor noted that the DPD method produces values consistently higher than the OTA method (i.e., 1.45 mg/l DPD and 1.25 mg/l OTA).

Chemical costs for ClO_2 production are incomplete. The cost of $NaClO_2$ is \$103/100 pounds (\$2.29/kg) and the estimated chemical cost for producing ClO_2 per 1,000 gallons of water is \$0.002 (0.06¢/cu m). The O/M costs are reportedly negligible. The plant sells the finished water to county customers at \$0.35/1,000 gallons (\$0.09/cu m).

WHEELING, WEST VIRGINIA

The 10 mgd (38,000 cu m) plant at Wheeling uses ClO_2 primarily for the control of phenolic compounds in its raw water source -- the Ohio River. Iron and manganese also pose problems. The source of the phenols, iron and manganese are attributed to both natural and industrial origin. The ClO_2 addition has been in operation since 1951.

The Wheeling plant has the following process steps:

- pre-settling
- aeration
- chemical addition - lime/ferric sulfate
- prechlorination - chlorine
- flash mix
- flocculation
- sedimentation
- chemical addition - sodium tripolyphosphate
- filtration
- chemical addition - sodium silicofluoride
  - chlorine dioxide
- clearwell - distribution system.

ClO_2 is generated by mixing aqueous Cl_2 and aqueous $NaClO_2$ in the ratio of 5:1 by weight [500 pounds (225 kg) of Cl_2/day and 100 pounds (45 kg) of $NaClO_2$/day]. This is a constant feed dosage year-round. The relatively high dosage of chlorine in the ClO_2 reactor reportedly is achieved to maintain a low pH for efficient ClO_2 production. Liquid chlorine is supplied to the site via rail tank car. The $NaClO_2$ is a technical grade manufactured by Olin Corporation and distributed by a local retailer. The anhydrous $NaClO_2$ is shipped in 100 pound (45 kg) drums.

The ClO_2 generation system consists of one Fischer and Porter reactor vessel for ClO_2 production, one Wallace and Tiernan diaphragm pump, $NaClO_2$ feed, and one Fischer and Porter gas evaporator and chlorinator. The piping for the aqueous $NaClO_2$ and ClO_2 presumably is schedule 80 PVC as specified by Fischer and Porter. The piping for both the liquid and gaseous chlorine is black steel. No leaks have been reported with the tubing materials. The Cl_2/$NaClO_2$ chemical feed equipment and ClO_2 reactor unit are housed in a separate building which adjoins the rail siding.

The building has a motor operated damper (MOD) exhaust system located at floor level. The air change rating is unknown but the unit appears to be of sufficient size. Makeup air is supplied through a louver on the opposite side of the room. The switches for the fan and interior lights are located on the exterior wall next to the door. An alarm switch is located on the interior wall inside the building. (The plant alarm system is tied into the city's fire station which is one block away.) The second of four alarm switches is located inside the entrance doorway. The chlorine leak detector, an ammonium hydroxide bottle, is located on top of a chlorinator. The entrance door has a glass window for visual inspection of the room interior. There are no floor drains inside the building.

The production of ClO_2 is monitored by the color of ClO_2 that appears inside the reactor vessel. Chemical feed rates of Cl_2 and $NaClO_2$ are also checked. The Fischer and Porter reactor vessel also has a sampling port on the immediate discharge port of the vessel. The plant does not analyze the ClO_2 product from the reactor vessel. Efficiency of ClO_2 addition is evaluated by the absence of taste and odor problems in the finished water. If the T/O problems persist, the $NaClO_2$ feed rate is increased, not the Cl_2 feed rate. Exact quantities or ranges of chemical feed for the operation of the ClO_2 generator are not available. The estimated concentration of $NaClO_2$ in the plant water is 1.2 mg/l.

The concentration of chlorine in the finished water is analyzed hourly by the laboratory. The plant uses the DPD method for the analysis of combined and free chlorine. Finished water has a free chlorine residual of 3 mg/l while the free chlorine residual at the extremities of the distribution system ranges from 0.5 to 1 mg/l.

The operator is concerned about the high cost of $NaClO_2$ as well as the potential hazard of storing dry $NaClO_2$. He is otherwise satisfied with the use of ClO_2 and intends to continue using it. The plant spends over $18,000/year for chlorine and $8,048/year for $NaClO_2$ (1976).

COVINGTON, KENTUCKY

The 10 mgd (38,000 cu m/day) Fort Thomas plant at Covington uses ClO_2 for control of taste and odor problems in its raw water source--the Ohio River. The cause of these taste and odor bodies is unknown. Addition of ClO_2 has been in operation since the early 1950s, although during the 1960s, the plant added ammonia with chlorine and discontinued ClO_2. Since 1972, the plant has resumed the addition of ClO_2 together with chlorine and terminated the use of ammonia because the chloramines were not holding the desired residual across the treatment basins. Chlorine has always been added to the plant water.

The Covington plant has the following process steps:

- chemical addition - lime/alum
  - chlorine - an injection point separate from the ClO_2 point of addition
  - chlorine dioxide
- flash mix
- coagulation
- sedimentation

- filtration
- chemical addition - chlorine
  - sodium silicofluoride
- clearwell - distribution system

The Covington plant pumps Ohio River water to two raw water reservoirs. During the warmer months, copper sulfate ($CuSO_4$) is added at the raw water station at approximately 200 pounds (90 kg) per day. ClO_2 is generated by mixing aqueous Cl_2 and 37% aqueous $NaClO_2$ in the ratio of 1.5:1 by weight [75 pounds (33.75 kg) of Cl_2 per day and 50 pounds (22.5 kg) of $NaClO_2$ per day]. The aqueous $NaClO_2$ is manufactured by Olin Corporation and delivered in 55 gallon dry weight drums by an area distributor. Chlorine is supplied in one ton (900 kg) cylinders.

The ClO_2 generation system consists of one Fischer and Porter ClO_2 reactor, one Fischer and Porter chlorinator and one Wallace and Tiernan diaphragm pump for liquid $NaClO_2$. The operator reported that the diaphragm pump develops cracks inside the plastic diaphragm casing. This causes the suction head to decrease and reduces the $NaClO_2$ feed rate. The reason for the cracks was not known. A second Fischer and Porter chlorinator is used for feeding chlorine into the plant water - separate from the ClO_2 feed system. Both Cl_2 and ClO_2 have the same application point in the chemical addition chamber. Piping from the $NaClO_2$ drum to the ClO_2 reactor vessel is Tygon tubing. Black steel pipe is used for conveying gaseous chlorine from the one ton (900 kg) cylinders to the chlorinator. A heavy-walled rubber hose is used for transporting aqueous chlorine from the chlorinator to the ClO_2 reactor. The same black rubber material is used for piping the ClO_2 product to the plant water. There were no reports of leakage from any of the piping material. The ClO_2 generation equipment as well as the lime/ferric feed equipment are located in one building. The one-ton (900 kg) chlorine cylinders are housed in a separate building.

Aqueous $NaClO_2$ is stored in 55 gallon (209 l) drums in the same room as the liquid alum feed equipment. There is no air exhaust system in this room although there are eight windows on the exterior wall. The floor drains inside this room empty to a french drain which empties beneath the building. $NaClO_2$ solution is fed undiluted from the drum to the ClO_2 reactor. When the $NaClO_2$ level is low, the operator removes the Tygon tubing from the spent drum and places it inside a full drum. Any $NaClO_2$ spills that occur are washed down the floor drain. This is done to prevent any potential fire hazard due to dry $NaClO_2$ that may remain on the floor. The working area around the $NaClO_2$ equipment is very close.

The chlorine storage room has a large exhaust fan mounted at floor level. Makeup air is provided by opening the entrance door or by an opened window. The switches for the exhaust fan and the ceiling lights are located on the interior wall next to the doorway. The entrance door has a glass panel for visual inspection of the room interior. The only leak detector for chlorine gas consists of two ammonium hydroxide bottles that are located in the center of the estimated 20' x 50' building. There is no alarm system on the plant grounds. Two self-contained breathing apparatuses and one MSA mask are stored in the control building -- about 125 feet away.

Production of ClO_2 is monitored visually by the color that appears inside the ClO_2 reactor. The chemical feed rates are also checked. Efficiency of ClO_2 production is monitored by the absence of T/O problems in the finished water. The estimated concentration of $NaClO_2$ in the plant water is 0.6 mg/l.

Chlorine in the plant water is analyzed by the DPD and the amperometric method. Free chlorine residual in the finished water varies from 2.5 to 4 mg/l but reportedly has been as high as 6 mg/l. The chlorine residual at the extremities of the distribution system at the plant was not immediately known. The State has required breakpoint chlorination since 1974. The plant reportedly uses at least one ton of chlorine every 2 to 3 days for water treatment (8 to 12 mg/l chlorine).

Chemical costs for the ClO_2 system were not available.

ANN ARBOR, MICHIGAN

The 16 mgd (60,800 cu m/day) plant at Ann Arbor uses ClO_2 for control of taste and odor problems in its raw water source -- the Huron River. Well water is used as an auxiliary source. The reported cause of the T/O problems is actinomycetes and is a year-round problem. The addition of ClO_2 has been in service since 1974.

The Ann Arbor plant has the following process steps:

- prechlorination - chlorine
- chemical addition - lime/soda ash
- flash mix
- coagulation
- sedimentation
- recarbonation
- chemical addition - powdered activated carbon
  - lime
- flash mix

- coagulation
- sedimentation
- chemical - chlorine dioxide
- filtration
- chemical addition - chlorine
  - fluoride
- clearwell - distribution system

ClO_2 is generated by mixing aqueous $NaClO_2$ and hydrochloric acid in the ratio of 6.1:1 by weight [32 pounds (14.4 kg) of $NaClO_2$ per day and 5.24 pounds (2.36 kg) of HCl per day dry weight]. HCl is delivered in 150 pound (67.5 kg) carboys as non-inhibited muriatic acid. The anhydrous $NaClO_2$ is delivered in 100 pound (45 kg) drums. Olin Corporation manufactures the technical grade $NaClO_2$ and an area distributor delivers the chemical to the plant.

Ann Arbor changed to HCl as a reagent for ClO_2 production because of the high costs of using liquid chlorine. The HCl also offers more reliable pH control than chlorine. Solution pH is measured periodically from the sample port on the discharge side of the reactor vessel.

The ClO_2 generation system consists of two Wallace and Tiernan ClO_2 reactor vessels (one as a standby) and one Wallace and Tiernan diaphragm pump for each of the HCl and $NaClO_2$ feed systems. There are also two plastic day tanks for liquid chemical feed storage. Plant personnel have installed a sampling port on the discharge end of the ClO_2 reactor. Heavy-walled Tygon tubing is used to transport liquid chemicals from the day tanks to chemical feed pumps. A semi-rigid plastic pipe carries the chemical reagents from the pumps to the reactor. On the ClO_2 unit, the effluent piping begins as schedule 80 PVC but then joins a semi-rigid plastic pipe which carries the aqueous ClO_2 product to the sedimentation tanks. The sampling valve on the ClO_2 unit is a PVC material. The piping network reportedly has had no problems with leakage. The plastic drums need replacing every two years or so; the Tygon tubing for the feed systems is replaced more frequently. The ClO_2 generation equipment is located in a hallway inside the filter gallery, which exits to the outside where the liquid HCl is stored in drums. $NaClO_2$ drums are stored along the wall of the hall. There are no floor drains immediately around the ClO_2 generation equipment.

Production of ClO_2 is monitored visually by the color that appears through the transparent reactor vessel. The ClO_2 solution had the richest red-brown color witnessed while visiting the various water plants, indicating that a higher concentration of ClO_2 is produced at Ann Arbor by the acid technique. Effluent from the ClO_2 reactor vessel is periodically analyzed for pH. The chemical feed pumps are also checked for proper feed rates.

Efficiency of ClO_2 addition is measured by the absence of taste and odor in the finished water.

The cost of HCl is approximately 28¢/lb (62¢/kg); $NaClO_2$ costs $78.70/100 pounds ($1.75/kg); liquid chlorine costs about 9¢/pound (20¢/kg). The chemical cost of producing ClO_2 reportedly is 0.16¢/1,000 gallons (0.06¢/cu m) of water treated.

In 1975, this plant stopped post-chlorinating with chlorine and changed to liquid sodium hypochlorite - NaOCl. Although this chemical is more expensive than chlorine, it is reportedly safer to handle than chlorine. Plant operators believe that less storage area is required for the liquid NaOCl and there is no longer a need to service the chlorinators. The NaOCl is 15% bleach and costs 37¢/gallon. Although liquid chlorine is no longer used at the plant, it is still used at the raw water lift station. However, $KMnO_4$ is frequently used at the lift station as a substitute for abating severe taste and odor problems. There is a 90 minute detention time from the raw water lift station to the filter plant.

The plant maintains a 1 mg/l residual of free chlorine in the clearwell. The free chlorine residual in the distribution system ranges from 0.3 to 0.4 mg/l. The DPD method is used to measure chlorine residual when reporting the results to the State. The OTA method is used around the plant to spot-check the chlorine residual of the process water.

The Ann Arbor plant is one of a few plants in the U.S. that recalcines the lime sludge for reuse.

APPENDIX F: CHLORINE DIOXIDE QUESTIONNAIRE SUMMARY TABLES

These tables summarize the questionnaire responses from the U.S. and Europe regarding use of chlorine dioxide in water treatment. As is the case with the ozone questionnaires, these data are reported as provided by the responding plants, without interpretation. In some cases, data were reported in incomplete form in a given category. Consequently, the reader is cautioned to discount data which vary widely from the norm, as the bases for those data may differ from the general case.

Also, note that Hamilton, Indiana is the only U.S. plant using ClO_2 strictly for residual disinfectant, even though other plants indicated this application on their questionnaire responses.

A process description is provided for each plant, beginning with the first process step and terminating at the clearwell. The method of ClO_2 preparation is also provided, where known.

In considering the dosages and/or residuals of ClO_2 reported, the reader should be aware that whenever excess chlorine is added to sodium chlorite, and the "chlorine dioxide" is measured by the OTA, DPD or amperometric titration methods, total oxidant is measured. This consists of chlorine plus chlorine dioxide, and the amount of chlorine present will depend upon the amount of excess chlorine added originally.

Many U.S. plants report dosages of ClO_2 as the amount of $NaClO_2$ added, and assume that it is all converted to ClO_2 with sufficient excess chlorine.

In the tables, the following abbreviations are used:

NA = not available

I = intermittent ClO_2 dosage

C = continuous dosage

sed. = sedimentation

exp. = expansion

W&T = Wallace & Tiernan

F&P = Fischer & Porter

T/O = taste and odor

GAC = granular activated carbon

PAC = powdered activated carbon

TABLE F-1. UNITED STATES ClO_2 QUESTIONNAIRE SUMMARY

Plant Name	Design Plant Capacity (MGD)	Year ClO_2 Use Initiated	Process Description	Purposes of ClO_2 Use	Average Dosage (mg/l)	Residual in Dist. System (mg/l)	Method of Preparation
Hartford, CT	NA	NA	NA	have capability to use ClO_2 but have not used it to date	NA	NA	NA
Lebanon, CT	9.5	NA	coagulation, sedimentation, filtration, chlorination	odor removal	I	NA	NA
Hamburg, NY	2.0	NA	air mixing, primary sed., secondary mixing, flocculation, secondary sed., filtration, storage	taste, odor, organics removal	I 1.0	NA	ClO_2 fed before primary sed. or after filtration as needed. ClO_2 from gaseous chlorine & $NaClO_2$. ClO_2 is discharged into clearwell or at raw water mixing chamber. Mathieson ClO_2 generator
Binghamton, NY	2.0	1951	coagulation, sed., filtration, ClO_2 addition	disinfection, taste, odor removal	C	0.03	ClO_2 fed into finished water. ClO_2 generated by gaseous chlorine & $NaClO_2$
Tonawanda, NY	24.0	1956	NA	phenols	NA	NA	ClO_2 generated with gaseous chlorine & $NaClO_2$. W&T ClO_2 system
New Brighton, PA	6.0	1969	mixing, flocculation, primary settling, mixing, secondary settling, filtration, chlorination	odor, organics, taste removal	I 0.18	NA	ClO_2 produced from gaseous chlorine & $NaClO_2$. W&T ClO_2 system

(continued)

TABLE F-1 (continued)

Plant Name	Design Plant Capacity (MGD)	Year ClO_2 Use Initiated	Process Description	Purposes of ClO_2 Use	Average Dosage (mg/l)	Residual in Dist. System (mg/l)	Method of Preparation
Oakmont, PA	10.0	1948 expd. 1974	chemical addition, rapid mixing, slow mixing, flocculation, settling, filtration, post chlorination & ClO_2 addition	disinfection, odor taste, organics removal	C 0.1-0.2	NA	ClO_2 produced from gaseous chlorine & $NaClO_2$. ClO_2 added before clearwell. W&T ClO_2 system
Midland, PA	4.4	1946	rapid mix, coagulation, sedimentation, filtration	NA	C 0.4	0.15	ClO_2 produced from gaseous chlorine & $NaClO_2$. ClO_2 added before clearwell. Mathieson ClO_2 generator
Pittsburgh, PA (Becks Rd.)	52.9	NA	chemical addition, rapid mixing, recarbonation, sedimentation, activated carbon filtration	odor, taste, organics removal	I 0.15	NA	ClO_2 produced from gaseous chlorine & $NaClO_2$. ClO_2 added before clearwell. pH of Cl_2 solution is adjusted to 3.5. W&T ClO_2 system
Charleroi, PA	10.0	1975	NA	odor, taste removal	C 0.023	NA	W&T ClO_2 system
Penn Hills, Wilksburg, PA	275.0	1947	flash mixing, powdered activated carbon, flocculation, settling, chlorination, filtration	odor, taste, organics removal	C 0.1-0.15	NA	ClO_2 produced from gaseous chlorine & $NaClO_2$ & injected before clearwell. W&T ClO_2 system
McKeesport, PA	10.0	1967	alum, lime, carbon, ClO_2 addition, flocculation, settling, chlorination, filtration	odor, taste, organics removal	I, but NA	NA	ClO_2 produced from gaseous chlorine & $NaClO_2$ & injected as needed before flocculation. W&T ClO_2 system

(continued)

TABLE F-1 (continued)

Plant Name	Design Plant Capacity (MGD)	Year ClO_2 Use Initiated	Process Description	Purposes of ClO_2 Use	Average Dosage (mg/l)	Residual in Dist. System (mg/l)	Method of Preparation
Washington, PA	5.2	1973	aeration, coagulation, pre-chlorination, settling, filtration, post chlorination, ClO_2 addition	odor, taste removal	C 0.20	NA	ClO_2 is produced from chlorine & $NaClO_2$ & added before clearwell. F&P ClO_2 generator
Charlottesville, VA	4.0	1966	aeration, ClO_2 addition, flash mix, coagulation, settling, filtration, chlorination	odor, taste removal	I, but NA	NA	ClO_2 produced from gaseous chlorine & $NaClO_2$ & added as needed before flash mixing. W&T ClO_2 system
Wheeling, WV	10.0	1951	presettling, aeration, chemical addition of lime & ferric sulfate, prechlorination, flash mixing, flocculation, settling, chemical addition of sodium tripolyphosphate, filtration, chemical addition of sodium silicofluoride, ClO_2 addition	organics, Fe & Mn control, taste & odor removal	C 3.0	1.0	ClO_2 produced from gaseous chlorine & $NaClO_2$ & added before clearwell
Newman, GA	4.5	1957	ClO_2 addition, flash mixing, flocculation, settling, filtration, chlorination	Fe, Mn removal	C, but NA	NA	ClO_2 produced from gaseous chlorine & $NaClO_2$. W&T ClO_2 system
Roswell, GA	0.6	1936	ClO_2 addition, coagulation, settling, filtration	Mn, taste removal, chlorine residual	C, but NA	0.5	ClO_2 produced from gaseous chlorine & $NaClO_2$. W&T ClO_2 system

(continued)

TABLE F-1 (continued)

Plant Name	Design Plant Capacity (MGD)	Year ClO_2 Use Initiated	Process Description	Purposes of ClO_2 Use	Average Dosage (mg/l)	Residual in Dist. System (mg/l)	Method of Preparation
Marietta, GA	24.0	1958	ClO_2 addition, coagulation, settling, filtration, flocculation, chlorination, pH adjustment	Fe, Mn, taste & odor removal, disinfection	C 0.32	0.5	ClO_2 produced from gaseous chlorine & $NaClO_2$. W&T ClO_2 system
Lincolnton, GA	0.375	1971	ClO_2 addition, rapid mixing, flocculation, settling, filtration, ClO_2 addition	Mn removal	I, but NA	NA	ClO_2 produced from gaseous chlorine & $NaClO_2$. W&T ClO_2 system
Forsyth, GA	1.0	1974	ClO_2 addition, coagulation, settling, Cl_2 addition, filtration	Mn, taste removal, chlorine residual	NA	NA	ClO_2 produced from gaseous chlorine & $NaClO_2$. W&T ClO_2 system
Atlanta, GA (Boulton Rd.)	80.0	1960	prechlorination (at intake), pH adjust. (lime), powd. act. carbon (as needed), chlorination, ClO_2 (as needed) addition, pH control (lime), PAC addition, alum coagulation, settling, filtration, post chlorination, lime addition for corrosion control, fluoride addition	Mn, Fe removal	I, but NA	NA	ClO_2 from gaseous chlorine & $NaClO_2$. W&T ClO_2 system
Griffin, GA	8.0	1966	ClO_2 addition, coagulation, settling, filtration, post chlorination	Mn removal	NA	NA	ClO_2 produced from gaseous chlorine & $NaClO_2$. W&T ClO_2 system

(continued)

TABLE F-1 (continued)

Plant Name	Design Plant Capacity (MGD)	Year ClO_2 Use Initiated	Process Description	Purposes of ClO_2 Use	Average Dosage (mg/l)	Residual in Dist. System (mg/l)	Method of Preparation
Carrollton, GA	8.0	1953	ClO_2 addition, rapid mixing, coagulation, settling, filtration, post chlorination, lime addition, fluoride addition	Fe, Mn removal	C, but NA	NA	ClO_2 produced from gaseous chlorine & $NaClO_2$. W&T ClO_2 system
Monroe, GA	4.0	1959	ClO_2 addition, flash mixing, coagulation, sedimentation, filtration, post chlorination	Mn removal	C, but NA	NA	$NaClO_2$ dissolved in water & pumped into chlorinated water line; average 70 lbs chlorine to 14 lbs $NaClO_2$. W&T ClO_2 system
East Point, GA	12.0	1963	prechlorination, ClO_2 addition, coagulation, settling, filtration, post chlorination	Fe, Mn, taste removal	C 0.5	NA	ClO_2 produced from gaseous chlorine & $NaClO_2$. W&T ClO_2 system
Fayetteville, GA	0.2	NA	prechlorination, ClO_2 addition, coagulation, settling, filtration, post chlorination	Fe, color, Mn, odor, taste removal, disinfection, chlorine residual	C 1.4	0.5	ClO_2 produced from gaseous chlorine & $NaClO_2$. W&T ClO_2 system
Warrenton, GA	0.75	1970	ClO_2 addition, coagulation, settling, filtration, chlorination	Fe, odor, color, Mn, taste removal	I 10.0	NA	ClO_2 produced from gaseous chlorine & $NaClO_2$. W&T ClO_2 system
Bremen, GA	0.8	NA	ClO_2 addition, flash mixing, alum & lime coagulation, settling, filtration, post chlorination, Cl_2 residual	Mn, odor, taste removal, disinfection	C 4.0-9.0	NA	ClO_2 produced from gaseous chlorine & $NaClO_2$. W&T ClO_2 system

(continued)

TABLE F-1 (continued)

Plant Name	Design Plant Capacity (MGD)	Year ClO_2 Use Initiated	Process Description	Purposes of ClO_2 Use	Average Dosage (mg/l)	Residual in Dist. System (mg/l)	Method of Preparation
Burford, GA	1.0	1967	ClO_2 addition, flash mixing, flocculation, settling, filtration, chlorination	Fe, Mn, odor, taste removal disinfection	C 0.5	NA	ClO_2 produced from gaseous chlorine & $NaClO_2$. W&T ClO_2 system
Monticello, GA	0.5	1946	ClO_2 addition, flash mixing, flocculation, settling, filtration	Fe, Mn removal, disinfection, chlorine residual	C, but NA	1.0	ClO_2 produced from gaseous chlorine & $NaClO_2$. W&T ClO_2 system
Buchanen, GA	2.5	1975	ClO_2 addition, flash mixing, flocculation, settling, filtration, chlorination	Fe, odor removal	C 2.0	NA	ClO_2 produced from gaseous chlorine & $NaClO_2$. W&T ClO_2 system
McDonald, GA	1.0	1959	ClO_2 & Cl_2 addition, coagulation, settling, filtration, pH adjust.	Fe, color, odor, organics, Mn, taste removal, disinfection, chlorine residual	C 5-6	0.8	ClO_2 produced from gaseous chlorine & $NaClO_2$. W&T ClO_2 system
Waynesboro, GA	1.2	1972	ClO_2 addition, flash mixing, coagulation, settling, filtration, final chlorination	scale control, corrosion control, disinfection	C, but NA	0.5	ClO_2 from gaseous chlorine & $NaClO_2$. W&T ClO_2 system
Louisville, KY	NA	1965	coagulation, settling, filtration, post chlorination & ClO_2 addition (as needed)	organic removal chlorine residual	NA	NA	ClO_2 use discontinued in 1968. Generated from gaseous chlorine & $NaClO_2$. W&T ClO_2 system

(continued)

TABLE F-1 (continued)

Plant Name	Design Plant Capacity (MGD)	Year ClO_2 Use Initiated	Process Description	Purposes of ClO_2 Use	Average Dosage (mg/l)	Residual in Dist. System (mg/l)	Method of Preparation
Covington, KY (Ft. Thomas)	18.0	1972	ClO_2 + Cl_2 addition, lime + alum addition, coagulation, settling, filtration, final chlorination, fluoride addition	odor, organics, taste removal	C 0.6 as $NaClO_2$	NA	ClO_2 from gaseous chlorine & $NaClO_2$
Portsmouth, OH	16.0	1950	prechlorination, chemical addition (alum + $KMnO_4$), flash mixing, coagulation, settling, chemical addition, (alum + lime), flash mixing, settling, filtration, final chlorination (Cl_2/ClO_2)	odor, taste removal	C 0.21	0.3 total; 0 as free chlorine	ClO_2 from gaseous chlorine and $NaClO_2$. W&T ClO_2 system
Campbell, OH	3.8	1975	chemical addition, flash mixing, coagulation, settling, filtration, Cl_2/ClO_2 addition into clearwell	taste, odor removal, disinfection	C 0.60	NA	ClO_2 from gaseous chlorine and $NaClO_2$. W&T ClO_2 system
Oregon, OH	0.8	1964	NA	taste, odor removal	C 0.1	NA	ClO_2 from gaseous chlorine & $NaClO_2$. W&T ClO_2 system with PVC header/diffuser
Elyria, OH	NA	NA	NA	haven't used in 9 yrs-was used for phenols but phenol problem now is corrected	NA	NA	NA

(continued)

TABLE F-1 (continued)

Plant Name	Design Plant Capacity (MGD)	Year ClO_2 Use Initiated	Process Description	Purposes of ClO_2 Use	Average Dosage (mg/l)	Residual in Dist. System (mg/l)	Method of Preparation
Ironton, OH	5.0	1919	NA	taste, odor removal, chlorine residual, disinfection	C 2 mg/l	NA	ClO_2 from gaseous chlorine & $NaClO_2$
Wellsville, OH	2.2	1937	chemical addition (lime, alum, $KMnO_4$), ClO_2 addition, carbon addition, flash mixing, coagulation, settling, filtration	Mn, taste, odor removal	C 0.50	NA	ClO_2 from gaseous chlorine & $NaClO_2$
Napolean, OH	4.5	1966	chemical addition (alum + lime), flash mixing, coagulation, settling, chemical addition (soda ash + $KMnO_4$), flash mixing, coagulation, settling, prechlorination, filtration, ClO_2 addition	taste, odor removal, Cl_2 residual, disinfection	C 0.25 I 0.5	2.0	ClO_2 from gaseous chlorine & $NaClO_2$. ClO_2 added before clearwell
Manawa, WI	8.5	NA	NA	taste, odor removal, Cl_2 residual, disinfection	C 1.2-1.5	0.2 free	ClO_2 from gaseous chlorine & $NaClO_2$. ClO_2 added into clearwell; contact 20 to 30 minutes
Steubenville, OH	12.0	1954	prechlorination, coagulation, settling, filtration, post chlorination, ClO_2 addition	taste, Cl_2 residual, disinfection	C 5.6	0.3	ClO_2 from gaseous chlorine & $NaClO_2$; added at clearwell entrance
Defiance, OH	5.0	1971	chemical addition, coagulation, settling, recarbonation, filtration, ClO_2 addition	taste, odor, organic compounds removal	I 0.5-2.0 (used twice since 1971)	NA	ClO_2 from gaseous chlorine & $NaClO_2$; added at clearwell entrance

(continued)

TABLE F-1 (continued)

Plant Name	Design Plant Capacity (MGD)	Year ClO_2 Use Initiated	Process Description	Purposes of ClO_2 Use	Average Dosage (mg/l)	Residual in Dist. System (mg/l)	Method of Preparation
Escanaba, MI	1.5	1970	pre ClO_2 addition, carbon addition, chemical addition (alum), flash mixing, flocculation, settling, chemical addition, filtration, ClO_2 addition into clearwell, final ClO_2 addition	taste, odor, organic compound removal, disinfection, Cl_2 residual	I 1-2	NA	ClO_2 from gaseous chlorine & $NaClO_2$
Rogers City, MI	1.5	1955	NA	taste, odor, color	C 6	NA	ClO_2 from gaseous chlorine & $NaClO_2$; added into distribution System
Menominee, MI	4.5	1971	carbon addition, chemical addition, coagulation, settling, ClO_2 addition, filtration	taste, odor removal, Cl_2 residual	I	NA	ClO_2 from gaseous chlorine & $NaClO_2$. ClO_2 unit fabricated by water treatment plant
Ontonagon, MI	0.3	1967	chemical addition,coagulation, settling, filtration, ClO_2 addition into clearwell	taste, odor, color removal, Cl_2 residual, disinfection	C 1.6	NA	ClO_2 from gaseous chlorine & $NaClO_2$
Marquette, MI	3.5	1945	chlorine, ClO_2 addition into clearwell	taste, odor removal	C 2.5 lbs/day	NA	ClO_2 from gaseous chlorine & $NaClO_2$
Gladstone, MI	3.0	1972	chlorine addition, coagulation, settling, filtration, ClO_2 addition into clearwell	taste, odor removal	I 0.5	NA	ClO_2 from gaseous chlorine & $NaClO_2$
Adrian, MI	10.0	1949	coagulation, ClO_2 addition, filtration, final chlorination	taste removal	I	NA	ClO_2 from gaseous chlorine & $NaClO_2$

(continued)

TABLE F-1 (continued)

Plant Name	Design Plant Capacity (MGD)	Year ClO_2 Use Initiated	Process Description	Purposes of ClO_2 Use	Average Dosage (mg/l)	Residual in Dist. System (mg/l)	Method of Preparation
Hamilton, OH	6.0	1956	preaeration, coagulation, settling, recarbonation, filtration, Cl_2/ClO_2 addition into clearwell	Cl_2 residual	NA	0.15 at plant 0.10 in system	ClO_2 from gaseous chlorine & $NaClO_2$
Bethesda, OH	0.216	1960	prechlorination, chemical addition, flash mixing, settling, filtration, ClO_2/Cl_2 addition	taste, odor removal, disinfection, chlorine residual	I 2.3-10.0	NA	ClO_2 from gaseous chlorine & $NaClO_2$
Newark, OH	15.0	1922	ClO_2 addition, chemical addition (lime, alum, carbon), flash mixing, coagulation, settling, recarbonation, chemical addition (lime), chlorination, filtration, ClO_2 addition	taste, odor removal	C 0.40	NA	ClO_2 from gaseous chlorine & $NaClO_2$
Columbus, OH	65.0	1975	chemical addition (lime, alum, $KMnO_4$), flash mixing, coagulation, settling, $Cl_2/NaCO_3$, and lime addition, settling, recarbonation, Cl_2 addition, filtration, ClO_2 addition into clearwell	taste, odor removal, disinfection	C 0.2	NA	ClO_2 from gaseous chlorine & $NaClO_2$
Toledo, OH	120.0	1942	carbon addition, lime/soda softening, settling, recarbonation, filtration, fluoride, ClO_2/Cl_2 addition	taste, odor removal, disinfection	C 0.2 (ranges from 0.2-0.4)	NA	ClO_2 from gaseous chlorine & $NaClO_2$

(continued)

TABLE F-1 (continued)

Plant Name	Design Plant Capacity (MGD)	Year ClO_2 Use Initiated	Process Description	Purposes of ClO_2 Use	Average Dosage (mg/l)	Residual in Dist. System (mg/l)	Method of Preparation
Ann Arbor, MI	50.0	1939	prechlorination, lime/soda softening, settling, recarbonation, carbon addition, coagulation, settling, ClO_2 addition, filtration, post chlorination, fluoride addition	taste, odor removal	C 0.24	NA	ClO_2 from hydrochloric acid & $NaClO_2$
Midland, TX	12.0	1970	ClO_2 addition when needed, coagulation, settling, filtration	taste, odor removal	I	NA	ClO_2 equipment installed but never used. W&T ClO_2 system
Augusta, KS	2.0	1976	addition of alum, lime, Cl_2 & ClO_2, coagulation, settling, filtration, final chlorination	taste, odor, organics removal, disinfection, chlorine residual, decrease Cl_2 demand	C 0.12 as $NaClO_2$	NA	ClO_2 from gaseous chlorine & $NaClO_2$. F&P ClO_2 system
Greenville, OH	3.0	1967	coagulation, settling, filtration, ClO_2/Cl_2 addition	taste, odor removal	C 0.5	NA	ClO_2 from gaseous chlorine & $NaClO_2$. Herberg Chemical ClO_2 system
Bowling Green, OH	6.0	1951	coagulation, settling, recarbonation, filtration, chlorination	taste removal	I 0.25	NA	ClO_2 used very infrequently

TABLE F-2. GREAT BRITAIN ClO_2 QUESTIONNAIRE SUMMARY

Plant Name	Design Plant Capacity cu m/day	Year ClO_2 Use Initiated	Process Description	Purposes of ClO_2	Average Dosage (mg/l)	Residual in Dist. System (mg/l)	Method of Preparation
Chester, G.B., (Huntington Plant) River Dee	288,000	1970	chemical addition, (lime, alum Cl_2, ClO_2), flash mixing, settling, sand filtration, pH adjustment & final chlorination	taste & odor control, bacterial disinfection, chlorine residual	C 0.2-0.4	NA	gaseous chlorine + sodium chlorite. Excess of chlorine
Great Sutton, G.B., (Sutton Hall Treatment works) River Dee	64,000	1961	chemical addition (alum polyelectrolyte), mixing, settling, filtration, lime & ClO_2, SO_2 addition	taste & odor removal, bacterial disinfection, residual oxidant	C 0.3-0.4	NA	gaseous chlorine + sodium chlorite. Excess of chlorine
Burham, G.B., (Rochester), Kent River Midway	45,000	1975	prechlorination, coagulation, sedimentation, filtration, pH adjustment, ClO_2 addition	taste & odor removal, bacterial disinfection, chlorine residual	C 0.05-0.1	NA	gaseous chlorine + sodium chlorite
Egham, G.B., North Surrey Water Co., River Thames	44,000	1966	prechlorination, coagulation settling, filtration, chlorination, ClO_2 addition	bacterial disinfection, chlorine residual, prevention of nuisance growth	C 0.3	0.05	gaseous chlorine + sodium chlorite combined in a Raschig ring reactor
Southampton, G.B., Testwood Pumping Station, River Tert	44,500	1968	prechlorination, coagulation, settling, ClO_2 addition, filtration	bacterial disinfection, phenolic taste	C 0.2-0.5	NA	gaseous chlorine + sodium chlorite, mixed in reactor

TABLE F-3. FRENCH AND AUSTRIAN ClO_2 QUESTIONNAIRE SUMMARY

Plant Name	Design Plant Capacity cu m/day	Year ClO_2 Use Initiated	Process Description	Purposes of ClO_2	Average Dosage (mg/l)	Residual in Dist. System (mg/l)	Method of Preparation
St. Oven l'Aumone, France, river water	2,400	1976	prechlorination, coagulation, settling, act. carbon filtration, final ClO_2 addition	bacterial disinfection, viral inactivation	C NA	0.2	gaseous chlorine + sodium chlorite-CIFEC "Bioxy" system
Vichy, France, river water	NA	NA	NA	NA	C	NA	used for pre ClO_2 addition and final ClO_2 addition. Only information given
Viry-Chatillion France	100,000	1951	prechlorination, coagulation, settling, filtration, ClO_2 addition	color, taste removal, bacterial disinfection	C 0.2-0.3	NA	gaseous chlorine + sodium chlorite, Degrémont system
Toulouse, France, surface water reservoir	110,000	1970	pre-ClO_2 addition, coagulation settling, filtration, O_3, sodium silicate addition	color, odor, taste, organics removal, bacterial disinfection	C 0.8	NA	gaseous chlorine + sodium chlorite-CIFEC "Bioxy" system
Rennes, France, surface water reservoir	24,000	1977	O_3, ClO_2 addition	bacterial disinfection	C NA	0.25	gaseous chlorine + sodium chlorite-CIFEC system
Vigneux-sur-Seine France, Seine River	57,600	1959	prechlorination, coagulation, settling, filtration, final ClO_2 addition	color, taste removal, bacterial disinfection	C 0.2-0.25	NA	gaseous chlorine + sodium chlorite. Degrémont system
Vienna, Austria, Ubergangskammer II	230,000	1973	NA	bacterial disinfection, chlorine residual	C 0.25	0.08	reaction of chlorine and sodium chlorite

(continued)

TABLE F-3 (continued)

Plant Name	Design Plant Capacity cu m/day	Year ClO_2 Use Initiated	Process Description	Purposes of ClO_2	Average Dosage (mg/l)	Residual in Dist. System (mg/l)	Method of Preparation
Vienna, Austria, ground waterworks Nussdorf	100,000	1967	NA	bacterial disinfection, chlorine residual	C 0.25	not measured	reaction of chlorine and sodium chlorite
Les Rousses, France, lake water	5,000	1959	pre-ClO_2 addition as needed, settling, filtration, final chlorination	odor, organics removal, bacterial disinfection	C 1.0	0.1	chlorine + sodium chlorite Degrémont equipment
Le Mans (Sarthe), France, river water	60,000	1970	prechlorination, coagulation, settling, filtration, ClO_2 addition, O_3, final ClO_2 addition	color, organics, odor, taste removal, bacterial disinfection, viral inactivation	C 2-3 (total)	trace	chlorine + sodium chlorite CIFEC system
Suresnes, France, Seine River	112,000	NA	line #1-chemical addition, powd. act. C., settling, filtration, ClO_2 addition, hypochlorite addition, line # 2-prechlorination, alum, powd. act. C., coagulation, settling, filtration, ClO_2 addition	bacterial disinfection, chlorine residual	NA	NA	hypochlorite + sodium. chlorite
Porticcio, France, river water	25,300	1966	coagulation, settling, filtration, ClO_2	bacterial disinfection	NA	0.1	gaseous chlorine + sodium chlorite
Bellerive-sur-Allier, France	6,600	NA	Cl_2, coagulation, sedimentation, filtration, O_3, Cl_2/ClO_2	disinfection	C 0.1-0.2	NA	Cl_2 + $NaClO_2$

(continued)

TABLE F-3 (continued)

Plant Name	Design Plant Capacity cu m/day	Year ClO_2 Use Initiated	Process Description	Purposes of ClO_2	Average Dosage (mg/l)	Residual in Dist. System (mg/l)	Method of Preparation
Le Poiroux, Aurille, France	16,000	NA	$KMnO_4$, ClO_2, Cl_2, coagulation, sedimentation, filtration, O_3, neutralization, Cl_2	Fe, color, odor, taste, organics, Mn removal, bacterial disinfection	1.0 C (disinf) 8.0 I (T/O)	0	Cl_2 + $NaClO_2$, CIFEC system
Chalandry, Elaire, France	10,000	1975	Cl_2/ClO_2, coagulation, sedimentation, filtration, O_3, Cl_2	Fe, Mn, color, odor, taste, organics removal, bacterial disinfection, viral inactivation	0.13 C (T/O) "trace" (disinf)	NA	Cl_2 + $NaClO_2$, CIFEC system
Warcq, France	8,400	1973	Cl_2/ClO_2, coagulation, sedimentation, filtration, O_3, Cl_2	Fe, Mn, taste, odor, taste, color, organics removal, disinfection, viral inactivation	"traces" (disinf) 0.2 C as Cl_2 (T/O)	NA	Cl_2 + $NaClO_2$, CIFEC system
Communante Urbaine de Lyon, France	600,000	1977	NA	disinfection	C 0.075	0	Cl_2 + $NaClO_2$, Degrémont system
Marseille, France	100,000	1953	Cl_2/ClO_2, coagulation, sedimentation, filtration, Cl_2/ClO_2	taste, odor, organics removal, disinfection, viral inactivation	NA	NA	Cl_2 + $NaClO_2$ (pretreatment by CIFEC)
Villefranche-sur-Saone, France	28,000	1975	Cl_2/ClO_2 addition only	taste, odor removal, disinfection, prevent formation of chlorophenols	C 0.15 (disinf) 0.25 (T/O)	0	Cl_2 + $NaClO_2$, CIFEC system

TABLE F-3 (continued)

Plant Name	Design Plant Capacity cu m/day	Year ClO_2 Use Initiated	Process Description	Purposes of ClO_2	Average Dosage (mg/l)	Residual in Dist. System (mg/l)	Method of Preparation
Paray le Monial, France, river water	6,000	1969	prechlorination, coagulation, flocculation, settling, filtration, ClO_2 addition	taste removal, bacterial disinfection	C 1.3	0.1	gaseous chlorine + sodium chlorite, Degrémont equipment
St. Etienne, France, lake water	20,000	1972	NA	bacterial disinfection	C 0.3	0.02	gaseous chlorine + sodium chlorite, CIFEC system
Carcassone, France river water	16,000	1946	settling, filtration, ClO_2 addition	organics, taste, odor removal, bacterial disinfection	C 0.3-0.5	0	gaseous chlorine + sodium chlorite. CIFEC system
La Ferte, Bernard, France	8,000	NA	prechlorination, coagulation, sedimentation, filtration, ClO_2	color, taste, odor removal, disinfection	0.7 C (disinf) 1.0 (T/O)	0	Cl_2 + $NaClO_2$ (Degrémont)
Agon Coutaimville, France	1,000	1972	ClO_2, coagulation, sedimentation, filtration, O_3, ClO_2	Fe, color, odor, organics, Mn, taste, turbidity removal, bacterial disinfection, viral inactivation	0.2 C (disinf)	0	Cl_2 + $NaClO_2$ (Degrémont)
Coutances, France	1,600	1957	ClO_2, coagulation, sedimentation, filtration, O_3, ClO_2	Fe, color, odor, organics, Mn, taste, turbidity removal, bacterial disinfection	0.3 C (disinf) 1.6 I (T/O)	0	Cl_2 + $NaClO_2$

(continued)

TABLE F-4. GERMAN ClO_2 QUESTIONNAIRE SUMMARY

Plant Name	Design Plant Capacity cu m/day	Year ClO_2 Use Initiated	Process Description	Purposes of ClO_2	Average Dosage (mg/l)	Residual in Dist. System (mg/l)	Method of Preparation
Einhausen (Hessen), Germany	50,000	1963	aeration, $KMnO_4$ addition, filtration, ClO_2 addition	chlorine residual	NA	NA	prepared by reacting $NaClO_2$ + chlorine gas
Wasserwerk III-Wittlaer, Duisburg, Germany	72,000	1962	O_3, filtration(anthracite and GAC), ClO_2 addition, neutralization	chlorine residual	C 0.20	0	prepared by reacting $NaClO_2$ + chlorine gas & water
Duisburg (46 plant), Germany	18,000	1968	pumping, pH adjustment, ClO_2 addition	bacterial disinfection	C 0.4	0.02	prepared by reacting $NaClO_2$ + Cl_2 gas & water
Duisburg (14 plant), Germany	22,000	1968	prechlorination, mixing basin pH adjustment, storage tank, ClO_2 addition	bacterial disinfection	C 0.4	0.02	prepared by reacting Cl_2 gas plus water + $NaClO_2$
Karlsruhe Waterworks, Germany	104,500	1977	aeration, coagulation, filtration, ClO_2 addition	bacterial disinfection	C 0.20	0	prepared by reacting $NaClO_2$ gas; capacity of 4.0 kg/day
Mülheim, Germany (Styrum)	60,000	1955	river sand bank filtration, storage well, ClO_2 addition	taste & odor removal bacterial disinfection, chlorine residual	C 0.4-0.5	0.1	prepared by reacting $NaClO_2$ + Cl_2 in a 1:1 ratio; analyze for the sum of Cl_2 + ClO_2
Wolfsburg City Works, Gem. Sassenburg, Germany	15,500	1967	coagulation, filtration, ClO_2 addition	chlorine residual	NA	NA	NA

(continued)

TABLE F-4 (continued)

Plant Name	Design Plant Capacity cu m/day	Year ClO_2 Use Initiated	Process Description	Purposes of ClO_2	Average Dosage (mg/l)	Residual in Dist. System (mg/l)	Method of Preparation
Mainz City Works, Mainz, Germany	25,000	1958	aeration, $KMnO_4$ addition, filtration, ClO_2 addition	organics, taste & odor removal, bacterial disinfection, chlorine residual	C 0.6	<0.1	prepared by reacting $NaClO_2$ + chlorine gas
City Works Mainz AG, Mainz, Germany Dr. Friedrich-Kirchhoff Street Plant	3,200	1960	ClO_2 addition is only treatment	organics, taste & odor removal, bacterial disinfection, chlorine residual	C 0.3	<0.1	prepared by reacting $NaClO_2$ + Cl_2 gas
City Works Mainz AG, Wormser Street Plant, Mainz, Germany	2,700	1967/ 1972	ClO_2 addition is only treatment	organics, taste & odor removal, bacterial disinfection, chlorine residual	C 0.3	<0.1	prepared by reacting chlorine gas + $NaClO_2$
Pforzheim City Works, Pforzheim, Germany	24,000	1962	wells, ClO_2 addition, holding tank, distribution	bacterial disinfection	C 0.25	NA	prepared by reacting $NaClO_2$ + chlorine gas; (Degussa-Chlorite system)
Düsseldorf City Works, Germany	403,000	1961-67	river sand bank filtration, O_3, filtration thru pre-activated carbon, GAC filtration, neutralization (NaOH), ClO_2 addition	bacterial disinfection, chlorine residual	C 0.1-0.2	<0.1	ClO_2 capacity is 415 kg/day. Prepared by reacting chlorine gas + $NaClO_2$. 0.25 kg Cl_2/1000 cu m + 0.68 kg $NaClO_2$/1000 cu m

(continued)

TABLE F-4 (continued)

Plant Name	Design Plant Capacity cu m/day	Year ClO_2 Use Initiated	Process Description	Purposes of ClO_2	Average Dosage (mg/l)	Residual in Dist. System (mg/l)	Method of Preparation
Wasserwerk das krelas, Aachen, Germany	172,800	1962	NA	taste & odor removal, bacterial disinfection	C 0.1	NA	ClO_2 capacity is 180 kg/day prepared by reacting 50 kg of $NaClO_2$ with 50 kg chlorine on a daily basis
Wuppertal City Works, Düsseldorf-Benrath, Germany	150,000	1967	river sand bank filtration, aeration, O_3, sand filtration, activated carbon, ClO_2 addition	bacterial disinfection	C 0.2	0.1	ClO_2 capacity is 72 kg/day prepared by reacting sodium chlorite + chlorine gas and water
Berlin, West Germany (Jungfernherde Plant)	170,000	1975	groundwater, aeration, filtration, ClO_2, clean water holding tank, ClO_2 measurement	bacterial disinfection	I 0.20	0	prepared by reacting 30% sodium chlorite + chlorine gas; capacity of 17.3 kg/day
Augsburg City Works, Bavaria, Germany	100,000	1963	ClO_2 addition only	bacterial disinfection, prevention of bacterial regrowth in distribution system	NA	NA	capacity of 740 kg/day. Prepared by reacting $NaClO_2$ + Cl_2 gas
Groundwater Pumpworks, Deggendorf, Germany	60,000	1970	magnesium & iron precipitation, filtration, ClO_2 addition, holding tank	chlorine residual	C 0.3	0.05-0.10	prepared by reacting chlorine gas + $NaClO_2$
Neuss City Works, Neuss, Germany	43,000	NA	ClO_2 addition only	taste & odor removal, bacterial disinfection	C 0.1	0.0	prepared by reacting $NaClO_2$ and chlorine gas

(continued)

TABLE F-4 (continued)

Plant Name	Design Plant Capacity cu m/day	Year ClO_2 Use Initiated	Process Description	Purposes of ClO_2	Average Dosage (mg/l)	Residual in Dist. System (mg/l)	Method of Preparation
Albstadt City Works, Albstadt, Germany	18,000	1977	pre-O_3, reaction tank, CO_2 stripping, two stage filtration, activated carbon, ClO_2 addition	chlorine residual	C 0.3	0.1	prepared by reacting Cl_2 gas & $NaClO_2$
Stolberger Waterworks, Stolberg, Germany	30,000	1970	ClO_2, aeration, filtration, activated carbon filtration, ClO_2 addition	Fe, Mn removal, bacterial disinfection, chlorine residual	C 0.01	0.0-0.02	prepared by reacting $NaClO_2$ with Cl_2 gas
Ansbach City Works (1), Ansbach, Germany	10,000	1973	aeration, iron & manganese filtration, ClO_2 addition	taste removal, bacterial disinfection	C 0.3	0.1	prepared by reacting 30% $NaClO_2$ + Cl_2 gas
Ansbach City Works (2), Ansbach, Germany	10,000	1967	aeration, iron & manganese filtration, ClO_2 addition	taste removal, bacterial disinfection	C 0.30	0.10	prepare ClO_2 by reacting 30% $NaClO_2$ + Cl_2 gas
Wittlich City Works, Wittlich, Germany	10,000	1967	NA	chlorine residual	C 0.2	0.05	prepared by reacting Cl_2 gas + $NaClO_2$
Zweckverband Keckquellen, Deisslingen, Germany	12,960	1974	O_3, coagulation, filtration, ClO_2 addition	post-chlorination	C 0.3	0.2	prepared by reacting $NaClO_2$ with chlorine gas

(continued)

TABLE F-4 (continued)

Plant Name	Design Plant Capacity cu m/day	Year ClO_2 Use Initiated	Process Description	Purposes of ClO_2	Average Dosage (mg/l)	Residual in Dist. System (mg/l)	Method of Preparation
Vallena Waterworks, Vallendar, Germany	5,700	1964	prechlorination (low dosage), oxygen addition, neutralization, activated carbon, ClO_2 addition, reaction basin	taste & odor removal, bacterial disinfection, chlorine residual	C 0.1 (disinf) I 0.3 (T/O)	0.0	prepare ClO_2 in two ways: $NaClO_2$ + Cl_2 gas and $NaClO_2$ + HCl in 2 sets of equipment
Wasserverband Dillkreis-Sud, Germany	8,000	1975	ozonation, neutralization, ClO_2 addition	bacterial disinfection	NA	NA	NA
Linnich City Waterworks, Linnich, Germany	7,200	1969	groundwater, ClO_2 addition ground storage	bacterial disinfection	C 0.1	0	prepared by Cl_2 gas reaction + $NaClO_2$
Lichtenfels City Works, Lichtenfels, Germany	4,500	1965	ClO_2 addition only	bacterial disinfection	C 0.2	0.04	NA
Herborn City Works, Herborn, Germany	2,500	1973	aeration, filtration, ClO_2 addition	iron & manganese removal, bacterial disinfection, to satisfy oxygen demand	C 0.5	0	prepared by reacting Cl_2 gas + $NaClO_2$
Eberbach City Works, Eberbach, Germany	3,970	1977	ClO_2 addition, neutralization	bacterial disinfection, chlorine residual	C 0.16	0.02	prepared by reacting Cl_2 gas + $NaClO_2$
Solingen City Works, Solingen, Germany	3,000	1965	CO_2 addition, $KMnO_4$ addition, 2-stage filtration, ClO_2 addition	chlorine residual	C 0.1	NA	prepared by reacting Cl_2 gas + $NaClO_2$

(continued)

TABLE F-4 (continued)

Plant Name	Design Plant Capacity (MGD)	Year ClO_2 Use Initiated	Process Description	Purposes of ClO_2 Use	Average Dosage (mg/l)	Residual in Dist. System (mg/l)	Method of Preparation
Plochingen City Works, Plochingen, Germany	1,800	1959	NA	odor removal, bacterial disinfection	C 0.2-0.3	0.03	prepared by reacting Cl_2 gas + $NaClO_2$
Spaichingen City Works, Spaichingen, Germany	1,500	1969	flocculation, ClO_2 addition, filtration	bacterial disinfection, chlorine residual	C 0.30	0.1	prepared by mixing bleach (sodium hypochlorite solution) and $NaClO_2$
Baumberg Waterworks, Solingen, Germany	2,000	1977	aeration, softening, double filtration, O_3, filtration plus flocculation, activated carbon, ClO_2 addition	chlorine residual	C 0.1	NA	prepared by reacting Cl_2 gas + $NaClO_2$
Braunschweig, Germany	24,000	1963	aeration, filtration, ClO_2 addition	bacterial disinfection	C 0.1	trace	prepared by reacting $NaClO_2$ + Cl_2 gas
Neumarkt, Germany	2,000	1974	ClO_2 addition only	bacterial disinfection	C 0.15	0.1	prepared by reacting Cl_2 gas and $NaClO_2$
Lohr on Main City Waterworks, Lohr on Main, Germany	5,200	1975	O_3, sedimentation, filtration, ClO_2 addition	NA	C 0.05	0.02	NA
Wollstein, Germany	1,200	1976	ClO_2 addition only	bacterial disinfection	C 0.3	0.1	NA

(continued)

TABLE F-4 (continued)

Plant Name	Design Plant Capacity (MGD)	Year ClO_2 Use Initiated	Process Description	Purposes of ClO_2 Use	Average Dosage (mg/l)	Residual in Dist. System (mg/l)	Method of Preparation
Ludwigsburg, Germany	12,000	1965	2 split streams: first stream is aerated, combined w/second stream in a holding basin prior to ClO_2 addition. Second stream is aerated and has biological removal of ammonia, Fe, Mn on a media filter of quartz, sand	bacterial disinfection, chlorine residual	C 0.12	0.05	prepared by reacting $NaClO_2 + Cl_2$ gas
Kitzingen Waterworks, Kitzingen, Germany	20,000	1961, 1966, 1967	ClO_2 addition only	bacterial disinfection	C 0.15	0.04	prepared by reacting $NaClO_2 + Cl_2$ gas
Friedsrichshafen, Germany	30,000	1971	O_3, alum flocculation, 2-stage filtration (sand, activated carbon), ClO_2 addition	bacterial disinfection	C 0.18	0	prepared by reacting $NaClO_2$ with Cl_2 gas
Ammerbuch-Poltringen, Germany	16,000	1967	NA	bacterial disinfection	C 0.25	NA	prepared by reacting $NaClO_2 + Cl_2$ gas
Bad Mergentheim, Germany	6,050	1976	flocculation, pre-filtration, O_3, post filtration, ClO_2 addition	chlorine residual	C 0.1	0	30% $NaClO_2$ reacted + Cl_2 gas
Schwabisch Hall City Works, Germany	8,600	1969	ClO_2 addition, flocculation, filtration, spraying, post-filtration, ClO_2 addition	chlorine residual	C 0.4-0.6	0.05	prepared by reacting $NaClO_2 + Cl_2$ gas